The CEA Study Guide: Securing Your Future in the International Electronics Marketplace

Compiled from works by
James L. Antonakos
Robert L. Boylestad
Thomas L. Floyd
Muhammed A. Mazidi
Janice G. Mazidi
Gary M. Miller

Prepared by

James L. Antonakos

Broome Community College

With assistance from
Bobby Rudder, Electronic Industries Alliance
Interactive Image Technologies

Prentice Hall

Upper Saddle River, New Jersey   Columbus, Ohio

To the mighty electron

Library of Congress Cataloging-in-Publication Data
The CEA Study Guide: Securing Your Future in the International
  Electronics Marketplace/compiled from works by James L.
  Antonakos et al.; prepared by James L. Antonakos, with assistance
  from Bobby Rudder.
      p.   cm.
    ISBN 0-13-081291-9
      1. Electronics Examinations, questions, etc.   2. Electronic
Industries Alliance—Examinations Study-guides.   3. Electronic
technicians—United States—Certification.   I. Antonakos, James L.
II. Rudder, Bobby.
TK7863.S78   1999
621.381′076—dc21   99-22291
                                                       CIP

Editor:	Scott Sambucci
Production Editor:	Stephen C. Robb
Design Coordinator:	Karrie M. Converse-Jones
Text Designer:	Elm Street Publishing Services, Inc.
Cover Designer:	Rod Harris
Production Manager:	Patricia A. Tonneman
Production Supervision:	Elm Street Publishing Services, Inc.
Marketing Manager:	Ben Leonard

This book was set in Times Roman by The Clarinda Company and was printed and bound by Banta Company. The cover was printed by Phoenix Color Corp.

©1999 by Prentice-Hall, Inc.
Pearson Education
Upper Saddle River, New Jersey 07458

Printed in the United States of America

10 9 8 7 6 5 4 3 2 1

ISBN: 0–13–081291–9

Prentice-Hall International (UK) Limited, *London*
Prentice-Hall of Australia Pty. Limited, *Sydney*
Prentice-Hall Canada, Inc., *Toronto*
Prentice-Hall Hispanoamericana, S. A., *Mexico*
Prentice-Hall of India Private Limited, *New Delhi*
Prentice-Hall of Japan, Inc., *Tokyo*
Prentice-Hall (Singapore) Pte. Ltd., *Singapore*
Editora Prentice-Hall do Brasil, Ltda., *Rio de Janeiro*

Preface

This study guide was developed to assist those individuals preparing for the Certified Electronics Associate Examination. The Electronic Industries Alliance (EIA), together with hundreds of individuals from industry, labor, education, and government, developed these skill standards to help provide the skilled, competent workers required by our highly technical industries, including consumer electronics, biomedical, automotive, and telecommunication products.

OUTLINE OF COVERAGE

Each section in this study guide provides information relevant to its associated topic area, with many true/false and multiple choice questions included to gauge retention. Figures, diagrams, and tables, along with typical examples and applications, are included to convey the required information. This study guide is intended to be used in conjunction with more intensive, formalized training, and thus does not cover every detail of every topic. Individuals wishing to delve deeper into a topic may wish to examine any of the following Prentice Hall textbooks, which contain material directly associated with the skill standards:

The Pentium Microprocessor by James L. Antonakos

An Introduction to the Intel Family of Microprocessors: A Hands-On Approach Utilizing the 80×86 Microprocessor Family, Third Edition, by James L. Antonakos

Microcomputer Repair, Third Edition, by James L. Antonakos

Introductory Circuit Analysis, Eighth Edition, by Robert L. Boylestad

Electronics Fundamentals: Circuits, Devices, and Applications, Fourth Edition, by Thomas L. Floyd

Principles of Electric Circuits: Electron-Flow Version, Fourth Edition, by Thomas L. Floyd

Digital Fundamentals, Sixth Edition, by Thomas L. Floyd

Electronic Devices: Electron-Flow Version, Second Edition, by Thomas L. Floyd

Modern Electronic Communication, Sixth Edition, by Gary M. Miller

The 80×86 IBM PC and Compatible Computers (Vols. I and II): Assembly Language, Design, and Interfacing, Second Edition, by Muhammed A. Mazidi and Janice G. Mazidi

For additional information, visit the Prentice Hall web site at www.prenhall.com.

The skill standards contain a mixture of theoretical and practical ("fabricate and demonstrate" and "troubleshooting") skills. The practical skills are designed to support the information presented in the theoretical skill sections.

In nearly all of the "fabricate and demonstrate" sections, a choice of using actual components or simulation software (such as *Electronics Workbench*) is provided. With the speed of computers increasing every day, realistic real-time simulation is now possible, with results as accurate as that of the actual circuit. Troubleshooting is also possible via software simulation, providing the "hands on" experience without the worry of damaging expensive components or equipment. Please note that the troubleshooting tips and techniques presented are detailed but not exhaustive; they provide a good stepping stone toward further troubleshooting experiences. Overall, troubleshooting is a skill learned by doing, not just reading.

The answers to the review questions are included at the end of the study guide to provide an immediate check on whether the material has been properly understood. Please note that the Review Questions may sometimes refer to material not reviewed directly in this Study Guide but still within the knowledge area of the topic.

This book includes a CD-ROM featuring a copy of Electronics Workbench™ (Version 5.12) software, a powerful software tool used to build and test electronic circuits. This software is provided by Interactive Image Technologies, the producers of Electronics Workbench™. Tips for using the software are available on the "README" section of the CD. Throughout the text, the "EWB" icon shown at left appears beside various figures. This indicates those figures are available electronically on the CD and are presented for the reader using the Electronics Workbench™ software. When reviewing material that includes an "EWB" labeled circuit, the reader should work through the example, then simulate the circuit to increase understanding of the particular topic. For more information about Electronics Workbench™, visit the Interactive Image Technologies website, www.electronicsworkbench.com.

HOW TO TAKE THE EXAM

Information on how to take the Certified Electronic Associate Examination can be found on the following web page:

http://www.cemacity.org

Look for the CEA logo and click on it. Sample exams, advice for taking the test, policy statements, and other useful information are provided.

ACKNOWLEDGMENTS

Many thanks to everyone at Prentice Hall, particularly Dave Garza and his assistant, Juanita Griffin, and Scott Sambucci and his assistant, Marcie Wademan, for all their help during the project. Thanks also to Bobby Rudder at CEMA for his valuable suggestions, and to everyone at Elm Street Publishing Services for their excellent work, especially Cathy Wacaser.

In addition, several authors have contributed material to the study guide and deserve recognition. These authors include James L. Antonakos, Robert L. Boylestad, Thomas L. Floyd, Muhammed A. Mazidi, Janice G. Mazidi, and Gary M. Miller. All of their contributions are appreciated.

James L. Antonakos
antonakos_j@sunybroome.edu
http://www.sunybroome.edu/~antonakos_j

Table of Contents

A General

A.01 Demonstrate an understanding of proper safety techniques for all types of circuits and components (DC circuits, AC circuits, analog circuits, digital circuits, discrete solid-state circuits, microprocessors)

A.02 Demonstrate an understanding of and comply with relevant OSHA safety standards

A.03 Demonstrate an understanding of proper troubleshooting techniques

A.04 Demonstrate an understanding of basic assembly skills using hand and power tools

A.05 Demonstrate an understanding of acceptable soldering/desoldering techniques, including through-hole and surface mount devices

A.06 Demonstrate an understanding of proper solderless connections

A.07 Demonstrate an understanding of use of data books and cross reference/technical manuals to specify and requisition electronic components

A.08 Demonstrate an understanding of the interpretation and creation of electronic schematics, technical drawings, and flow diagrams

A.09 Demonstrate an understanding of design curves, tables, graphs, and recording of data

A.10 Demonstrate an understanding of color codes and other component descriptors

A.11 Demonstrate an understanding of site electrical and environmental survey

A.12 Demonstrate the use of listening skills or assistive devices to assess signs and symptoms of malfunctions

A.01 Demonstrate an understanding of proper safety techniques for all types of circuits and components (DC circuits, AC circuits, analog circuits, digital circuits, discrete solid-state circuits, microprocessors)

INTRODUCTION

The safety techniques involved in working with the various components and circuits covered in sections B through F are, at the minimum, composed of the following:

☐ Double-check all wiring before power is applied.

☐ Wear appropriate clothing and other protection (safety glasses, gloves, earplugs).

☐ Use the appropriate antistatic procedure when working with static-sensitive parts. This may involve wearing an antistatic wrist strap or working on an anti-static bench.

☐ Be as neat as possible. From the circuit setup to your troubleshooting notes, having everything in order helps reduce confusion and irritation when working with the circuit.

☐ In a high voltage circuit, make measurements using the one-hand rule.

The environment where the circuit is being operated may have its own unique safety rules and requirements. Familiarize yourself with these prior to beginning your work.

REVIEW QUESTIONS

True/False

1. Turn power on as soon as possible so a circuit can warm up.
2. Eliminate any static charge before working on a circuit.

Multiple Choice

3. When working in a new environment,
    a. Do whatever you want.
    b. Follow the specific rules of the environment.
    c. Learn by your mistakes.
4. Neatness helps
    a. Eliminate confusion.
    b. Reduce irritation.
    c. Both a and b.
5. When making high-voltage measurements,
    a. Use only one hand.
    b. Use both hands.
    c. Use both hands for DC measurements only.

A.02 Demonstrate an understanding of and comply with relevant OSHA safety standards

INTRODUCTION

OSHA (Occupational Safety and Health Association) is the U.S. Department of Labor's main watchdog, overseeing the safety of more than 100 million workers. The goal of OSHA and the standards it enforces is to ensure the safety and health of the U.S. workforce. You are encouraged to spend time looking around the OSHA home page at http://www.osha.gov. Figure A.02-1 shows the main OSHA home page. There you can find on-line documents (safety manuals and regulations, workforce statistics), news, publications, links to other organizations, and many other useful categories.

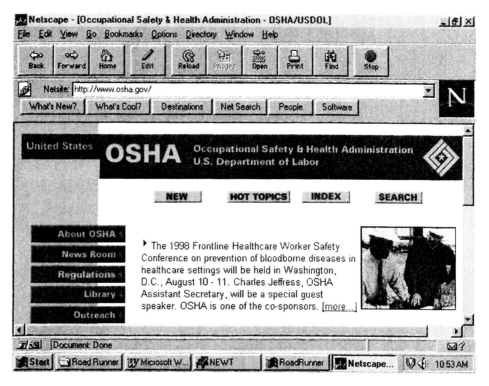

FIGURE A.02-1
OSHA home page.

Let us examine some typical safety rules. Safety in the microcomputer laboratory requires observance of the commonsense safety rules that you should follow in any situation when working with or on electrical and mechanical equipment. The following safety rules should be observed at all times.

1. *Do not allow "horseplay" in the lab.* Many lab injuries are caused by students playing jokes or "booby-trapping" equipment. This practice can cause serious permanent injuries to yourself and others. This kind of behavior should not be tolerated in any laboratory situation.

2. *Always get instructor approval.* Your instructor is there to help. Always ask for instructor approval before starting any new task. Doing this can save valuable lab time and also help prevent injuries to you and/or damage to equipment.

3. *Report any injuries immediately.* Always report any injury to your lab instructor. You should do this no matter how small the injury. What may appear to be a small cut, mild shock, or minor bruise could lead to serious complications if not properly treated.

4. *Use safety glasses.* When any mechanical or electrical equipment is used, there is always the chance of sparks or particles being ejected. This can happen in an electrical circuit when a part such as an electrolytic capacitor has been installed incorrectly. Remember that it takes only a very small particle to cause permanent eye damage.

5. *Use tools correctly.* The improper use of tools can result in injuries to you or others as well as permanent damage to the tools. Never attempt to use a tool that is damaged. Never use a tool that you do not know how to use. Never use a tool for a purpose other than that for which it was designed.

6. *Use equipment correctly.* What applies to tools applies equally well to electrical or mechanical equipment. If you are not sure about the operation of any piece of equipment, ask your instructor before attempting to use it.

7. *Do not distract others.* Do not talk to or otherwise distract someone who is in the process of using electrical or mechanical equipment. Doing so could lead to personal injury and/or damage to the equipment.

8. *Use correct lifting techniques.* Always use the proper method for lifting or pushing heavy objects. Ask for help in lifting very heavy objects. If you cannot lift or move an object for any reason, let your instructor know.

9. *Remove jewelry.* Remove all rings, watches, chains, and other jewelry; these are all capable of conducting electricity and causing a shock.

REVIEW QUESTIONS

True/False

1. It is safe to wear jewelry around equipment.
2. A few practical jokes are good for morale.

Multiple Choice

3. When in doubt about operating a piece of equipment, you should
   a. Ask your lab partner for a demonstration.
   b. Try it yourself first so you don't appear stupid.
   c. Ask your instructor.
4. If you accidentally cut your finger while using a small hand tool, you should
   a. Report the accident immediately to your instructor.
   b. Wait until after class and then report the accident to your instructor.
   c. Quietly leave the lab in order not to disturb anyone, and seek first aid.
5. If you find that the tool you are using in your lab experiment is damaged, you should
   a. Try to repair it to save the school money.
   b. Not use it and let your instructor know that the tool is damaged.
   c. Use it so that you don't waste time in the lab.

A.03 Demonstrate an understanding of proper troubleshooting techniques

INTRODUCTION

Troubleshooting is the art of examining a problem, finding its cause, and correcting it. For example, perhaps the speaker volume of an amplifier is very low, even with a good signal applied to the amplifier. What could be the cause?

When troubleshooting a problem with an electrical circuit, do the following:

☐ Check the power connections. Many times a missing ground wire is all that is keeping the circuit from operating. On rare occasions, the only problem is that someone forgot to turn power on. It pays to check even the most obvious things.

☐ Give the circuit a good visual. Are any components missing or installed improperly? Have any components been damaged (cracked, burned, bent, or shorted pins)? Has a paper clip fallen into the circuit?

☐ Examine the operating voltages and currents. Use multimeters, oscilloscopes, or logic analyzers to gather data. Does the input signal do anything to the circuit? If there are several stages to the circuit, are any of the stages operating correctly?

☐ Many AC circuits respond to one frequency or a range of frequencies. Be sure to examine the circuit at the appropriate frequencies.

☐ If the circuit has been setup on a breadboard or wire-wrapped, check the wiring with a new schematic. Even one missing connection can prevent the entire circuit from operating correctly.

☐ Has the circuit ever worked? If so, the most probable cause of failure is a bad component or some form of mechanical problem (a broken wire, bent connector pin). Maybe the circuit was disconnected, moved to another location, and reconnected improperly. It pays to know the history of the problem and the circuit.

☐ Talk to other individuals about the circuit. They may have a fresh viewpoint or explanation that you have not thought of.

Finally, if you've tried everything and the problem still exists, walk away from it for a while. Take some time off. Relax. Sometimes a break can help organize your thoughts. When you look at the circuit again, if only to repeat previous steps, you may discover something you missed the first time, or you may try something slightly different. The answer may even come to you when you least expect it, such as when watching television or taking a shower. Overall, a little patience will go a long way in helping you solve many of your troubleshooting problems.

REVIEW QUESTIONS

True/False

1. Always assume that power is on when troubleshooting.

2. A visual inspection is not very helpful.

Multiple Choice

3. A new circuit has never worked. The problem could be
   a. Bad wiring or improperly installed or missing components.
   b. A bad design that could never work.
   c. Both a and b.

4. A once-working circuit has stopped working. The problem could be
   a. Bad wiring or improperly installed or missing components.
   b. A bad design that could never work.
   c. A damaged component.

5. When troubleshooting a circuit,
   a. Try a few things and then give up.
   b. Replace every component and see if it works.
   c. Patiently go through a checklist of items to examine.

A.04 Demonstrate an understanding of basic assembly skills using hand and power tools

INTRODUCTION

Hand tools are instruments used with the hands to extend their working capabilities. In the microcomputer lab, hand tools aid in the disassembly and reassembly of microcomputer equipment and parts. They are not intended to serve as any kind of electrical testing devices; attempting to use them in this manner is very dangerous.

Figure A.04-1 illustrates the hand tools that will be used in this exercise. All the hand tools shown in Figure A.04-1 are made of metal. Because metal is a good conductor of electricity, such tools present a potential shock hazard when used in working with electrical equipment.

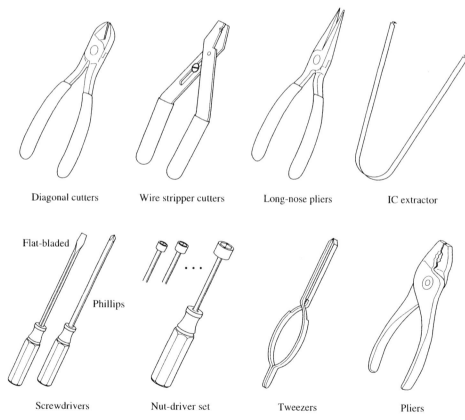

| Diagonal cutters | Wire stripper cutters | Long-nose pliers | IC extractor |

| Screwdrivers | Nut-driver set | Tweezers | Pliers |

Flat-bladed
Phillips

FIGURE A.04-1
Hand tools for microcomputer repair.

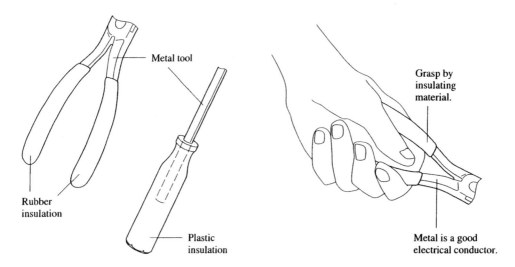

FIGURE A.04-2
Insulated hand tools.

Properly made hand tools for working on electrical equipment have metal handles that are insulated with a rubber coating or are cast in an insulating plastic. This is illustrated in Figure A.04-2.

Note from Figure A.04-2 that when using hand tools on electrical equipment, you do not touch the metal part of the hand tool with any part of your body. By grasping the tool only on the insulating material, you reduce your chance of electrical shock. Recall that an insulator is not a good conductor of electricity. Hand tools on which the insulation is frayed or has been removed should not be used; you should discard such tools and replace them with a new set.

Diagonal Cutters

Figure A.04-3 illustrates a typical set of **diagonal cutters** and the correct way to use them. As illustrated, diagonal cutters can be damaged by cutting thick material close to the ends of the cutting tips.

FIGURE A.04-3
Diagonal cutters.

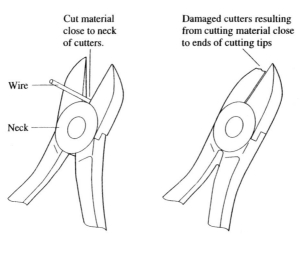

Proper method   Damaged cutters

Wire Stripper Cutters

Figure A.04-4 illustrates a typical set of **wire stripper cutters.** As shown in the figure, these instruments are used to strip the insulation from wire to prepare the wire for use in an electrical connection.

Observe from Figure A.04-4 that the minimum size of the wire stripper opening must be small enough to cut through the insulation of the wire completely, but not so small as to cut or nick the wire itself.

FIGURE A.04-4
Wire stripper cutters.

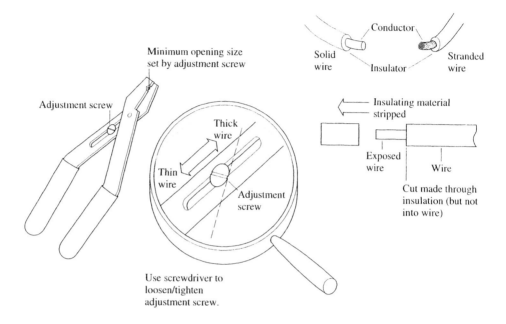

Long-Nose Pliers

Figure A.04-5 shows a typical set of **long-nose pliers.** As shown in the figure, these pliers are not intended for use in removing hardware such as nuts and bolts. Such misuse can result in permanent damage to these tools.

Long-nose pliers come in various sizes. Keep in mind that these are delicate instruments intended for delicate work, not for removing nuts and bolts.

FIGURE A.04-5
Long-nose pliers.

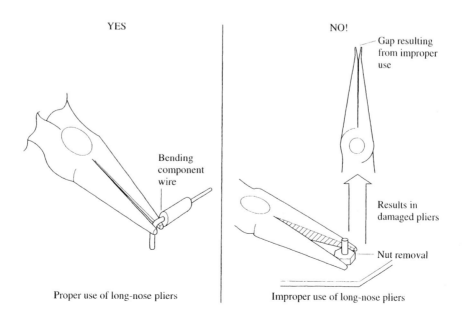

Flat-Bladed and Phillips Screwdrivers

A typical set of **flat-bladed and Phillips screwdrivers** is shown in Figure A.04-6. These screwdrivers come in various sizes and may use a carbide tip for strength. They are intended to be used for the insertion and removal of screws. Remember to use a screwdriver of the right size for the job at hand.

IC Extractors

Figure A.04-7 shows typical **IC** (integrated circuit) **extractors.** As shown in the figure, these instruments are intended for the removal of IC packages. It's important that all power be disconnected from the system before an IC is removed.

FIGURE A.04-6
Flat-bladed and Phillips screwdrivers.

Tips come in various widths and thicknesses.

Slot screw

Tips come in various sizes.

Phillips screw

Blade shafts come in various lengths.

Carbide-tipped for strength

Flat-bladed screwdriver

Phillips screwdriver

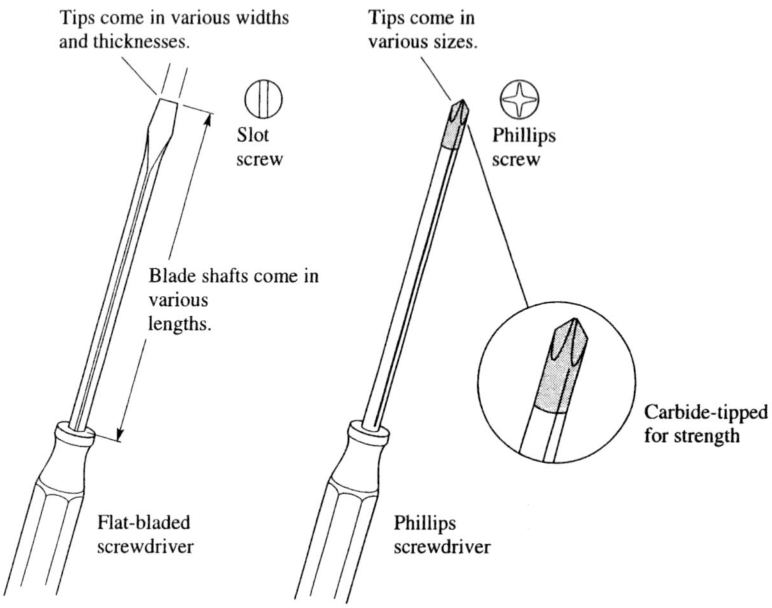

FIGURE A.04-7
IC extractors.

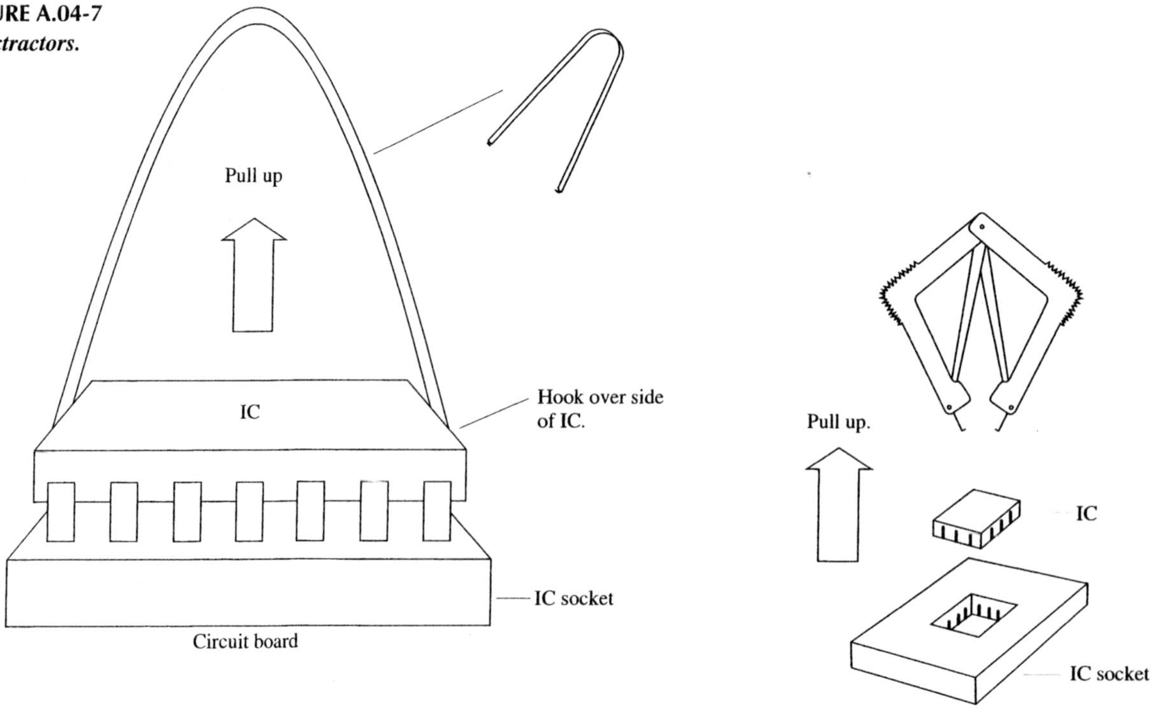

Pull up

IC

Hook over side of IC.

IC socket

Circuit board

(a) DIP (Dual Inline Package) extractor

Pull up.

IC

IC socket

(b) PLCC (Plastic-Leaded Chip Carrier) extractor

Antistatic Wrist Strap

Static electricity can damage many types of integrated circuits, including CPU and memory devices. To avoid static damage during handling, it is common to wear an antistatic wrist strap. The strap wraps around your wrist and connects to a ground terminal via an attached cable, as indicated in Figure A.04-8. The cable provides a path to ground for any static electricity encountered. A resistor built in to the cable is typically used to reduce the current flow during a discharge.

Antistatic mats or pads may also be used. These are placed on top of the workbench and provide a safe surface for equipment and electronics.

Tweezers

A typical pair of **tweezers** used in microcomputer repair is shown in Figure A.04-9. Tweezers may also be used as a soldering aid, as you will see in a later exercise. Be careful of static problems introduced by tweezers.

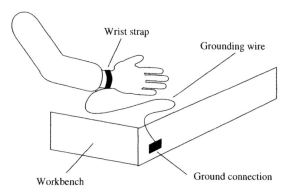

FIGURE A.04-8
Using a wrist strap.

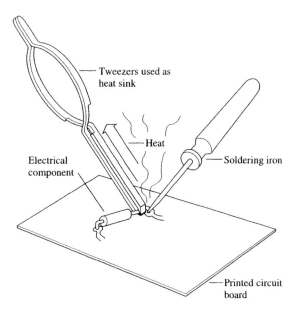

FIGURE A.04-9
Tweezers.

Nut Drivers

Figure A.04-10 illustrates a typical **nut driver.** As shown in the figure, nut drives are not adjustable and therefore come in various sizes to accommodate different-size nuts. Note that these instruments present a small surface area, thus causing minimum marring of computer surfaces. Nut drivers are always the preferred instrument for loosening or tightening of nuts.

Pliers

Common **pliers** are shown in Figure A.04-11. Because pliers of this type can easily cause damage to computer surfaces, they should be the last instrument of choice for use on computer equipment.

FIGURE A.04-10
Nut driver.

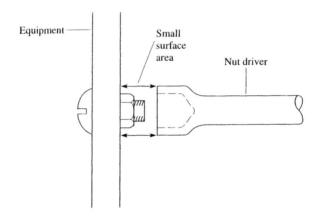

FIGURE A.04-11
Common pliers.

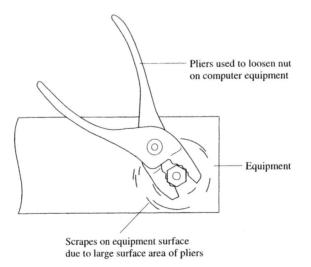

REVIEW QUESTIONS

True/False

1. Hand tools are instruments used to extend the capabilities of the hands.

2. Hand tools may be used to make electrical tests.

Multiple Choice

3. When using hand tools around electrical equipment, you should
   a. Touch only the metal parts of the tools.
   b. Avoid touching the metal parts of the tools.
   c. Not use hand tools around electrical equipment.

4. Insulating material on hand tools is
   a. A good conductor of electricity.
   b. Used to prevent electrical shock.
   c. There to prevent the hand tool from rusting.

5. Hand tools on which the insulation is frayed or missing
   a. Should be discarded and replaced by new hand tools.
   b. Are dangerous to use.
   c. Both a and b.

A.05 Demonstrate an understanding of acceptable soldering/desoldering techniques, including through-hole and surface mount devices

INTRODUCTION

Soldering is the process used to secure the wire connections of electrical components. Soldering is the least expensive, fastest, most reliable, and simplest method of making electrical connections between electronic components. The process requires three things:

1. A metal alloy, called solder.

2. A material to clean the connection.

3. A source of heat.

Figure A.05-1 illustrates the process of soldering.

FIGURE A.05-1
The process of soldering.

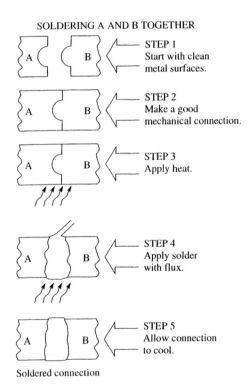

Why Soldering Works

Why Soldering Works

The metal alloy called **solder** frequently consists of a combination of 60% lead and 40% tin, which is referred to as 60/40 solder. This combination, called an **alloy,** has a melting point, 370°F, that is lower than the melting point of either metal by itself. Solder has the ability to form an interface that embeds itself into the metals to be connected, forming an excellent electrical connection. This is shown in Figure A.05-2.

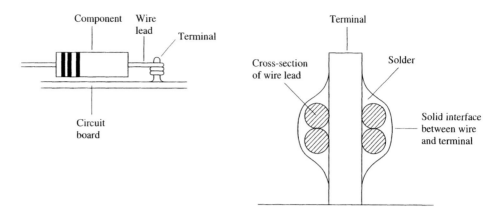

FIGURE A.05-2
The soldered electrical connection.

FIGURE A.05-3
Three types of solder.

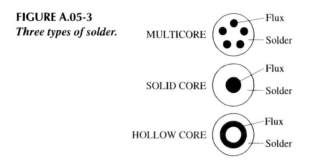

Because wire leads, which are made of copper, have a tendency to oxidize when heated, a method of keeping them clean is needed in order for the solder to make the required metal bonding. A chemical called a **rosin,** or **flux,** is used. The flux chemical comes contained within the core of the solder. Solder is available in a variety of shapes; the most common type is in the form of a long wire that is about 1/16″ in diameter (good for most electrical work) or 1/32″ in diameter (good for small, detailed work). Three types of solder are shown in Figure A.05-3.

Soldering Irons

The tool most commonly used as a source of heat for the melting and application of solder is a **soldering iron.** Figure A.05-4 illustrates the basic construction of a soldering iron.

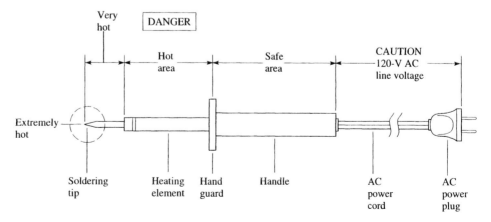

FIGURE A.05-4
Basic construction of a soldering iron.

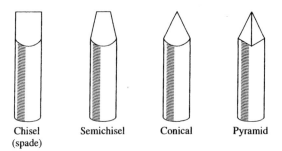

FIGURE A.05-5
Various soldering tips.

Soldering irons come in different wattage ratings, from 10 to 250 W. For most computer work, an iron with a wattage rating of between 10 and 50 W is used. In fact, the most common type used for computer work is an iron with a rating of 25 to 35 W. Soldering irons with higher wattage ratings get so hot that they can easily damage delicate components and printed circuit boards. These higher-wattage soldering irons are used only for larger and more rugged electrical soldering jobs, not for work on PCs.

The working part of a soldering iron is its **tip.** The tip is the part of the iron that melts the solder. Various soldering tips are shown in Figure A.05-5.

Tinning and Cleaning a Soldering Tip

The tip of a soldering iron must be capable of providing a concentrated area of heat. This is best achieved when the tip surface is bright and shiny. However, soldering tips will oxidize and turn black from the heat they generate, causing the amount of heat leaving the surface to be reduced, as shown in Figure A.05-6.

In order to ensure that the tip of the soldering iron stays bright and shiny, a process called **tinning** is performed. Tinning is done by carefully melting solder on the tip of the soldering iron when the iron is first turned on for use. As soon as a small amount of solder melts on the tip, the tip is wiped clean with a small, damp cloth or damp sponge. The process of tinning the tip is shown in Figure A.05-7.

As you are using the soldering iron, you want to ensure that the tip is kept clean. This is achieved by wiping it quickly with a damp cloth or damp sponge. It is important that the tip remain clean throughout the entire soldering process.

FIGURE A.05-6
Heat on the surface of a soldering tip.

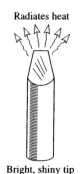

Radiates heat

Bright, shiny tip

Reduced heat flow

Dull, dark tip

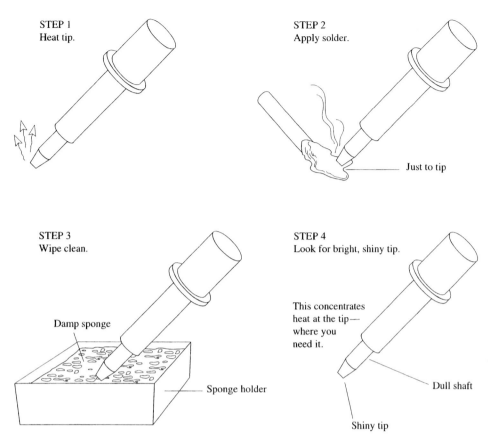

STEP 1
Heat tip.

STEP 2
Apply solder.

Just to tip

STEP 3
Wipe clean.

Damp sponge

Sponge holder

STEP 4
Look for bright, shiny tip.

This concentrates
heat at the tip—
where you
need it.

Dull shaft

Shiny tip

FIGURE A.05-7
Tinning the tip of a soldering iron.

How to Solder

Soldering is a four-step process:

1. Make a good mechanical connection.

2. Heat both parts to be connected.

3. Apply solder to both parts at the same time.

4. Remove the iron and allow parts to cool slowly.

This four-step process is shown in Figure A.05-8.

There are times when you may not have an actual physical connection on a printed circuit board to which to attach the lead of a component. In this case, the lead must be placed on the foil surface of the circuit board. When you must do this, be sure that the lead is absolutely still while the solder cools and hardens. This type of connection is shown in Figure A.05-9.

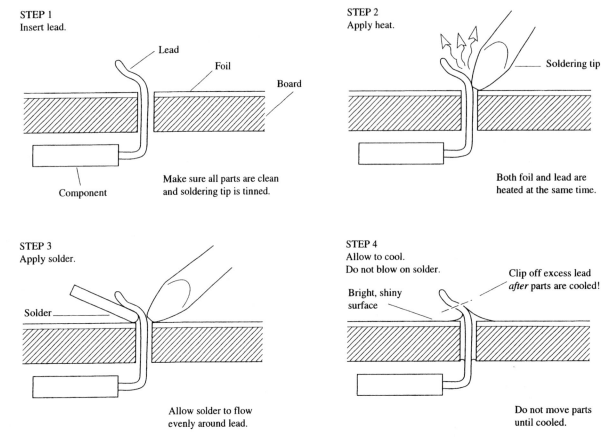

FIGURE A.05-8
The four-step process of soldering.

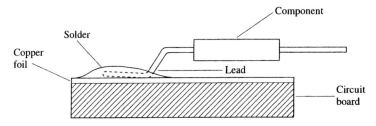

FIGURE A.05-9
Soldering to a foil surface.

Cold Solder Joints

A **cold solder joint** is an undesirable condition that results in an unreliable electrical connection. Cold solder joints are caused by

1. Dirty soldering tips.
2. Heat not uniformly distributed to both parts.
3. Movement of the connection during cooling.
4. Blowing on the connection or using an external source to cause rapid cooling of the connection.

Some of the different types of cold solder joints and their causes are shown in Figure A.05-10. A good soldered connection is smooth and shiny. A poor soldered connection (a cold solder joint) looks dull and pitted. If you find that you have created a cold solder joint,

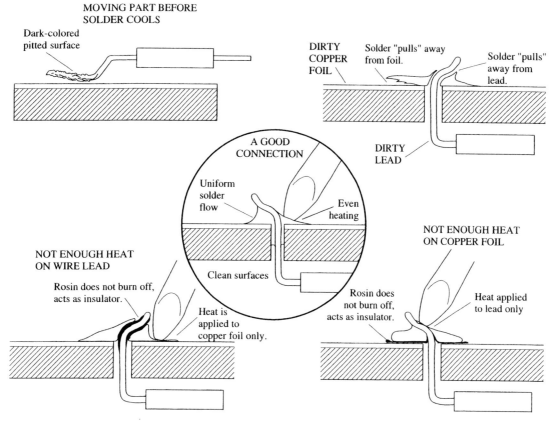

FIGURE A.05-10
Cold solder joints and their causes.

simply reheat the joint to remove the old solder from both surfaces and, with a clean tip, resolder.

Solder Bridges

A solder bridge is an undesirable condition that can result in severe electrical damage to a computer system. Solder bridges are caused by solder accidentally allowing an electrical connection between two adjoining parts of a printed circuit board. A solder bridge is shown in Figure A.05-11.

FIGURE A.05-11
A solder bridge.

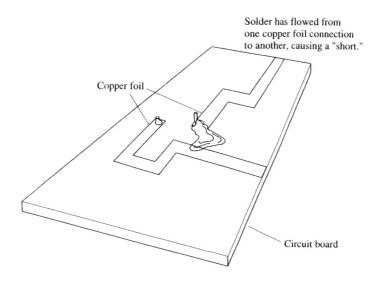

Always inspect your soldering work closely to ensure that no solder bridges have formed. If you find one, heat the solder carefully to remove it. The section on desoldering techniques will show you how to remove unwanted solder.

Tinning Wire

Insulated wire is either solid or stranded. Stranded wire is more flexible and not as breakable as solid wire. Figure A.05-12 illustrates the two types of wire.

When stranded wire is used to make an electrical connection, the tips of the strands of wire should be soldered together. The process of doing this is called *tinning* the wire and is shown in Figure A.05-13.

FIGURE A.05-12
Two types of wire.

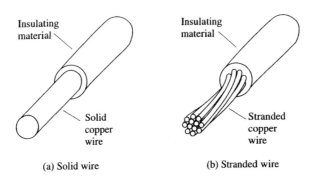

(a) Solid wire   (b) Stranded wire

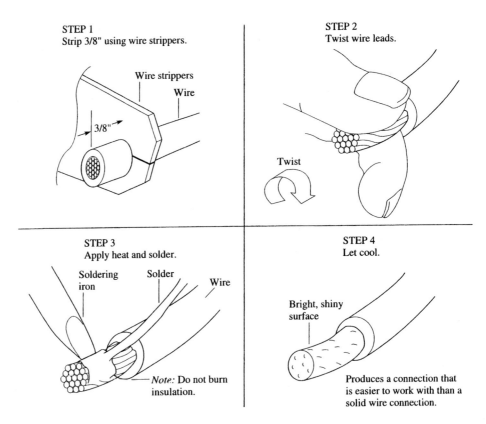

STEP 1
Strip 3/8" using wire strippers.

STEP 2
Twist wire leads.

STEP 3
Apply heat and solder.

STEP 4
Let cool.

FIGURE A.05-13
Process of tinning wire.

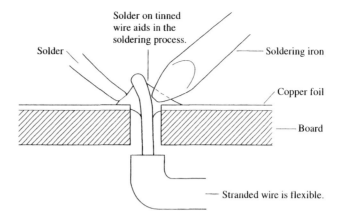

FIGURE A.05-14
Connecting tinned wire.

After the wire is properly tinned, it may then be used to make electrical connections. Tinning the wire helps make soldering it to other parts of the circuit an easy task, as shown in Figure A.05-14.

Using a Heat Sink

Many of the components with which you will be working are very sensitive to heat. This is especially true of solid-state components such as integrated circuits, diodes, and transistors; other small components may also be sensitive. In order to help prevent heat from destroying these devices, a tool is placed between the heat source and the device to help conduct away the heat. Such a tool is called a **heat sink** and is illustrated in Figure A.05-15.

It is important to use a heat sink to avoid damaging good electrical parts. If in doubt, use a heat sink; it can never hurt, but if you don't use it, you could wind up with a damaged part.

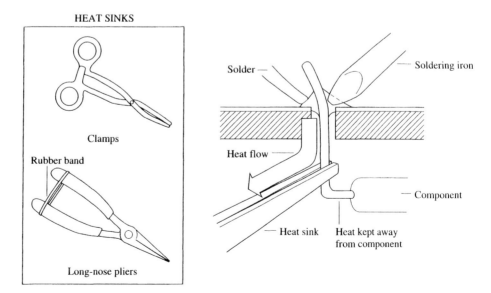

FIGURE A.05-15
The use of a heat sink during soldering.

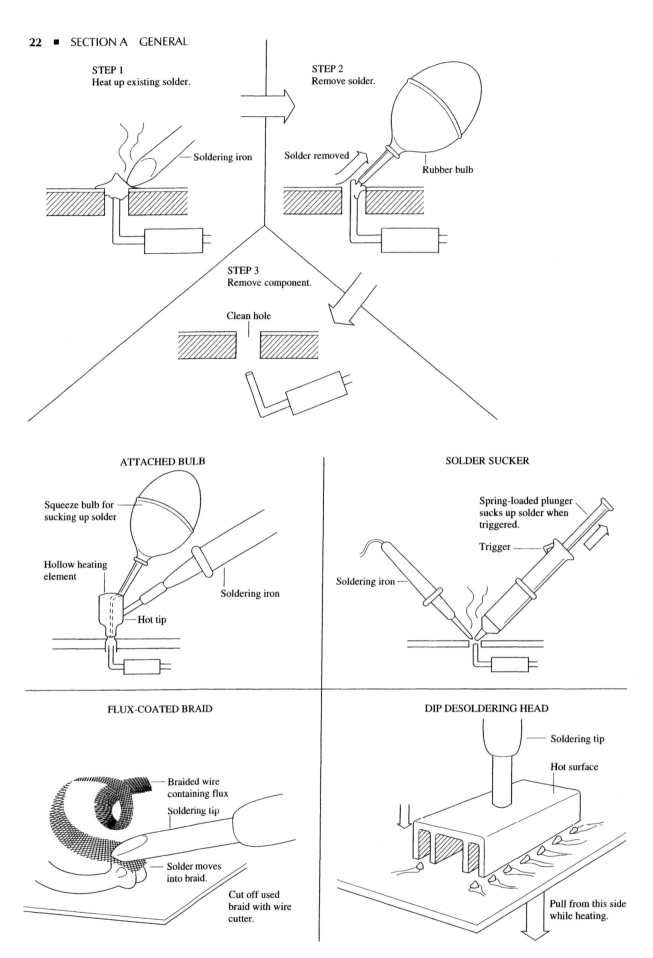

STEP 1
Heat up existing solder.

Soldering iron

STEP 2
Remove solder.

Solder removed

Rubber bulb

STEP 3
Remove component.

Clean hole

ATTACHED BULB

Squeeze bulb for sucking up solder

Hollow heating element

Soldering iron

Hot tip

SOLDER SUCKER

Spring-loaded plunger sucks up solder when triggered.

Trigger

Soldering iron

FLUX-COATED BRAID

Braided wire containing flux

Soldering tip

Solder moves into braid.

Cut off used braid with wire cutter.

DIP DESOLDERING HEAD

Soldering tip

Hot surface

Pull from this side while heating.

FIGURE A.05-16
The process of desoldering, and various desoldering tools.

Desoldering Techniques

Desoldering is the process of removing a soldered component or simply removing unwanted solder (such as from a solder bridge).

The desoldering process consists of the following steps:

1. Heating the solder.
2. Removing the solder.
3. Removing the component.
4. Cleaning the surface.

The process of desoldering, along with various desoldering tools, is shown in Figure A.05-16. In this exercise, you will have an opportunity to practice soldering and desoldering.

Surface-Mount Components

Technological advances in printed circuit board design and fabrication now allow components (resistors, capacitors, integrated circuits) to be soldered directly onto the surface of the printed circuit board. No holes need to be drilled, since the connecting pins or pads of the component do not go through the board. Figure A.05-17 illustrates the difference between a surface-mounted capacitor and a disk capacitor. Note the size difference between the two capacitors, even though they both have ratings of 0.001 μF. Surface-mounted components are much smaller than their through-hole-mounted counterparts. This allows more components to be placed on a board, or the same number of parts to be placed on a smaller board.

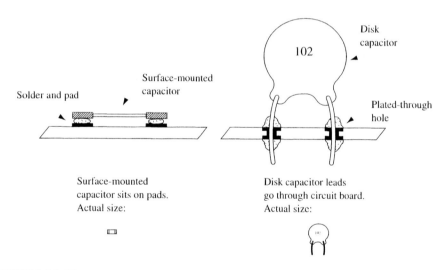

FIGURE A.05-17
Mounting capacitors two different ways.

Surface mounting requires a solder paste to be applied to the pads on the surface of the printed circuit board. Then the component is carefully placed on the pads. There is no room for error when placing a component. If the component wiggles around on the pad, the solder paste smears, resulting in a bad connection when the board is heated up in a special machine used to melt the solder paste. This tight control of placement is especially important when dealing with a surface-mounted integrated circuit, whose leads are very close together (as shown in Figure A.05-18). For this reason, surface-mounted components are rarely placed by human hands. Instead, large industrial pick-and-place machines are used to mechanically position each part on a surface-mount printed circuit board. These parts often need to be placed with an accuracy of one ten-thousandth of an inch!

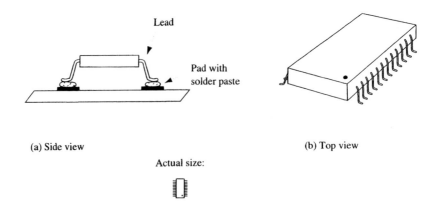

Lead

Pad with
solder paste

(a) Side view

(b) Top view

Actual size:

FIGURE A.05-18
Surface-mounted integrated circuit.

If you are willing to spend a good amount of money, you can set up a surface-mount station that will allow you to experiment with the technology, rather than try to build 1000 motherboards a day. For a few thousand dollars you can buy a **hot air rework system** that holds printed circuit boards and uses hot air blown through a special tip to heat up the surface-mount pads when soldering/desoldering. For the same price, a **manual placement machine** allows precise placement (to 0.0005″) of components onto a circuit board.

For several thousand dollars more a **high-resolution vision system** can be added, allowing visual inspection of surface-mount components and circuit boards.

On a smaller cost scale, many small tools and other items are available for surface-mount applications. These include **vacuum pickup** instruments (Figure A.05-19), which can pick up small components using a suction cup and vacuum, pin straighteners, miniature soldering-iron tips, and surface-mount components of all varieties, from resistors to ICs, LEDs, and connectors.

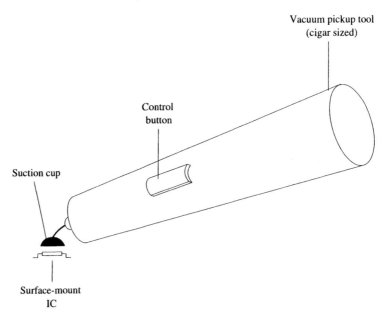

Vacuum pickup tool
(cigar sized)

Control
button

Suction cup

Surface-mount
IC

FIGURE A.05-19
Vacuum pickup tool.

True/False

1. Solder is a metal alloy consisting of zinc and copper.
2. The material used to clean the connections to be soldered is called a flux.

Multiple Choice

3. The process of soldering requires
    a. A metal alloy called the solder.
    b. A source of heat.
    c. Both a and b.
4. The metal alloy called solder consists of
    a. 50% zinc and 50% copper.
    b. 60% lead and 40% tin.
    c. 40% copper and 60% silver.
5. When soldering, the proper process is to
    a. Heat both surfaces at the same time and apply the solder.
    b. Heat one surface at a time, applying solder to each.
    c. Heat the solder and let it drip onto the connection.

A.06 Demonstrate an understanding of proper solderless connections

INTRODUCTION

A solderless connection is one that relies on mechanical contact to provide the conductivity. Examples of solderless connections are components that plug into sockets, such as integrated circuits, connectors (RCA, bananna, BNC, DB25, etc.) and cards with edge connectors (such as adapter cards plugged into a PC's motherboard). In this section we will examine the different types of IC sockets and their associated connections.

Identifying an IC

An integrated circuit (IC) consists of many different electronic devices (such as transistors, resistors, capacitors, diodes, etc.), all connected together in a single small package. Figure A.06-1 shows some typical integrated circuits.

Note from Figure A.06-1 that each IC contains several pins along its sides. These pins are used to make electrical connections to external circuits, usually through an IC socket. Figure A.06-2 shows some typical IC sockets.

FIGURE A.06-1
Typical integrated circuits (courtesy of Motorola).

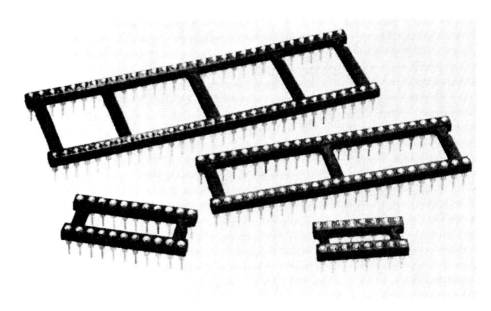

FIGURE A.06-2
Typical IC DIP sockets (courtesy of Aries Electronics, Inc.).

Connecting an IC

Integrated circuits can be connected to external circuits by many methods. The following are three of the most common methods.

1. Soldering directly to the PC board (through-hole and surface-mount).

2. Insertion into an IC socket, where the socket itself is soldered directly to the PC board.

3. Connection with wire, such as wire-wrapped connections.

These three different methods of connecting integrated circuits are shown in Figure A.06-3.

For this exercise, you will be working with ICs that are inserted into sockets.

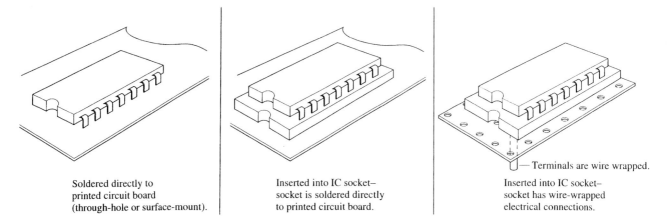

Soldered directly to
printed circuit board
(through-hole or surface-mount).

Inserted into IC socket–
socket is soldered directly
to printed circuit board.

— Terminals are wire wrapped.

Inserted into IC socket–
socket has wire-wrapped
electrical connections.

FIGURE A.06-3
Three methods of connecting ICs.

Pin Numbering

One advantage of connecting an IC to a circuit through the use of sockets is that the IC may be easily removed for replacement. One disadvantage is that manufacturing costs are higher, since the socket must first be soldered to the board and then the IC inserted into the socket.

As a technician, you must ensure that you place an IC in the correct position when replacing it in a socket. All IC pins have distinct numbers, as do the corresponding sockets. The numbering of integrated circuits is shown in Figure A.06-4. Figure A.06-5 illustrates the numbering of IC sockets.

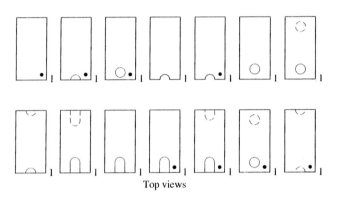

Top views

FIGURE A.06-4
Pin numbers of various ICs.

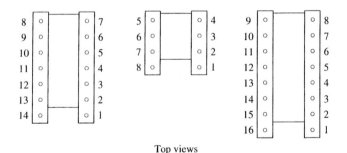

Top views

FIGURE A.06-5
Pin numbers of IC sockets.

Reason for IC Failures

Integrated circuits are normally very reliable devices. One of the major causes of initial IC failure is improper insertion of the ICs into their sockets. Figure A.06-6 shows the common insertion errors that may cause IC failures.

Because computer owners may have tried to repair their own computers before bringing them in for service, it is always good practice to give all printed circuit boards with IC sockets a complete visual inspection. Do this to ensure that the customer or another technician has not inserted an IC incorrectly into the board.

Types of ICs

Figure A.06-7 shows the nomenclature used to identify an IC. It is important that you always use an exact replacement, as recommended by the manufacturer.

Many different functions are performed by ICs in your computer. One common troubleshooting technique is to replace a suspect IC with a known good one. It is important that you practice the technique of IC removal and insertion.

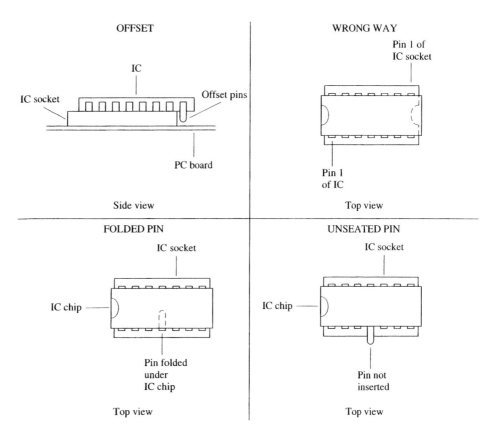

FIGURE A.06-6
Common IC insertion errors.

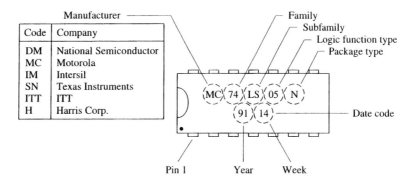

FIGURE A.06-7
IC identification.

Removing an IC

Figure A.06-8 shows the proper procedure for removing an IC using an IC removal tool.

When removing an IC, it is important to pull it straight up in order to prevent bending the pins. There are times when the IC is so large that the removal tool will not work properly. If this is the case, use the method shown in Figure A.06-9.

If you find it necessary to use the method shown in Figure A.06-9, be sure to pry up, gently, first one end and then the other. Again, you are trying to pull the IC straight up in order to avoid bending the pins. Do *not* pull an IC out with your fingers. One slip, and the tiny metal pins of the IC can easily pierce your fingertips.

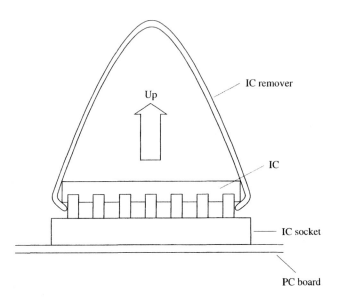

FIGURE A.06-8
Proper method of removing an IC.

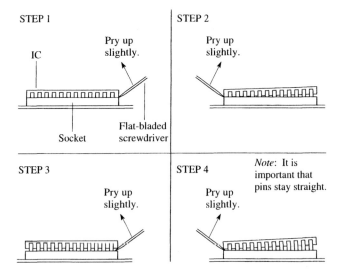

FIGURE A.06-9
Removing a large IC.

Inserting an IC

Figure A.06-10 shows the correct method of inserting an IC into an IC socket. Note that the pins of an IC tend to spread out to the side.

The reason an IC comes from the factory with its pins spread out is to help form a better electrical connection within the IC socket. Because of the mechanical pressure caused by the spreading of the pins, a more reliable electrical connection is made between each IC pin and its corresponding socket connection.

Figure A.06-11 shows how an insertion tool is used to insert an IC. The adapter holds the pins of the IC in grooves, which helps guide them into the socket holes.

It is important to note that when you are replacing an IC, you must always have the power off. If the power is on as you replace an IC, you could destroy the new IC. In a like manner, if power is still on when you remove an IC, you could destroy an otherwise good IC. Also, be sure to wear a protective wrist strap to avoid static damage to sensitive integrated circuits.

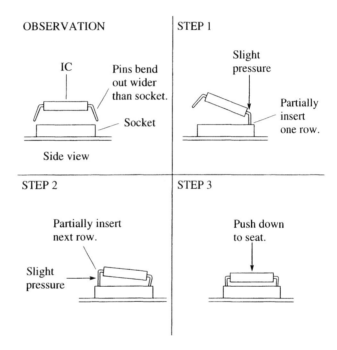

FIGURE A.06-10
Inserting an IC into its socket.

FIGURE A.06-11
Using an insertion tool.

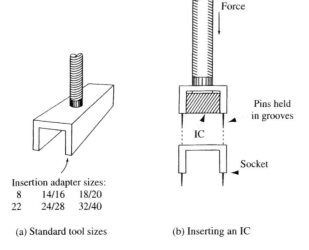

(a) Standard tool sizes   (b) Inserting an IC

Advanced Chips and Packages

Early 16-bit microprocessors, such as Intel's 8088, were victims of the 40-pin DIP, which allowed the CPU to utilize only 40 external connections. This forced the designers to use eight pins for dual purposes (multiplexed address and data information) and reduced the performance of the processor. By the time the 64-pin DIP package became available, microprocessors had graduated to 32-bit data and 32-bit address buses and required sockets with even larger numbers of pins. Figure A.06-12 shows two types of advanced packages—the **PGA (Pin Grid Array)** and the **PLCC (Plastic Leaded Chip Carrier)**—that offer solutions to the connection problem. Both packages allow more pins to be connected in a smaller surface area of board space. In the PGA-style chip, all of the pins are arranged in a two-dimensional structure, protruding from the bottom of a square, ceramic chip housing, with up to 168 connections available. Considerable force is needed to push a PGA chip into its socket, making for very good connections.

FIGURE A.06-12
PGA and PLCC chips and sockets.

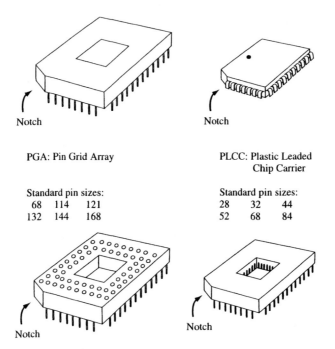

Notch

Notch

PGA: Pin Grid Array

PLCC: Plastic Leaded
Chip Carrier

Standard pin sizes:
68	114	121
132	144	168

Standard pin sizes:
28	32	44
52	68	84

Notch

Notch

The PLCC-style chip consists of springy pins mounted on all four sides of a square-shaped integrated circuit housing. Up to 84 pins are available, enough for a coprocessor chip or other suitable device. The PLCC is pushed straight down into its socket, with equal pressure applied to all four sides. The springy pins push against the connectors on the inside of the PLCC socket with plenty of force, resulting in good connections and the ability to lock the chip in place.

If you examine the motherboard of a newer microcomputer, the odds are good that you will find one, or both, of these types of sockets and chip styles.

ZIF Sockets

New motherboards are equipped with **ZIF (zero insertion force)** sockets that allow you to easily replace the processor. A small handle on the ZIF socket is lifted, releasing the processor by removing pressure on its pins (as indicated in Figure A.06-13). The Pentium processors use ZIF sockets 5 and 7 (320 and 321 pins, respectively), the Pentium Pro uses the 387-pin socket 8, and the Pentium II uses a new connector design called *Slot 1,* a 242-contact rectangular cartridge that plugs into a slot on the motherboard.

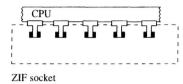

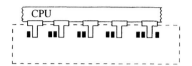

ZIF socket

(a) Lever down, pins locked in place

(b) Lever up, pins are released

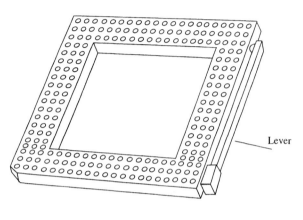

(c) ZIF socket

FIGURE A.06-13
Operation of a ZIF socket.

REVIEW QUESTIONS

True/False

1. When an IC is inserted into a socket, it may be inserted in any direction and still operate properly.
2. To remove an IC, use a pair of long-nose pliers and remove one pin at a time.

Multiple Choice

3. An IC may be connected to a printed circuit board by
   a. Soldering it directly to the board.
   b. Inserting it into an IC socket, where the socket is soldered to the board.
   c. Both a and b.
4. Figure A.06-14 shows some ICs with corresponding pin numbers. The ICs on which the first halves of the pin numbers are correct are
   a. A, C, and D.
   b. B and C.
   c. All have the first half correct.

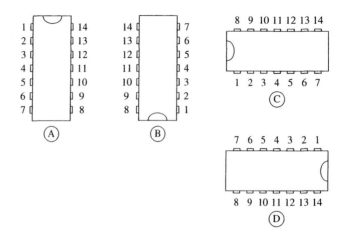

FIGURE A.06-14
IC pin numbers.

5. Referring again to Figure A.06-14, the ICs on which all of the pin numbers are correct are

a. A, C, and D.

b. B and C.

c. A and D.

A.07 Demonstrate an understanding of use of data books and cross reference/technical manuals to specify and requisition electronic components

INTRODUCTION

There are several ways to obtain electronic components. Here are a few of them:

☐ Fill out an order form included in the component catalog.

☐ Place an order with the company over the phone.

☐ FAX a purchase order to the company.

☐ Order the part over the Web.

If you have no product literature at all, the Web is a good place to start. Just search for the component or product you need, and browse the search sites until you find what you are looking for.

If you have a data manual, find the part you require by matching your requirements with the manufacturer's specifications. For example, say you need to order 1000 rectifier diodes. Should you buy 1N4001 or 1N4004 diodes? The answer depends on what the diodes will be used for, thus requiring knowledge of the diode's application. Once you decide on a component, simply call the appropriate sales office listed in the data manual.

Many electronic catalogs come with order forms included (as well as directions on how to order over the phone or through the Web). The order form can be mailed or FAXed to the company, with FAX orders often shipping the same day.

Cross references, such as the line of D.A.T.A. references, list many different part numbers (with their manufacturer) for each type of component, so that the same part may be found by looking in several different locations. Often an equivalent part that is acceptable as a substitution in cases where the original part is no longer available is also listed.

REVIEW QUESTIONS

True/False

1. Components may only be ordered over the phone.

2. Without a data manual, you cannot find the part you need.

Multiple Choice

3. A part is out of date and not manufactured anymore. You should

   a. Redesign the circuit with an available part.

   b. Try to find a replacement part by cross reference.

   c. Either a or b.

4. FAX orders
   a. Typically ship the same day.
   b. Are processed once a week by most companies.
   c. Add unnecessary processing charges.
5. How would you decide on what type of transistor to buy?
   a. Pick the first one in the catalog.
   b. Match specifications with requirements.
   c. Always buy the most expensive component.

A.08 Demonstrate an understanding of the interpretation and creation of electronic schematics, technical drawings, and flow diagrams

This material is covered in detail in every section of topics B through H.

A.09 Demonstrate an understanding of design curves, tables, graphs, and recording of data

This material is covered in detail in every section of topics B through H.

A.10 Demonstrate an understanding of color codes and other component descriptors

INTRODUCTION

In this section we examine the markings and parameters of two common electrical components: the resistor and the capacitor.

Resistors

Figure A.10-1 shows several kinds of fixed **resistors.** As shown in the figure, resistors come in different sizes. The physical size of the resistor determines its wattage rating, which is a measure of how much electrical power it is capable of handling. The larger the wattage rating in watts, the more electrical power the resistor can dissipate. You should never replace a resistor with one that has a *lower* wattage rating.

FIGURE A.10-1
Various fixed resistors (courtesy of SEI Electronics, Inc., Raleigh, NC).

The purpose of a resistor is to *resist,* or limit, the flow of current in a circuit. The value of a resistor is measured in ohms. The more ohms a resistor has, the more it will limit the flow of electrical current. The symbol for ohms is the Greek letter omega (Ω). Thus, 24 ohms is written as 24 Ω.

The value of a resistor in ohms is indicated by the colors of the bands on the resistor. Table A.10-1 shows the meanings of the color-coded resistor bands. Figure A.10-2 shows a diagram of the color-coded bands.

Resistor networks are shown in Figure A.10-3. A resistor network consists of several resistors connected in a specified way. These networks are used to simplify the construction of digital circuits so that only one network needs to be inserted into the circuit rather than several separate resistors.

The schematic symbols for resistors are shown in Figure A.10-4, which illustrates several different circuit connections of resistors. Note that a variable resistor has an arrow drawn through it to indicate that its value can be changed.

Most resistor values are quite large, usually in thousands or even millions of ohms. A notation called **metric notation** is used to represent these values. Table A.10-2 lists the metric notation used for typical resistors.

TABLE A.10-1
Resistor color code.

Digit* (Bands 1 and 2)	Color
0	Black
1	Brown
2	Red
3	Orange
4	Yellow
5	Green
6	Blue
7	Violet
8	Gray
9	White

Tolerance (Band 4)	Color
5%	Gold
10%	Silver
20%	No band

*Or multiplier for band 3.

FIGURE A.10-2
Color-coded resistor bands.

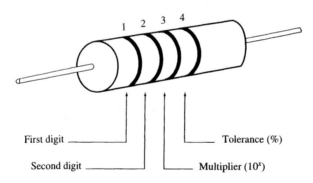

First digit —

Second digit —

Tolerance (%)

Multiplier (10^x)

FIGURE A.10-3
Resistor networks (courtesy of Bourns, Inc.).

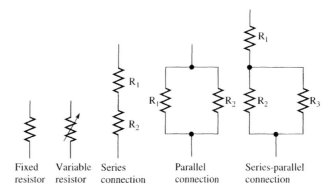

FIGURE A.10-4
Resistor symbols and typical circuit connections.

TABLE A.10-2
Metric notation for resistors

Metric Prefix	Meaning	Examples
Kilo (k)	×1000	$12 \text{ k}\Omega = 12 \times 1000 = 12,000 \ \Omega$ $4.3 \text{ k}\Omega = 4.3 \times 1000 = 4300 \ \Omega$ $280 \text{ k}\Omega = 280 \times 1000 = 280,000 \ \Omega$
Mega (M)	×1,000,000	$18 \text{ M}\Omega = 18 \times 1,000,000$ $= 18,000,000 \ \Omega$ $8.2 \text{ M}\Omega = 8.2 \times 1,000,000$ $= 8,200,000 \ \Omega$ $750 \text{ M}\Omega = 750 \times 1,000,000$ $= 750,000,000 \ \Omega$

Capacitors

As shown in Figure A.10-5, **capacitors** come in a variety of shapes and sizes. The purpose of a capacitor is to store an electrical charge. Because of this capability, one of the uses of capacitors is to help maintain a steady supply of voltage to circuits within the computer.

The value of a capacitor is measured by farad (F). A capacitor also has a voltage rating. When replacing a capacitor, you should use a capacitor with the same value in farads and never one with a lower voltage rating.

FIGURE A.10-5
Capacitors (courtesy of KEMET Electronics Corp.).

Capacitor values are usually quite small. Typical values for a capacitor are in the range of millionths of a farad. As with resistors, metric notation is used to represent the value of a capacitor. Table A.10-3 lists the metric notation used for typical capacitors.

TABLE A.10-3
Metric notation for capacitors.

Metric Prefix	Meaning	Examples
Micro (μ)	×0.000001	12 μF = 0.000012 F 3.5 μF = 0.0000035 F 0.1 μF = 0.0000001 F
Pico (p)	×0.000000000001	12 pF = 0.000000000012 F 3.5 pF = 0.0000000000035 F 100 pF = 0.000000000100 F

REVIEW QUESTIONS

True/False

1. A resistor opposes the flow of current.

2. A resistor with a color code of *red-black-orange* would have a value of 2 kΩ.

Multiple Choice

3. The value of a capacitor is measured in
   a. Ohms.
   b. Henries.
   c. Farads.

4. The important consideration of electrolytic capacitors is that they
   a. Are polarized.
   b. Have small values.
   c. Should not be used in digital circuits.

5. A capacitor will
   a. Store an electrical charge.
   b. Cause current to flow in one direction.
   c. Oppose a change in current.

A.11 Demonstrate an understanding of site electrical and environmental survey

INTRODUCTION

It is important to be familiar with your surroundings when placed in a new environment. Where are the emergency exits? How is a fire reported? What are the workplace rules? Where is the nearest first-aid kit?

In general, spend some time exploring a new laboratory or office complex. Look at everything, from the plumbing to the number of elevators. In an educational environment, it is not uncommon to find hazardous chemicals, radioactive material, large industrial-type equipment, high voltage, or other sensitive equipment or locations. Know your way around. Stay out of areas you have no business in, if only to prevent an embarrassing false alarm when you accidentally go through a protected exit.

REVIEW QUESTIONS

Regarding your workplace, educational setting, or other appropriate environment:

1. What are the emergency phone numbers?

2. Where is first aid available?

3. How many exits are there in your building?

4. Can you sketch an overhead view of your environment?

5. What is the fire-alarm code?

A.12 Demonstrate the use of listening skills or assistive devices to assess signs and symptoms of malfunctions

INTRODUCTION

Your senses all play a role in an effective troubleshooting experience. For example, you may smell a burned component, feel an excessive amount of heat coming from a component, or hear an electrolytic capacitor hissing as it gets ready to explode. Knowing how components work and their operating characteristics assists your troubleshooting by allowing you to match what you are experiencing with your senses with what you should expect. For instance, if a capacitor feels warm it is probably bad, since capacitors normally run cool (they use no real power). A low-frequency hum in the audio output of an amplifier may lead you to examine the power supply, finding a significant amount of 120 Hz ripple due to an overworked filter capacitor.

In general, it pays to be a good observer when troubleshooting. This means observing with all your senses. Just looking at a circuit may not be enough.

REVIEW QUESTIONS

True/False

1. A capacitor should feel cool to the touch.

2. You can sometimes smell a bad component.

Multiple Choice

3. Which sense is used when locating a burned component?

   a. Touch.

   b. Smell.

   c. Sight.

4. Which sense is used when locating a missing component?

   a. Touch.

   b. Smell.

   c. Sight.

5. Which sense is used when locating an improperly installed component?

   a. Touch.

   b. Smell.

   c. Sight.

B DC Circuits

B.01 Demonstrate an understanding of sources of electricity in DC circuits

INTRODUCTION

All electrical circuits require a voltage source for proper operation. The voltage source provides the ability to create current in the circuit components. A DC voltage source creates a constant current in the components. In this section we will examine several types of DC voltage sources.

Sources of Voltage

The Battery   A voltage source is a **source** of potential energy that is also called *electromotive force* (emf). The **battery** is one type of voltage source that converts chemical energy into electrical energy. A voltage exists between the electrodes (terminals) of a battery, as shown by a voltaic cell in Figure B.01-1. One electrode is positive and the other negative as a result of the separation of charges caused by the chemical action when two different conducting materials are dissolved in the electrolyte.

FIGURE B.01-1
A voltaic cell converts chemical energy into electrical energy.

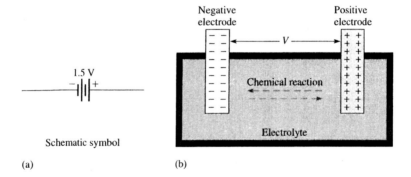

Negative electrode   Positive electrode

1.5 V

Schematic symbol

(a)   (b)

Battery cells are generally classified as *primary cells,* which cannot be recharged, and *secondary cells,* which can be recharged by reversal of the chemical action. Examples of primary cells are carbon-zinc cells and alkaline cells. Examples of secondary cells are lead-acid cells and nickel-cadmium (Ni-Cd) cells. The amount of voltage provided by a cell varies depending on the materials used in the construction of the cell. For example, a carbon-zinc cell (flashlight battery) produces 1.5 V, whereas a lead-acid cell used in automobile batteries produces 2 V. Individual cells are combined to produce higher voltages. A 9 V battery is made up of six 1.5 V cells and a 12 V car battery consists of six 2 V cells.

The Electronic Power Supply   These voltage sources convert the AC voltage from the wall outlet to a constant (DC) voltage which is available across two output terminals, as indicated in Figure B.01-2(a). Typical commercial **power supplies** are shown in part (b).

FIGURE B.01-2
Electronic power supplies.
Part (b) courtesy of B & K
Precision Corp.

(a)

(b)

Power supplies use a technique called *rectification* to convert an AC input voltage into a DC output voltage. Special circuitry is typically included to allow voltage and/or current *regulation* (varying or limiting the amount of voltage or current available at the output).

The Solar Cell The operation of solar cells is based on the principle of *photovoltaic action,* which is the process whereby light energy is converted directly into electrical energy. A basic solar cell consists of two layers of different semiconductive materials joined together to form a junction. When one layer is exposed to light, many electrons acquire enough energy to break away from their parent atoms and cross the junction. This process forms negative ions on one side of the junction and positive ions on the other; thus, a potential difference (voltage) is developed.

The Electric Generator **Generators** convert mechanical energy into electrical energy using a principle called *electromagnetic induction*. A conductor is rotated through a magnetic field, and a voltage is produced across the conductor. A typical generator is pictured in Figure B.01-3.

FIGURE B.01-3
Cutaway view of a typical
DC generator.

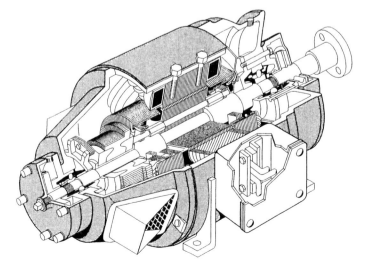

True/False

1. Batteries rely on chemical reactions to create voltage.
2. Primary cells can be recharged.

Multiple Choice

3. A 6 V battery can be made using
   a. Four 1.5 V cells.
   b. Three 2 V cells.
   c. Either a or b.
4. Rectification is the process of
   a. Converting AC to DC.
   b. Converting DC to AC.
   c. Maintaining a constant output voltage.
5. A solar cell creates voltage through the process of
   a. Electromagnetic induction.
   b. Regulation.
   c. Photovoltaic action.

B.02 Demonstrate an understanding of principles and operation of batteries

INTRODUCTION

Basically, an electric **circuit** consists of a voltage source, a load, and a path for current between the source and the **load.** Figure B.02-1 shows an example of a simple electric circuit: a battery connected to a lamp with two conductors (wires). The battery is the voltage source, the lamp is the load on the battery because it draws current from the battery, and the two wires provide the current path from the negative terminal of the battery to the lamp and back to the positive terminal of the battery, as shown in part (b). There is current through the filament of the lamp (which has a resistance), causing it to emit visible light. Current through the battery is produced by chemical action. In many practical cases, one terminal of the battery is connected to a **ground** point. For example, in automobiles, the negative battery terminal is connected to the metal chassis of the car. The chassis is the ground for the automobile electrical system. An electrical schematic of the circuit in part (a) is shown in part (c).

FIGURE B.02-1
A simple electric circuit.

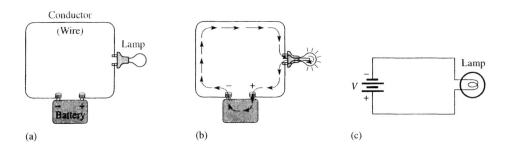

(a) (b) (c)

Ampere-hour Ratings of Batteries

Batteries convert chemical energy into electrical energy. Because of their limited source of chemical energy, batteries have a certain capacity that limits the amount of time over which they can produce a given power level. This capacity is measured in ampere-hours (Ah). The **ampere-hour rating** determines the length of time that a battery can deliver a certain amount of current to a load at the rated voltage.

A rating of one ampere-hour means that a battery can deliver one ampere of current to a load for one hour at the rated voltage output. This same battery can deliver two amperes for one-half hour. The more current the battery is required to deliver, the shorter is the life of the battery. In practice, a battery usually is rated for a specified current level and output voltage. For example, a 12 V automobile battery may be rated for 70 Ah at 3.5 A. This means that it can produce 3.5 A for 20 h.

REVIEW QUESTIONS

True/False

1. A battery uses a chemical reaction to generate voltage.
2. Batteries have unlimited life.

Multiple Choice

3. A rechargeable battery is called a
   a. Primary cell.
   b. Secondary cell.
   c. Bipolar cell.

4. A battery is rated at 200 mAh. How long can it supply a current of 5 mA?
   a. 4 hours.
   b. 40 hours.
   c. 1000 hours.

5. A battery rated at 50 Ah must last for 10 days. What current can be drawn from the battery?
   a. 20.8 mA.
   b. 208 mA.
   c. 2.08 A.

B.03 Demonstrate an understanding of the meaning of and relationships among and between voltage, current, resistance, and power in DC

INTRODUCTION

Ohm's law is the foundation of all circuit analysis and design, relating voltage, current, and resistance. In this section we examine these relationships and the associated power requirements.

EXAMPLE B.03-1

Using the Ohm's law formula, verify that the current through a 10 Ω resistor increases when the voltage is increased from 5 V to 20 V.

Solution   The following calculations show that the current increases from 0.5 A to 2 A.

For $V = 5$ V,

$$I = \frac{V}{R} = \frac{5 \text{ V}}{10 \text{ } \Omega} = 0.5 \text{ A}$$

For $V = 20$ V,

$$I = \frac{V}{R} = \frac{20 \text{ V}}{10 \text{ } \Omega} = 2 \text{ A}$$

EXAMPLE B.03-2

Use the Ohm's law formula to calculate the voltage across a 100 Ω resistor when the current is 2 A.

Solution $V = IR = (2 \text{ A})(100 \text{ } \Omega) = 200 \text{ V}$

EXAMPLE B.03-3

Use the Ohm's law formula to calculate the resistance in a circuit when the voltage is 12 V and the current is 0.5 A.

Solution $R = \frac{V}{I} = \frac{12 \text{ V}}{0.5 \text{ A}} = 24 \text{ } \Omega$

EXAMPLE B.03-4

Calculate the power in each of the three circuits of Figure B.03-1.

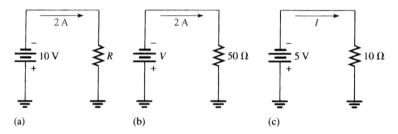

(a)   (b)   (c)

FIGURE B.03-1

Solution   In circuit (a), V and I are known. The power is determined as follows:

$$P = VI = (10 \text{ V})(2 \text{ A}) = 20 \text{ W}$$

In circuit (b), I and R are known. Therefore,

$$P = I^2R = (2 \text{ A})^2(50 \text{ }\Omega) = 200 \text{ W}$$

In circuit (c), V and R are known. Therefore,

$$P = \frac{V^2}{R} = \frac{(5 \text{ V})^2}{10 \text{ }\Omega} = 2.5 \text{ W}$$

REVIEW QUESTIONS

True/False

1. When voltage is increased and resistance is kept constant, the resulting current decreases.
2. A current of 2 A flowing through a 10 ohm resistor produces a power of 40 W.

Multiple Choice

3. If $V = 10$ V and $I = 2$ A then R equals
   a. 20 ohms.
   b. 5 ohms.
   c. 12 ohms.
4. What is the voltage across a 5 ohm resistor that is using 20 W of power?
   a. 100 V.
   b. 20 V.
   c. 10 V.
5. What is the current through a 10 ohm resistor that has 50 V across it?
   a. 0.2 A.
   b. 5 A.
   c. 500 A.

B.04 Demonstrate an understanding of measurement of resistance of conductors and insulators and the computation of conductance

INTRODUCTION

In this section we look at two instruments designed to measure resistance, and we review the relationship between resistance and conductance.

Measuring Resistance with an Ohmmeter

To measure resistance, connect the ohmmeter across the resistor. *The resistor must first be removed or disconnected from the circuit.* This procedure is shown in Figure B.04-1.

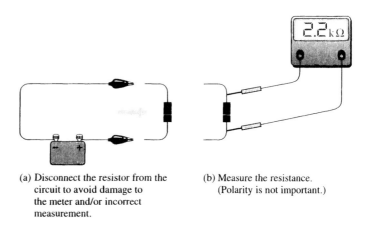

(a) Disconnect the resistor from the   (b) Measure the resistance.
circuit to avoid damage to   (Polarity is not important.)
the meter and/or incorrect
measurement.

FIGURE B.04-1
Example of using an ohmmeter to measure resistance.

Measuring Resistance with the Multimeter

The meter must be set to the ohmmeter function to measure current. Before the ohmmeter is connected, the resistance to be measured must be disconnected from the circuit. Before disconnecting any component, first turn the power supply off. Refer to Figure B.04-2.

☐ For which component is the resistance measured?

☐ For which measurement (A or B) will the lamp be brighter when the circuit is reconnected and the power turned on? Explain.

Conductance   The reciprocal of resistance is **conductance,** symbolized by *G*. It is a measure of the ease with which current is established. The formula is

$$G = \frac{1}{R}$$

The unit of conductance is the siemens, abbreviated S. Occasionally, the earlier unit of mho is still used.

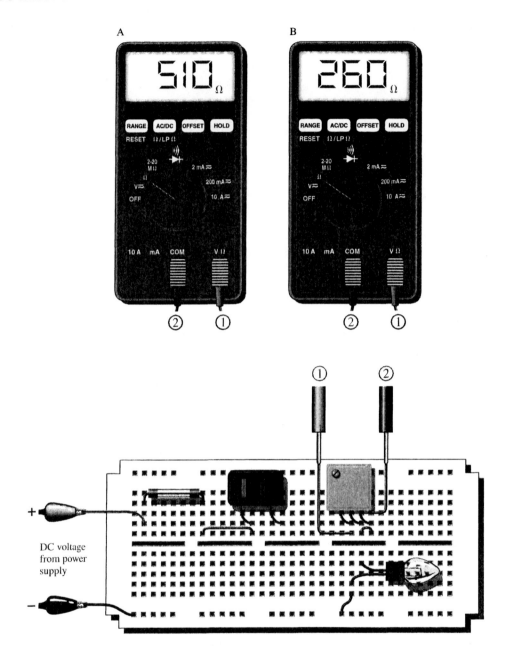

FIGURE B.04-2
Resistance measurements.

REVIEW QUESTIONS

True/False

1. Conductance is proportional to resistance.
2. Resistance measurements can be made in circuit.

Multiple Choice

3. When measuring resistance,

    a. The order of the meter leads placed on the resistor is important.

    b. The order of the meter leads is unimportant.

    c. Only the red meter lead is used.

4. Given 10 ohms of resistance, what is the associated conductance?

    a. 0.1 mhos.

    b. 1 mho.

    c. 10 mhos.

5. Which has more resistance?

    a. A 2 mho conductor.

    b. A 5 mho conductor.

    c. Neither.

B.05 Demonstrate an understanding of application of Ohm's law to series, parallel, and series-parallel circuits

INTRODUCTION

This section introduces the formulas frequently used during series, parallel, and series-parallel circuit analysis. Additional details are provided in B.08, B.11, and B.14.

EXAMPLE B.05-1

Find the current in the circuit of Figure B.05-1.

FIGURE B.05-1

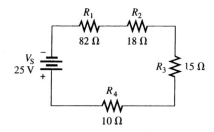

Solution   The current is determined by the voltage and the total resistance. First, calculate the total resistance.

$$R_T = R_1 + R_2 + R_3 + R_4 = 82 \ \Omega + 18 \ \Omega + 15 \ \Omega + 10 \ \Omega = 125 \ \Omega$$

Next, use Ohm's law to calculate the current.

$$I = \frac{V_S}{R_T} = \frac{25 \text{ V}}{125 \ \Omega} = 0.2 \text{ A} = 200 \text{ mA}$$

Remember, the same current exists at all points in the circuit. Thus, each resistor has 200 mA through it.

EXAMPLE B.05-2

The current in the circuit of Figure B.05-2 is 1 mA. For this amount of current, what must the source voltage V_S be?

FIGURE B.05-2

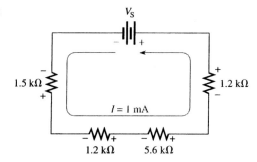

Solution   In order to calculate V_S, first determine R_T.

$$R_T = 1.2 \text{ k}\Omega + 5.6 \text{ k}\Omega + 1.2 \text{ k}\Omega + 1.5 \text{ k}\Omega = 9.5 \text{ k}\Omega$$

Now use Ohm's law to get V_S.

$$V_S = IR_T = (1 \text{ mA})(9.5 \text{ k}\Omega) = 9.5 \text{ V}$$

Additional details are provided in B.08.

EXAMPLE B.05-3

Find the total current produced by the battery in Figure B.05-3.

FIGURE B.05-3

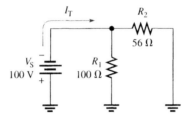

Solution   The battery "sees" a total parallel resistance that determines the amount of current that it generates. First, calculate R_T.

$$R_T = \frac{R_1 R_2}{R_1 + R_2} = \frac{(100 \ \Omega)(56 \ \Omega)}{100 \ \Omega + 56 \ \Omega} = \frac{5600 \ \Omega^2}{156 \ \Omega} = 35.9 \ \Omega$$

The battery voltage is 100 V. Use Ohm's law to find I_T.

$$I_T = \frac{100 \text{ V}}{35.9 \ \Omega} = 2.79 \text{ A}$$

EXAMPLE B.05-4

Determine the current through each resistor in the parallel circuit of Figure B.05-4.

FIGURE B.05-4

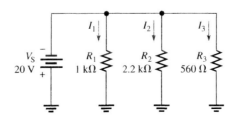

Solution   The voltage across each resistor (branch) is equal to the source voltage. That is, the voltage across R_1 is 20 V, the voltage across R_2 is 20 V, and the voltage across R_3 is 20 V. The current through each resistor is determined as follows:

$$I_1 = \frac{V_S}{R_1} = \frac{20\ V}{1\ k\Omega} = 20\ mA$$

$$I_2 = \frac{V_S}{R_2} = \frac{20\ V}{2.2\ k\Omega} = 9.1\ mA$$

$$I_3 = \frac{V_S}{R_3} = \frac{20\ V}{560\ \Omega} = 35.7\ mA$$

Additional details are provided in B.11.

OHM'S LAW IN SERIES-PARALLEL CIRCUITS

A series-parallel circuit is nothing more than two or more individual series or parallel sub-circuits connected together. Through the use of series and parallel resistor simplification, eventually all series-parallel circuits may be reduced to simple series or parallel combinations. Thus, Ohm's law applies to series-parallel analysis as it does for series and parallel analysis.

Additional details are provided in B.14.

REVIEW QUESTIONS

True/False

1. In a series circuit, an increase in resistance causes a decrease in current.

2. In a parallel circuit, lowering one of the resistors increases the total circuit current.

Multiple Choice

3. Two 47 ohm resistors are connected across a 12 V source. What is the voltage across each resistor?

   a. 4.7 V.

   b. 6 V.

   c. 12 V.

4. Two 100 ohm resistors are connected in parallel across a 5 V source. The total circuit current is

   a. 25 mA.

   b. 50 mA.

   c. 100 mA.

5. A series-parallel circuit requires

   a. Ordinary series and parallel calculations.

   b. Special series-parallel calculations.

   c. Parallel first, then series calculations.

B.06 Demonstrate an understanding of magnetic properties of circuits and devices

INTRODUCTION

A permanent magnet, such as the bar magnet shown in Figure B.06-1, has a magnetic field surrounding it. The magnetic field consists of lines of force, or flux lines, that radiate from the north pole (N) to the south pole (S) and back to the north pole through the magnetic material. For clarity, only a few lines of force are shown in the figure. Imagine, however, that many lines surround the magnet in three dimensions. The lines shrink to the smallest possible size and blend together, although they do not touch. This effectively forms a continuous magnetic field surrounding the magnet.

FIGURE B.06-1
Magnetic lines of force around a bar magnet.

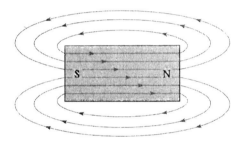

Gray lines represent a few magnetic lines of force in the magnetic field.

Attraction and Repulsion of Magnetic Poles

When unlike poles of two permanent magnets are placed close together, an attractive force is produced by the magnetic fields, as indicated in Figure B.06-2 on the next page. When two like poles are brought close together, they repel each other, as shown in part (b).

Altering a Magnetic Field

When a nonmagnetic material such as paper, glass, wood, or plastic is placed in a magnetic field, the lines of force are unaltered, as shown in Figure B.06-3(a). However, when a magnetic material such as iron is placed in the magnetic field, the lines of force tend to change course and pass through the iron rather than through the surrounding air. They do so because the iron provides a magnetic path that is more easily established than that of air. Figure B.06-3(b) illustrates this principle.

FIGURE B.06-2
Magnetic attraction and repulsion.

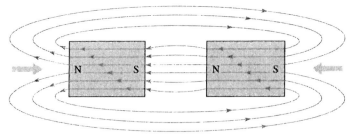

(a) Unlike poles attract.

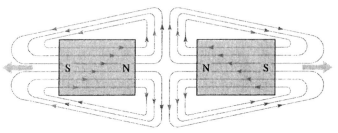

(b) Like poles repel.

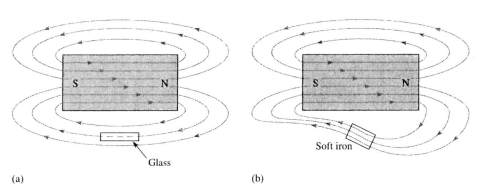

(a)   (b)

FIGURE B.06-3
Effect of (a) nonmagnetic and (b) magnetic materials on a magnetic field.

Magnetic Flux (ϕ)

The group of force lines going from the north pole to the south pole of a magnet is called the **magnetic flux,** symbolized by ϕ (the Greek letter phi). The number of lines of force in a magnetic field determines the value of the flux. The more lines of force, the greater the flux and the stronger the magnetic field.

The unit of magnetic flux is the **weber** (Wb). One weber equals 10^8 lines. The weber is a very large unit; thus, in most practical situations, the microweber (μWb) is used. One microweber equals 100 lines of magnetic flux.

ELECTROMAGNETISM

Current produces a magnetic field, called an **electromagnetic field,** around a conductor, as illustrated in Figure B.06-4. The invisible lines of force of the magnetic field form a concentric circular pattern around the conductor and are continuous along its length.

Although the magnetic field cannot be seen, it is capable of producing visible effects. For example, if a current-carrying wire is inserted through a sheet of paper in a perpendicular direction, iron filings placed on the surface of the paper arrange themselves along the magnetic lines of force in concentric rings, as illustrated in Figure B.06-5(a). Part (b) of the figure illustrates that the north pole of a compass placed in the electromagnetic field will point in the direction of the lines of force. The field is stronger closer to the conductor and becomes weaker with increasing distance from the conductor.

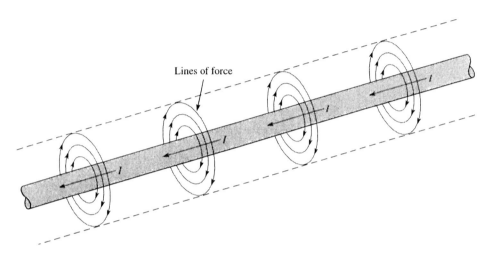

FIGURE B.06-4
Magnetic field around a current-carrying conductor.

FIGURE B.06-5
Visible effects of an electromagnetic field.

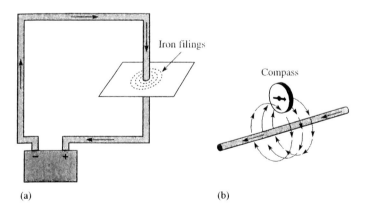

(a) (b)

Direction of the Lines of Force

The direction of the lines of force surrounding the conductor is indicated in Figure B.06-6. When the direction of current is right to left, as in part (a), the lines are in a clockwise direction. When current is left to right, as in part (b), the lines are in a counterclockwise direction.

FIGURE B.06-6
Magnetic lines of force around a current-carrying conductor.

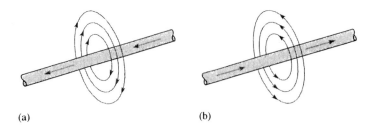

(a)   (b)

FIGURE B.06-7
Illustration of left-hand rule.

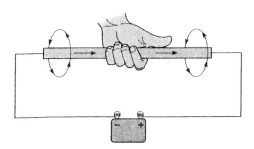

Left-Hand Rule   An aid to remembering the direction of the lines of force is illustrated in Figure B.06-7. Imagine that you are grasping the conductor with your left hand, with your thumb pointing in the direction of current. Your fingers point in the direction of the magnetic lines of force.

The Electromagnet

An electromagnet is based on the properties that you have just learned. A basic electromagnet is simply a coil of wire wound around a core material that can be easily magnetized.

The shape of the electromagnet can be designed for various applications. For example, Figure B.06-8 shows a U-shaped magnetic core. When the coil of wire is connected to a battery and there is current, as shown in part (a), a magnetic field is established as indicated. If the current is reversed, as shown in part (b), the direction of the magnetic field is also reversed. The closer the north and south poles are brought together, the smaller the air gap between them becomes, and the easier it becomes to establish a magnetic field, because the reluctance is lessened.

FIGURE B.06-8
Reversing the current in the coil causes the electromagnetic field to reverse.

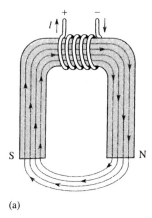
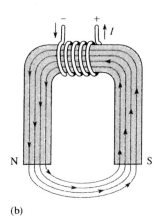

(a)   (b)

A good example of one application of an electromagnet is the process of recording on magnetic tape. In this situation, the recording head is an electromagnet with a narrow air gap, as shown in Figure B.06-9. Current sets up a magnetic field across the air gap, and as the recording head passes over the magnetic tape, the tape is permanently magnetized. In digital recording, for example, the tape is magnetized in one direction for a binary 1 and in the other direction for a binary 0, as illustrated in the figure. This magnetization in different directions is accomplished by reversing the coil current in the recording head.

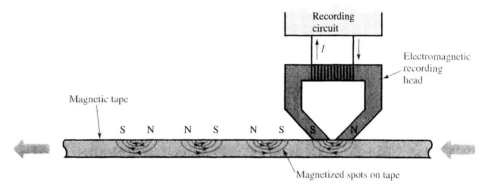

FIGURE B.06-9
An electromagnetic recording head recording information on magnetic tape by magnetizing the tape as it passes by.

The Relay

Relays differ from solenoids in that the electromagnetic action is used to open or close electrical contacts rather than to provide mechanical movement. Figure B.06-10 shows the basic operation of a relay with one normally open (NO) contact and one normally closed (NC) contact (single-pole–double-throw). When there is no coil current, the armature is held against the upper contact by the spring, thus providing continuity from terminal 1 to terminal 2, as shown in part (a) of the figure. When energized with coil current, the armature is pulled down by the attractive force of the electromagnetic field and makes connection with the lower contact to provide continuity from terminal 1 to terminal 3, as shown in Figure B.06-10(b).

A typical relay and its schematic symbol are shown in Figure B.06-11.

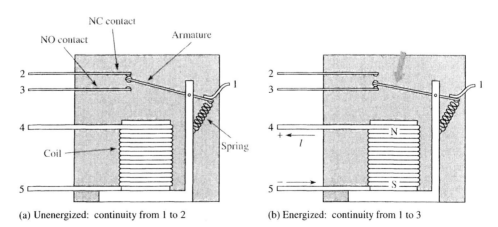

(a) Unenergized: continuity from 1 to 2   (b) Energized: continuity from 1 to 3

FIGURE B.06-10
Basic structure of a single-pole–double-throw relay.

FIGURE B.06-11
A typical relay.

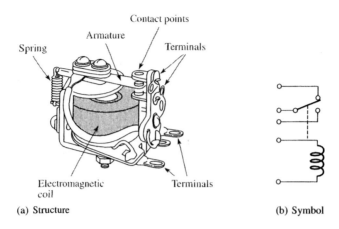

(a) Structure   (b) Symbol

FIGURE B.06-11
A typical relay.

The Speaker

Permanent-magnet speakers are commonly used in stereos, radios, and TVs, and their operation is based on the principle of electromagnetism. A typical speaker is constructed with a permanent magnet and an electromagnet, as shown in Figure B.06-12(a). The cone of the speaker consists of a paper-like diaphragm to which is attached a hollow cylinder with a coil around it, forming an electromagnet. One of the poles of the permanent magnet is positioned within the cylindrical coil. When there is current through the coil in one direction, the interaction of the permanent magnetic field with the electromagnetic field causes the cylinder to move to the right, as indicated in Figure B.06-12(b). Current through the coil in the other direction causes the cylinder to move to the left, as shown in part (c).

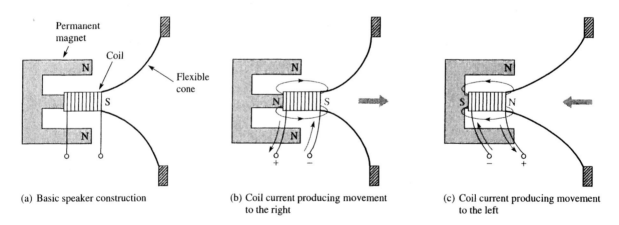

(a) Basic speaker construction   (b) Coil current producing movement   (c) Coil current producing movement
                                 to the right   to the left

FIGURE B.06-12
Basic speaker operation.

The movement of the coil cylinder causes the flexible diaphragm also to move in or out, depending on the direction of the coil current. The amount of coil current determines the intensity of the magnetic field, which controls the amount that the diaphragm moves.

As shown in Figure B.06-13, when an audio signal (voice or music) is applied to the coil, the current varies in both direction and amount. In response, the diaphragm will vibrate in and out by varying amounts and at varying rates corresponding to the audio signal. Vibration in the diaphragm causes the air that is in contact with it to vibrate in the same manner. These air vibrations move through the air as sound waves.

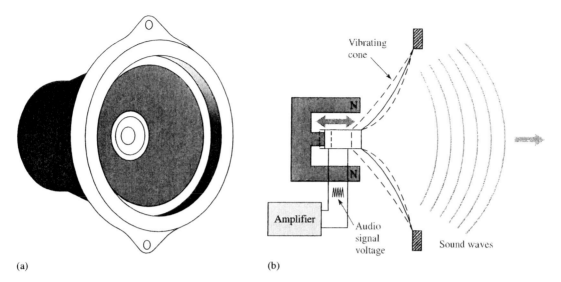

FIGURE B.06-13
The speaker converts audio signal voltages into sound waves.

True/False

1. The unit of magnetic flux is the weber.
2. The unit of magnetic flux density is the Tesla.

Multiple Choice

3. Current flowing through a wire produces a(n)
    a. Atomic field.
    b. Electromagnetic field.
    c. Density field.
4. To determine the direction of a magnetic field, use the
    a. Left-hand rule.
    b. Right-hand rule.
    c. Flux density rule.
5. A curve that relates magnetizing force and flux density is the
    a. Lenz curve.
    b. Faraday curve.
    c. Hysteresis curve.

B.07 Demonstrate an understanding of the physical, electrical characteristics of capacitors and inductors

INTRODUCTION

In this section we examine the properties of capacitors and inductors. Additional details are provided in B.21.

THE BASIC CAPACITOR

Basic Construction

In its simplest form, a **capacitor** is an electrical device constructed of two parallel conductive plates separated by an insulating material called the **dielectric.** Connecting leads are attached to the parallel plates. A basic capacitor is shown in Figure B.07-1, and the schematic symbol is shown in part (b).

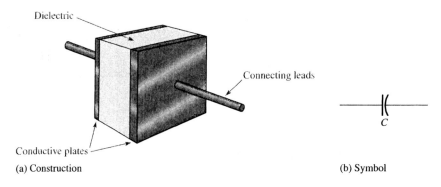

Dielectric

Connecting leads

C

Conductive plates

(a) Construction

(b) Symbol

FIGURE B.07-1
The basic capacitor.

How a Capacitor Stores Charge

In the neutral state, both plates of a capacitor have an equal number of free electrons, as indicated in Figure B.07-2(a). When the capacitor is connected to a voltage source through a resistor, as shown in part (b), electrons (negative charge) are removed from plate *A,* and an equal number are deposited on plate *B.* As plate *A* loses electrons and plate *B* gains electrons, plate *A* becomes positive with respect to plate *B.* During this charging process, electrons flow only through the connecting leads and the source. No electrons flow through the dielectric of the capacitor because it is an insulator. The movement of electrons ceases when the voltage across the capacitor equals the source voltage, as indicated in Figure B.07-2(c). If the capacitor is disconnected from the source, it retains the stored charge for a long period of time (the length of time depends on the type of capacitor) and still has the voltage across it, as shown in Figure B.07-2(d). Actually, the charged capacitor can be considered a temporary battery.

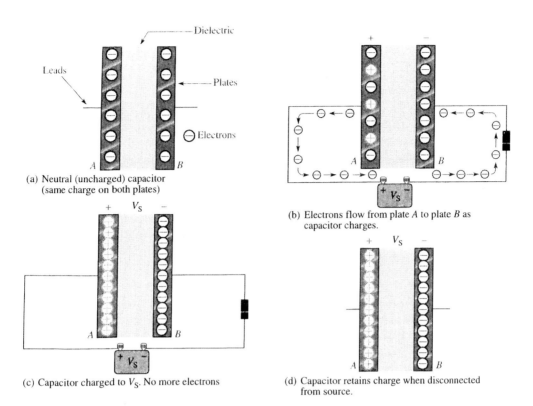

(a) Neutral (uncharged) capacitor
(same charge on both plates)

(b) Electrons flow from plate *A* to plate *B* as
capacitor charges.

(c) Capacitor charged to V_S. No more electrons

(d) Capacitor retains charge when disconnected
from source.

FIGURE B.07-2
Illustration of a capacitor storing charge.

Capacitance

The amount of charge that a capacitor can store per unit of voltage across its plates is its **capacitance,** designated *C.* That is, capacitance is a measure of a capacitor's ability to store charge. The more charge per unit of voltage a capacitor can store, the greater its capacitance, as expressed by the following formula:

$$C = \frac{Q}{V}$$

where *C* is capacitance, *Q* is charge, and *V* is voltage. By rearranging the equation, you can obtain two other forms for calculating *Q* and *V.*

$$Q = CV$$

$$V = \frac{Q}{C}$$

The Unit of Capacitance   The **farad (F)** is the basic unit of capacitance, and the **coulomb (C)** is the unit of electrical charge.

> **One farad is the amount of capacitance when one coulomb of charge is stored with one volt across the plates.**

Most capacitors that are used in electronics work have capacitance values in microfarads (μF) and picofarads (pF).

> **A microfarad is one-millionth of a farad ($1 \ \mu$F $= 1 \times 10^{-6}$ F).**

> **A picofarad is one-trillionth of a farad (1 pF $= 1 \times 10^{-12}$ F).**

Conversions for farads, microfarads, and picofarads are given in Table B.07-1.

TABLE B.07-1
Conversions for farads, microfarads, and picofarads.

To Convert From	To	Move the Decimal Point
Farads	Microfarads	6 places to right ($\times 10^6$)
Farads	Picofarads	12 places to right ($\times 10^{12}$)
Microfarads	Farads	6 places to left ($\times 10^{-6}$)
Microfarads	Picofarads	6 places to right ($\times 10^6$)
Picofarads	Farads	12 places to left ($\times 10^{-12}$)
Picofarads	Microfarads	6 places to left ($\times 10^{-6}$)

EXAMPLE B.07-1

(a) A certain capacitor stores 50 microcoulombs (50 μC) when 10 V are applied across its plates. What is its capacitance?

(b) A 2 μF capacitor has 100 V across its plates. How much charge does it store?

(c) Determine the voltage across a 100 pF capacitor that is storing 2 μC of charge.

Solution

(a) $C = \dfrac{Q}{V} = \dfrac{50\ \mu C}{10\ V} = 5\ \mu F$

(b) $Q = CV = (2\ \mu F)(100\ V) = 200\ \mu C$

(c) $V = \dfrac{Q}{C} = \dfrac{2\ \mu C}{100\ pF} = 0.02 \times 10^6 = 20 \times 10^3\ V = 20\ kV$

EXAMPLE B.07-2

Convert the following values to microfarads:

(a) 0.00001 F   (b) 0.005 F   (c) 1000 pF   (d) 200 pF

Solution

(a) $0.00001\ F \times 10^6 = 10\ \mu F$ (b) $0.005\ F \times 10^6 = 5000\ \mu F$

(c) $1000\ pF \times 10^{-6} = 0.001\ \mu F$ (d) $200\ pF \times 10^{-6} = 0.0002\ \mu F$

Capacitor Labeling

Capacitor values are indicated on the body of the capacitor either by typographical labels or by color codes. Typographical labels consist of letters and numbers that indicate various parameters such as capacitance, voltage rating, and tolerance.

Some capacitors carry no unit designation for capacitance. In these cases, the units are implied by the value indicated and are recognized by experience. For example, a ceramic capacitor marked .001 or .01 has units of microfarads because picofarad values that small are not available. As another example, a ceramic capacitor labeled 50 or 330 has units of picofarads because microfarad units that large normally are not available in this type. In some cases a 3-digit designation is used. The first two digits are the first two digits of the capacitance value. The third digit is the number of zeros after the second digit. For example, 103 means 10,000 pF. In some instances, the units are labeled as pF or μF; sometimes the microfarad unit is labeled as MF or MFD.

A voltage rating appears on some types of capacitors with WV or WVDC and is omitted on others. When it is omitted, the voltage rating can be determined from information supplied by the manufacturer. The tolerance of the capacitor is usually labeled as a percentage, such as ±10%. The temperature coefficient is indicated by a *parts per million* marking. This type of label consists of a P or N followed by a number. For example, N750 means a negative temperature coefficient of 750 ppm/C°, and P30 means a positive temperature coefficient of 30 ppm/C°. An NP0 designation means that the positive and negative coefficients are zero; thus, the capacitance does not change with temperature. Certain types of capacitors are color coded.

THE BASIC INDUCTOR

When a length of wire is formed into a coil, as shown in Figure B.07-3, it becomes a basic **inductor.** Current through the coil produces an electromagnetic field. The magnetic lines of force around each loop (turn) in the **winding** of the coil effectively add to the lines of force around the adjoining loops, forming a strong magnetic field within and around the coil, as shown. The net direction of the total magnetic field creates a north and a south pole.

FIGURE B.07-3
A coil of wire forms an inductor. When there is current through it, a three-dimensional electromagnetic field is created, surrounding the coil in all directions.

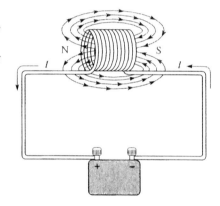

Self-Inductance

When there is current through an inductor, an electromagnetic field is established. When the current changes, the electromagnetic field also changes. An increase in current expands the field, and a decrease in current reduces it. Therefore, a changing current produces a changing electromagnetic field around the inductor (also known as **coil** or **choke**). In turn, the changing electromagnetic field causes an **induced voltage** across the coil in a direction to oppose the change in current. This property is called *self-inductance,* but it is usually referred to as simply **inductance.** Inductance is symbolized by *L*.

> **Inductance is a measure of a coil's ability to establish an induced voltage as a result of a change in its current, and that induced voltage is in a direction to oppose that change in current.**

The Unit of Inductance   The **henry,** symbolized by H, is the basic unit of inductance. By definition, the inductance is one henry when current through the coil, changing at the rate of one ampere per second, induces one volt across the coil. In many practical applications, millihenries (mH) and microhenries (μH) are the more common units. A schematic symbol for the inductor is shown in Figure B.07-4.

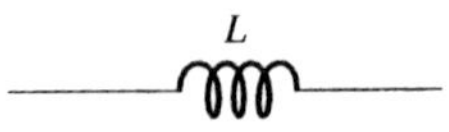

FIGURE B.07-4
A symbol for the inductor.

$$L$$

Energy Storage

An inductor stores energy in the magnetic field created by the current. The energy stored is expressed as follows:

$$W = \frac{1}{2}LI^2$$

As you can see, the energy stored is proportional to the inductance and the square of the current. When current (I) is in amperes and inductance (L) is in henries, the energy (W) is in joules.

Physical Characteristics of Inductors

The following characteristics are important in establishing the inductance of a coil: the permeability of the core material, the number of turns of wire, the length, and the cross-sectional area of the core.

Core Material As discussed earlier, an inductor is basically a coil of wire. The material around which the coil is formed is called the **core.** Coils are wound on either nonmagnetic or magnetic materials. Examples of nonmagnetic materials are air, wood, copper, plastic, and glass. The permeabilities of these materials are the same as for a vacuum. Examples of magnetic materials are iron, nickel, steel, cobalt, or alloys. These materials have permeabilities that are hundreds or thousands of times greater than that of a vacuum and are classified as *ferromagnetic.* A ferromagnetic core provides a better path for the magnetic lines of force and thus permits a stronger magnetic field.

The permeability (μ) of the core material determines how easily a magnetic field can be established. *The inductance is directly proportional to the permeability of the core material.*

Physical Parameters As indicated in Figure B.07-5, the number of turns of wire, the length, and the cross-sectional area of the core are factors in setting the value of inductance. The inductance is inversely proportional to the length of the core and directly proportional to the cross-sectional area. Also, the inductance is directly related to the number of turns squared. This relationship is as follows:

$$L = \frac{N^2 \mu A}{l}$$

where L is the inductance in henries, N is the number of turns, μ is the permeability, A is the cross-sectional area in meters squared, and l is the core length in meters.

FIGURE B.07-5
Factors that determine the inductance of a coil.

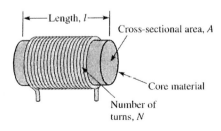

EXAMPLE B.07-3

Determine the inductance of the coil in Figure B.07-6. The permeability of the core is 0.25×10^{-3}.

FIGURE B.07-6

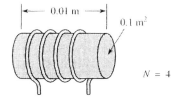

Solution
$$L = \frac{N^2 \mu A}{l} = \frac{(4)^2(0.25 \times 10^{-3})(0.1)}{0.01} = 40 \text{ mH}$$

Winding Resistance   When a coil is made of a certain material, for example, insulated copper wire, that wire has a certain resistance per unit of length. When many turns of wire are used to construct a coil, the total resistance may be significant. This inherent resistance is called the *DC resistance* or the *winding resistance* (R_W). Although this resistance is distributed along the length of the wire, as shown in Figure B.07-7(a), it is sometimes indicated in a schematic as resistance appearing in series with the inductance of the coil, as shown in Figure B.07-7(b). In many applications, the winding resistance can be ignored and the coil considered as an ideal inductor. In other cases, the resistance must be considered.

FIGURE B.07-7
Winding resistance of a coil.

(a) The wire has resistance distributed along its length.

(b) Equivalent circuit

Winding Capacitance

When two conductors are placed side by side, there is always some capacitance between them. Thus, when many turns of wire are placed close together in a coil, a certain amount of stray capacitance, called *winding capacitance* (C_W), is a natural side effect. In many applications, this winding capacitance is very small and has no significant effect. In other cases, particularly at high frequencies, it may become quite important.

The equivalent circuit for an inductor with both its winding resistance (R_W) and its winding capacitance (C_W) is shown in Figure B.07-8. The capacitance effectively acts in parallel. The total of the stray capacitances between each loop of the winding is indicated in a schematic as a capacitance appearing in parallel with the coil and its winding resistance, as shown in Figure B.07-8(b).

FIGURE B.07-8
Winding capacitance of a coil.

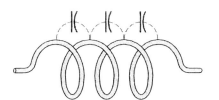

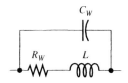

(a) Stray capacitance between each loop appears as a total parallel capacitance (C_W).

(b) Equivalent circuit

Review of Faraday's Law

The principle of **Faraday's law** is reviewed here because of its importance in the study of inductors. Faraday found that by moving a magnet through a coil of wire, a voltage was induced across the coil, and that when a complete path was provided, the induced voltage caused an induced current.

> **The amount of induced voltage is directly proportional to the rate of change of the magnetic field with respect to the coil.**

This principle is illustrated in Figure B.07-9, where a bar magnet is moved through a coil of wire. An induced voltage is indicated by the voltmeter connected across the coil. The faster the magnet is moved, the greater is the induced voltage.

FIGURE B.07-9
Induced voltage is created by a changing magnetic field.

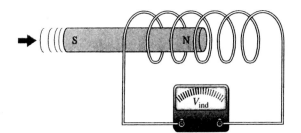

When a wire is formed into a certain number of loops or turns and is exposed to a changing magnetic field, a voltage is induced across the coil. The induced voltage is proportional to the number of turns of wire in the coil, N, and to the rate at which the magnetic field changes.

Lenz's Law

Lenz's law adds to Faraday's law by defining the direction of induced voltage as follows:

> **When the current through a coil changes and an induced voltage is created as a result of the changing magnetic field, the direction of the induced voltage is such that it always opposes the change in current.**

Figure B.07-10 illustrates Lenz's law. In part (a), the current is constant and is limited by R_1. There is no induced voltage because the magnetic field is unchanging. In part (b), the switch suddenly is closed, placing R_2 in parallel with R_1 and thus reducing the resistance. Naturally, the current tries to increase and the magnetic field begins to expand, but the induced voltage opposes this attempted increase in current for an instant.

In Figure B.07-10(c), the induced voltage gradually decreases, allowing the current to increase. In part (d), the current has reached a constant value as determined by the parallel resistors, and the induced voltage is zero. In part (e), the switch has been suddenly opened, and, for an instant, the induced voltage prevents any decrease in current. In part (f), the induced voltage gradually decreases, allowing the current to decrease back to a value determined by R_1. Notice that the induced voltage has a polarity that opposes any current change. The polarity of the induced voltage is opposite that of the battery voltage for an increase in current and aids the battery voltage for a decrease in current.

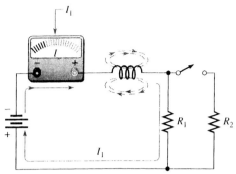

(a) Switch open: Constant current and constant magnetic field; no induced voltage.

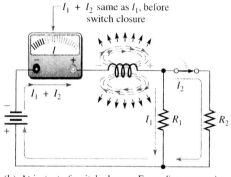

(b) At instant of switch closure: Expanding magnetic field induces voltage, which opposes increase in total current.

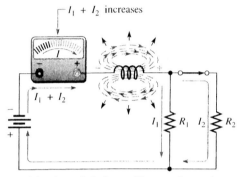

(c) Right after switch closure: The rate of expansion of the magnetic field decreases, allowing the current to increase exponentially as induced voltage decreases.

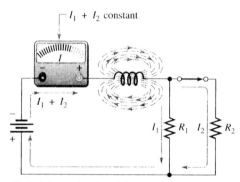

(d) Switch remains closed: Current and magnetic field reach constant value.

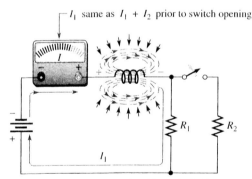

(e) At instant of switch opening: Magnetic field begins to collapse, creating an induced voltage, which opposes decrease in current.

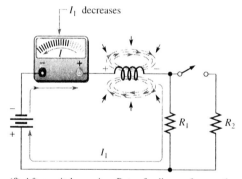

(f) After switch opening: Rate of collapse of magnetic field decreases, allowing current to decrease exponentially back to original value.

FIGURE B.07-10
Demonstration of Lenz's law: When the current tries to change suddenly, the electromagnetic field changes and induces a voltage in a direction that opposes that change in current.

REVIEW QUESTIONS

True/False

1. A capacitor stores energy in a magnetic field.
2. An inductor stores energy in an electric field.

Multiple Choice

3. The value 0.00001 microfarads is the same as
   a. 1 picofarad.
   b. 10 picofarads.
   c. 100 picofarads.

4. The unit of inductance is the
   a. Farad.
   b. Henry.
   c. Lenz.

5. Mica, ceramic, and electrolytic are three types of
   a. Capacitors.
   b. Inductors.
   c. Insulators.

B.08 Understand principles and operations of DC series circuits

INTRODUCTION

The first basic type of electrical connection is the series connection. In this section we will examine the operation of resistors connected in series.

EXAMPLE B.08-1

Calculate the voltage across each resistor in Figure B.08-1, and find the value of V_S. To what value can V_S be raised before the 5 mA fuse blows?

FIGURE B.08-1

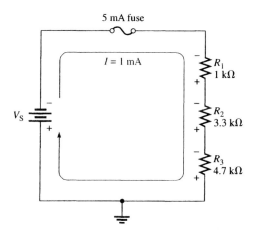

Solution   By Ohm's law, the voltage across each resistor is equal to its resistance multiplied by the current through it. Use the Ohm's law formula $V = IR$ to determine the voltage across each of the resistors. Keep in mind that there is the same current through each series resistor. The voltage across R_1 (designated V_1) is

$$V_1 = IR_1 = (1 \text{ mA})(1 \text{ k}\Omega) = 1 \text{ V}$$

The voltage across R_2 is

$$V_2 = IR_2 = (1 \text{ mA})(3.3 \text{ }\Omega) = 3.3 \text{ V}$$

The voltage across R_3 is

$$V_3 = IR_3 = (1 \text{ mA})(4.7 \text{ k}\Omega) = 4.7 \text{ V}$$

To calculate the value of V_S, first determine the total resistance.

$$R_T = 1 \text{ k}\Omega + 3.3 \text{ k}\Omega + 4.7 \text{ k}\Omega = 9 \text{ k}\Omega$$

The source voltage V_S is equal to the current times the total resistance.

$$V_S = IR_T = (1 \text{ mA})(9 \text{ k}\Omega) = 9 \text{ V}$$

Notice that if you add the voltage drops of the resistors, they total 9 V, which is the same as the source voltage.

The fuse can handle a maximum current of 5 mA. The maximum value of V_S is therefore

$$V_S = IR_T = (5 \text{ mA})(9 \text{ k}\Omega) = 45 \text{ V}$$

EXAMPLE B.08-2

Some resistors are not color coded with bands but have the values stamped on the resistor body. When the portion of the circuit board shown in Figure B.08-2 was assembled, someone mounted the resistors with the labels turned down, and there is no documentation showing the resistor values. Assume that there are a voltmeter, ammeter, and power supply immediately available, but no ohmmeter. Without removing the resistors from the board, use Ohm's law to determine the resistance of each one.

FIGURE B.08-2

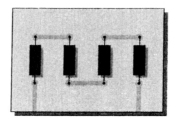

Solution The resistors are all in series, so the current is the same through each one. Measure the current by connecting a 12 V source (arbitrary value) and an ammeter as shown in Figure B.08-3. Measure the voltage across each resistor by placing the voltmeter across the first resistor (R_1). Then repeat this measurement for the other three resistors. For illustration, the voltage values indicated are assumed to be the measured values.

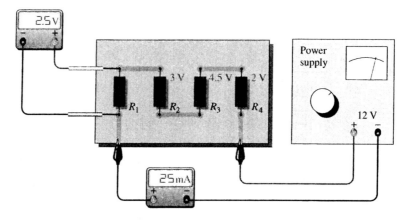

FIGURE B.08-3
The voltmeter readings across each resistor are indicated.

Determine the resistance of each resistor by substituting the measured values of current and voltage into the Ohm's law formula.

$$R_1 = \frac{V_1}{I} = \frac{2.5 \text{ V}}{25 \text{ mA}} = 100 \text{ }\Omega$$

$$R_2 = \frac{V_2}{I} = \frac{3 \text{ V}}{25 \text{ mA}} = 120 \text{ }\Omega$$

$$R_3 = \frac{V_3}{I} = \frac{4.5 \text{ V}}{25 \text{ mA}} = 180 \ \Omega$$

$$R_4 = \frac{V_4}{I} = \frac{2 \text{ V}}{25 \text{ mA}} = 80 \ \Omega$$

VOLTAGE SOURCES IN SERIES

When batteries are placed in a flashlight, they are connected series-aiding to produce a larger voltage, as illustrated in Figure B.08-4. In this example, three 1.5 V batteries are placed in series to produce a total voltage ($V_{S(tot)}$) as follows:

$$V_{S(tot)} = V_{S1} + V_{S2} + V_{S3} = 1.5 \text{ V} + 1.5 \text{ V} + 1.5 \text{ V} = 4.5 \text{ V}$$

Series voltage sources are added when their polarities are in the same direction and are subtracted when their polarities are in opposite directions, or series-opposing. For example, if one of the batteries in the flashlight is turned around, as indicated in the schematic of Figure B.08-5, its voltage subtracts, because it has a negative value, and reduces the total voltage.

$$V_{S(tot)} = V_{S1} - V_{S2} + V_{S3} = 1.5 \text{ V} - 1.5 \text{ V} + 1.5 \text{ V} = 1.5 \text{ V}$$

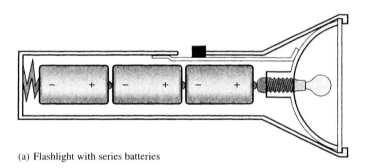

(a) Flashlight with series batteries

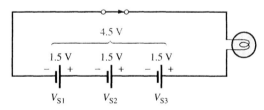

(b) Schematic of flashlight circuit

FIGURE B.08-4
Example of series-aiding voltage sources.

FIGURE B.08-5
Opposite polarities subtract.

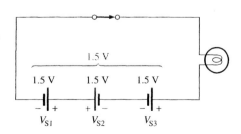

Determine the source voltage V_S in Figure B.08-6 where the two voltage drops are given.

FIGURE B.08-6

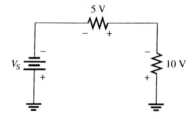

Solution By Kirchhoff's voltage law, the source voltage (applied voltage) must equal the sum of the voltage drops. Adding the voltage drops gives the value of the source voltage.

$$V_S = 5 \text{ V} + 10 \text{ V} = 15 \text{ V}$$

REVIEW QUESTIONS

True/False

1. The current is the same everywhere in a series circuit.
2. The sum of the rises and the drops in a closed loop is zero.

Multiple Choice

3. Two resistors connected in series across a 12 V source have a total resistance of 100 ohms. The voltage across one resistor is 3 V. What is the value of the other resistor?
   a. 25 ohms.
   b. 75 ohms.
   c. 100 ohms.

4. The current in a series circuit containing three resistors is 10 mA. One resistor has a value of 100 ohms. Another resistor has 6.5 V across it. If the source voltage equals 10 V, what is the value of the third resistor?
   a. 100 ohms.
   b. 250 ohms.
   c. 650 ohms.

5. What is the total resistance of eight 20-ohm resistors connected in series?
   a. 2.5 ohms.
   b. 20 ohms.
   c. 160 ohms.

B.09 Fabricate and demonstrate DC series circuits

INTRODUCTION

In this section you will set up several series circuits and make the necessary measurements to validate your own calculations.

Test Circuits

Set up the circuits shown in Figure B.09-1 one at a time, using actual components or a software package such as *Electronics Workbench*. Measure the circuit current and the individual resistor voltages. Verify by calculation that the measurements are correct.

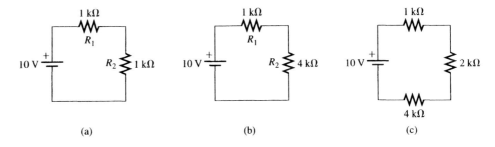

(a)   (b)   (c)

FIGURE B.09-1
Series circuits.

B.10 Troubleshoot and repair DC series circuits

INTRODUCTION

Open resistors or contacts and shorts between conductors are common problems in all circuits including series circuits.

Open Circuit

The most common failure in a series circuit is an open. For example, when a resistor or a lamp burns out, it causes a break in the current path and creates an **open circuit** as illustrated in Figure B.10-1.

An open in a series circuit prevents current.

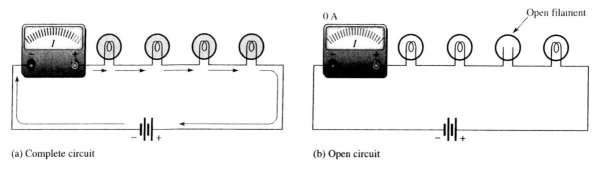

(a) Complete circuit   (b) Open circuit

FIGURE B.10-1
An open circuit prevents current.

Checking for an Open Element   Sometimes a visual check will reveal a charred resistor or an open lamp filament. However, it is possible for a resistor to open without showing visible signs of damage. In this situation, a voltage check of the series circuit is required. The general procedure is as follows: *Measure the voltage across each resistor in series. The voltage across all of the good resistors will be zero. The voltage across the open resistor will equal the total voltage across the series combination.*

This condition occurs because an open resistor will prevent current through the series circuit. With no current, there can be no voltage drop across any of the good resistors. Since $IR = (0 A)R = 0$ V, in accordance with Ohm's law, the voltage on each side of a good resistor is the same. The total voltage must then appear across the open resistor in accordance with Kirchhoff's voltage law, as illustrated in Figure B.10-2. To fix the circuit, replace the open resistor.

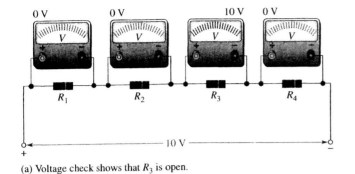

(a) Voltage check shows that R_3 is open.

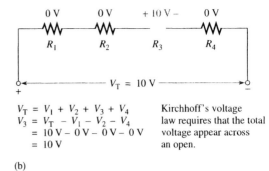

$$V_T = V_1 + V_2 + V_3 + V_4$$
$$V_3 = V_T - V_1 - V_2 - V_4$$
$$= 10\ V - 0\ V - 0\ V - 0\ V$$
$$= 10\ V$$

Kirchhoff's voltage law requires that the total voltage appear across an open.

(b)

FIGURE B.10-2
Troubleshooting a series circuit for an open element.

Short Circuit

Sometimes an unwanted **short circuit** occurs when two conductors touch or a foreign object such as solder or a wire clipping accidentally connects two sections of a circuit together. This situation is particularly common in circuits with a high component density. Three potential causes of short circuits are illustrated on the PC board in Figure B.10-3.

When there is a short, a portion of the series resistance is bypassed (all of the current goes through the short), thus reducing the total resistance as illustrated in Figure B.10-4. Notice that the current increases as a result of the short.

A short in a series circuit causes more current.

FIGURE B.10-3
Examples of shorts on a PC board.

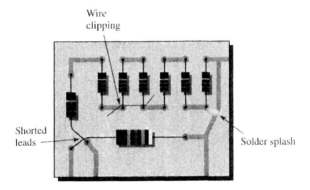

FIGURE B.10-4
The effect of a short in a series circuit.

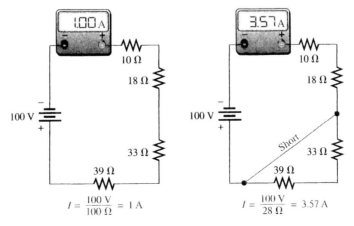

(a) Before short

(b) After short

B.11 Understand principles and operations of DC parallel circuits

INTRODUCTION

The parallel connection is the second basic electrical connection. In this section we examine the operation of parallel resistive circuits.

Kirchhoff's current law is stated as follows:

The sum of the currents into a junction (total current in) is equal to the sum of the currents out of that junction (total current out).

A **junction** is any point in a circuit where two or more components are connected. In a parallel circuit, a junction is a point where the parallel branches come together. For example, in the circuit of Figure B.11-1, point A is one junction and point B is another. Let's start at the negative terminal of the source and follow the current. The total current I_T from the source is *into* the junction at point A. At this point, the current splits up among the three branches as indicated. Each of the three branch currents (I_1, I_2 and I_3) is *out of* junction A. Kirchhoff's current law says that the total current into junction A is equal to the total current out of junction A; that is,

$$I_T = I_1 + I_2 + I_3$$

Now, following the currents in Figure B.11-1 through the three branches, you see that they come back together at point B. Currents I_1, I_2, and I_3 are into junction B, and I_T is out of junction B. Kirchhoff's current law formula at this junction is therefore the same as at junction A.

$$I_T = I_1 + I_2 + I_3$$

FIGURE B.11-1
Kirchhoff's current law: The current into a junction equals the current out of that junction.

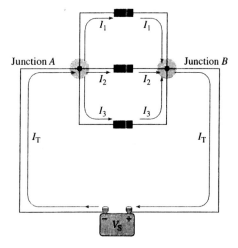

Junction A Junction B

General Formula for Kirchhoff's Current Law

The previous discussion used a specific case to illustrate Kirchhoff's current law, often abbreviated KCL. Now let's look at the general case. Figure B.11-2 shows a generalized circuit junction where a number of branches are connected to a point in the circuit. Currents

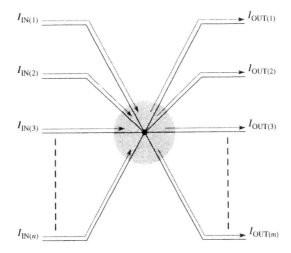

$$I_{IN(1)} + I_{IN(2)} + I_{IN(3)} + \cdots + I_{IN(n)} = I_{OUT(1)} + I_{OUT(2)} + I_{OUT(3)} + \cdots I_{OUT(m)}$$

FIGURE B.11-2
Generalized circuit junction illustrates Kirchhoff's current law.

$I_{IN(1)}$ through $I_{IN(n)}$ are into the junction (n can be any number). Currents $I_{OUT(1)}$ through $I_{OUT(m)}$ are out of the junction (m can be any number but not necessarily equal to n).

By Kirchhoff's current law, the sum of the currents into a junction must equal the sum of the currents out of the junction. With reference to Figure B.11-2, the general formula for Kirchhoff's current law is

$$I_{IN(1)} + I_{IN(2)} + \cdots + I_{IN(n)} = I_{OUT(1)} + I_{OUT(2)} + \cdots + I_{OUT(m)}$$

If all of the terms on the right side of the equation are brought over to the left side, their signs change to negative, and a zero is left on the right side.

$$I_{IN(1)} + I_{IN(2)} + \cdots + I_{IN(n)} - I_{OUT(1)} - I_{OUT(2)} - \cdots - I_{OUT(m)} = 0$$

Kirchhoff's current law can also be stated in this way:

The algebraic sum of all the currents entering and leaving a junction is equal to zero.

You can verify Kirchhoff's current law by connecting a circuit and measuring each branch current and the total current from the source, as illustrated in Figure B.11-3. When the branch currents are added together, their sum will equal the total current. This rule applies for any number of branches.

FIGURE B.11-3
Experimental verification of Kirchhoff's current law.

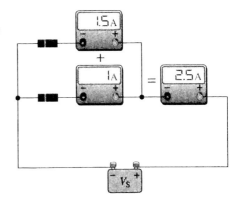

EXAMPLE B.11-1

The branch currents in the circuit of Figure B.11-4 are known. Determine the total current entering junction A and the total current leaving junction B.

FIGURE B.11-4

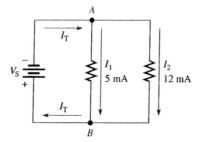

Solution   The total current out of junction A is the sum of the two branch currents. So the total current into A is

$$I_T = I_1 + I_2 = 5 \text{ mA} + 12 \text{ mA} = 17 \text{ mA}$$

The total current entering point B is the sum of the two branch currents. So the total current out of B is

$$I_T = I_1 + I_2 = 5 \text{ mA} + 12 \text{ mA} = 17 \text{ mA}$$

EXAMPLE B.11-2

Use Kirchhoff's current law to find the current measured by ammeters A3 and A5 in Figure B.11-5.

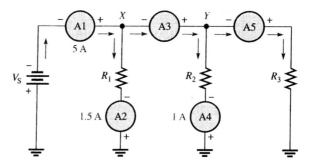

FIGURE B.11-5

Solution   The total current into junction X is 5 A. Two currents are out of junction X: 1.5 A through resistor R_1 and the current through A3. Kirchhoff's current law at junction X is

$$5 \text{ A} = 1.5 \text{ A} + I_{A3}$$

Solving for I_{A3} yields

$$I_{A3} = 5 \text{ A} - 1.5 \text{ A} = 3.5 \text{ A}$$

The total current into junction Y is $I_{A3} = 3.5$ A. Two currents are out of junction Y: 1 A through resistor R_2 and the current through A5 and R_3. Kirchhoff's current law applied to junction Y gives

$$3.5 \text{ A} = 1 \text{ A} + I_{A5}$$

Solving for I_{A5} yields

$$I_{A5} = 3.5 \text{ A} - 1 \text{ A} = 2.5 \text{ A}$$

The Case of Two Resistors in Parallel

It is often useful to consider only two resistors in parallel because this situation occurs commonly in practice. Also, any number of resistors in parallel can be broken down into pairs as an alternate way to find the R_T.

The formula for two resistors in parallel is

$$R_T = \frac{R_1 R_2}{R_1 + R_2}$$

That is,

The total resistance of two resistors in parallel is equal to the product of the two resistors divided by the sum of the two resistors.

This equation is sometimes referred to as the "product over the sum" formula. Example B.11-3 illustrates how to use it.

EXAMPLE B.11-3

Calculate the total resistance connected to the voltage source of the circuit in Figure B.11-6.

FIGURE B.11-6

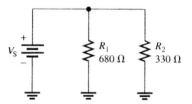

Solution $\qquad R_T = \dfrac{R_1 R_2}{R_1 + R_2} = \dfrac{(680\ \Omega)(330\ \Omega)}{680\ \Omega + 330\ \Omega} = \dfrac{224{,}000\ \Omega^2}{1{,}010\ \Omega} = 222\ \Omega$

The Case of Equal-Value Resistors in Parallel

Another special case of parallel circuits is the parallel connection of several resistors having the same value. The following is a shortcut method of calculating R_T when this case occurs:

$$R_T = \frac{R}{n}$$

When any number of resistors (n), all having the same resistance (R), are connected in parallel, R_T is equal to the resistance divided by the number of resistors in parallel.

Notation for Parallel Resistors

Sometimes, for convenience, parallel resistors are designated by two parallel vertical marks. For example, R_1 in parallel with R_2 can be written as $R_1 \| R_2$. Also, when several resistors are in parallel with each other, this notation can be used. For example,

$$R_1 \| R_2 \| R_3 \| R_4 \| R_5$$

indicates that R_1 through R_5 are all in parallel.

This notation is also used with resistance values. For example,

$$10 \text{ k}\Omega \| 5 \text{ k}\Omega$$

means that a 10 kΩ resistor is in parallel with a 5 kΩ resistor.

REVIEW QUESTIONS

True/False

1. At most, three resistors may be connected in parallel.
2. The equivalent resistance of a parallel circuit is smaller than the smallest resistor in the circuit.

Multiple Choice

3. Two resistors, 10 ohms and 40 ohms, are connected in parallel. Their equivalent resistance is
   a. 8 ohms.
   b. 30 ohms.
   c. 50 ohms.

4. If 40 V is applied to the two resistors in Question 3, the total current is
   a. 1 A.
   b. 4 A.
   c. 5 A.

5. What resistor, in parallel with 20 ohms, results in an equivalent resistance of 10 ohms?
   a. 10 ohms.
   b. 20 ohms.
   c. 25 ohms.

B.12 Fabricate and demonstrate DC parallel circuits

INTRODUCTION

In this section you will set up several parallel circuits and make the necessary measurements to validate your own calculations.

Test Circuits

Set up the circuits shown in Figure B.12-1 one at a time, using actual components or a software package such as *Electronics Workbench*. Measure the total circuit current and the individual resistor currents. Verify by calculation that the measurements are correct.

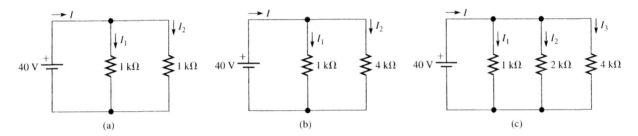

FIGURE B.12-1
Parallel circuits.

B.13 Troubleshoot and repair DC parallel circuits

INTRODUCTION

In this section, you will see how an open in a parallel branch affects the parallel circuit.

Open Branches

Recall that an open circuit is one in which the current path is interrupted and there is no current. In this section we examine what happens when a branch of a parallel circuit opens.

If a switch is connected in a branch of a parallel circuit, as shown in Figure B.13-1, an open or a closed path can be made by the switch. When the switch is closed, as in Figure B.13-1(a), R_1 and R_2 are in parallel. The total resistance is 50 Ω (two 100 Ω resistors in parallel). Current is through both resistors. If the switch is opened, as in Figure B.13-1(b), R_1 is effectively removed from the circuit, and the total resistance is 100 Ω. Current is now only through R_2.

In general,

> **When an open circuit occurs in a parallel branch, the total resistance increases, the total current decreases, and the same current continues through each of the remaining parallel paths.**

The decrease in total current equals the amount of current that was previously in the open branch. The other branch currents remain the same.

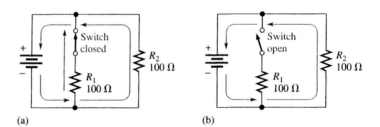

(a)   (b)

FIGURE B.13-1
When the switch opens, total current decreases and current through R_2 remains unchanged.

Consider the lamp circuit in Figure B.13-2. There are four bulbs in parallel with a 120 V source. In part (a), there is current through each bulb. Now suppose that one of the bulbs burns out, creating an open path as shown in Figure B.13-2(b). This light will go out because there is no current through the open path. Notice, however, that current continues through all the other parallel bulbs, and they continue to glow. The open branch does not change the voltage across the parallel branches; it remains at 120 V.

You can see that a parallel circuit has an advantage over a series connection in lighting systems because if one or more of the parallel bulbs burns out, the others will stay on. In a series circuit, when one bulb goes out, all of the others go out also because the current path is completely interrupted.

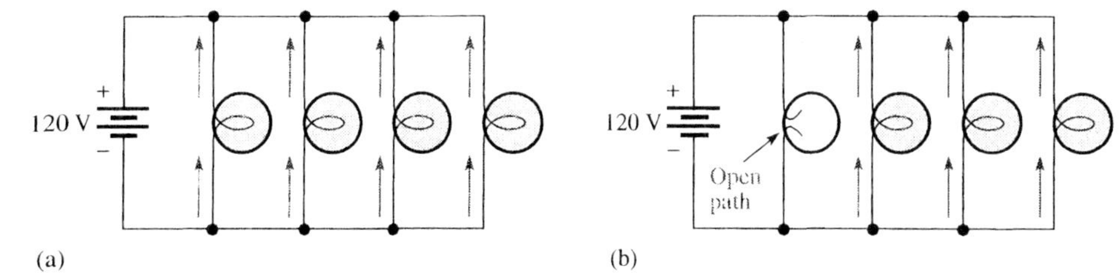

(a)   (b)

FIGURE B.13-2
When a lamp filament opens, total current decreases and the other branch currents remain unchanged.

When a resistor in a parallel circuit opens, the open resistor cannot be located by measurement of the voltage across the branches because the same voltage exists across all the branches. Thus, there is no way to tell which resistor is open by simply measuring voltage. The good resistors will always have the same voltage as the open one, as illustrated in Figure B.13-3 (note that the middle resistor is open).

If a visual inspection does not reveal the open resistor, it can be located by current measurements. In practice, measuring current is more difficult than measuring voltage because you must insert the ammeter in series to measure the current. Thus, a wire or a printed circuit connection must be cut or disconnected, or one end of a component must be lifted off the circuit board, in order to connect the ammeter in series. This procedure, of course, is not required when voltage measurements are made because the meter leads are simply connected across a component.

FIGURE B.13-3
All parallel branches (open or not) have the same voltage.

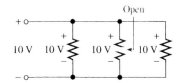

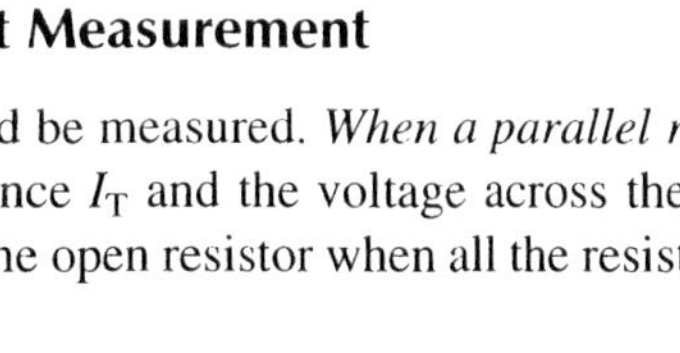

Finding an Open Branch by Current Measurement

In a parallel circuit, the total current should be measured. *When a parallel resistor opens, I_T is always less than its normal value.* Once I_T and the voltage across the branches are known, a few calculations will determine the open resistor when all the resistors are of different values.

Consider the two-branch circuit in Figure B.13-4(a). If one of the resistors opens, the total current will equal the current in the good resistor. Ohm's law quickly tells you what the current in each resistor should be.

$$I_1 = \frac{50 \text{ V}}{560 \text{ } \Omega} = 89.3 \text{ mA}$$

$$I_2 = \frac{50 \text{ V}}{100 \text{ } \Omega} = 500 \text{ mA}$$

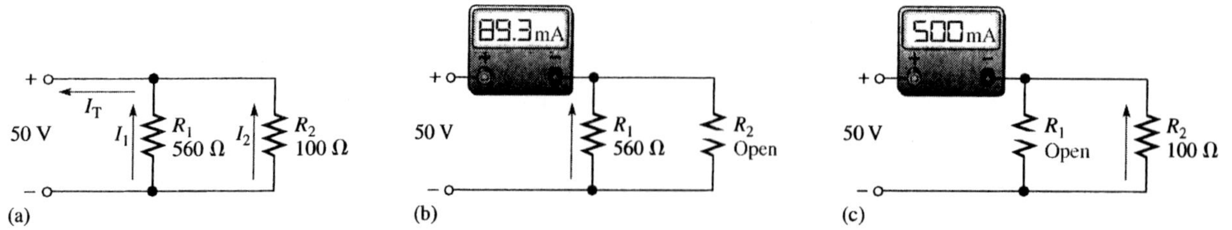

(a)   (b)   (c)

FIGURE B.13-4
Finding an open path by current measurement.

If R_2 is open, the total current is 89.3 mA, as indicated in Figure B.13-4(b). If R_1 is open, the total current is 500 mA, as indicated in Figure B.13-4(c).

This procedure can be extended to any number of branches having unequal resistances. If the parallel resistances are all equal, the current in each branch must be checked until a branch is found with no current. This branch has the open resistor.

B.14 Understand the principles and operations of DC series-parallel and bridge circuits

INTRODUCTION

When three or more resistors are used to build a circuit, many different series-parallel combinations are possible. In this section we examine the operation of several series-parallel circuits.

EXAMPLE B.14-1

Determine R_T between points A and B of the circuit in Figure B.14-1.

FIGURE B.14-1

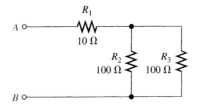

Solution   Resistors R_2 and R_3 are in parallel, and this parallel combination is in series with R_1. First find the parallel resistance of R_2 and R_3. Since R_2 and R_3 are equal in value, divide the value by 2.

$$R_{2\text{-}3} = \frac{R}{n} = \frac{100 \ \Omega}{2} = 50 \ \Omega$$

Now, since R_1 is in series with $R_{2\text{-}3}$, add their values.

$$R_T = R_1 + R_{2\text{-}3} = 10 \ \Omega + 50 \ \Omega = 60 \ \Omega$$

EXAMPLE B.14-2

Find R_T of the circuit in Figure B.14-2.

FIGURE B.14-2

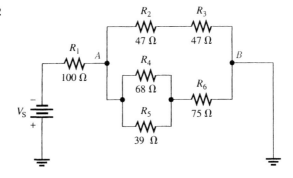

Solution   In the upper branch, R_2 is in series with R_3. The series combination is designated $R_{2\text{-}3}$ and is equal to $R_2 + R_3$.

$$R_{2\text{-}3} = R_2 + R_3 = 47\ \Omega + 47\ \Omega = 94\ \Omega$$

In the lower branch, R_4 and R_5 are in parallel with each other. This parallel combination is designated $R_{4\text{-}5}$.

$$R_{4\text{-}5} = \frac{R_4 R_5}{R_4 + R_5} = \frac{(68\ \Omega)(39\ \Omega)}{68\ \Omega + 39\ \Omega} = 24.8\ \Omega$$

Also in the lower branch, the parallel combination of R_4 and R_5 is in series with R_6. This series-parallel combination is designated $R_{4\text{-}5\text{-}6}$.

$$R_{4\text{-}5\text{-}6} = R_6 + R_{4\text{-}5} = 75\ \Omega + 24.8\ \Omega = 99.8\ \Omega$$

Figure B.14-3 shows the original circuit in a simplified equivalent form.

FIGURE B.14-3

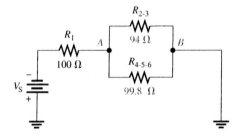

Now you can find the resistance between points A and B. It is $R_{2\text{-}3}$ in parallel with $R_{4\text{-}5\text{-}6}$. The equivalent resistance is calculated as follows:

$$R_{AB} = \frac{1}{\left(\dfrac{1}{94\ \Omega}\right) + \left(\dfrac{1}{99.8\ \Omega}\right)} = 48.4\ \Omega$$

Finally, the total resistance is R_1 in series with R_{AB}.

$$R_T = R_1 + R_{AB} = 100\ \Omega + 48.4\ \Omega = 148.4\ \Omega$$

Total Current

Once the total resistance and the source voltage are known, you can find total current in a circuit by applying Ohm's law. Total current is the total source voltage divided by the total resistance.

$$I_T = \frac{V_S}{R_T}$$

For example, let's find the total current in the circuit of Example B.14-2 (Figure B.14-2). Assume that the source voltage is 30 V. The calculation is

$$I_T = \frac{V_S}{R_T} = \frac{30\ \text{V}}{148.4\ \Omega} = 202\ \text{mA}$$

Branch Currents

Using the current-divider formula, Kirchhoff's current law, Ohm's law, or combinations of these, you can find the current in any branch of a series-parallel circuit. In some cases it may take repeated application of the formula to find a given current.

EXAMPLE B.14-3

Determine the current through R_4 in Figure B.14-4 if $V_S = 50$ V.

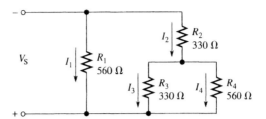

FIGURE B.14-4

Solution First find the current (I_2) into the junction of R_3 and R_4. Once you know this current, you can use the current-divider formula to find I_4, the current through R_4.

Notice that there are two main branches in the circuit. The left-most branch consists of only R_1. The right-most branch has R_2 in series with the parallel combination of R_3 and R_4. The voltage across both of these main branches is the same and equal to 50 V. Find the current (I_2) into the junction of R_3 and R_4 by calculating the equivalent resistance ($R_{2\text{-}3\text{-}4}$) of the right-most main branch and then applying Ohm's law; I_2 is the total current through this main branch. Thus,

$$R_{2\text{-}3\text{-}4} = R_2 + \frac{R_3 R_4}{R_3 + R_4} = 330\ \Omega + \frac{(330\ \Omega)(560\ \Omega)}{890\ \Omega} = 538\ \Omega$$

$$I_2 = \frac{V_S}{R_{2\text{-}3\text{-}4}} = \frac{50\ \text{V}}{538\ \Omega} = 93\ \text{mA}$$

Use the current-divider formula to calculate I_4.

$$I_4 = \left(\frac{R_3}{R_3 + R_4}\right) I_2 = \left(\frac{330\ \Omega}{890\ \Omega}\right) 93\ \text{mA} = 34.5\ \text{mA}$$

You can find the current through R_3 by applying Kirchhoff's law ($I_2 = I_3 + I_4$) and subtracting I_4 from I_2 ($I_3 = I_2 - I_4$).

You can find the current through R_1 by using Ohm's law ($I_1 = V_S/R_1$).

Voltage Relationships

The circuit in Figure B.14-5 illustrates voltage relationships in a series-parallel circuit. Voltmeters are connected to measure each of the resistor voltages, and the readings are indicated.

Some general observations about Figure B.14-5 are as follows:

1. V_{R1} and V_{R2} are equal because R_1 and R_2 are in parallel. (Recall that voltages across parallel branches are the same.) V_{R1} and V_{R2} are the same as the voltage from A to B.

2. V_{R3} is equal to $V_{R4} + V_{R5}$ because R_3 is in parallel with the series combination of R_4 and R_5. (V_{R3} is the same as the voltage from B to C.)

3. V_{R4} is about one-third of the voltage from B to C because R_4 is about one-third of the resistance $R_4 + R_5$ (by the voltage-divider principle).

4. V_{R5} is about two-thirds of the voltage from B to C because R_5 is about two-thirds of $R_4 + R_5$.

5. $V_{R1} + V_{R3}$ equals V_S because, by Kirchhoff's voltage law, the sum of the voltage drops must equal the source voltage.

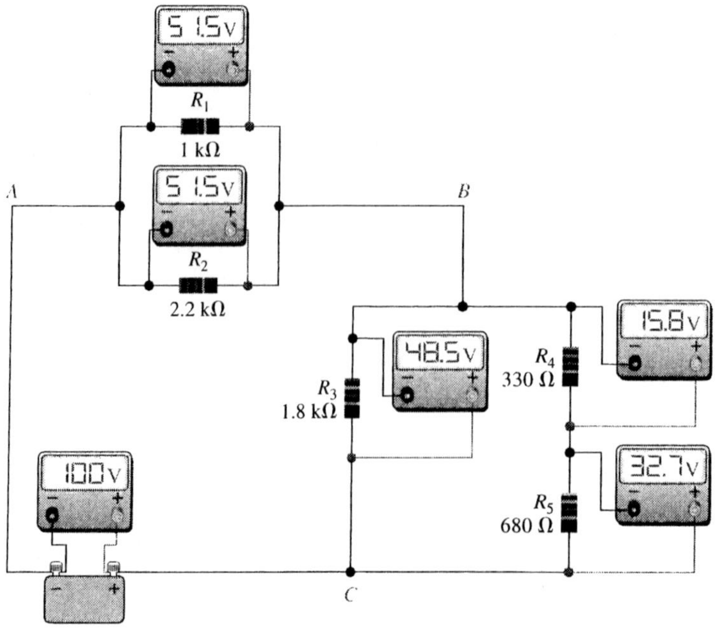

FIGURE B.14-5
Illustration of voltage relationships.

REVIEW QUESTIONS

True/False

1. The total resistance of a series-parallel circuit depends on how it is drawn.
2. Series and parallel analysis techniques are used to analyze series-parallel circuits.

Multiple Choice

3. What does $Ra + Rb \parallel Rc$ mean?
   a. *Ra* is in series with *Rb*. The series pair is in parallel with *Rc*.
   b. *Ra* is in series with the parallel combination of *Rb* and *Rc*.
   c. All three resistors are in parallel.
4. Series-parallel connections are often made clearer when
   a. A good guess is made at the beginning.
   b. The circuit is redrawn.
   c. All the resistors are the same value.
5. The voltages in a series-parallel circuit must still obey
   a. Ohm's law.
   b. Kirchhoff's voltage law.
   c. Both a and b.

B.15 Fabricate and demonstrate DC series-parallel and bridge circuits

INTRODUCTION

In this section you will set up two series-parallel circuits and make the necessary measurements to validate your own calculations.

Test Circuits

Set up the circuits shown in Figure B.15-1 one at a time, using actual components or a software package such as *Electronics Workbench*. Measure the total circuit current, the individual resistor currents, and the individual resistor voltages. For Figure B.15-1(b) also measure V_{ab}. Verify by calculation that the measurements are correct.

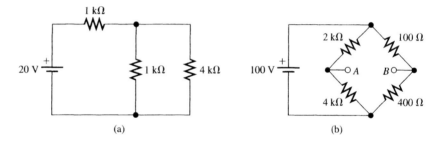

(a)   (b)

FIGURE B.15-1
Series-parallel circuits.

B.16 Troubleshoot and repair DC series-parallel and bridge circuits

INTRODUCTION

Opens and shorts are typical problems that occur in electric circuits. If a resistor burns out, it will normally produce an open circuit. Bad solder connections, broken wires, and poor contacts can also be causes of open paths. Pieces of foreign material, such as solder splashes, broken insulation on wires, and so on, can often lead to shorts in a circuit. A short is a zero resistance path between two points.

In addition to complete opens or shorts, partial opens or partial shorts can develop in a circuit. A partial open would be a much higher than normal resistance, but not infinitely large. A partial short would be a much lower than normal resistance, but not zero.

The following examples illustrate troubleshooting series-parallel circuits.

EXAMPLE B.16-1

From the indicated voltmeter reading in Figure B.16-1, determine if there is a fault. If there is a fault, identify it as either a short or an open.

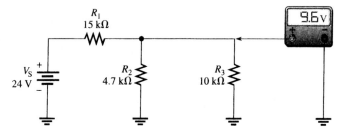

FIGURE B.16-1

Solution First determine what the voltmeter should be indicating. Since R_2 and R_3 are in parallel, their combined resistance is

$$R_{2\text{-}3} = \frac{R_2 R_3}{R_2 + R_3} = \frac{(4.7 \text{ k}\Omega)(10 \text{ k}\Omega)}{14.7 \text{ k}\Omega} = 3.20 \text{ k}\Omega$$

The voltage across the parallel combination is determined by the voltage-divider formula.

$$V_{2\text{-}3} = \left(\frac{R_{2\text{-}3}}{R_1 + R_{2\text{-}3}}\right)V_S = \left(\frac{3.2 \text{ k}\Omega}{18.2 \text{ k}\Omega}\right)24 \text{ V} = 4.22 \text{ V}$$

This calculation shows that 4.22 V is the voltage reading that you should get on the meter. But the meter reads 9.6 V instead. This value is incorrect, and, because it is higher than it should be, R_2 or R_3 is probably open. Why? Because if either R_2 or R_3 is open, the resistance across which the meter is connected is larger than expected. A higher resistance will drop a higher voltage in this circuit, which is, in effect, a voltage divider.

Start by assuming that R_2 is open. If it is, the voltage across R_3 is

$$V_3 = \left(\frac{R_3}{R_1 + R_3}\right)V_S = \left(\frac{10 \text{ k}\Omega}{25 \text{ k}\Omega}\right)24 \text{ V} = 9.6 \text{ V}$$

Since the measured voltage is also 9.6 V, this calculation shows that R_2 is open. Replace R_2 with a new resistor.

EXAMPLE B.16-2

Suppose that you measure 24 V with the voltmeter in Figure B.16-2. Determine if there is a fault, and, if there is, identify it.

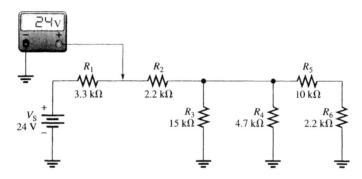

FIGURE B.16-2

Solution   There is no voltage drop across R_1 because both sides of the resistor are at +24 V. Either there is no current through R_1 from the source, which tells us that R_2 is open in the circuit, or R_1 is shorted.

The most probable failure is an open R_2. If it is open, then there will be no current from the source. To verify this, measure across R_2 with the voltmeter as shown in Figure B.16-3. If R_2 is open, the meter will indicate 24 V. The right side of R_2 will be at zero volts because there is no current through any of the other resistors to cause a voltage drop across them.

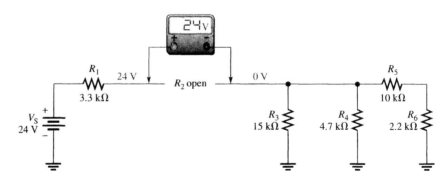

FIGURE B.16-3

B.17 Understand the principles and operations of the Wheatstone bridge

INTRODUCTION

Bridge circuits are widely used in measurement devices and other applications that you will learn later. In this section, you will study the balanced resistive bridge, which can be used to measure unknown resistance values.

The circuit shown in Figure B.17-1(a) is known as a *Wheatstone bridge*. Figure B.17-1(b) is the same circuit electrically, but it is drawn in a different way.

A bridge is said to be balanced when the voltage (V_{OUT}) across the output terminals A and B is zero; that is, $V_A = V_B$. In Figure B.17-1(b), if V_A equals V_B, then $V_{R1} = V_{R2}$, since the top sides of both R_1 and R_2 are connected to the same point. Also $V_{R3} = V_{R4}$, since the bottom sides of both R_3 and R_4 connect to the same point. The voltage ratios can be written as

$$\frac{V_1}{V_3} = \frac{V_2}{V_4}$$

Substituting by Ohm's law yields

$$\frac{I_1 R_1}{I_1 R_3} = \frac{I_2 R_2}{I_2 R_4}$$

The currents cancel to give

$$\frac{R_1}{R_3} = \frac{R_2}{R_4}$$

Solving for R_1 yields the following formula:

$$R_1 = R_3 \left(\frac{R_2}{R_4} \right)$$

This formula can be used to determine an unknown resistance. First, make R_3 a variable resistor and call it R_V. Also, set the ratio R_2/R_4 to a known value. If R_V is adjusted until the bridge is balanced, the product of R_V and the ratio R_2/R_4 is equal to R_1, which is the unknown resistor (R_{UNK}).

$$R_{UNK} = R_V \left(\frac{R_2}{R_4} \right)$$

FIGURE B.17-1
Wheatstone bridge.

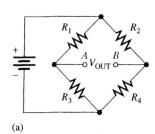

(a)

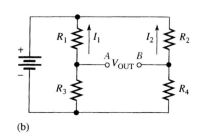

(b)

FIGURE B.17-2
Balanced Wheatstone bridge.

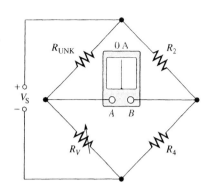

The bridge is balanced when the voltage across the output terminals equals zero $(V_A = V_B)$. A *galvanometer* (a meter that measures small currents in either direction and is zero at center scale) is connected between the output terminals. Then R_V is adjusted until the galvanometer shows zero current $(V_A = V_B)$, indicating a balanced condition. The setting of R_V multiplied by the ratio R_2/R_4 gives the value of R_{UNK}. Figure B.17-2 shows this arrangement. For example, if $R_2/R_4 = \frac{1}{10}$ and $R_V = 680\ \Omega$, then $R_{UNK} = (680\ \Omega)(\frac{1}{10}) = 68\ \Omega$.

EXAMPLE B.17-1

What is R_{UNK} under the balanced bridge conditions shown in Figure B.17-3?

FIGURE B.17-3

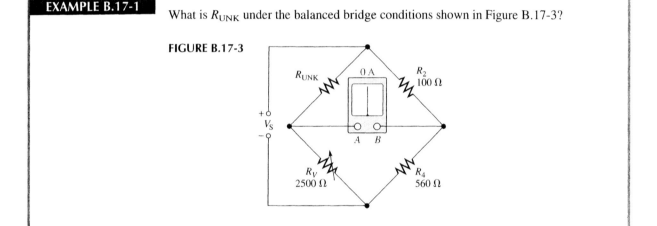

Solution $R_{UNK} = R_V \left(\dfrac{R_2}{R_4}\right) = 2500\ \Omega \left(\dfrac{100\ \Omega}{560\ \Omega}\right) = 446\ \Omega$

REVIEW QUESTIONS

True/False

1. A bridge is balanced when its output voltage equals one-half the supply voltage.
2. A galvanometer is used to sense the output of a bridge.

Multiple Choice

3. If R_V equals 10 ohms, R_2 equals 470 ohms, and R_4 equals 100 ohms, what is R_{UNK}?
   a. 4.7ohms.
   b. 47 ohms.
   c. 470 ohms.

4. A component used to measure temperature with a bridge is the
   a. Thermistor.
   b. Tempister.
   c. Galvanometer.
5. A Wheatstone bridge whose output voltage is 2 volts is
   a. Balanced.
   b. Unbalanced.
   c. Defective.

B.18 Understand the principles and operations of DC voltage divider circuits (loaded and unloaded)

INTRODUCTION

To illustrate how a series string of resistors acts as a voltage divider, we will examine Figure B.18-1 where there are two resistors in series. There are two voltage drops: one across R_1 and one across R_2. These voltage drops are V_1 and V_2, respectively, as indicated in the diagram.

Since each resistor has the same current, the voltage drops are proportional to the resistance values. For example, if the value of R_2 is twice that of R_1, then the value of V_2 is twice that of V_1. In other words, *the total voltage drop divides among the series resistors in amounts directly proportional to the resistance values.*

For example, in Figure B.18-1, if V_S is 10 V, R_1 is 50 Ω, and R_2 is 100 Ω, then V_1 is one-third the total voltage, or 3.33 V, because R_1 is one-third the total resistance (150 Ω). Likewise, V_2 is two-thirds V_S, or 6.67 V.

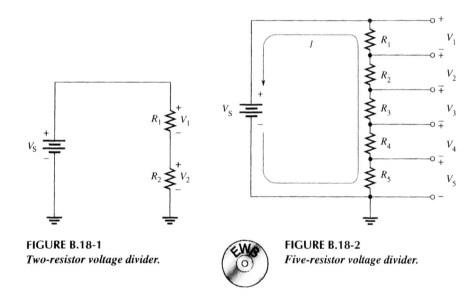

FIGURE B.18-1
Two-resistor voltage divider.

FIGURE B.18-2
Five-resistor voltage divider.

Voltage-Divider Formula

With a few steps, a formula for determining how the voltages divide among series resistors can be developed. Let's assume that we have several resistors in series as shown in Figure B.18-2. This figure shows five resistors as an example, but there can be any number.

Let's call the voltage drop across any one of the resistors V_x, where x represents the number of a particular resistor (1, 2, 3, and so on). By Ohm's law, the voltage drop across any of the resistors in Figure B.18-2 can be written as follows:

$$V_x = IR_x$$

where $x = 1, 2, 3, 4,$ or 5.

The current is equal to the source voltage divided by the total resistance ($I = V_S/R_T$). For the example circuit of Figure B.18-2, the total resistance is $R_1 + R_2 + R_3 + R_4 + R_5$. Substituting V_S/R_T for I in the expression for V_x results in

$$V_x = \left(\frac{V_S}{R_T}\right)R_x$$

Rearranging the terms yields

$$V_x = \left(\frac{R_x}{R_T}\right)V_S$$

This equation is the general voltage-divider formula. It can be stated as follows:

The voltage drop across any resistor or combination of resistors in a series circuit is equal to the ratio of that resistance value to the total resistance, multiplied by the source voltage.

The following three examples illustrate use of the voltage-divider formula.

EXAMPLE B.18-1

Determine V_1 (the voltage across R_1) and V_2 (the voltage across R_2) in the voltage divider in Figure B.18-3.

FIGURE B.18-3

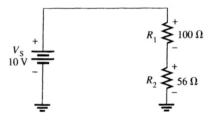

Solution   To determine V_1, use the voltage-divider formula, $V_x = (R_x/R_T)V_S$, where $x = 1$. The total resistance is

$$R_T = R_1 + R_2 = 100\ \Omega + 56\ \Omega = 156\ \Omega$$

R_1 is 100 Ω and V_S is 10 V. Substitute these values into the voltage-divider formula.

$$V_1 = \left(\frac{R_1}{R_T}\right)V_S = \left(\frac{100\ \Omega}{156\ \Omega}\right)10\ V = 6.41\ V$$

There are two ways to find the value of V_2: Kirchhoff's voltage law or the voltage-divider formula. If you use Kirchhoff's voltage law ($V_S = V_1 + V_2$), substitute the values for V_S and V_1 as follows:

$$V_2 = V_S - V_1 = 10\ V - 6.41\ V = 3.59\ V$$

A second way to find V_2 is to use the voltage-divider formula where $x = 2$.

$$V_2 = \left(\frac{R_2}{R_T}\right)V_S = \left(\frac{56\ \Omega}{156\ \Omega}\right)10\ V = 3.59\ V$$

You get the same result either way.

EXAMPLE B.18-2

Calculate the voltage drop across each resistor in the voltage divider of Figure B.18-4.

FIGURE B.18-4

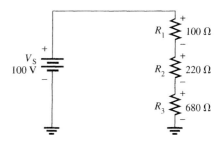

Solution Look at the circuit for a moment and consider the following: The total resistance is 1000 Ω. Ten percent of the total voltage is across R_1 because it is 10% of the total resistance (100 Ω is 10% of 1000 Ω). Likewise, 22% of the total voltage is dropped across R_2 because it is 22% of the total resistance (220 Ω is 22% of 1000 Ω). Finally, R_3 drops 68% of the total voltage because 680 Ω is 68% of 1000 Ω.

Because of the convenient values in this problem, it is easy to figure the voltages mentally. ($V_1 = 0.10 \times 100$ V = 10 V, $V_2 = 0.22 \times 100$ V = 22 V, and $V_3 = 0.68 \times 100$ V = 68 V). Such is not always the case, but sometimes a little thinking will produce a result more efficiently and eliminate some calculating. This is also a good way to roughly estimate what your results should be so that you will recognize an unreasonable answer as a result of a calculation error.

Although you have already reasoned through this problem, the calculations will verify your results.

$$V_1 = \left(\frac{R_1}{R_T}\right)V_S = \left(\frac{100\ \Omega}{1000\ \Omega}\right)100\ \text{V} = 10\ \text{V}$$

$$V_2 = \left(\frac{R_2}{R_T}\right)V_S = \left(\frac{220\ \Omega}{1000\ \Omega}\right)100\ \text{V} = 22\ \text{V}$$

$$V_3 = \left(\frac{R_3}{R_T}\right)V_S = \left(\frac{680\ \Omega}{1000\ \Omega}\right)100\ \text{V} = 68\ \text{V}$$

Notice that the sum of the voltage drops is equal to the source voltage, in accordance with Kirchhoff's voltage law. This check is a good way to verify your results.

EXAMPLE B.18-3

Determine the voltages between the following points in the voltage divider of Figure B.18-5:
(a) A to B   **(b)** A to C   **(c)** B to C   **(d)** B to D   **(e)** C to D

FIGURE B.18-5

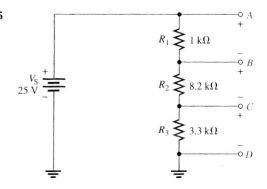

Solution First determine R_T.

$$R_T = R_1 + R_2 + R_3 = 1 \text{ k}\Omega + 8.2 \text{ k}\Omega + 3.3 \text{ k}\Omega = 12.5 \text{ k}\Omega$$

Now apply the voltage-divider formula to obtain each required voltage.

(a) The voltage A to B is the voltage drop across R_1.

$$V_{AB} = \left(\frac{R_1}{R_T}\right)V_S = \left(\frac{1 \text{ k}\Omega}{12.5 \text{ k}\Omega}\right)25 \text{ V} = 2 \text{ V}$$

(b) The voltage from A to C is the combined voltage drop across both R_1 and R_2. In this case, R_x in the general formula is $R_1 + R_2$.

$$V_{AC} = \left(\frac{R_1 + R_2}{R_T}\right)V_S = \left(\frac{9.2 \text{ k}\Omega}{12.5 \text{ k}\Omega}\right)25 \text{ V} = 18.4 \text{ V}$$

(c) The voltage from B to C is the voltage drop across R_2.

$$V_{BC} = \left(\frac{R_2}{R_T}\right)V_S = \left(\frac{8.2 \text{ k}\Omega}{12.5 \text{ k}\Omega}\right)25 \text{ V} = 16.4 \text{ V}$$

Verify for yourself that the answers to parts (d) and (e) are 23 V and 6.6 V, respectively.

REVIEW QUESTIONS

True/False

1. An unloaded voltage divider is analyzed the same as a series circuit.

2. A loaded voltage divider has the same voltage across each resistor as an unloaded divider.

Multiple Choice

3. What is the voltage across R_2 if R_1 equals 1 k ohms and R_2 equals 4 k ohms, with an applied voltage of 10 V?

    a. 1 V.

    b. 4 V.

    c. 8 V.

4. The potentiometer is actually a

    a. Fixed output voltage divider.

    b. Variable output voltage divider.

    c. Variable output current divider.

5. When a voltage divider is loaded, its output voltage

    a. Goes down.

    b. Goes up.

    c. Stays the same.

B.19 Fabricate and demonstrate DC voltage-divider circuits (loaded and unloaded)

INTRODUCTION

A voltage divider with three output voltages has been designed and constructed on a PC board. The voltage divider is to be used as part of a portable power supply unit for supplying up to three different reference voltages to measuring instruments in the field. The power supply unit contains a battery pack combined with a voltage regulator that produces a constant +12 V to the voltage-divider circuit board. In this section, you will apply your knowledge of loaded voltage dividers, Kirchhoff's laws, and Ohm's law to determine the operating parameters of the voltage divider in terms of voltages and currents for all possible load configurations. You will also troubleshoot the circuit for various malfunctions.

The Schematic of the Voltage Divider

☐ Draw the schematic and label the resistor values for the circuit board in Figure B.19-1.

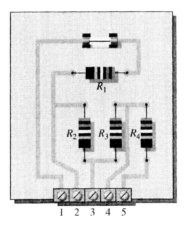

FIGURE B.19-1
Voltage-divider circuit board.

The 12 V Power Supply

☐ Specify how to connect a 12 V power supply to the circuit board so that all resistors are in series and pin 2 has the highest output voltage.

The Unloaded Output Voltages

☐ Calculate each of the output voltages with no loads connected. Add these voltage values to a copy of the table in Figure B.19-2.

FIGURE B.19-2
Table of operating parameters for the power supply voltage divider.

10 MΩ Load	$V_{OUT\,(2)}$	$V_{OUT\,(3)}$	$V_{OUT\,(4)}$	% Deviation	$I_{LOAD\,(2)}$	$I_{LOAD\,(3)}$	$I_{LOAD\,(4)}$
None							
Pin 2							
Pin 3							
Pin 4							
Pins 2 and 3				2 3			
Pins 2 and 4				2 4			
Pins 3 and 4				3 4			
Pins 2, 3, and 4				2 3 4			

The Loaded Output Voltages

The instruments to be connected to the voltage divider each have a 10 MΩ input resistance. This means that when an instrument is connected to a voltage-divider output there is effectively a 10 MΩ resistor from that output to ground (negative side of source).

☐ Determine the output voltage across each load for the following load configurations and add these voltage values to a copy of the table in Figure B.19-2.

1. A 10 MΩ load connected to pin 2.
2. A 10 MΩ load connected to pin 3.
3. A 10 MΩ load connected to pin 4.
4. 10 MΩ loads connected to pin 2 and pin 3.
5. 10 MΩ loads connected to pin 2 and pin 4.
6. 10 MΩ loads connected to pin 3 and pin 4.
7. 10 MΩ loads connected to pin 2, pin 3, and pin 4.

Percent Deviation of the Output Voltages

☐ Calculate how much each loaded output voltage deviates from its unloaded value for each of the load configurations in the previous list and express each as a percentage using the following formula:

$$\text{Percent deviation} = \left(\frac{V_{OUT(unloaded)} - V_{OUT(loaded)}}{V_{OUT(unloaded)}} \right) 100\%$$

Add the values to a copy of the table of Figure B.19-2.

The Load Currents

☐ Calculate the current to each 10 MΩ load for each of the load configurations listed previously. Add these values to a copy of the table in Figure B.19-2.

Troubleshooting

The voltage-divider circuit board is connected to a 12 V power supply and to the three instruments to which it provides reference voltages, as shown in Figure B.19-3. Voltages at each of the numbered test points are measured with a voltmeter in each of eight different cases.

☐ For each case in Figure B.19-3, determine the fault indicated by the voltage measurements.

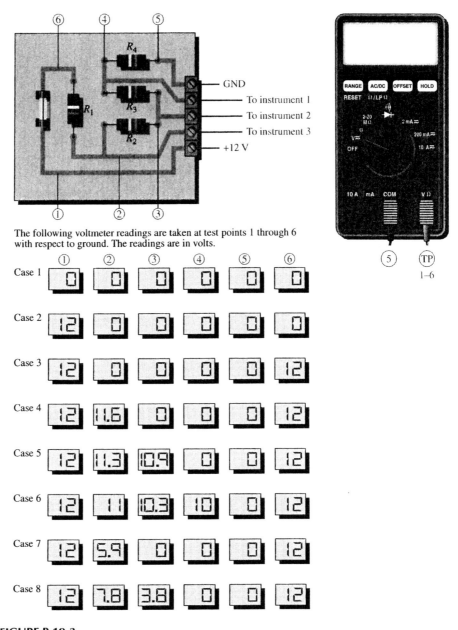

The following voltmeter readings are taken at test points 1 through 6 with respect to ground. The readings are in volts.

FIGURE B.19-3
Test-point readings.

B.20 Troubleshoot and repair DC voltage-divider circuits (loaded and unloaded)

This material is covered in B.19.

B.21 Understand principles and operations of DC *RC* and *RL* circuits

INTRODUCTION

This section provides details concerning the use of capacitors and inductors in DC circuits.

Charging from Zero The formula for the special case in which an increasing exponential voltage curve begins at zero ($V_i = 0$), is given here. It is developed as follows, starting with the general formula.

$$
\begin{aligned}
v &= V_F + (V_i - V_F)e^{-t/\tau} \\
&= V_F + (0 - V_F)e^{-t/RC} \\
&= V_F - V_F e^{-t/RC}
\end{aligned}
$$

$$v = V_F(1 - e^{-t/RC})$$

Using this equation, you can calculate the value of the charging voltage of a capacitor at any instant of time if it is initially uncharged. The same is true for an increasing current.

EXAMPLE B.21-1

In Figure B.21-1, determine the capacitor voltage 50 μs after the switch is closed if the capacitor initially is uncharged. Sketch the charging curve.

FIGURE B.21-1

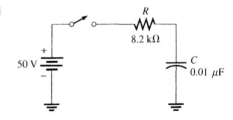

Solution The time constant is $RC = (8.2 \text{ k}\Omega)(0.01 \ \mu\text{F}) = 82 \ \mu\text{s}$. The voltage to which the capacitor will fully charge is 50 V (this is V_F). The initial voltage is zero. Notice that 50 μs is less than one time constant, so the capacitor will charge less than 63% of the full voltage in that time.

$$
\begin{aligned}
v_C &= V_F(1 - e^{-t/RC}) = 50 \text{ V}(1 - e^{-50\mu s/82\mu s}) \\
&= 50 \text{ V}(1 - e^{-0.61}) = 50 \text{ V}(1 - 0.543) = 22.8 \text{ V}
\end{aligned}
$$

Determine the value of $e^{-0.61}$ on the calculator by entering -0.61 and then pressing $\boxed{e^x}$ or $\boxed{\text{INV}}$ $\boxed{\text{ln} x}$.

The charging curve for the capacitor is shown in Figure B.21-2.

FIGURE B.21-2

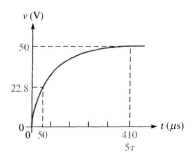

Discharging to Zero The formula for the special case in which a decreasing exponential voltage curve ends at zero ($V_F = 0$) is derived from the general formula as follows:

$$v = V_F + (V_i - V_F)e^{-t/\tau}$$
$$= 0 + (V_i - 0)e^{-t/RC}$$

$$v = V_i e^{-t/RC}$$

where V_i is the voltage at the beginning of the discharge. You can use this formula to calculate the discharging voltage at any instant.

Energizing Current in an Inductor

In a series *RL* circuit, the current will increase to 63% of its full value in one time-constant interval after the switch is closed. This buildup of current is analogous to the buildup of capacitor voltage during the charging in an *RC* circuit; they both follow an exponential curve and reach the approximate percentages of final value as indicated in Table B.21-1 and as illustrated in Figure B.21-3.

TABLE B.21-1
Percentage of final current after each time-constant interval during current buildup.

Number of Time Constants	% of Final Value
1	63
2	86
3	95
4	98
5	99 (considered 100%)

FIGURE B.21-3
Energizing current in an inductor.

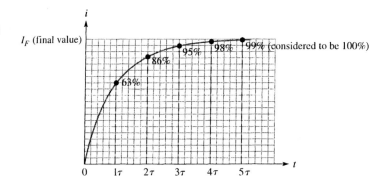

The change in current over five time-constant intervals is illustrated in Figure B.21-4. When the current reaches its final value at approximately 5τ, it ceases to change. At this time, the inductor acts as a short (except for winding resistance) to the constant current. The final value of the current is

$$I_F = \frac{V_S}{R_W} = \frac{10 \text{ V}}{1 \text{ k}\Omega} = 10 \text{ mA}$$

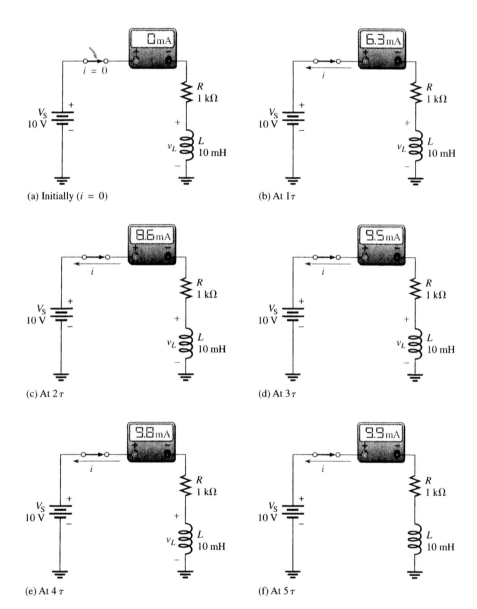

(a) Initially $(i = 0)$

(b) At 1τ

(c) At 2τ

(d) At 3τ

(e) At 4τ

(f) At 5τ

FIGURE B.21-4

Illustration of the exponential buildup of current in an inductor. The current increases another 63% during each time-constant interval. A voltage (v_L) is induced in the coil that tends to oppose the increase in current.

EXAMPLE B.21-2

Calculate the time constant for Figure B.21-5. Then determine the current and the time at each time-constant interval, measured from the instant the switch is closed.

FIGURE B.21-5

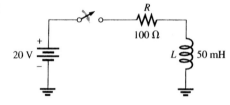

Solution   The time constant is

$$\tau = \frac{L}{R} = \frac{50 \text{ mH}}{100 \text{ }\Omega} = 500 \text{ }\mu s$$

The final current is

$$I_F = \frac{V_S}{R} = \frac{20 \text{ V}}{100 \text{ }\Omega} = 200 \text{ mA}$$

Using the time-constant percentage values from Table B.21-1.

$$\begin{aligned}
\text{At } 1\tau = 500 \text{ }\mu s: \quad & i = 0.63(200 \text{ mA}) = 126 \text{ mA} \\
\text{At } 2\tau = 1 \text{ ms}: \quad & i = 0.86(200 \text{ mA}) = 172 \text{ mA} \\
\text{At } 3\tau = 1.5 \text{ ms}: \quad & i = 0.95(200 \text{ mA}) = 190 \text{ mA} \\
\text{At } 4\tau = 2 \text{ ms}: \quad & i = 0.98(200 \text{ mA}) = 196 \text{ mA} \\
\text{At } 5\tau = 2.5 \text{ ms}: \quad & i = 0.99(200 \text{ mA}) = 198 \text{ mA} \cong 200 \text{ mA}
\end{aligned}$$

Deenergizing Current in an Inductor

Current in an inductor decreases exponentially according to the approximate percentage values shown in Table B.21-2 and in Figure B.21-6.

TABLE B.21-2
Percentage of initial current after each time-constant interval while current is decreasing.

Number of Time Constants	% of Initial Value
1	37
2	14
3	5
4	2
5	1 (considered 0)

FIGURE B.21-6
Deenergizing current in an inductor.

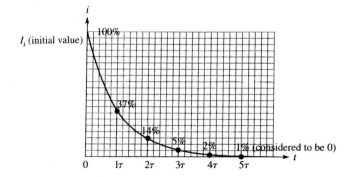

True/False

1. Capacitors charge instantly to the maximum available voltage.

2. Inductors discharge exponentially.

Multiple Choice

3. How long does a 10 μF capacitor take to fully charge through a 470 ohm resistor?

    a. 4.7 μS.

    b. 4.7 mS.

    c. 23.5 mS.

4. How long does an inductor take to discharge?

    a. One time constant.

    b. Five time constants.

    c. Always less than 10 μS.

5. What is the capacitor charging voltage after 8 seconds in a circuit with a supply voltage of 100 V and a time constant of 1.5 seconds?

    a. 36.6 V.

    b. 63.4 V.

    c. 100 V.

B.22 Fabricate and demonstrate DC *RC* and *RL* circuits

INTRODUCTION

Your supervisor has handed you two unmarked coils and wants to know their inductance values. You cannot find an inductance bridge, which is an instrument for measuring inductance directly. After scratching your head for a while, you decide to use the time-constant characteristic of inductive circuits to determine the unknown inductances. A test setup consisting of a square wave generator and an oscilloscope is used to make the measurements.

The method used is to place the coil in series with a resistor of known value and measure the time constant by applying a square wave to the circuit and observing the resulting voltage across the resistor with an oscilloscope. Knowing the time constant and the resistance value, you can calculate the inductance *L*.

Each time the square wave input voltage goes high, the inductor is energized; and each time the square wave goes back to zero, the inductor is deenergized. The time it takes for the exponential resistor voltage to increase to approximately its final value equals five time constants. This operation is illustrated in Figure B.22-1. To make sure that the winding resistance of the coil can be neglected, it must be measured; and the value of the resistor used in the circuit must be selected to be considerably larger than the winding resistance.

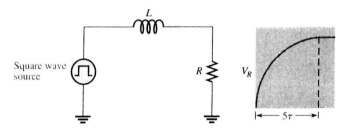

FIGURE B.22-1
Circuit for time-constant measurement.

Step 1: Measure the Coil Resistance and Select a Series Resistor

Assume that the winding resistance has been measured with an ohmmeter and found to be 85 Ω. To make the winding resistance negligible, a 10 kΩ series resistor is used in the circuit.

Step 2: Determine the Inductance of Coil 1

Refer to Figure B.22-2. To determine the inductance, a 10 V square wave is applied to the breadboarded circuit. The frequency of the square wave is adjusted so that the inductor has

10 V square wave input

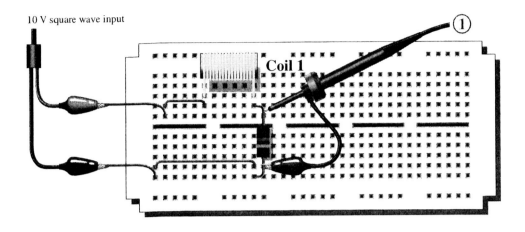

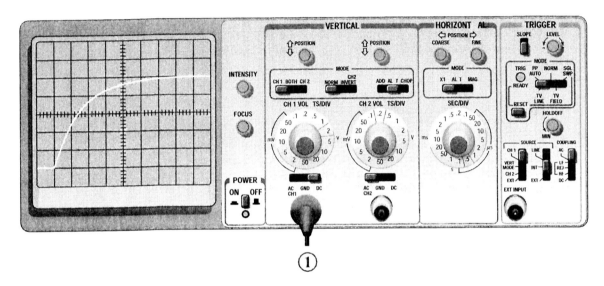

FIGURE B.22-2
Testing coil 1.

time to fully energize during each square wave pulse; the scope is set to view a complete energizing curve as shown. Determine the approximate circuit time constant from the scope display and calculate the inductance of coil 1.

Step 3: Determine the Inductance of Coil 2

Refer to Figure B.22-3. To determine the inductance, a 10 V square wave is applied to the breadboarded circuit. The frequency of the square wave is adjusted so that the inductor has time to fully energize during each square wave pulse; the scope is set to view a complete energizing curve as shown. Determine the approximate circuit time constant from the scope display and calculate the inductance of coil 2. Discuss any difficulty you find with this method.

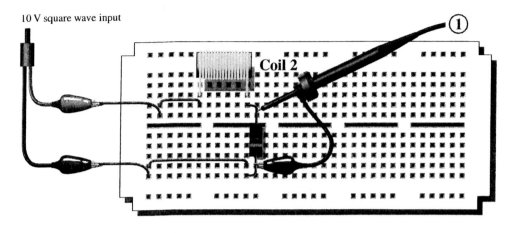

10 V square wave input

Coil 2

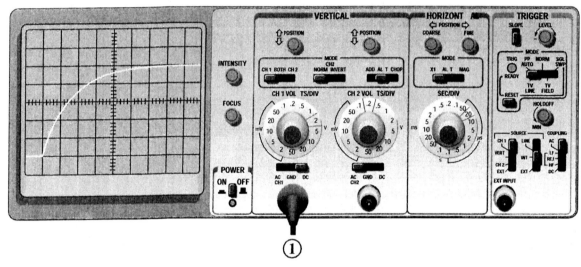

FIGURE B.22-3
Testing coil 2.

Step 4: Determine Another Way to Find Unknown Inductance

Determination of the time constant is not the only way that you can find an unknown inductance. Specify a method using a sinusoidal input voltage instead of the square wave.

A Note Concerning Capacitors

The same type of test may be performed on an unknown capacitor. Choose a suitable series resistance and use the *RC* time constant relationship to determine the unknown capacitance value once measurements from the oscilloscope are available.

B.23 Troubleshoot and repair DC *RC* and *RL* circuits

INTRODUCTION

In this section we will examine how to test inductors and capacitors.

TESTING INDUCTORS

The most common failure in an inductor is an open coil. To check for an open, remove the coil from the circuit. If there is an open, an ohmmeter check will indicate infinite resistance, as shown in Figure B.23-1(a). If the coil is good, the ohmmeter will show the winding resistance. The value of the winding resistance depends on the wire size and length of the coil. It can be anywhere from one ohm to several hundred ohms. Figure B.23-1(b) shows a good reading.

FIGURE B.23-1
Checking a coil by measuring the resistance.

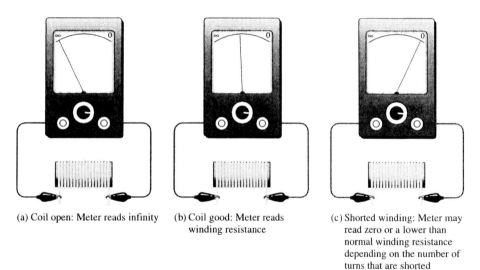

(a) Coil open: Meter reads infinity

(b) Coil good: Meter reads winding resistance

(c) Shorted winding: Meter may read zero or a lower than normal winding resistance depending on the number of turns that are shorted

Occasionally, when an inductor is overheated with excessive current, the wire insulation will melt, and some coils will short together. This must be tested on an *LC* meter because, with two shorted turns (or even several), an ohmmeter check may show the coil to be perfectly good from a resistance standpoint. Two shorted turns occur more frequently because the turns are adjacent and can easily short across from poor insulation, voltage breakdown, or simple wear if something is rubbing on them.

TESTING CAPACITORS

Capacitor failures can be categorized into two areas: catastrophic and degradation. The catastrophic failures are usually a short circuit caused by dielectric breakdown or an open circuit caused by connection failure. Degradation usually results in a gradual decrease in leakage resistance, hence an increase in leakage current or an increase in equivalent series resistance or dielectric absorption.

Ohmmeter Check

When there is a suspected problem, the capacitor can be removed from the circuit and checked with an ohmmeter. First, to be sure that the capacitor is discharged, short its leads, as indicated in Figure B.23-2(a). Connect the meter—set on a high ohms range such as R × 1M—to the capacitor, as shown in Figure B.23-2(b), and observe the needle. It should initially indicate near zero ohms. Then it should begin to move toward the high-resistance end of the scale as the capacitor charges from the ohmmeter's battery, as shown in Figure B.23-2(c). When the capacitor is fully charged, the meter will indicate an extremely high resistance, as shown in Figure B.23-2(d).

(a) Discharge by shorting leads

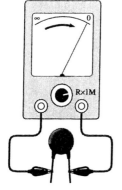

(b) Initially, when the meter is first connected, the pointer jumps immediately to zero.

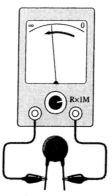

(c) The needle moves slowly toward infinity as the capacitor charges from the ohmmeter's battery.

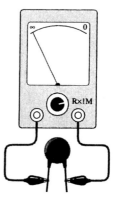

(d) When the capacitor is fully charged the pointer is at infinity.

FIGURE B.23-2
Checking a capacitor with an analog ohmmeter. This check shows a good capacitor.

As mentioned, the capacitor charges from the internal battery of the ohmmeter, and the meter responds to the charging current. The larger the capacitance value, the more slowly the capacitor will charge, as indicated by the needle movement. For small pF values, the meter response may be too slow to indicate the fast charging action.

If the capacitor is internally shorted, the meter will go to zero and stay there. If it is leaky, the final meter reading will be much less than normal. Most capacitors have a resistance of several hundred megohms. The exception is the electrolytic capacitor, which may normally have less than one megohm of leakage resistance. If the capacitor is open, no charging action will be observed, and the meter will indicate an infinite resistance.

Testing for Capacitance Value and Other Parameters with an *LC* Meter

An *LC* meter such as the one shown in Figure B.23-3 can be used to check the value of a capacitor or inductor. All capacitors change value over a period of time, some more than others. Ceramic capacitors, for example, often exhibit a 10% to 15% change in value

FIGURE B.23-3
*A typical LC meter
(courtesy of SENCORE).*

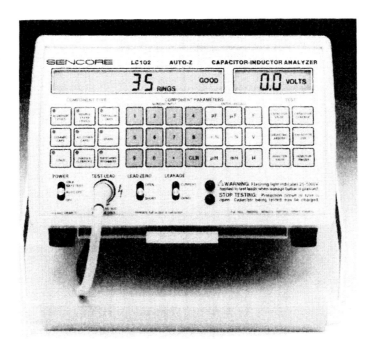

during the first year. Electrolytic capacitors are particularly subject to value change due to drying of the electrolyte. In other cases, capacitors may be labeled incorrectly or the wrong value might have been installed in the circuit. Although a value change represents fewer than 25% of defective capacitors, a value check should be made to quickly eliminate this as a source of trouble when troubleshooting a circuit.

Typically, values from 1 pF to 200,000 μF can be measured by simply connecting the capacitor, pushing the appropriate button, and reading the value of the display.

Many *LC* meters can also be used to check for leakage current in capacitors. In order to check for leakage, a sufficient voltage must be applied across the capacitor to simulate operating conditions. This is automatically done by the test instrument. Over 40% of all defective capacitors have excessive leakage current and electrolytic capacitors are particularly susceptible to this problem.

The problem of dielectric absorption occurs mostly in electrolytic capacitors when they do not completely discharge during use and retain a residual charge. Approximately 25% of defective capacitors have exhibited this condition.

B.24 Demonstrate an understanding of measurement of power in DC circuits

INTRODUCTION

No matter what type of circuit is used (series, parallel, or series-parallel), the total power supplied is equal to the sum of the powers required by the individual components. In this section we review several power formulas and their application.

EXAMPLE B.24-1

Determine the total amount of power in the series circuit in Figure B.24-1.

FIGURE B.24-1

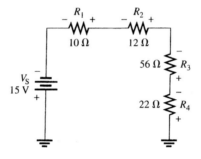

Solution The source voltage is 15 V. The total resistance is

$$R_T = 10 \ \Omega + 12 \ \Omega + 56 \ \Omega + 22 \ \Omega = 100 \ \Omega$$

The easiest formula to use is $P_T = V_S^2/R_T$, since you know both V_S and R_T.

$$P_T = \frac{V_S^2}{R_T} = \frac{(15 \ \text{V})^2}{100 \ \Omega} = \frac{225 \ \text{V}^2}{100 \ \Omega} = 2.25 \ \text{W}$$

If you determine the power of each resistor separately and add all these powers, you obtain the same result. Another calculation will illustrate. First, find the current.

$$I = \frac{V_S}{R_T} = \frac{15 \ \text{V}}{100 \ \Omega} = 0.15 \ \text{A}$$

Next, calculate the power for each resistor using $P = I^2R$.

$$P_1 = (0.15 \ \text{A})^2(10 \ \Omega) = 0.225 \ \text{W}$$
$$P_2 = (0.15 \ \text{A})^2(12 \ \Omega) = 0.270 \ \text{W}$$
$$P_3 = (0.15 \ \text{A})^2(56 \ \Omega) = 1.260 \ \text{W}$$
$$P_4 = (0.15 \ \text{A})^2(22 \ \Omega) = 0.495 \ \text{W}$$

Now, add these powers to get the total power.

$$P_T = 0.225 \ \text{W} + 0.270 \ \text{W} + 1.260 \ \text{W} + 0.495 \ \text{W} = 2.25 \ \text{W}$$

This result shows that the sum of the individual powers is equal to the total power as determined by the formula $P_T = V_S^2/R_T$.

EXAMPLE B.24-2

Determine the total amount of power in the parallel circuit in Figure B.24-2.

FIGURE B.24-2

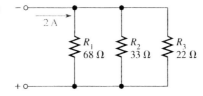

Solution   The total current is 2 A. The total resistance is

$$R_T = \cfrac{1}{\left(\cfrac{1}{68\ \Omega}\right) + \left(\cfrac{1}{33\ \Omega}\right) + \left(\cfrac{1}{22\ \Omega}\right)} = 11.1\ \Omega$$

The easiest formula to use is $P_T = I_T^2 R_T$ since you know both I_T and R_T. Thus,

$$P_T = I_T^2 R_T = (2\ \text{A})^2(11.1\ \Omega) = 44.4\ \text{W}$$

To demonstrate that if the power in each resistor is determined and if all of these values are added together you get the same result, work through another calculation. First, find the voltage across each branch of the circuit.

$$V_S = I_T R_T = (2\ \text{A})(11.1\ \Omega) = 22.2\ \text{V}$$

Remember that the voltage across all branches is the same.
   Next, use $P = V^2/R$ to calculate the power for each resistor.

$$P_1 = \frac{(22.2\ \text{V})^2}{68\ \Omega} = 7.25\ \text{W}$$

$$P_2 = \frac{(22.2\ \text{V})^2}{33\ \Omega} = 14.9\ \text{W}$$

$$P_3 = \frac{(22.2\ \text{V})^2}{22\ \Omega} = 22.4\ \text{W}$$

Now, add these powers to get the total power.

$$P_T = 7.25\ \text{W} + 14.9\ \text{W} + 22.4\ \text{W} = 44.6\ \text{W}$$

This calculation shows that the sum of the individual powers is equal (approximately) to the total power as determined by one of the power formulas. Rounding to three significant figures accounts for the difference.

Wattmeters

Analog and digital Wattmeters are available that measure the amount of power used by a circuit. Older Wattmeters used two sets of terminals, one for measuring voltage and the other for measuring current. Newer Wattmeters also have two sets of terminals, but these are organized as power-in (source) and power-out (circuit) terminals. Both types of Wattmeter are illustrated in Figure B.24-3.

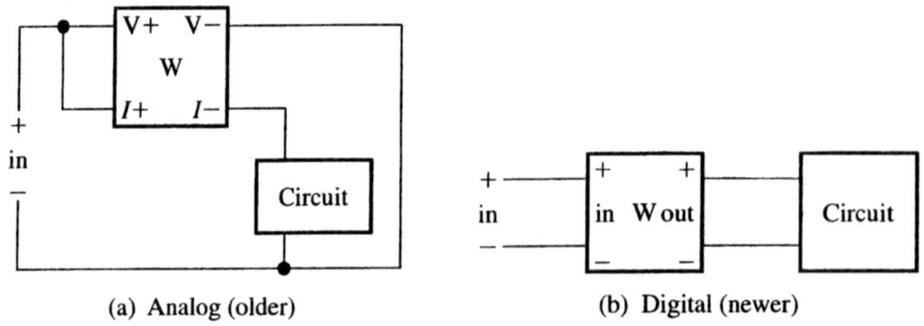
V+ V−
W
I+ I−
+
in
−
Circuit
(a) Analog (older)
+
in
−
+ +
in W out
− −
Circuit
(b) Digital (newer)

FIGURE B.24-3
Wattmeters.

REVIEW QUESTIONS

True/False

1. Power is always additive.
2. Power supplied to a circuit must equal power used by the circuit.

Multiple Choice

3. Two 10 ohm resistors are connected in series. What is the power supplied to the resistors from a 20 V source?
   a. 10 W each.
   b. 20 W each.
   c. 200 W each.
4. Repeat Question 3 for the same two resistors connected in parallel.
   a. 5 W.
   b. 20 W.
   c. 40 W.
5. A 2 W resistor has 2 V across it. What is the maximum current possible?
   a. 1 A.
   b. 1.414 A.
   c. 2 A.

C AC Circuits

C.01 Demonstrate an understanding of sources of electricity in AC circuits

C.02 Demonstrate an understanding of the properties of an AC signal

C.03 Demonstrate an understanding of the principles of operation and characteristics of sinusoidal and nonsinusoidal wave forms

C.04 Demonstrate an understanding of basic motor/generator theory and operation

C.05 Demonstrate an understanding of measurement of power in AC circuits

C.06 Demonstrate an understanding of the principle and operation of various power conditioning: (isolation transformers, surge suppressors, uninterruptable power systems)

C.07 Demonstrate an understanding of the principle and operation of safety grounding systems: (lightning arresters, ground fault interrupters, etc.)

C.08 Understand principles and operations of AC capacitive circuits

C.09 Fabricate and demonstrate AC capacitive circuits

C.10 Troubleshoot and repair AC capacitive circuits

C.11 Understand principles and operations of AC inductive circuits

C.12 Fabricate and demonstrate AC inductive circuits

C.13 Troubleshoot and repair AC inductive circuits

C.14 Understand principles and operations of AC circuits using transformers

C.15 Demonstrate an understanding of impedance matching theory

C.16 Fabricate and demonstrate AC circuits using transformers

C.17 Troubleshoot and repair AC circuits using transformers

C.18 Understand principles and operations of AC differentiator and integrator circuits (determine RC and RL time constants)

C.19 Fabricate and demonstrate AC differentiator and integrator circuits

C.20 Troubleshoot and repair AC differentiator and integrator circuits

C.21 Understand principles and operations of AC series and parallel resonant circuits

C.22 Fabricate and demonstrate AC series and parallel resonant circuits

C.23 Troubleshoot and repair AC series and parallel resonant circuits

C.24 Understand principles and operations of AC RC, RL, and RLC circuits

C.25 Fabricate and demonstrate AC RC, RL, and RLC circuits

C.26 Troubleshoot and repair AC RC, RL, and RLC circuits

C.27 Understand principles and operations of AC frequency selective filter circuits

C.28 Fabricate and demonstrate AC frequency selective filter circuits

C.29 Troubleshoot and repair AC frequency selective filter circuits

C.30 Understand principles and operations of AC polyphase circuits

C.31 Understand principles and operations of AC phase-locked loop circuits

C.32 Troubleshoot and repair AC phase-locked loop circuits

C.01 Demonstrate an understanding of sources of electricity in AC circuits

INTRODUCTION

Two basic methods of generating sine wave voltages are electromagnetic and electronic. Sine waves are produced electromagnetically by AC generators and electronically by oscillator circuits.

The AC Generator

Figure C.01-1 is a cutaway view of one type of AC generator. To illustrate the basic operation, Figure C.01-2 shows a greatly simplified AC **generator** consisting of a single loop of wire in a permanent magnetic field. Notice that each end of the loop is connected to a separate solid conductive ring called a *slip ring*. As the loop rotates in the magnetic field, the slip rings also rotate and rub against the brushes which connect the loop to an external load.

FIGURE C.01-1
The AC generator, cutaway view (courtesy of General Electric).

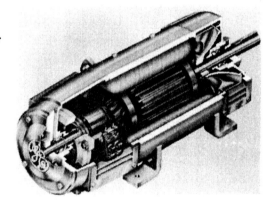

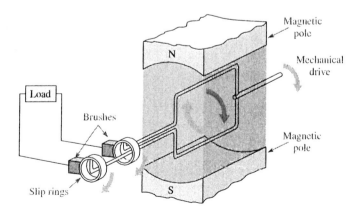

FIGURE C.01-2
Basic AC generator operations.

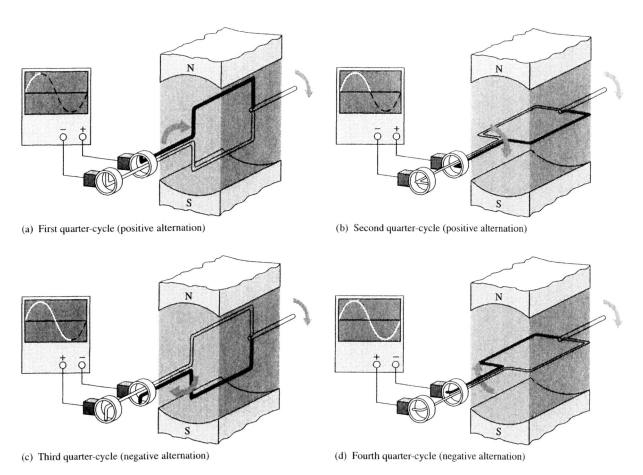

(a) First quarter-cycle (positive alternation)

(b) Second quarter-cycle (positive alternation)

(c) Third quarter-cycle (negative alternation)

(d) Fourth quarter-cycle (negative alternation)

FIGURE C.01-3
One revolution of the loop generates one cycle of the sine wave voltage.

When a conductor moves through a magnetic field, a voltage is induced. Figure C.01-3 illustrates how a sine wave voltage is produced by the basic AC generator as the loop rotates. An oscilloscope is used to display the voltage waveform.

To begin, Figure C.01-3(a) shows the loop rotating through the first quarter of a revolution. It goes from an instantaneous horizontal position, where the induced voltage is zero, to an instantaneous vertical position, where the induced voltage is maximum. At the horizontal position, the loop is instantaneously moving parallel with the flux lines; thus, no lines are being cut and the voltage is zero. As the loop rotates through the first quarter-cycle, it cuts through the flux lines at an increasing rate until it is instantaneously moving perpendicular to the flux lines at the vertical position and cutting through them at a maximum rate. Thus, the induced voltage increases from zero to a peak during the quarter-cycle. As shown on the display in part (a), this part of the rotation produces the first quarter of the sine wave cycle as the voltage builds up from zero to its positive maximum.

Figure C.01-3(b) shows the loop completing the first half of a revolution. During this part of the rotation, the voltage decreases from its positive maximum back to zero.

During the second half of the revolution, illustrated in Figure C.01-3, parts (c) and (d), the loop is cutting through the magnetic field in the opposite direction, so the voltage produced has a polarity opposite of that produced during the first half of the revolution. After one complete revolution of the loop, one full cycle of the sine wave voltage has been produced. As the loop continues to rotate, repetitive cycles of the sine wave are generated.

Frequency You have seen that one revolution of the conductor through the magnetic field in the basic AC generator (also called an *alternator*) produces one cycle of induced sinusoidal voltage. It is obvious that the rate at which the conductor is rotated determines the

FIGURE C.01-4
*Frequency is directly
proportional to the rate of
rotation of the generator
loop.*

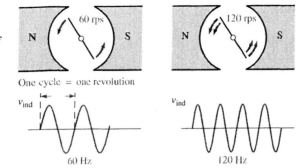

time for completion of one cycle. For example, if the conductor completes 60 revolutions in one second (rps), the period of the resulting sine wave is 1/60 s, corresponding to a frequency of 60 Hz. Thus, the faster the conductor rotates, the higher the resulting frequency of the induced voltage, as illustrated in Figure C.01-4.

Another way of achieving a higher frequency is to increase the number of magnetic poles. In the previous discussion, two magnetic poles were used to illustrate the AC generator principle. During one revolution, the conductor passes under a north pole and a south pole, thus producing one cycle of a sine wave. When four magnetic poles are used instead of two, as shown in Figure C.01-5, one cycle is generated during one-half a revolution. This doubles the frequency for the same rate of rotation.

FIGURE C.01-5
*Four poles achieve a
higher frequency than
two for the same rps.*

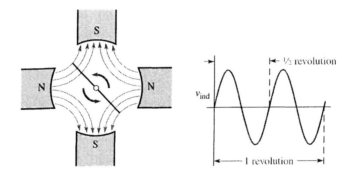

An expression for frequency in terms of the number of poles and the number of revolutions per second (rps) is as follows:

$$f = \text{(number of pole pairs) (rps)}$$

EXAMPLE C.01-1

A four-pole generator has a rotation speed of 100 rps. Determine the frequency of the output voltage.

Solution $f = \text{(number of pole pairs) (rps)} = 2(100) = 200$ Hz

Voltage Amplitude   The amount of voltage induced in a conductor depends on the number of turns (N) and the rate of change with respect to the magnetic field. Therefore, when the speed of rotation of the conductor is increased, not only does the frequency of the induced voltage increase, so also does the amplitude. Since the frequency value normally is fixed, the most practical method of increasing the amount or amplitude of induced voltage is to increase the number of loops.

Electronic Signal Generators

The signal generator is an instrument that electronically produces sine waves for use in testing or controlling electronic circuits and systems. There are a variety of signal generators, ranging from special-purpose instruments that produce only one type of waveform in a limited frequency range, to programmable instruments that produce a wide range of frequencies and a variety of waveforms. All signal generators consist basically of an **oscillator**, which is an electronic circuit that produces sine wave voltages whose amplitude and frequency can be adjusted. Typical signal generators are shown in Figure C.01-6.

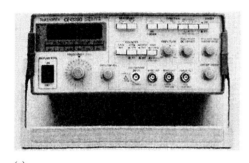

(a)

(b)

FIGURE C.01-6
Typical signal generators. Part (a) copyright 1999 Tektronix, Inc. All rights reserved—reproduced by permission. Part (b) courtesy of B & K Precision Corp.

REVIEW QUESTIONS

True/False

1. A sine wave may be produced by rotational action.
2. A 4-pole generator contains four north poles.

Multiple Choice

3. An electronic circuit that produces repetitive waveforms is called a(n)
   a. Wave generator.
   b. Oscillator.
   c. Amplifier.
4. AC generators are also called
   a. Alternators.
   b. Rotors.
   c. Stators.
5. Frequency in a generator is proportional to
   a. Amplitude.
   b. Induction.
   c. Rate of rotation.

C.02 Demonstrate an understanding of the properties of an AC signal

INTRODUCTION

The sine wave is a common type of alternating current (AC) and alternating voltage. It is also referred to as a sinusoidal waveform, or, simply, sinusoid. The electrical service provided by the power companies is in the form of sinusoidal voltage and current. In addition, other types of waveforms are composites of many individual sine waves called **harmonics.**

Sine waves are produced by two types of sources: rotating electrical machines (AC generators) or electronic oscillator circuits, which are in instruments known as electronic signal generators. Figure C.02-1 shows the symbol used to represent either source of sine wave voltage.

Figure C.02-2 is a graph showing the general shape of a sine wave, which can be either an **alternating current** or voltage. Voltage (or current) is displayed on the vertical axis, and time (*t*) is displayed on the horizontal axis. Notice how the voltage (or current) varies with time. Starting at zero, the voltage (or current) increases to a positive maximum (peak), returns to zero, and then increases to a negative maximum (peak) before returning again to zero, thus completing one full cycle.

FIGURE C.02-1
Symbol for a sine wave (sinusoidal) voltage source.

FIGURE C.02-2
Graph of one cycle of a sine wave.

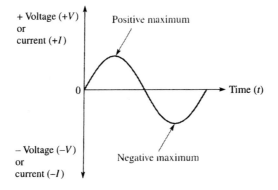

Polarity of a Sine Wave

As you have seen, a sine wave changes polarity at its zero value; that is, it alternates between positive and negative values. When a sine wave voltage is applied to a resistive circuit, as in Figure C.02-3, an alternating sine wave current results. When the voltage changes polarity, the current correspondingly changes direction as indicated.

FIGURE C.02-3
Alternating current and voltage.

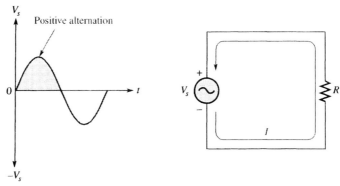

(a) Positive voltage: current direction as shown

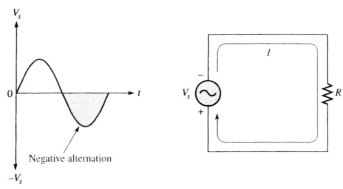

(b) Negative voltage: current reverses direction

During the positive alternation of the applied voltage V_s, the current is in the direction shown in Figure C.02-3(a). During a negative alternation of the applied voltage, the current is in the opposite direction, as shown in Figure C.02-3(b). The combined positive and negative alternations make up one **cycle** of a sine wave.

Period of a Sine Wave

The time required for a given sine wave to complete one full cycle is called the period (T).

Figure C.02-4(a) illustrates the **period** of a sine wave. Typically, a sine wave continues to repeat itself in identical cycles, as shown in Figure C.02-4(b). Since all cycles of a repetitive sine wave are the same, the period is always a fixed value for a given sine wave. The period of a sine wave can be measured from any peak in a given cycle to the corresponding peak in the next cycle.

FIGURE C.02-4
The period of a given sine wave is the same for each cycle.

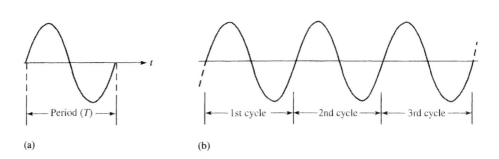

(a)   (b)

EXAMPLE C.02-1

What is the period of the sine wave in Figure C.02-5?

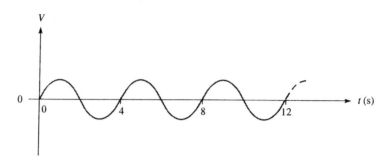

FIGURE C.02-5

Solution   As shown in Figure C.02-5, it takes four seconds (4 s) to complete each cycle. Therefore, the period is 4 s.

$$T = 4 \text{ s}$$

EXAMPLE C.02-2

Show three possible ways to measure the period of the sine wave in Figure C.02-6. How many cycles are shown?

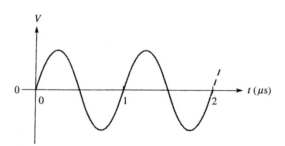

FIGURE C.02-6

Solution

Method 1:   The period can be measured from one zero crossing to the corresponding zero crossing in the next cycle.

Method 2:   The period can be measured from the positive peak in one cycle to the positive peak in the next cycle.

Method 3:   The period can be measured from the negative peak in one cycle to the negative peak in the next cycle.

These measurements are indicated in Figure C.02-7, where two cycles of the sine wave are shown. Keep in mind that you obtain the same value for the period no matter which corresponding peaks on the waveform you use.

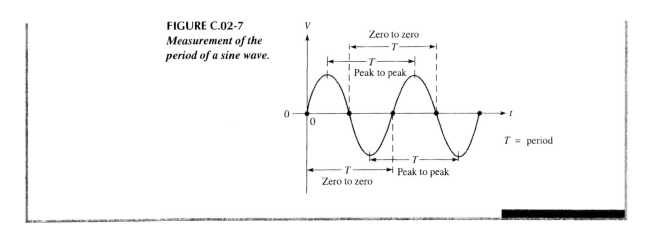

FIGURE C.02-7
Measurement of the period of a sine wave.

Frequency of a Sine Wave

Frequency is the number of cycles that a sine wave completes in one second.

The more cycles completed in one second, the higher the frequency. **Frequency** (f) is measured in units of **hertz.** One hertz (Hz) is equivalent to one cycle per second; for example, 60 Hz is 60 cycles per second. Figure C.02-8 shows two sine waves. The sine wave in part (a) completes two full cycles in one second. The one in part (b) completes four cycles in one second. Therefore, the sine wave in part (b) has twice the frequency of the one in part (a).

FIGURE C.02-8
Illustration of the frequency of a sine wave.

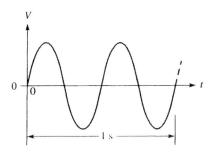

(a) Lower frequency: fewer cycles per second

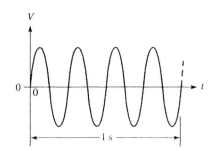

(b) Higher frequency: more cycles per second

Relationship of Frequency and Period

The relationship between frequency and period is very important. The formulas for this relationship are as follows:

$$f = \frac{1}{T}$$

$$T = \frac{1}{f}$$

There is a reciprocal relationship between f and T. Knowing one, you can calculate the other with the $\boxed{\text{2nd}}$ $\boxed{\text{1/x}}$ keys on your calculator. On some calculators, the reciprocal key is not a secondary function. This inverse relationship makes sense because a sine wave with a longer period goes through fewer cycles in one second than one with a shorter period.

EXAMPLE C.02-3

Which sine wave in Figure C.02-9 has the higher frequency? Determine the period and the frequency of both waveforms.

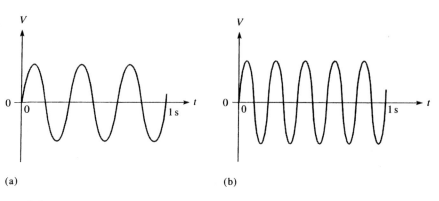

(a)   (b)

FIGURE C.02-9

Solution   The sine wave in Figure C.02-9(b) has the higher frequency because it completes more cycles in 1 s than does the one in part (a).

In Figure C.02-9(a), three cycles take 1 s. Therefore, one cycle takes 0.333 s (one-third second), and this is the period.

$$T = 0.333 \text{ s} = 333 \text{ ms}$$

The frequency is

$$f = \frac{1}{T} = \frac{1}{333 \text{ ms}} = 3 \text{ Hz}$$

In Figure C.02-9(b), five cycles take 1 s. Therefore, one cycle takes 0.2 s (one-fifth second), and this is the period.

$$T = 0.2 \text{ s} = 200 \text{ ms}$$

The frequency is

$$f = \frac{1}{T} = \frac{1}{200 \text{ ms}} = 5 \text{ Hz}$$

REVIEW QUESTIONS

True/False

1. The time required for a single cycle is called its period.
2. Period and frequency are proportional.

Multiple Choice

3. If the period of a sine wave is 1 mS, what is its frequency?
   a. 1 mHz.
   b. 100 Hz.
   c. 1000 Hz.

FIGURE C.03-5
Half-cycle average value.

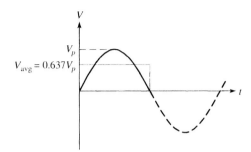

EXAMPLE C.03-1

Determine V_p, V_{pp}, V_{rms}, and the half-cycle V_{avg} for the sine wave in Figure C.03-6.

FIGURE C.03-6 $+V$ (V)

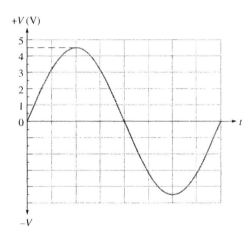

Solution As taken directly from the graph, V_p = 4.5 V. From this value calculate the other values.

$$V_{pp} = 2V_p = 2(4.5 \text{ V}) = 9 \text{ V}$$
$$V_{rms} = 0.707V_p = 0.707(4.5 \text{ V}) = 3.18 \text{ V}$$
$$V_{avg} = 0.637V_p = 0.637(4.5 \text{ V}) = 2.87 \text{ V}$$

Phase of a Sine Wave

The **phase** of a sine wave is an angular measurement that specifies the position of that sine wave relative to a reference. Figure C.03-7 shows one cycle of a sine wave to be used as

FIGURE C.03-7
Phase reference.

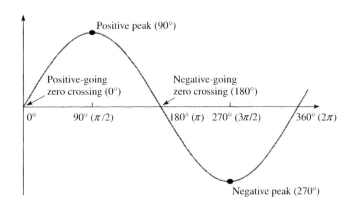

the reference. Note that the first positive-going crossing of the horizontal axis (zero cross-ing) is at 0° (0 rad), and the positive peak is at 90° (π/2 rad). The negative-going zero crossing is at 180° (π rad), and the negative peak is at 270° (3π/2 rad). The cycle is com-pleted at 360° (2π rad). When the sine wave is shifted left or right with respect to this ref-erence, there is a phase shift.

Figure C.03-8 illustrates phase shifts of a sine wave. In part (a), sine wave *B* is shifted to the right by 90° (π/2 rad). Thus, there is a phase angle of 90° between sine wave *A* and sine wave *B*. In terms of time, the positive peak of sine wave *B* occurs later than the positive peak of sine wave *A* because time increases to the right along the horizontal axis. In this case, sine wave *B* is said to **lag** sine wave *A* by 90° or π/2 radians. In other words, sine wave *A* leads sine wave *B* by 90°.

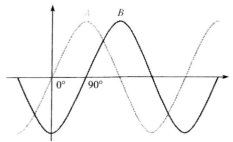

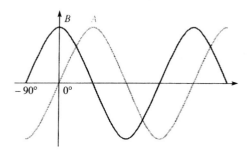

(a) *A* leads *B* by 90°, or *B* lags *A* by 90°. (b) *B* leads *A* by 90°, or *A* lags *B* by 90°.

FIGURE C.03-8
Illustration of a phase shift.

In Figure C.03-8(b), sine wave *B* is shown shifted left by 90°. Thus, again there is a phase angle of 90° between sine wave *A* and sine wave *B*. In this case, the positive peak of sine wave *B* occurs earlier in time than that of sine wave *A*; therefore, sine wave *B* is said to **lead** by 90°. In both cases there is a 90° phase angle between the two waveforms.

EXAMPLE C.03-2

What are the phase angles between the two sine waves in Figure C.03-9(a) and C.03-9(b)?

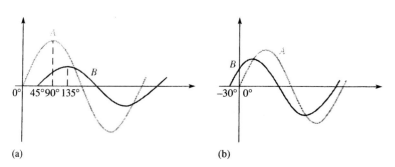

(a)   (b)

FIGURE C.03-9

Solution In Figure C.03-9(a) the zero crossing of sine wave *A* is at 0°, and the corre-sponding zero crossing of sine wave *B* is at 45°. There is a 45° phase angle between the two waveforms with sine wave *A* leading.

In Figure C.03-9(b) the zero crossing of sine wave *B* is at −30°, and the corre-sponding zero crossing of sine wave *A* is at 0°. There is a 30° phase angle between the two waveforms with sine wave *B* leading.

EXAMPLE C.03-3

Determine the instantaneous value at the 90° reference point on the horizontal axis for each sine wave voltage in Figure C.03-10.

FIGURE C.03-10

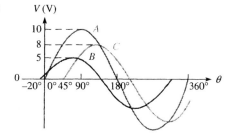

Solution Sine wave A is the reference. Sine wave B is shifted left 20° with respect to A, so it leads. Sine wave C is shifted right 45° with respect to A, so it lags.

$$v_A = V_p\sin\theta = 10\sin 90° = 10(1) = 10 \text{ V}$$
$$v_B = V_p\sin(\theta + \phi_B)$$
$$= 5\sin(90° + 20°) = 5\sin 110° = 5(0.9397) = 4.7 \text{ V}$$
$$v_C = V_p\sin(\theta - \phi_C)$$
$$= 8\sin(90° - 45°) = 8\sin 45° = 8(0.707) = 5.66 \text{ V}$$

EXAMPLE C.03-4

Determine the average value of each of the positive-going waveforms in Figure C.03-11.

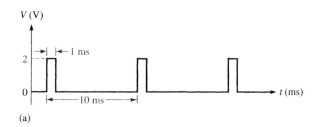

(a)

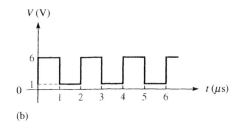

(b)

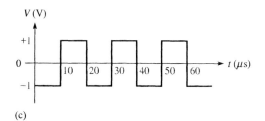

(c)

FIGURE C.03-11

Solution In Figure C.03-11(a), the baseline is at 0 V, the amplitude is 2 V, and the duty cycle is 10%. The average value is

$$V_{avg} = \text{baseline} + (\text{duty cycle})(\text{amplitude})$$
$$= 0 \text{ V} + (0.1)(2 \text{ V}) = 0.2 \text{ V}$$

The waveform in Figure C.03-11(b) has a baseline of +1 V, an amplitude of 5 V, and a duty cycle of 50%. The average value is

$$V_{avg} = \text{baseline} + (\text{duty cycle})(\text{amplitude})$$
$$= 1\text{ V} + (0.5)(5\text{ V}) = 1\text{ V} + 2.5\text{ V} = 3.5\text{ V}$$

The waveform in Figure C.03-11(c) is a square wave with a baseline of −1 V and an amplitude of 2 V. The duty cycle is 50%. The average value is

$$V_{avg} = \text{baseline} + (\text{duty cycle})(\text{amplitude})$$
$$= -1\text{ V} + (0.5)(2\text{ V}) = -1\text{ V} + 1\text{ V} = 0\text{ V}$$

This is an alternating square wave, and, like an alternating sine wave, it has an average value of zero over a full cycle.

Triangular and Sawtooth Waveforms

Triangular and sawtooth waveforms are formed by voltage or current ramps. A **ramp** is a linear increase or decrease in the voltage or current. Figure C.03-12 shows both positive- and negative-going ramps. In part (a) the ramp has a positive slope; in part (b), the ramp has a negative slope. The slope of a voltage ramp is $\pm\ V/t$ and is expressed in units of V/s. The slope of a current ramp is $\pm\ I/t$ and is expressed in units of A/s.

FIGURE C.03-12
Ramps.

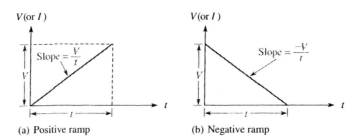

(a) Positive ramp   (b) Negative ramp

Triangular Waveforms   Figure C.03-13 shows that a **triangular waveform** is composed of positive-going and negative-going ramps having equal slopes. The period of this waveform is measured from one peak to the next corresponding peak, as illustrated. This particular triangular waveform is alternating and has an average value of zero.

Figure C.03-14 depicts a triangular waveform with a nonzero average value. The frequency for triangular waves is determined in the same way as for sine waves, that is, $f = 1/T$.

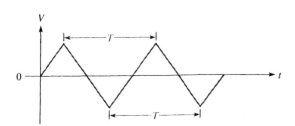

FIGURE C.03-13
Alternating triangular waveforms.

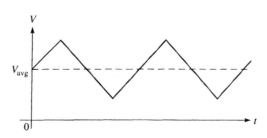

FIGURE C.03-14
Nonalternating triangular waveform.

Sawtooth Waveforms The **sawtooth waveform** is actually a special case of the triangular waveform consisting of two ramps, one of much longer duration than the other. Sawtooth waveforms are commonly used in many electronic systems. For example, the electron beam that sweeps across the screen of your TV receiver, creating the picture, is controlled by sawtooth voltages and currents. One sawtooth wave produces the horizontal beam movement, and the other produces the vertical beam movement. A sawtooth voltage is sometimes called a *sweep voltage*.

Figure C.03-15 is an example of a sawtooth waveform. Notice that it consists of a positive-going ramp of relatively long duration, followed by a negative-going ramp of relatively short duration.

FIGURE C.03-15
Alternating sawtooth waveform.

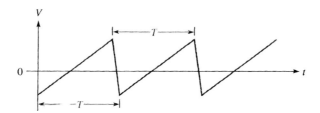

REVIEW QUESTIONS

True/False

1. Average volts and volts rms are the same thing.
2. Pulse width is measured between the 10% and 90% levels of a pulse.

Multiple Choice

3. If the peak voltage of a sine wave is 100 V, the rms voltage is
   a. 70.7 V.
   b. 100 V.
   c. 141 V.
4. 3.14159 radians is the same as
   a. 45 degrees.
   b. 90 degrees.
   c. 180 degrees.
5. A pulse has an on time of 4 mS and an off time of 12 mS. What is its duty cycle?
   a. 25%.
   b. 50%.
   c. 75%.

C.04 Demonstrate an understanding of basic motor/generator theory and operation

Refer to C.01.

C.05 Demonstrate an understanding of measurement of power in AC circuits

INTRODUCTION

There are three types of power in an AC circuit: true power (measured in Watts), reactive power (in VARs, for volt-amperes reactive), and apparent power (in VA, or volt-amperes). In this section we examine how each power may be determined.

Determine the power factor and the true power in the circuit of Figure C.05-1.

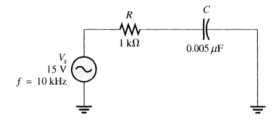

FIGURE C.05-1

Solution Calculate the capacitive reactance and phase angle as follows:

$$X_C = \frac{1}{2\pi f C} = \frac{1}{2\pi(10 \text{ kHz})(0.005 \ \mu\text{F})} = 3.18 \text{ k}\Omega$$

$$\theta = \tan^{-1}\left(\frac{X_C}{R}\right) = \tan^{-1}\left(\frac{3.18 \text{ k}\Omega}{1 \text{ k}\Omega}\right) = 72.5°$$

The power factor is

$$PF = \cos\theta = \cos(72.5°) = 0.301$$

The impedance is

$$Z = \sqrt{R^2 + X_C^2} = \sqrt{(1 \text{ k}\Omega)^2 + (3.18 \text{ k}\Omega)^2} = 3.33 \text{ k}\Omega$$

Therefore, the current is

$$I = \frac{V_s}{Z} = \frac{15 \text{ V}}{3.33 \text{ k}\Omega} = 4.50 \text{ mA}$$

The true power is

$$P_{\text{true}} = V_s I \cos\theta = (15 \text{ V})(4.50 \text{ mA})(0.301) = 20.3 \text{ mW}$$

EXAMPLE C.05-2

For the circuit in Figure C.05-2, find the true power, the reactive power, and the apparent power. X_C has been determined to be 2 kΩ.

FIGURE C.05-2

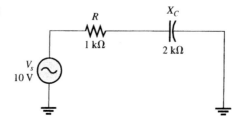

Solution First find the total impedance so that the current can be calculated.

$$Z_{tot} = \sqrt{R^2 + X_C^2} = \sqrt{(1 \text{ k}\Omega)^2 + (2 \text{ k}\Omega)^2} = 2.24 \text{ k}\Omega$$

$$I = \frac{V_s}{Z} = \frac{10 \text{ V}}{2.24 \text{ k}\Omega} = 4.46 \text{ mA}$$

The phase angle, θ, is

$$\theta = \tan^{-1}\left(\frac{X_C}{R}\right) = \tan^{-1}\left(\frac{2 \text{ k}\Omega}{1 \text{ k}\Omega}\right) = 63.4°$$

The true power is

$$P_{\text{true}} = V_s I \cos\theta = (10 \text{ V})(4.46 \text{ mA}) \cos(63.4°) = 20 \text{ mW}$$

Note that the same result is realized using the formula $P_{\text{true}} = I^2 R$.
    The reactive power is

$$P_r = I^2 X_C = (4.46 \text{ mA})^2(2 \text{ k}\Omega) = 39.8 \text{ mVAR}$$

The apparent power is

$$P_a = I^2 Z = (4.46 \text{ mA})^2(2.24 \text{ k}\Omega) = 44.6 \text{ mVA}$$

The apparent power is also the phasor sum of P_{true} and P_r.

$$P_a = \sqrt{P_{\text{true}}^2 + P_r^2} = 44.6 \text{ mVA}$$

EXAMPLE C.05-3

Determine the power factor, the true power, the reactive power, and the apparent power in Figure C.05-3.

FIGURE C.05-3

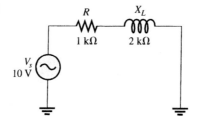

Solution The impedance of the circuit is

$$Z = \sqrt{R^2 + X_L^2} = \sqrt{(1 \text{ k}\Omega)^2 + (2 \text{ k}\Omega)^2} = 2.24 \text{ k}\Omega$$

The current is

$$I = \frac{V_s}{Z} = \frac{10 \text{ V}}{2.24 \text{ k}\Omega} = 4.46 \text{ mA}$$

The phase angle is

$$\theta = \tan^{-1}\left(\frac{X_L}{R}\right) = \tan^{-1}\left(\frac{2 \text{ k}\Omega}{1 \text{ k}\Omega}\right) = 63.4°$$

Therefore, the power factor is

$$PF = \cos\theta = \cos(63.4°) = 0.448$$

The true power is

$$P_{true} = V_s I \cos\theta = (10 \text{ V})(4.46 \text{ mA})(0.448) = 20 \text{ mW}$$

The reactive power is

$$P_r = I^2 X_L = (4.46 \text{ mA})^2(2 \text{ k}\Omega) = 39.8 \text{ mVAR}$$

The apparent power is

$$P_a = I^2 Z = (4.46 \text{ mA})^2(2.24 \text{ k}\Omega) = 44.6 \text{ mVA}$$

REVIEW QUESTIONS

True/False

1. Power in a resistor is called *reactive power.*
2. The power factor of a purely resistive circuit is 1.0.

Multiple Choice

3. If a source supplies 4 A at 12 V and the angle between voltage and current is 60 degrees, the true power is
   a. 12 W.
   b. 24 W.
   c. 48 W.
4. The power factor of an inductive load is called a(n)
   a. Leading power factor.
   b. Lagging power factor.
   c. In-phase power factor.
5. To correct the power factor of an inductive load, add additional
   a. Capacitance.
   b. Inductance.
   c. Resistance.

C.06 Demonstrate an understanding of the principle and operation of various power conditioning: (isolation transformers, surge suppressors, uninterruptable power systems)

INTRODUCTION

Transformers are useful in providing electrical isolation between the primary circuit and the secondary circuit because there is no electrical connection between the two windings. In a transformer, energy is transferred entirely by magnetic coupling.

DC Isolation

If there is a nonchanging direct current through the primary circuit of a transformer, nothing happens in the secondary circuit, as indicated in Figure C.06-1(a). The reason is that a changing current in the primary winding is necessary in order to create a changing magnetic field. This will cause voltage to be induced in the secondary circuit, as indicated in Figure C.06-1(b). Therefore, the transformer isolates the secondary circuit from any DC voltage in the primary circuit.

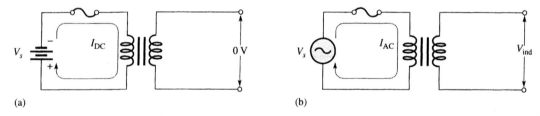

(a)   (b)

FIGURE C.06-1
DC isolation and AC coupling.

In a typical application, a transformer can be used to keep the DC voltage on the output of an amplifier stage from affecting the DC bias of the next amplifier. Only the AC signal is coupled through the transformer from one stage to the next, as Figure C.06-2 illustrates.

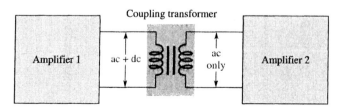

FIGURE C.06-2
Audio amplifier stages with transformer coupling for DC isolation.

Power Line Isolation

Transformers are often used to electrically isolate electronic equipment from the 60 Hz, 110 V AC power line. The reason for using an isolation transformer to couple the 60 Hz AC to an instrument is to prevent a possible shock hazard if the 110 V line is connected to the metal chassis of the equipment. This condition is possible if the line cord plug can be inserted into an outlet either way. Incidentally, to prevent this situation, many plugs have keyed prongs so that they can be plugged in only one way.

Figure C.06-3 illustrates how a transformer can prevent the metal chassis from being connected to the 110 V line rather than to neutral (ground), no matter how the cord is plugged into the outlet. When an isolation transformer is used, the secondary circuit is said to be "floating" because it is not referenced to the power line ground. Should a person come in contact with the secondary voltage, there is no complete current path back to ground, and therefore there is no shock hazard. There must be current through your body in order for you to receive an electrical shock.

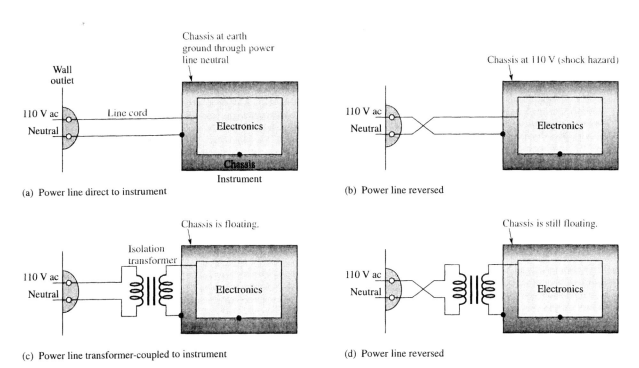

(a) Power line direct to instrument

(b) Power line reversed

(c) Power line transformer-coupled to instrument

(d) Power line reversed

FIGURE C.06-3
The use of an isolation transformer to prevent a shock hazard.

Surge Supressors

Surge supressors are used to protect sensitive electronic circuitry from high-voltage spikes on power or signal inputs/outputs. In its simplest form, a surge supressor may be a series element (in series with the protected signal) or a shunt element (connected between the signal and ground). The resistance of the element changes during a high-voltage spike, either blocking the spike (via high resistance) or shorting it to ground (with a low resistance).

Surge supressors are used to protect practically every electrical product from personal computers to FAX machines, VCR's, and stereos. Surge supressors can be used on telephone lines to protect against damage from electrical storms.

Since surge supressors for use around the home or office are relatively inexpensive, it is well worth the cost to protect your equipment.

UPS (Uninterruptable Power Source) Systems

An uninterruptable power source is a power supply that is able to provide power to its load for a short period of time in the event of a power failure. Typically, a UPS system will contain one or more batteries, which are normally kept in a charging mode when main power is available. During a power outage the batteries supply power to the load, either as regulated DC or as AC. Power protection may last for only a few minutes or up to several hours (with a proportional increase in cost).

For critical power needs (such as equipment in a hospital) UPS systems are a must. Can you think of other places where UPS systems may be required?

REVIEW QUESTIONS

True/False

1. A transformer provides electrical isolation of both DC and AC signals.

2. An isolation transformer causes its associated circuit to *float.*

Multiple Choice

3. A surge supressor

   a. Is only used on telephone lines.

   b. Only operates when equipment is turned on.

   c. Protects against high-voltage spikes.

4. UPS stands for

   a. Up surge.

   b. Unknown power source.

   c. Uninterruptable power source.

5. UPS systems provide backup

   a. DC.

   b. AC.

   c. Both a and b.

C.07 Demonstrate an understanding of the principle and operation of safety grounding systems: (lightning arresters, ground fault interrupters, etc.)

INTRODUCTION

In this section we examine the operation of common safety devices used in electrical circuits.

Circuit Breakers, GFCIs, and Fuses

The incoming power to any large industrial plant, heavy equipment, simple circuit in the home, or meters used in the laboratory must be limited to ensure that the current through the lines is not above the rated value. Otherwise, the conductors or the electrical or electronic equipment may be damaged or dangerous side effects such as fire or smoke may result. To limit the current level, fuses or circuit breakers are installed where the power enters the installation, such as in the panel in the basement of most homes at the point where the outside feeder lines enter the dwelling. The fuses of Figure C.07-1 have an internal metallic conductor through which the current will pass; it will begin to melt if the current through the system exceeds the rated value printed on the casing. Of course, if it melts through, the current path is broken and the load in its path is protected.

FIGURE C.07-1
Fuses: (a) CC-TRON®
(0–10 A); (b) subminiature
solid matrix; (c) Semitron
(0–600 A). (Courtesy of
Cooper Bussman)

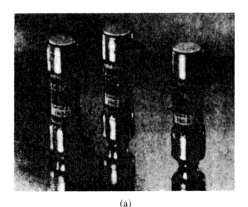

(a)

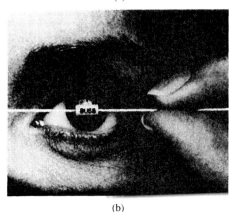

(b)

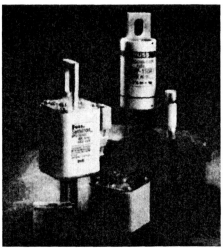

(c)

FIGURE C.07-2
Circuit breakers. (Courtesy of Siemens
Electrochemical Components, Inc.)

In homes built in recent years, fuses have been replaced by circuit breakers such as those appearing in Figure C.07-2. When the current exceeds rated conditions, an electromagnet in the device will have sufficient strength to draw the connecting metallic link in the breaker out of the circuit and open the current path. When conditions have been corrected, the breaker can be reset and used again.

The current National Electrical Code requires that outlets in the bathroom and other sensitive areas be of the ground fault current interrupt (GFCI) variety. They are designed to trip more quickly than the standard circuit breaker. The units in Figure C.07-3 trip in 500 ms (½ s). It has been determined that 6 mA is the maximum level that most individuals should be exposed to for a short period of time and not be seriously injured. A current higher than 11 mA can cause involuntary muscle contractions that could prevent a person from letting go of the conductor and possibly cause him or her to enter a state of shock. Higher currents lasting more than a second can cause the heart to go into fibrillation and possibly cause death in a few minutes. The GFCI is able to react as quickly as it does by sensing the difference between the input and output currents to the outlet. They should be the same if everything is working properly. An errant path such as through an individual establishes a difference in the two current levels and causes the breaker to trip and disconnect the power source.

FIGURE C.07-3
Ground fault current interrupter (GFCI)
125 V AC, 60 Hz, 15-A outlet. (Courtesy of
Leviton Manufacturing Co., Inc.)

REVIEW QUESTIONS

True/False

1. Only current supplied to an industrial plant needs to be limited.

2. Fuses have been replaced by circuit breakers in many homes.

Multiple Choice

3. Circuit breakers

   a. Break the connection once and must be replaced.

   b. Can break the connection many times.

   c. Must never be manually tripped.

4. GFCI stands for
    a. Ground frequency interrupt.
    b. Ground fault capacitance.
    c. Ground fault current interrupt.
5. An individual can withstand brief currents of
    a. 6 mA.
    b. 11 mA.
    c. 500 mA.

C.08 Understand principles and operations of AC capacitive circuits

INTRODUCTION

In this section we examine the reactance, phase, and power characteristics of the capacitor.

EXAMPLE C.08-1

A sinusoidal voltage is applied to a capacitor, as shown in Figure C.08-1. The frequency of the sine wave is 1 kHz. Determine the capacitive reactance.

FIGURE C.08-1

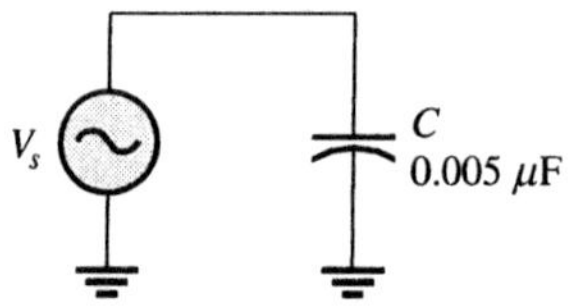

V_s C 0.005 μF

Solution $X_C = \dfrac{1}{2\pi f C} = \dfrac{1}{2\pi(1 \times 10^3 \text{ Hz})(0.005 \times 10^{-6} \text{ F})} = 31.8 \text{ k}\Omega$

EXAMPLE C.08-2

Determine the rms current in Figure C.08-2.

FIGURE C.08-2

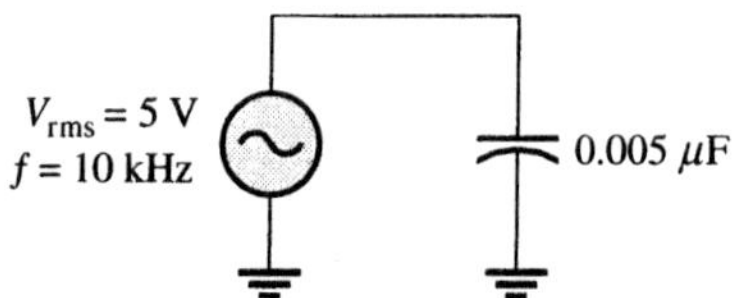

$V_{rms} = 5$ V
$f = 10$ kHz   0.005 μF

Solution First, find X_C.

$$X_C = \frac{1}{2\pi f C} = \frac{1}{2\pi(10 \times 10^3 \text{ Hz})(0.005 \times 10^{-6} \text{ F})} = 3.18 \text{ k}\Omega$$

Then, apply Ohm's law.

$$V_{rms} = I_{rms}X_C$$

$$I_{rms} = \frac{V_{rms}}{X_C} = \frac{5 \text{ V}}{3.18 \text{ k}\Omega} = 1.57 \text{ mA}$$

EXAMPLE C.08-3

Determine the true power and the reactive power in Figure C.08-3.

FIGURE C.08-3

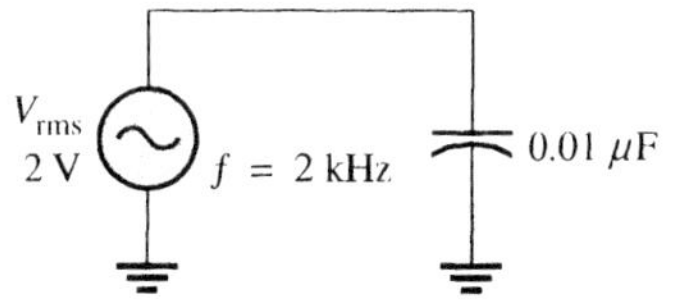

Solution The true power P_{true} is always zero for an ideal capacitor. The reactive power is determined by first finding the value for the capacitive reactance.

$$X_C = \frac{1}{2\pi fC} = \frac{1}{2\pi(2 \times 10^3 \text{ Hz})(0.01 \times 10^{-6} \text{ F})} = 7.96 \text{ k}\Omega$$

$$P_r = \frac{V_{\text{rms}}^2}{X_C} = \frac{(2 \text{ V})^2}{7.96 \text{ k}\Omega} = 503 \times 10^{-6} \text{ VAR} = 503 \text{ }\mu\text{VAR}$$

REVIEW QUESTIONS

True/False

1. The reactance of a capacitor can be thought of as its AC resistance.

2. X_C goes up as frequency goes down.

Multiple Choice

3. What is the reactance of a .047 μF capacitor when operated at 1 KHz?

    a. 338 ohms.

    b. 3386 ohms.

    c. 33.8 K ohms.

4. Current in a capacitor

    a. Lags voltage by 90 degrees.

    b. Leads voltage by 90 degrees.

    c. Is in-phase with voltage.

5. A capacitor with 100 ohms of reactance has 20 V across it. What is the reactive power?

    a. 4 Watts.

    b. 4 Vars.

    c. 4 Joules.

C.09 Fabricate and demonstrate AC capacitive circuits

INTRODUCTION

Capacitors are used in certain types of amplifiers for coupling AC signals and blocking DC voltages. In this section, an amplifier circuit board contains two coupling capacitors. Your assignment is to check certain voltages on three identical amplifier circuit boards to determine if the capacitors are working properly. A knowledge of amplifier circuits is not necessary for this assignment.

All amplifier circuits contain transistors that require DC voltages to establish proper operating conditions for amplifying AC signals. These DC voltages are referred to as bias voltages. As indicated in Figure C.09-1(a), a common type of DC bias circuit used in amplifiers is the voltage divider formed by R_1 and R_2, which sets up the proper DC voltage at the input to the amplifier.

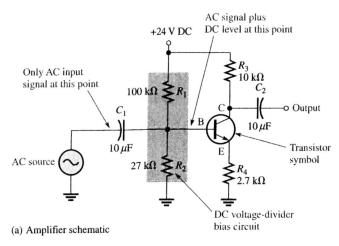

(a) Amplifier schematic

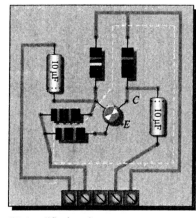

(b) Amplifier board

FIGURE C.09-1
Capacitively coupled amplifier.

When an AC signal voltage is applied to the amplifier, the input coupling capacitor, C_1, prevents the internal resistance of the AC source from changing the DC bias voltage. Without the capacitor, the internal source resistance would appear in parallel with R_2 and drastically change the value of the DC voltage.

The coupling capacitance is chosen so that its reactance (X_C) at the frequency of the AC signal is very small compared to the bias resistor values. The coupling capacitance therefore efficiently couples the AC signal from the source to the input of the amplifier. On the source side of the input coupling capacitor there is only AC, but on the amplifier side there is AC plus DC (the signal voltage is riding on the DC bias voltage set by the voltage divider), as indicated in Figure C.09-1(a). Capacitor C_2 is the output coupling capacitor which couples the amplified AC signal to another amplifier stage that would be connected to the output.

You will check three amplifier boards like the one in Figure C.09-1(b) for the proper input voltages using an oscilloscope. If the voltages are incorrect, you will determine the most likely fault. For all measurements, assume the amplifier has no DC loading effect on the voltage-divider bias circuit.

Step 1: Compare the Printed Circuit Board with the Schematic

Check the printed circuit board in Figure C.09-1(b) to make sure it agrees with the amplifier schematic in part (a).

Step 2: Test the Input to Board 1

The oscilloscope probe is connected from channel 1 to the board as shown in Figure C.09-2.

The input signal from a sinusoidal voltage source is connected to the board and set to a frequency of 5 kHz with an amplitude of 1 V rms. Determine if the voltage and frequency displayed on the scope in Figure C.09-2 are correct. If the scope measurement is incorrect, specify the most likely fault in the circuit.

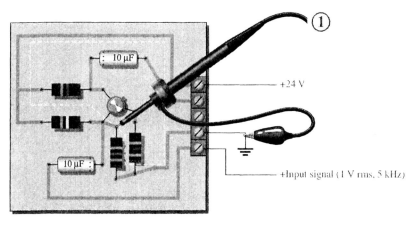

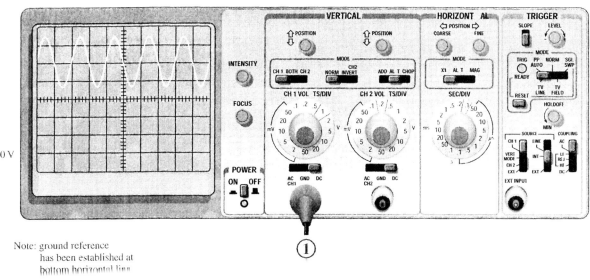

Note: ground reference
has been established at
bottom horizontal line

FIGURE C.09-2
Testing board 1.

Step 3: Test the Input to Board 2

The oscilloscope probe is connected from channel 1 to board 2 the same as was shown in Figure C.09-2 for board 1.

The input signal from the sinusoidal voltage source is the same as Step 2. Determine if the scope display in Figure C.09-3 is correct. If the scope measurement is incorrect, specify the most likely fault in the circuit.

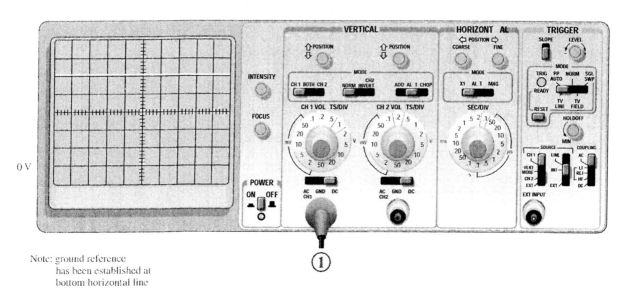

Note: ground reference
 has been established at
 bottom horizontal line

FIGURE C.09-3
Testing board 2.

Step 4: Test the Input to Board 3

The oscilloscope probe is connected from channel 1 to board 3 the same as was shown in Figure C.09-2 for board 1.

The input signal from the sinusoidal voltage source is the same as Step 3. Determine if the scope display in Figure C.09-4 is correct. If the scope measurement is incorrect, specify the most likely fault in the circuit.

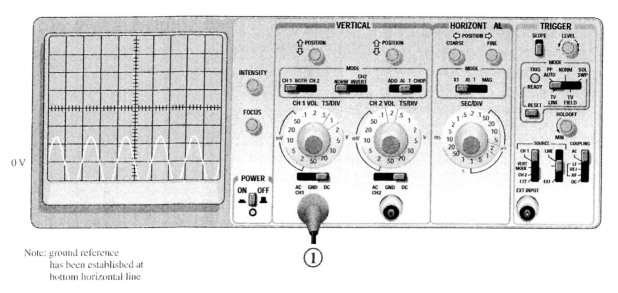

Note: ground reference
 has been established at
 bottom horizontal line

FIGURE C.09-4
Testing board 3.

C.10 Troubleshoot and repair AC capacitive circuits

INTRODUCTION

Here are a few things to examine when troubleshooting AC capacitive circuits:

☐ Is the operating frequency correct? The capacitor values used in the circuit were chosen so that they have a specific reactance at the operating frequency (or throughout the range of operating frequencies).

☐ Verify the capacitive reactance by measuring the voltage across and current through the capacitor and dividing the measured values. How does the calculation compare with the value predicted by the capacitive reactance formula?

☐ Measure the actual capacitance using a digital capacitance meter. If the capacitor is old or has been stressed, it may have changed value.

☐ Carefully touch the capacitor while it is on. If it is warm it is leaky and not working properly.

C.11 Understand principles and operations of AC inductive circuits

INTRODUCTION

In this section the reactance, phase, and power characteristics of an inductor are examined.

EXAMPLE C.11-1

A sinusoidal voltage is applied to the circuit in Figure C.11-1. The frequency is 1 kHz. Determine the inductive reactance.

FIGURE C.11-1

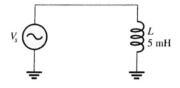

Solution   Convert 1 kHz to 1×10^3 Hz and 5 mH to 5×10^{-3} H. Then, the inductive reactance is

$$X_L = 2\pi fL = 2\pi(1 \times 10^3 \text{ Hz})(5 \times 10^{-3} \text{ H}) = 31.4 \ \Omega$$

EXAMPLE C.11-2

Determine the rms current in Figure C.11-2.

FIGURE C.11-2

Solution   Convert 10 kHz to 10×10^3 Hz and 100 mH to 100×10^{-3} H. Then calculate X_L.

$$X_L = 2\pi fL = 2\pi(10 \times 10^3 \text{ Hz})(100 \times 10^{-3} \text{ H}) = 6283 \ \Omega$$

Apply Ohm's law as follows:

$$I_{\text{rms}} = \frac{V_{\text{rms}}}{X_L} = \frac{5 \text{ V}}{6283 \ \Omega} = 796 \ \mu\text{A}$$

EXAMPLE C.11-3

A 10 V rms signal with a frequency of 1 kHz is applied to a 10 mH coil with a winding resistance of 5 Ω. Determine the reactive power (P_r) and the true power (P_{true}).

Solution   First, calculate the inductive reactance and current values.

$$X_L = 2\pi f L = 2\pi(1 \text{ kHz})(10 \text{ mH}) = 62.8 \ \Omega$$

$$I = \frac{V_S}{X_L} = \frac{10 \text{ V}}{62.8 \ \Omega} = 159 \text{ mA}$$

$$P_r = I^2 X_L = (159 \text{ mA})^2(62.8 \ \Omega) = 1.59 \text{ VAR}$$

The true power is

$$P_{\text{true}} = I^2 R_W = (159 \text{ mA})^2(5 \ \Omega) = 126 \text{ mW}$$

The Quality Factor (Q) of a Coil

The **quality factor** (*Q*) is the ratio of the reactive power in the inductor to the true power in the winding resistance of the coil or the resistance in series with the coil. It is a ratio of the power in *L* to the power in R_W. The quality factor is very important in resonant circuits. A formula for *Q* is developed as follows:

$$Q = \frac{\text{reactive power}}{\text{true power}} = \frac{P_r}{P_{\text{true}}} = \frac{I^2 X_L}{I^2 R_W}$$

In a series circuit, *I* is the same in *L* and *R;* thus, the I^2 terms cancel, leaving

$$Q = \frac{X_L}{R_W}$$

When the resistance is just the winding resistance of the coil, the circuit *Q* and the coil *Q* are the same. Note that *Q* is a ratio of like units and, therefore, has no unit itself.

REVIEW QUESTIONS

True/False

1. The reactance of an inductor can be thought of as its AC resistance.

2. X_L goes up as frequency goes down.

Multiple Choice

3. What is the reactance of a 10 mH inductor when operated at 1 KHz?
   a. 62.8 ohms.
   b. 628 ohms.
   c. 6280 ohms.

4. Current in an inductor
   a. Lags voltage by 90 degrees.
   b. Leads voltage by 90 degrees.
   c. Is in-phase with voltage.

5. The ratio of reactive power to true power in an inductor is called
   a. *P.*
   b. *Q.*
   c. *R.*

C.12 Fabricate and demonstrate AC inductive circuits

INTRODUCTION

Using actual components (or simulation software such as *Electronics Workbench*), set up and examine the circuit in Figure C.12-1.

Adjust the signal frequency to 500 Hz. Measure the voltage across the resistor, the voltage across the inductor, and the current. Divide the inductor voltage by the current to find the inductive reactance. How does the reactance compare with the value predicted by the formula? Is Kirchhoff's voltage law satisfied? What is the phase shift between the source voltage and source current?

Repeat for a frequency of 1 KHz.

FIGURE C.12-1
RL test circuit.

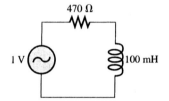

C.13 Troubleshoot and repair AC inductive circuits

INTRODUCTION

Many of the points made in Section C.10 for troubleshooting capacitors applies to the inductor as well. Here are a few additional things to examine when troubleshooting AC inductive circuits:

☐ What is the resistance of the inductor? Is it small compared to other resistance in series with the inductor? For example, in a series resonant circuit, a resistive load in series with the inductor and capacitor might be 470 ohms. If the inductor has a winding resistance of 5 ohms, its effect on the circuit will be extremely small. If the load is small as well, say 50 ohms or less, the resistance of the inductor becomes significant and must be taken into account.

☐ Are there any other inductors present, providing the possibility of mutual inductance? Placing the inductors at right angles to each other should eliminate this type of problem.

C.14 Understand principles and operations of AC circuits using transformers

INTRODUCTION

In this section we look at several examples of step-up and step-down transformers, and their loading properties.

EXAMPLE C.14-1

The transformer in Figure C.14-1 has a 200 turn primary winding and a 600 turn secondary winding. What is the voltage across the secondary?

FIGURE C.14-1

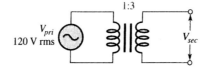

Solution   The secondary voltage is

$$V_{sec} = \left(\frac{N_{sec}}{N_{pri}}\right)V_{pri} = \left(\frac{600}{200}\right)120 \text{ V} = 3(120 \text{ V}) = 360 \text{ V}$$

Note that the turns ratio of 3 is indicated on the schematic as 1:3, meaning that there are three secondary turns for each primary turn.

EXAMPLE C.14-2

The transformer in Figure C.14-2 is part of a laboratory power supply and has 50 turns in the primary winding and 10 turns in the secondary winding. What is the secondary voltage?

FIGURE C.14-2

Solution   The secondary voltage is

$$V_{sec} = \left(\frac{N_{sec}}{N_{pri}}\right)V_{pri} = \left(\frac{10}{50}\right)120 \text{ V} = 0.2(120 \text{ V}) = 24 \text{ V}$$

EXAMPLE C.14-3

The transformers in Figures C.14-3(a) and C.14-3(b) have loaded secondaries. If the primary current is 100 mA in each case, what is the current through the load?

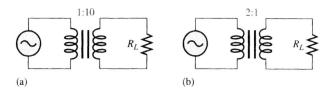

(a)   (b)

FIGURE C.14-3

Solution   In Figure C.14-3(a), the secondary load current is

$$I_L = I_{sec} = \left(\frac{N_{pri}}{N_{sec}}\right)I_{pri} = 0.1(100 \text{ mA}) = 10 \text{ mA}$$

In Figure C.14-3(b), the secondary load current is

$$I_L = I_{sec} = \left(\frac{N_{pri}}{N_{sec}}\right)I_{pri} = 2(100 \text{ mA}) = 200 \text{ mA}$$

EXAMPLE C.14-4

Figure C.14-4 shows a source that is transformer-coupled to a load resistor of 100 Ω. The transformer has a turns ratio of 4. What is the reflected resistance seen by the source?

FIGURE C.14-4

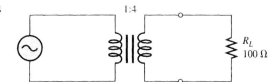

Solution   The reflected resistance is determined as follows:

$$R_{pri} = \left(\frac{N_{pri}}{N_{sec}}\right)^2 R_L = \left(\frac{1}{4}\right)^2 100 \text{ Ω} = \left(\frac{1}{16}\right)100 \text{ Ω} = 6.25 \text{ Ω}$$

The source sees a resistance of 6.25 Ω just as if it were connected directly, as shown in the equivalent circuit of Figure C.14-5.

FIGURE C.14-5

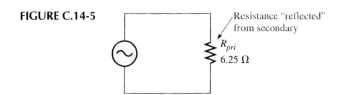

True/False

1. A transformer contains two primary windings.

2. There is no electrical connection between the windings of a transformer.

Multiple Choice

3. A transformer with a turns ratio of 1:10 is a
   a. Step-down transformer.
   b. Step-up transformer.
   c. Decade transformer.

4. A transformer with a turns ratio of 1:10 has a 4.7 K ohm load resistance. What is the reflected resistance?
   a. 47 ohms.
   b. 4.7 K ohms.
   c. 470 K ohms.

5. What is the output voltage of a transformer with a turns ratio of 4:1 if the input voltage is 120 V?
   a. 30 V.
   b. 120 V.
   c. 480 V.

C.15 Demonstrate an understanding of impedance matching theory

INTRODUCTION

One application of transformers is in the matching of a load resistance to a source resistance in order to achieve maximum transfer of power. This technique is called impedance matching. In audio systems, transformers are often used to get the maximum amount of power from the amplifier to the speaker by proper selection of the turns ratio.

The concept of power transfer is illustrated in the basic circuit of Figure C.15-1. Part (a) shows an AC voltage source with a series resistance representing its internal resistance. Some internal resistance is inherent in all sources due to their internal circuitry or physical makeup. When the source is connected directly to a load, as shown in part (b), the objective is generally to transfer as much of the power produced by the source to the load as possible. However, a certain amount of the power produced by the source is lost in its internal resistance, and the remaining power goes to the load.

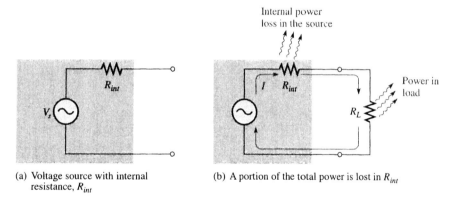

(a) Voltage source with internal resistance, R_{int}

(b) A portion of the total power is lost in R_{int}

FIGURE C.15-1
Power transfer from a nonideal voltage source to a load.

Maximum Power Transfer Theorem

The **maximum power transfer theorem** is important when you need to know the value of the load at which the most power is delivered from the source. The theorem can be stated as follows:

> **When a source is connected to a load, maximum power is delivered to the load when the load resistance is equal to the internal source resistance.**

We will demonstrate this theorem by finding the load power for various values of load resistance and a specific internal source resistance in Example C.15-1.

EXAMPLE C.15-1

The source in Figure C.15-2 has an internal source resistance of 75 Ω. Determine the load power for each of the following values of load resistance:
(a) 25 Ω (b) 50 Ω (c) 75 Ω (d) 100 Ω (e) 125 Ω
Draw a graph showing the load power versus the load resistance.

FIGURE C.15-2

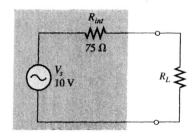

Solution   Use Ohm's law ($I = V/R$) and the power formula ($P = I^2R$) to find the load power for each value of load resistance.
(a) For $R_L = 25$ Ω,

$$I = \frac{V_s}{R_{int} + R_L} = \frac{10 \text{ V}}{100 \text{ }\Omega} = 100 \text{ mA}$$
$$P_L = I^2R_L = (100 \text{ mA})^2(25 \text{ }\Omega) = 250 \text{ mW}$$

(b) For $R_L = 50$ Ω,

$$I = \frac{V_s}{R_{int} + R_L} = \frac{10 \text{ V}}{125 \text{ }\Omega} = 80 \text{ mA}$$
$$P_L = I^2R_L = (80 \text{ mA})^2(50 \text{ }\Omega) = 320 \text{ mW}$$

(c) For $R_L = 75$ Ω,

$$I = \frac{V_s}{R_{int} + R_L} = \frac{10 \text{ V}}{150 \text{ }\Omega} = 66.7 \text{ mA}$$
$$P_L = I^2R_L = (66.7 \text{ mA})^2(75 \text{ }\Omega) = 334 \text{ mW}$$

(d) For $R_L = 100$ Ω,

$$I = \frac{V_s}{R_{int} + R_L} = \frac{10 \text{ V}}{175 \text{ }\Omega} = 57.1 \text{ mA}$$
$$P_L = I^2R_L = (57.1 \text{ mA})^2(100 \text{ }\Omega) = 326 \text{ mW}$$

(e) For $R_L = 125$ Ω,

$$I = \frac{V_s}{R_{int} + R_L} = \frac{10 \text{ V}}{200 \text{ }\Omega} = 50 \text{ mA}$$
$$P_L = I^2R_L = (50 \text{ mA})^2(125 \text{ }\Omega) = 313 \text{ mW}$$

Notice that the load power is greatest when $R_L = 75$ Ω, which is the same as the internal source resistance. When the load resistance is less than or greater than this value, the load power drops off, as the curve in Figure C.15-3 graphically illustrates.

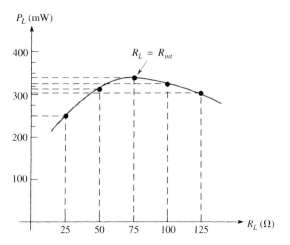

FIGURE C.15-3
Curve showing that the load power is maximum when $R_L = R_{int}$.

In most practical situations, the internal source resistance of various types of sources is fixed. Also, in many cases, the resistance of a device that acts as a load is fixed and cannot be altered. If you need to connect a given source to a given load, remember that only by chance will their resistances match. In this situation a transformer comes in handy. You can use the reflected-resistance characteristic of a transformer to make the load resistance appear to have the same value as the source resistance, thereby "fooling" the source into "thinking" that there is a match. This technique is called **impedance matching.**

Let's take a practical, everyday situation to illustrate. The typical resistance of the input to a TV receiver is 300 Ω. An antenna must be connected to this input by a lead-in cable in order to receive TV signals. In this situation, the antenna and the lead-in act as the source, and the input resistance of the TV receiver is the load, as illustrated in Figure C.15-4.

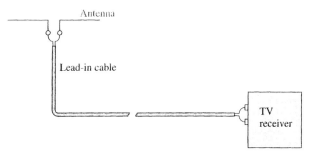

(a) The antenna/lead-in is the source; the TV input is the load.

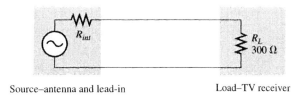

(b) Circuit equivalent of antenna and TV receiver system

FIGURE C.15-4
An antenna directly coupled to a TV receiver.

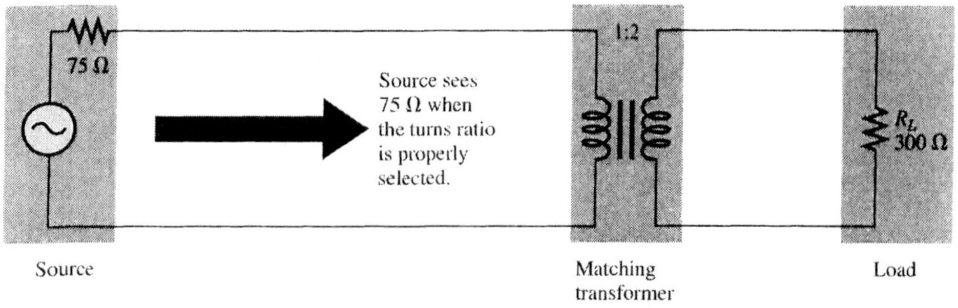

FIGURE C.15-5
Example of a load matched to a source by transformer coupling for maximum power transfer.

It is common for an antenna system to have a characteristic resistance of 75 Ω. Thus, if the 75 Ω source (antenna and lead-in) is connected directly to the 300 Ω TV input, maximum power will not be delivered to the input to the TV, and you will have poor signal reception. The solution is to use a matching transformer, connected as indicated in Figure C.15-5, in order to match the 300 Ω load resistance to the 75 Ω source resistance.

To match the resistances, that is, to reflect the load resistance (R_L) into the primary circuit so that it appears to have a value equal to the internal source resistance (R_{int}), you must select a proper value of turns ratio. You want the 300 Ω load resistance to look like 75 Ω to the source. First, obtain a formula to determine the turns ratio (N_{sec}/N_{pri}) when you know the values for R_L and R_{pri}.

$$R_{pri} = \left(\frac{N_{pri}}{N_{sec}}\right)^2 R_L$$

Transpose terms and divide both sides by R_L.

$$\left(\frac{N_{pri}}{N_{sec}}\right)^2 = \frac{R_{pri}}{R_L}$$

Then take the square root of both sides.

$$\frac{N_{pri}}{N_{sec}} = \sqrt{\frac{R_{pri}}{R_L}}$$

Invert both sides to get the following formula for the turns ratio.

$$n = \frac{N_{sec}}{N_{pri}} = \sqrt{\frac{R_L}{R_{pri}}}$$

Finally, solve for this particular turns ratio.

$$n = \frac{N_{sec}}{N_{pri}} = \sqrt{\frac{300\ \Omega}{75\ \Omega}} = \sqrt{4} = 2$$

Therefore, a matching transformer with a turns ratio of 2 must be used in this application.

EXAMPLE C.15-2

An amplifier has an 800 Ω internal resistance. In order to provide maximum power to an 8 Ω speaker, what turns ratio must be used in the coupling transformer?

Solution   The reflected resistance must equal 800 Ω. Thus, the turns ratio can be determined.

$$n = \frac{N_{sec}}{N_{pri}} = \sqrt{\frac{R_L}{R_{pri}}} = \sqrt{\frac{8\ \Omega}{800\ \Omega}} = \sqrt{0.01} = 0.1$$

There must be ten primary turns for each secondary turn because $n = 0.1$. The diagram and its equivalent reflected circuit are shown in Figure C.15-6.

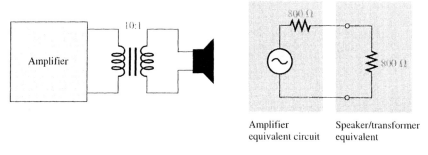

Amplifier   Speaker/transformer
equivalent circuit   equivalent

FIGURE C.15-6

REVIEW QUESTIONS

True/False

1. Maximum power transfer occurs when the load resistance equals the source resistance.
2. Transformers are used to match impedance.

Multiple Choice

3. What is the maximum load power available from a 10 V source with an internal resistance of 5 ohms?

   a. 1 W.
   b. 5 W.
   c. 10 W.

4. A step-down transformer provides

   a. A smaller reflected impedance than the load resistance.
   b. A larger reflected impedance than the load resistance.
   c. The same impedance as the load resistance.

5. What turns ratio is required to reflect 5 ohms into 500 ohms?

   a. 1:10.
   b. 10:1.
   c. 100:1.

C.16 Fabricate and demonstrate AC circuits using transformers

INTRODUCTION

Using actual components (or simulation software such as *Electronics Workbench*), set up and examine the circuit in Figure C.16-1.

The half-wave rectifier in Figure C.16-1 uses a 10:1 step-down transformer to rectify 12 Vrms into pulsating DC. View the output voltage using an oscilloscope. Measure the DC output voltage. How does it compare to the value predicted by this formula:

$$V_o = 0.318V_p$$

where V_p is the peak voltage at the cathode of the diode.

FIGURE C.16-1
Half-wave rectifier.

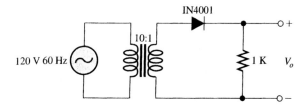

The rectifier shown in Figure C.16-2 is a full-wave center-tapped rectifier. Examine its output with an oscilloscope. How does it differ from the half-wave output? Measure the DC output voltage. How does it compare to the value predicted by

$$V_o = 0.636V_p$$

where V_p is once again the peak output voltage at the cathode (of either diode). What effect does the center tap introduce into the circuit?

FIGURE C.16-2
Full-wave center-tapped rectifier.

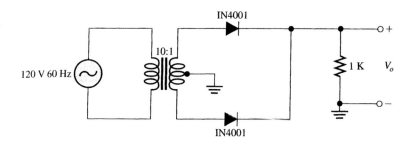

C.17 Troubleshoot and repair AC circuits using transformers

INTRODUCTION

The common failures in transformers are opens, shorts, or partial shorts in either the primary or the secondary windings. One cause of such failures is the operation of the device under conditions that exceed its ratings. A few transformer failures and the associated symptoms are covered in this section.

Open Primary Winding

When there is an open primary winding, there is no primary current and, therefore, no induced voltage or current in the secondary. This condition is illustrated in Figure C.17-1(a), and the method of checking with an ohmmeter is shown in part (b).

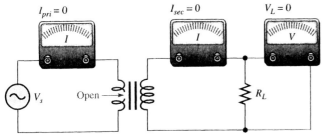

(a) Conditions when the primary winding is open

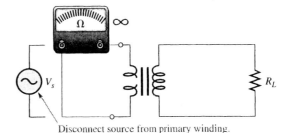

Disconnect source from primary winding.

(b) Checking the primary winding with an ohmmeter

FIGURE C.17-1
Open primary winding.

Open Secondary Winding

When there is an open secondary winding, there is no current in the secondary circuit and, as a result, no voltage across the load. Also, an open secondary winding causes the primary current to be very small (there is only a small magnetizing current). In fact, the primary current may be practically zero. This condition is illustrated in Figure C.17-2(a), and the ohmmeter check is shown in part (b).

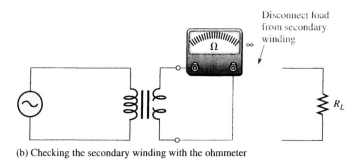

(a) Conditions when the secondary winding is open

(b) Checking the secondary winding with the ohmmeter

FIGURE C.17-2
Open secondary winding.

Shorted or Partially Shorted Primary Winding

A completely shorted primary winding will draw excessive current from the source and, unless there is a breaker or a fuse in the circuit, either the source or the transformer or both will burn out. A partial short in the primary winding can cause higher than normal or even excessive primary current.

Shorted or Partially Shorted Secondary Winding

In this case, there is an excessive primary current because of the low reflected resistance due to the short. Often, this excessive current will burn out the primary winding and result in an open. The short-circuit current in the secondary winding causes the load current to be zero (full short) or smaller than normal (partial short), as demonstrated in Figure C.17-3(a) and C.17-3(b). The ohmmeter check for this condition is shown in part (c).

Normally, when a transformer fails, it is very difficult to repair, and therefore the simplest procedure is to replace it.

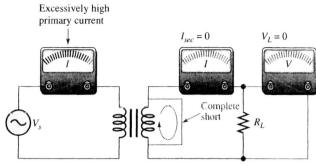

(a) Secondary winding completely shorted

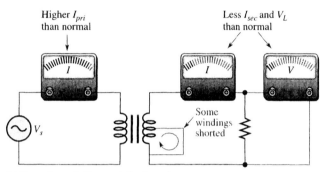

(b) Secondary winding partially shorted

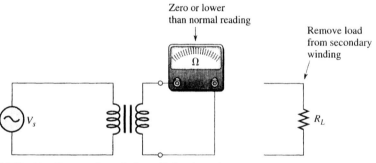

(c) Checking the secondary winding with the ohmmeter

FIGURE C.17-3
Shorted secondary winding.

C.18 Understand principles and operations of AC differentiator and integrator circuits (determine *RC* and *RL* time constants)

INTRODUCTION

In addition to the reactive properties of capacitors and inductors, we must also examine their *transient* response. What happens when a nonsinusoidal input is applied (such as a square wave) to an *RC* or *RL* circuit? In this section we will look at the transient response of two useful circuits: the integrator and the differentiator.

EXAMPLE C.18-1

A single 10 V pulse with a width of 100 μs is applied to the integrator in Figure C.18-1.
(a) To what voltage will the capacitor charge?
(b) How long will it take the capacitor to discharge?
(c) Sketch the output voltage.

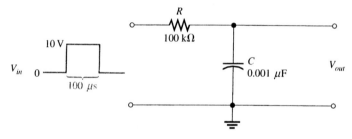

FIGURE C.18-1

Solution
(a) The circuit time constant is

$$\tau = RC = (100 \text{ k}\Omega)(0.001 \text{ } \mu\text{F}) = 100 \text{ } \mu\text{s}$$

Notice that the pulse width is exactly equal to one time constant. Thus, the capacitor will charge 63% of the full input amplitude in one time constant, so the output will reach a maximum voltage of 6.3 V.

(b) The capacitor discharges back through the source when the pulse ends. The total discharge time is

$$5\tau = 5(100 \text{ } \mu\text{s}) = 500 \text{ } \mu\text{s}$$

(c) The output charging and discharging curve is shown in Figure C.18-2.

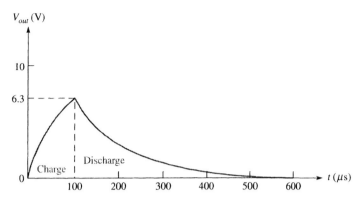

FIGURE C.18-2

Determine the output voltage waveform for the first two pulses applied to the integrator circuit in Figure C.18-3. Assume that the capacitor is initially uncharged.

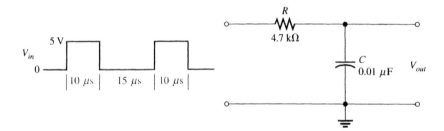

FIGURE C.18-3

Solution First calculate the circuit time constant.

$$\tau = RC = (4.7 \text{ k}\Omega)(0.01 \text{ }\mu\text{F}) = 47 \text{ }\mu\text{s}$$

Obviously, the time constant is much longer than the input pulse width or the interval between pulses (notice that the input is not a square wave). Thus, in this case, the exponential formulas must be applied, and the analysis is relatively difficult. Follow the solution carefully.

1. *Calculation for first pulse:* Note that V_F is 5 V, and t equals the pulse width of 10 μs. Therefore,

$$v_C = V_F(1 - e^{-t/RC}) = (5 \text{ V})(1 - e^{-10\mu s/47\mu s})$$
$$= (5 \text{ V})(1 - 0.808) = 960 \text{ mV}$$

This result is plotted in Figure C.18-4(a).

2. *Calculation for interval between first and second pulse:* Note that V_i is 960 mV because C begins to discharge from this value at the end of the first pulse. The discharge time is 15 μs. Therefore,

$$v_C = V_i e^{-t/RC} = (960 \text{ mV})^{-15\mu s/47\mu s}$$
$$= (960 \text{ mV})(0.727) = 698 \text{ mV}$$

This result is shown in Figure C.18-4(b).

3. *Calculation for second pulse:* At the beginning of the second pulse, the output voltage is 698 mV. During the second pulse, the capacitor will again charge. In this case it does not begin at zero volts. It already has 698 mV from the previous charge and discharge.

$$v = V_F + (V_i - V_F)e^{-t/\tau}$$

Using this equation, you can calculate the voltage across the capacitor at the end of the second pulse as follows:

$$
\begin{aligned}
v_C &= V_F + (V_i - V_F)e^{-t/RC} \\
&= 5\text{ V} + (698\text{ mV} - 5\text{ V})e^{-10\mu s/47\mu s} \\
&= 5\text{ V} + (-4.30\text{ V})(0.808) \\
&= 5\text{ V} - 3.47\text{ V} = 1.53\text{ V}
\end{aligned}
$$

This result is shown in Figure C.18-4(c).

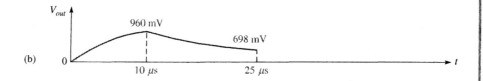

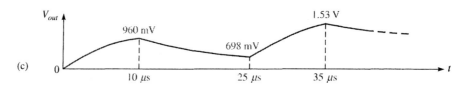

FIGURE C.18-4

Notice that the output waveform builds up on successive input pulses. After approximately 5τ, it will reach its steady state and will fluctuate between a constant maximum and a constant minimum, with an average equal to the average value of the input. You can demonstrate this pattern by carrying the analysis in this example further.

EXAMPLE C.18-3

Sketch the output voltage for the circuit in Figure C.18-5.

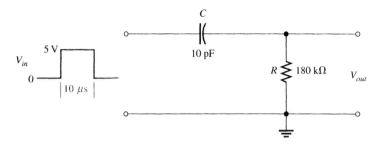

FIGURE C.18-5

Solution   First calculate the time constant.

$$\tau = RC = (180 \text{ k}\Omega)(10 \text{ pF}) = 1.8 \ \mu s$$

In this case, $t_W > 5\tau$, so the capacitor reaches full charge in 9 μs (before the end of the pulse).

On the rising edge, the resistor voltage jumps to $+5$ V and then decreases exponentially to zero before the end of the pulse. On the falling edge, the resistor voltage jumps to -5 V and then goes back to zero exponentially. The resistor voltage is, of course, the output, and its shape is shown in Figure C.18-6.

FIGURE C.18-6

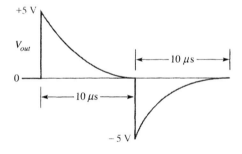

EXAMPLE C.18-4

Determine the maximum output voltage for the integrator in Figure C.18-7 when a single pulse is applied as shown.

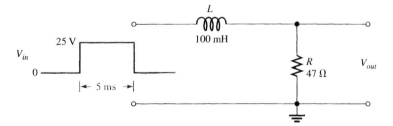

FIGURE C.18-7

Solution   Calculate the time constant.

$$\tau = \frac{L}{R} = \frac{100 \text{ mH}}{47 \ \Omega} = 2.13 \text{ ms}$$

Because the pulse width is 5 ms, the inductor charges for approximately 2.35τ; $(5 \text{ ms}/2.13 \text{ ms} = 2.35)$. Use the exponential formula to calculate the voltage.

$$v_{out} = V_F(1 - e^{-t/\tau}) = 25(1 - e^{-5\text{ms}/2.13\text{ms}})$$
$$= 25(1 - e^{-2.35}) = 25(1 - 0.095) = 25(0.905) = 22.6 \text{ V}$$

EXAMPLE C.18-5

Sketch the output voltage for the circuit in Figure C.18-8.

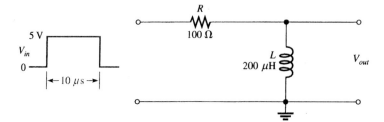

FIGURE C.18-8

Solution   First calculate the time constant.

$$\tau = \frac{L}{R} = \frac{200 \ \mu\text{H}}{100 \ \Omega} = 2 \ \mu\text{s}$$

In this case, $t_W = 5\tau$, so the output will decay to zero at the end of the pulse.

On the rising edge, the inductor voltage jumps to $+5$ V and then decays exponentially to zero. It reaches approximately zero at the instant of the falling edge. On the falling edge of the input, the inductor voltage jumps to -5 V and then goes back to zero. The output waveform is shown in Figure C.18-9.

FIGURE C.18-9

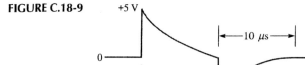

REVIEW QUESTIONS

True/False

1. The integrator takes its output from the capacitor.

2. The differentiator never reaches a steady state of operation.

Multiple Choice

3. The transient time is typically
   a. One time constant.
   b. Five time constants.
   c. Ten time constants.

4. An integrator uses a 4.7 K ohm resistor and a 10 μF capacitor. If the pulse width is 100 mS, the capacitor

   a. Does not fully charge.

   b. Fully charges.

   c. Stays at the same voltage.

5. A differentiator detects

   a. Rising edges.

   b. Falling edges.

   c. Both a and b.

C.19 Fabricate and demonstrate AC differentiator and integrator circuits

INTRODUCTION

Your supervisor has given you the assignment of breadboarding and testing a time-delay circuit that will provide five switch-selectable delay times. An *RC* integrator is selected for this application. The input is a 5 V pulse of long duration which begins the time delay. The exponential output voltage goes to a threshold trigger circuit that is used to turn the power on to a portion of a system after any one of the five selectable time delays.

A schematic of the selectable time-delay integrator is shown in Figure C.19-1. The *RC* integrator is driven by a positive pulse input, and the output is the exponential voltage across the selected capacitor. The output voltage triggers a threshold circuit at the 3.5 V level, which then turns on the power to a portion of a system. The basic concept of operation is shown in Figure C.19-2. In this application, the delay time of the integrator is spec-

FIGURE C.19-1
Integrator delay circuit.

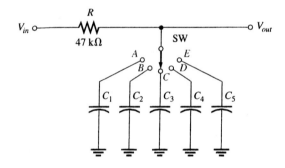

FIGURE C.19-2
Illustration of the time-delay operation.

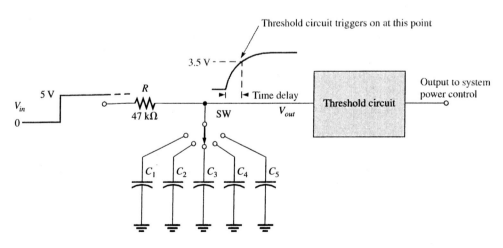

TABLE C.19-1

Switch Position	Delay Time
A	10 ms
B	25 ms
C	40 ms
D	65 ms
E	85 ms

ified as the time from the rising edge of the input pulse to the point where the output voltage reaches 3.5 V. The specified time delays are listed in Table C.19-1.

Step 1: Capacitor Values

Determine a value for each of the five capacitors in the time-delay circuit that will provide a delay time within 10% of the specified delay time. Select from the following list of standard values: 0.1 μF, 0.12 μF, 0.15 μF, 0.18 μF, 0.22 μF, 0.27 μF, 0.33 μF, 0.39 μF, 0.47 μF, 0.56 μF, 0.68 μF, 0.82 μF, 1.0 μF, 1.2 μF, 1.5 μF, 1.8 μF, 2.2 μF, 2.7 μF, 3.3 μF, 3.9 μF, 4.7 μF, 5.6 μF, 6.8 μF, and 8.2 μF.

Step 2: Circuit Connections

Refer to Figure C.19-3 on the next page. The components for the time-delay circuit are assembled, but not connected, on the breadboard. Using the circled numbers, develop a point-to-point wiring list to show how the circuit and measurement instruments should be connected.

Step 3: Test Procedure and Instrument Settings

Develop a procedure for fully testing the time-delay circuit. Specify the amplitude, frequency, and duty cycle settings for the function generator in order to test each delay time. Specify the oscilloscope sec/div settings for measuring each of the five specified delay times.

Step 4: Measurement

Explain how you will verify that each switch setting produces the proper output delay time.

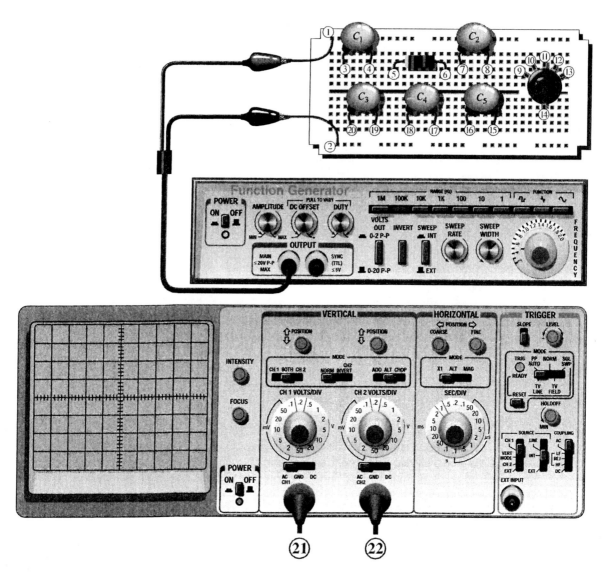

FIGURE C.19-3
Time-delay circuit breadboard and test instruments.

C.20 Troubleshoot and repair AC differentiator and integrator circuits

INTRODUCTION

Open Capacitor   If the capacitor in an integrator opens, the output has the same wave-shape as the input, as shown in Figure C.20-1(a). If the capacitor in a differentiator opens, the output is zero because it is held at ground through the resistor, as illustrated in part (b).

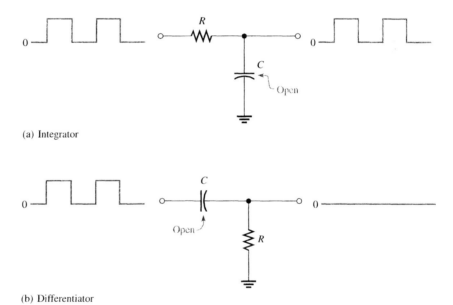

(a) Integrator

(b) Differentiator

FIGURE C.20-1
Examples of the effect of an open capacitor.

Leaky Capacitor   If the capacitor in an integrator becomes leaky, the time constant will be effectively reduced by the leakage resistance. Using Thevenin analysis, looking from C, the leakage resistance R_{leak} appears in parallel with R through the zero resistance of the input source. The waveshape of the output voltage (across C) is altered from normal by a shorter charging time. Also, the amplitude of the output is reduced because R and R_{leak} effectively act as a voltage divider. These effects are illustrated in Figure C.20-2(a).

   If the capacitor in a differentiator becomes leaky, the time constant is reduced, just as in the integrator (they are both simply series RC circuits). When the capacitor reaches full charge, the output voltage (across R) is set by the effective voltage-divider action of R and R_{leak}, as shown in Figure C.20-2(b).

183

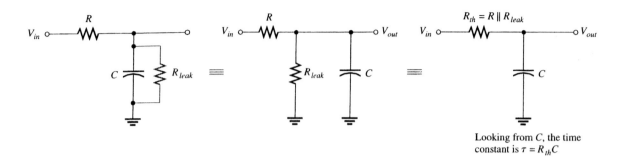

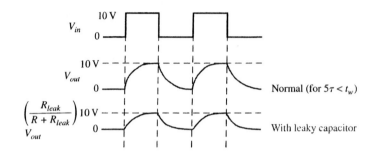

(a)

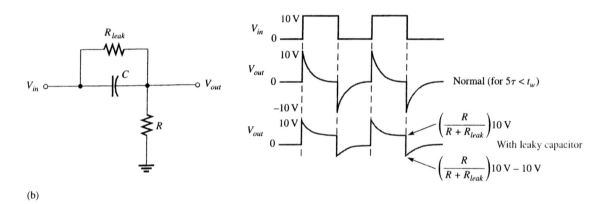

(b)

FIGURE C.20-2
Examples of the effect of a leaky capacitor.

Shorted Capacitor   If the capacitor in an integrator shorts, the output is at ground as shown in Figure C.20-3(a). If the capacitor in a differentiator shorts, the output is the same as the input, as shown in part (b).

Open Resistor   If the resistor in an integrator opens, the capacitor has no discharge path, and, ideally, it will hold its charge. In an actual situation, the charge will gradually leak off or the capacitor will discharge slowly through a measuring instrument connected to the output. This is illustrated in Figure C.20-4(a).
     If the resistor in a differentiator opens, the output looks like the input except for the DC level because the capacitor now must charge and discharge through the extremely high resistance of the oscilloscope, as shown in Figure C.20-4(b).

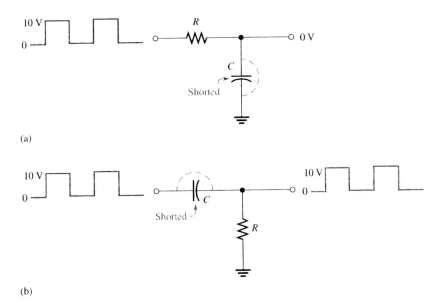

FIGURE C.20-3
Examples of the effect of a shorted capacitor.

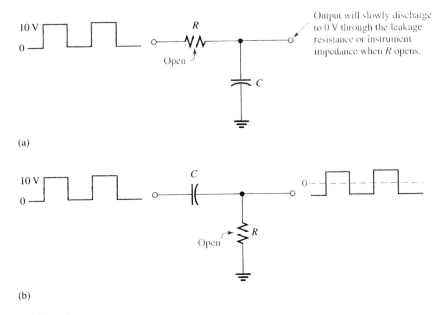

FIGURE C.20-4
Examples of the effect of an open resistor.

C.21 Understand principles and operations of AC series and parallel resonant circuits

INTRODUCTION

Every *LC* combination (series or parallel) has an associated *resonant* frequency, a frequency that causes the inductive and capacitive reactances to cancel each other. In this section we examine the operation of series and parallel resonant circuits.

EXAMPLE C.21-1

For the series *RLC* circuit in Figure C.21-1, determine X_C and Z at resonance.

FIGURE C.21-1

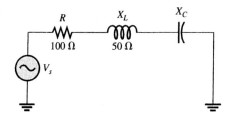

Solution X_L equals X_C at resonance, therefore,

$$X_L = X_C = 50 \ \Omega$$

Since the reactances cancel each other, the impedance equals the resistance.

$$Z_r = R = 100 \ \Omega$$

EXAMPLE C.21-2

Find the series resonant frequency for the circuit in Figure C.21-2.

FIGURE C.21-2

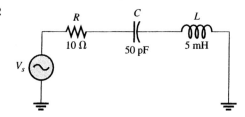

Solution   The resonant frequency is

$$f_r = \frac{1}{2\pi\sqrt{LC}} = \frac{1}{2\pi\sqrt{(5 \text{ mH})(50 \text{ pF})}} = 318 \text{ kHz}$$

EXAMPLE C.21-3

Find I, V_R, V_L, and V_C at resonance in Figure C.21-3.

FIGURE C.21-3

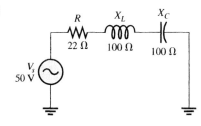

Solution   At resonance, I is maximum and equal to V_s/R.

$$I = \frac{V_s}{R} = \frac{50 \text{ V}}{22 \text{ } \Omega} = 2.27 \text{ A}$$

Applying Ohm's law, obtain the following voltages:

$$V_R = IR = (2.27 \text{ A})(22 \text{ } \Omega) = 50 \text{ V}$$
$$V_L = IX_L = (2.27 \text{ A})(100 \text{ } \Omega) = 227 \text{ V}$$
$$V_C = IX_C = (2.27 \text{ A})(100 \text{ } \Omega) = 227 \text{ V}$$

Notice that all of the source voltage is dropped across the resistor. Also, of course, V_L and V_C are equal in magnitude but opposite in phase. This causes the voltages to cancel, making the total reactive voltage zero.

EXAMPLE C.21-4

For the circuit in Figure C.21-4, determine the impedance at the following frequencies:
(a) f_r **(b)** 1 kHz below f_r **(c)** 1 kHz above f_r

FIGURE C.21-4

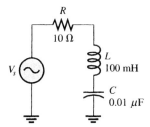

Solution
(a) At f_r, the impedance is equal to R.

$$Z = R = 10 \text{ } \Omega$$

To determine the impedance above and below f_r, first calculate the resonant frequency as follows:

$$f_r = \frac{1}{2\pi\sqrt{LC}} = \frac{1}{2\pi\sqrt{(100 \text{ mH})(0.01 \text{ } \mu\text{F})}} = 5.03 \text{ kHz}$$

(b) At 1 kHz below f_r, the frequency and reactances are as follows:

$$f = f_r - 1 \text{ kHz} = 5.03 \text{ kHz} - 1 \text{ kHz} = 4.03 \text{ kHz}$$

$$X_C = \frac{1}{2\pi fC} = \frac{1}{2\pi(4.03 \text{ kHz})(0.01 \ \mu\text{F})} = 3.95 \text{ k}\Omega$$

$$X_L = 2\pi fL = 2\pi(4.03 \text{ kHz})(100 \text{ mH}) = 2.53 \text{ k}\Omega$$

Therefore, the impedance at $f_r - 1$ kHz is

$$Z = \sqrt{R^2 + (X_L - X_C)^2} = \sqrt{(10 \ \Omega)^2 + (2.53 \text{ k}\Omega - 3.95 \text{ k}\Omega)^2} = 1.42 \text{ k}\Omega$$

(c) At 1 kHz above f_r,

$$f = 5.03 \text{ kHz} + 1 \text{ kHz} = 6.03 \text{ kHz}$$

$$X_C = \frac{1}{2\pi(6.03 \text{ kHz})(0.01 \ \mu\text{F})} = 2.64 \text{ k}\Omega$$

$$X_L = 2\pi(6.03 \text{ kHz})(100 \text{ mH}) = 3.79 \text{ k}\Omega$$

Therefore, the impedance at $f_r + 1$ kHz is

$$Z = \sqrt{(10 \ \Omega)^2 + (3.79 \text{ k}\Omega - 2.64 \text{ k}\Omega)^2} = 1.15 \text{ k}\Omega$$

In part (b), Z is capacitive, and in part (c) Z is inductive.

EXAMPLE C.21-5

Find the resonant frequency and the branch currents in the ideal (winding resistance neglected) parallel LC circuit of Figure C.21-5.

FIGURE C.21-5

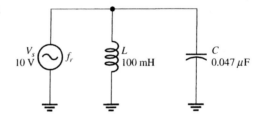

Solution   The resonant frequency is

$$f_r = \frac{1}{2\pi\sqrt{LC}} = \frac{1}{2\pi\sqrt{(100 \text{ mH})(0.047 \ \mu\text{F})}} = 2.32 \text{ kHz}$$

Determine the branch currents as follows:

$$X_L = 2\pi f_r L = 2\pi(2.32 \text{ kHz})(100 \text{ mH}) = 1.46 \text{ k}\Omega$$

$$X_C = X_L = 1.46 \text{ k}\Omega$$

$$I_L = \frac{V_s}{X_L} = \frac{10 \text{ V}}{1.46 \text{ k}\Omega} = 6.85 \text{ mA}$$

$$I_C = I_L = 6.85 \text{ mA}$$

The total current into the parallel LC circuit is

$$I_{tot} = |I_C - I_L| = 0 \text{ A}$$

EXAMPLE C.21-6

Determine the impedance of the circuit in Figure C.21-6 at the resonant frequency ($f_r \cong 17{,}794$ Hz).

FIGURE C.21-6

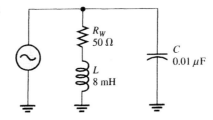

Solution First calculate the quality factor Q as follows:

$$X_L = 2\pi f_r L = 2\pi (17{,}794 \text{ Hz})(8 \text{ mH}) = 894 \ \Omega$$

$$Q = \frac{X_L}{R_W} = \frac{894 \ \Omega}{50 \ \Omega} = 17.9$$

The impedance is

$$Z_r = R_W(Q^2 + 1) = 50 \ \Omega(17.9^2 + 1) = 16.1 \text{ k}\Omega$$

EXAMPLE C.21-7

Find the frequency, impedance, and total current at resonance for the circuit in Figure C.21-7.

FIGURE C.21-7

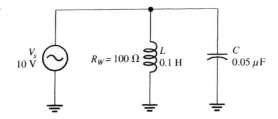

Solution The precise resonant frequency is

$$f_r = \frac{\sqrt{1 - (R_W^2 C/L)}}{2\pi\sqrt{LC}} = \frac{\sqrt{1 - [(100 \ \Omega)^2(0.05 \ \mu F)/0.1 \text{ H}]}}{2\pi\sqrt{(0.05 \ \mu F)(0.1 \text{ H})}} = 2.25 \text{ kHz}$$

Calculate impedance as follows:

$$X_L = 2\pi f_r L = 2\pi(2.25 \text{ kHz})(0.1 \text{ H}) = 1.4 \text{ k}\Omega$$

$$Q = \frac{X_L}{R_W} = \frac{1.4 \text{ k}\Omega}{100 \ \Omega} = 14$$

$$Z_r = R_W(Q^2 + 1) = 100 \ \Omega(14^2 + 1) = 19.7 \text{ k}\Omega$$

The total current is

$$I_{tot} = \frac{V_s}{Z_r} = \frac{10 \text{ V}}{19.7 \text{ k}\Omega} = 508 \ \mu A$$

True/False

1. A series resonant circuit has minimum impedance at resonance.
2. A parallel resonant conducts a minimum current at resonance.

Multiple Choice

3. A series resonant circuit uses a 35 mH inductor and a 0.1 μF capacitor. What is its resonant frequency?

   a. 2690 Hz.

   b. 16.9 KHz.

   c. 45.4 MHz.

4. Increasing the resistance in a series resonant circuit

   a. Lowers the resonant frequency.

   b. Raises the resonant frequency.

   c. Has no effect on the resonant frequency.

5. When a series resonant circuit is operated below resonance, it is

   a. Inductive.

   b. Capacitive.

   c. Purely resistive.

C.22 Fabricate and demonstrate AC series and parallel resonant circuits

INTRODUCTION

Using actual components (or simulation software such as *Electronics Workbench*), set up and examine the circuit in Figure C.22-1.

Vary the input frequency until you locate the resonant frequency. Explain how you found the resonant frequency. Measure all component voltages and the circuit current. Lower the frequency until 0.707 of the original current is found. Record this frequency as F_1. Raise the frequency (above F_r) until the current again drops to 0.707 of its original value. This is F_2. Subtract F_2 and F_1 to find the bandwidth. Divide F_r by the bandwidth to find the Q. Change the 100 ohm resistor to 470 ohms and repeat all measurements.

The parallel resonant circuit shown in Figure C.22-2 should have a similar resonant frequency.

Repeat the various measurements performed on the series resonant circuit. Does the parallel resistance have any effect on F_r?

FIGURE C.22-1
Series resonant circuit.

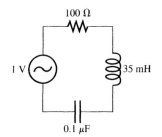

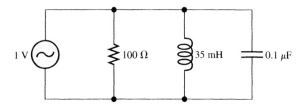

FIGURE C.22-2
Parallel resonant circuit.

C.23 Troubleshoot and repair AC series and parallel resonant circuits

INTRODUCTION

What could cause a resonant circuit to have a different resonant frequency than expected? Or a different bandwidth? Or a different Q? Here are some of the things we can check:

☐ Are all component values correct? Check this via observation and direct measurement.

☐ If the inductor has a tunable coil, has it been adjusted properly?

☐ Has the internal resistance of the inductor coil been taken into account during design?

☐ Has the internal resistance of the voltage source been taken into account? Many function generators look like a 50 ohm impedance internally. Is this enough additional resistance to significantly affect the overall resistance of the resonant circuit (thus affecting Q and bandwidth)?

C.24 Understand principles and operations of AC *RC, RL,* and *RLC* circuits

INTRODUCTION

In this section we examine the voltage, current, phase, and impedance properties of AC *RC, RL,* and *RLC* series and parallel circuits.

EXAMPLE C.24-1

If the current in Figure C.24-1 is 0.2 mA, determine the source voltage and the phase angle.

FIGURE C.24-1

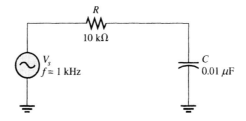

Solution The capacitive reactance is

$$X_C = \frac{1}{2\pi f C} = \frac{1}{2\pi(1000 \text{ Hz})(0.01 \ \mu\text{F})} = 15.9 \text{ k}\Omega$$

The impedance is

$$Z = \sqrt{R^2 + X_C^2} = \sqrt{(10 \text{ k}\Omega)^2 + (15.9 \text{ k}\Omega)^2} = 18.8 \text{ k}\Omega$$

Applying Ohm's law yields

$$V_s = IZ = (0.2 \text{ mA})(18.8 \text{ k}\Omega) = 3.76 \text{ V}$$

The phase angle is

$$\theta = \tan^{-1}\left(\frac{X_C}{R}\right) = \tan^{-1}\left(\frac{15.9 \text{ k}\Omega}{10 \text{ k}\Omega}\right) = 57.8°$$

The source voltage has a magnitude of 3.76 V and lags the current by 57.8°.

EXAMPLE C.24-2

Determine the current in the circuit of Figure C.24-2.

FIGURE C.24-2

Solution   The capacitive reactance is

$$X_C = \frac{1}{2\pi f C} = \frac{1}{2\pi(1.5 \text{ kHz})(0.02 \ \mu\text{F})} = 5.3 \text{ k}\Omega$$

The impedance is

$$Z = \sqrt{R^2 + X_C^2} = \sqrt{(2.2 \text{ k}\Omega)^2 + (5.3 \text{ k}\Omega)^2} = 5.74 \text{ k}\Omega$$

Applying Ohm's law yields

$$I = \frac{V}{Z} = \frac{10 \text{ V}}{5.74 \text{ k}\Omega} = 1.74 \text{ mA}$$

EXAMPLE C.24-3

Determine the source voltage and the phase angle in Figure C.24-3.

FIGURE C.24-3

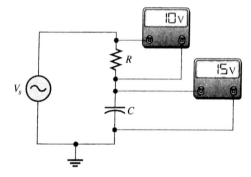

Solution   Since V_R and V_C are 90° out of phase, they cannot be added directly. The source voltage is the phasor sum of V_R and V_C.

$$V_s = \sqrt{V_R^2 + V_C^2} = \sqrt{(10 \text{ V})^2 + (15 \text{ V})^2} = 18 \text{ V}$$

The phase angle between the current and the source voltage is

$$\theta = \tan^{-1}\left(\frac{V_C}{V_R}\right) = \tan^{-1}\left(\frac{15 \text{ V}}{10 \text{ V}}\right) = 56.3°$$

EXAMPLE C.24-4

For the series *RC* circuit in Figure C.24-4, determine the impedance and phase angle for each of the following values of frequency:
(a) 10 kHz   **(b)** 20 kHz   **(c)** 30 kHz

FIGURE C.24-4

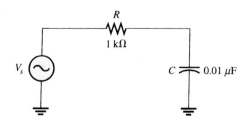

Solution

(a) For $f = 10$ kHz, the impedance is calculated as follows:

$$X_C = \frac{1}{2\pi f C} = \frac{1}{2\pi(10 \text{ kHz})(0.01 \ \mu\text{F})} = 1.59 \text{ k}\Omega$$

$$Z = \sqrt{R^2 + X_C^2} = \sqrt{(1 \text{ k}\Omega)^2 + (1.59 \text{ k}\Omega)^2} = 1.88 \text{ k}\Omega$$

The phase angle is

$$\theta = \tan^{-1}\left(\frac{X_C}{R}\right) = \tan^{-1}\left(\frac{1.59 \text{ k}\Omega}{1 \text{ k}\Omega}\right) = 57.8°$$

(b) For $f = 20$ kHz,

$$X_C = \frac{1}{2\pi(20 \text{ kHz})(0.01 \ \mu\text{F})} = 796 \ \Omega$$

$$Z = \sqrt{(1 \text{ k}\Omega)^2 + (796 \ \Omega)^2} = 1.28 \text{ k}\Omega$$

$$\theta = \tan^{-1}\left(\frac{796 \ \Omega}{1 \text{ k}\Omega}\right) = 38.5°$$

(c) For $f = 30$ kHz,

$$X_C = \frac{1}{2\pi(30 \text{ kHz})(0.01 \ \mu\text{F})} = 531 \ \Omega$$

$$Z = \sqrt{(1 \text{ k}\Omega)^2 + (531 \ \Omega)^2} = 1.13 \text{ k}\Omega$$

$$\theta = \tan^{-1}\left(\frac{531 \ \Omega}{1 \text{ k}\Omega}\right) = 28.0°$$

Notice that as the frequency increases, X_C, Z, and θ decrease.

For convenience in the analysis of parallel circuits, the Ohm's law formulas using impedance can be rewritten for admittance using the relation $Y = 1/Z$.

$$V = \frac{I}{Y}$$

$$I = VY$$

$$Y = \frac{I}{V}$$

EXAMPLE C.24-5

Determine the total current and the phase angle in Figure C.24-5.

FIGURE C.24-5

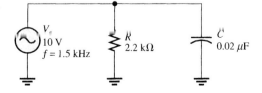

Solution   First determine the total admittance. The capacitive reactance is

$$X_C = \frac{1}{2\pi f C} = \frac{1}{2\pi(1.5 \text{ kHz})(0.02 \ \mu F)} = 5.31 \text{ k}\Omega$$

The capacitive susceptance is

$$B_C = \frac{1}{X_C} = \frac{1}{5.31 \text{ k}\Omega} = 188 \ \mu S$$

The conductance is

$$G = \frac{1}{R} = \frac{1}{2.2 \text{ k}\Omega} = 455 \ \mu S$$

Therefore, the total admittance is

$$Y_{tot} = \sqrt{G^2 + B_C^2} = \sqrt{(455 \ \mu S)^2 + (188 \ \mu S)^2} = 492 \ \mu S$$

Use Ohm's law to calculate the total current.

$$I_{tot} = V Y_{tot} = (10 \text{ V})(492 \ \mu S) = 4.92 \text{ mA}$$

The phase angle is

$$\theta = \tan^{-1}\left(\frac{R}{X_C}\right) = \tan^{-1}\left(\frac{2.2 \text{ k}\Omega}{5.31 \text{ k}\Omega}\right) = 22.5°$$

The total current is 4.92 mA, and it leads the applied voltage by 22.5°.

EXAMPLE C.24-6

Determine the value of each current in Figure C.24-6, and describe the phase relationship of each with the applied voltage.

FIGURE C.24-6

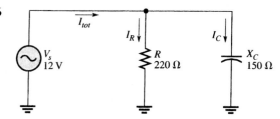

Solution   The resistor current, the capacitor current, and the total current are expressed as follows:

$$I_R = \frac{V_s}{R} = \frac{12 \text{ V}}{220 \ \Omega} = 54.5 \text{ mA}$$

$$I_C = \frac{V_s}{X_C} = \frac{12 \text{ V}}{150 \ \Omega} = 80 \text{ mA}$$

$$I_{tot} = \sqrt{I_R^2 + I_C^2} = \sqrt{(54.5 \text{ mA})^2 + (80 \text{ mA})^2} = 96.8 \text{ mA}$$

The phase angle is

$$\theta = \tan^{-1}\left(\frac{I_C}{I_R}\right) = \tan^{-1}\left(\frac{80 \text{ mA}}{54.5 \text{ mA}}\right) = 55.7°$$

I_R is in phase with the voltage, I_C leads the voltage by 90°, and I_{tot} leads the voltage by 55.7°.

EXAMPLE C.24-7

The current in Figure C.24-7 is 200 μA. Determine the source voltage.

FIGURE C.24-7

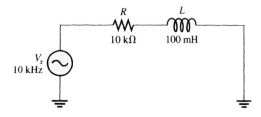

Solution   The inductive reactance is

$$X_L = 2\pi f L = 2\pi(10 \text{ kHz})(100 \text{ mH}) = 6.28 \text{ k}\Omega$$

The impedance is

$$Z = \sqrt{R^2 + X_L^2} = \sqrt{(10 \text{ k}\Omega)^2 + (6.28 \text{ k}\Omega)^2} = 11.8 \text{ k}\Omega$$

Applying Ohm's law yields

$$V_s = IZ = (200 \text{ }\mu\text{A})(11.8 \text{ k}\Omega) = 2.36 \text{ V}$$

EXAMPLE C.24-8

Determine the source voltage and the phase angle in Figure C.24-8.

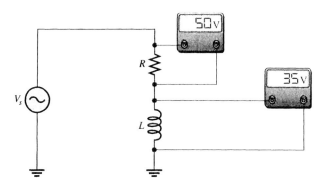

FIGURE C.24-8

Solution   Since V_R and V_L are 90° out of phase, they cannot be added directly and must be added as phasor quantities. The source voltage is

$$V_s = \sqrt{V_R^2 + V_L^2} = \sqrt{(50 \text{ V})^2 + (35 \text{ V})^2} = 61 \text{ V}$$

The phase angle between the current and the source voltage is

$$\theta = \tan^{-1}\left(\frac{V_L}{V_R}\right) = \tan^{-1}\left(\frac{35 \text{ V}}{50 \text{ V}}\right) = 35°$$

EXAMPLE C.24-9

For the series *RL* circuit in Figure C.24-9, determine the impedance and the phase angle for each of the following frequencies:
(a) 10 kHz   **(b)** 20 kHz   **(c)** 30 kHz

FIGURE C.24-9

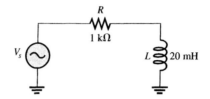

Solution
(a) For $f = 10$ kHz, the impedance is calculated as follows:

$$X_L = 2\pi fL = 2\pi(10 \text{ kHz})(20 \text{ mH}) = 1.26 \text{ k}\Omega$$
$$Z = \sqrt{R^2 + X_L^2} = \sqrt{(1 \text{ k}\Omega)^2 + (1.26 \text{ k}\Omega)^2} = 1.61 \text{ k}\Omega$$

The phase angle is

$$\theta = \tan^{-1}\left(\frac{X_L}{R}\right) = \tan^{-1}\left(\frac{1.26 \text{ k}\Omega}{1 \text{ k}\Omega}\right) = 51.6°$$

(b) For $f = 20$ kHz,

$$X_L = 2\pi(20 \text{ kHz})(20 \text{ mH}) = 2.51 \text{ k}\Omega$$
$$Z = \sqrt{(1 \text{ k}\Omega)^2 + (2.51 \text{ k}\Omega)^2} = 2.70 \text{ k}\Omega$$
$$\theta = \tan^{-1}\left(\frac{2.51 \text{ k}\Omega}{1 \text{ k}\Omega}\right) = 68.3°$$

(c) For $f = 30$ kHz,

$$X_L = 2\pi(30 \text{ kHz})(20 \text{ mH}) = 3.77 \text{ k}\Omega$$
$$Z = \sqrt{(1 \text{ k}\Omega)^2 + (3.77 \text{ k}\Omega)^2} = 3.90 \text{ k}\Omega$$
$$\theta = \tan^{-1}\left(\frac{3.77 \text{ k}\Omega}{1 \text{ k}\Omega}\right) = 75.1°$$

Notice that as the frequency increases, X_L, Z, and θ also increase.

EXAMPLE C.24-10

Determine the total current and the phase angle in the circuit of Figure C.24-10.

FIGURE C.24-10

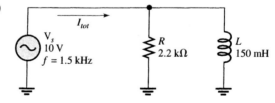

Solution First determine the total admittance. The inductive reactance is

$$X_L = 2\pi fL = 2\pi(1.5 \text{ kHz})(150 \text{ mH}) = 1.41 \text{ k}\Omega$$

The inductive susceptance is

$$B_L = \frac{1}{X_L} = \frac{1}{1.41 \text{ k}\Omega} = 709 \text{ }\mu\text{S}$$

The conductance is

$$G = \frac{1}{R} = \frac{1}{2.2 \text{ k}\Omega} = 455 \text{ }\mu\text{S}$$

Therefore, the total admittance is

$$Y_{tot} = \sqrt{G^2 + B_L^2} = \sqrt{(455 \ \mu S)^2 + (709 \ \mu S)^2} = 842 \ \mu S$$

Use Ohm's law to get the total current.

$$I_{tot} = VY_{tot} = (10 \ V)(842 \ \mu S) = 8.42 \ mA$$

The phase angle is

$$\theta = \tan^{-1}\left(\frac{R}{X_L}\right) = \tan^{-1}\left(\frac{2.2 \ k\Omega}{1.41 \ k\Omega}\right) = 57.3°$$

The total current lags the source voltage by 57.3°.

EXAMPLE C.24-11

Determine the value of each current in Figure C.24-11, and describe the phase relationship of each with the applied voltage.

FIGURE C.24-11

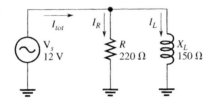

Solution The resistor current, the inductor current, and the total current are calculated as follows:

$$I_R = \frac{V_s}{R} = \frac{12 \ V}{220 \ \Omega} = 54.5 \ mA$$

$$I_L = \frac{V_s}{X_L} = \frac{12 \ V}{150 \ \Omega} = 80 \ mA$$

$$I_{tot} = \sqrt{I_R^2 + I_L^2} = \sqrt{(54.5 \ mA)^2 + (80 \ mA)^2} = 96.8 \ mA$$

The phase angle is

$$\theta = \tan^{-1}\left(\frac{R}{X_L}\right) = \tan^{-1}\left(\frac{220 \ \Omega}{150 \ \Omega}\right) = 55.7°$$

The resistor current is 54.5 mA and is in phase with the applied voltage. The inductor current is 80 mA and lags the applied voltage by 90°. The total current is 96.8 mA and lags the voltage by 55. 7°.

EXAMPLE C.24-12

For each of the following frequencies of the source voltage, find the impedance and the phase angle for the circuit in Figure C.24-12. Note how the impedance and the phase angle change with frequency.
(a) $f = 1$ kHz   **(b)** $f = 3.5$ kHz   **(c)** $f = 5$ kHz

FIGURE C.24-12

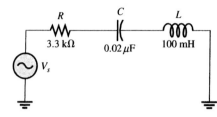

Solution

(a) At $f = 1$ kHz,

$$X_C = \frac{1}{2\pi fC} = \frac{1}{2\pi(1\ \text{kHz})(0.02\ \mu\text{F})} = 7.96\ \text{k}\Omega$$

$$X_L = 2\pi fL = 2\pi(1\ \text{kHz})(100\ \text{mH}) = 628\ \Omega$$

The circuit is highly capacitive because X_C is much larger than X_L. The impedance is

$$Z = \sqrt{R^2 + (X_L - X_C)^2} = \sqrt{(3.3\ \text{k}\Omega)^2 + (628\ \Omega - 7.96\ \text{k}\Omega)^2} = 8.04\ \text{k}\Omega$$

The phase angle is

$$\theta = \tan^{-1}\left(\frac{X_{tot}}{R}\right) = \tan^{-1}\left(\frac{|X_L - X_C|}{R}\right) = \tan^{-1}\left(\frac{7.33\ \text{k}\Omega}{3.3\ \text{k}\Omega}\right) = 65.8°$$

I leads V_s by 65.8°.

(b) At $f = 3.5$ kHz,

$$X_C = \frac{1}{2\pi(3.5\ \text{kHz})(0.02\ \mu\text{F})} = 2.27\ \text{k}\Omega$$

$$X_L = 2\pi(3.5\ \text{kHz})(100\ \text{mH}) = 2.20\ \text{k}\Omega$$

The circuit is very close to being purely resistive but is still slightly capacitive because X_C is slightly larger than X_L. The impedance and phase angle are

$$Z = \sqrt{(3.3\ \text{k}\Omega)^2 + (2.20\ \text{k}\Omega - 2.27\ \text{k}\Omega)^2} = 3.30\ \text{k}\Omega$$

$$\theta = \tan^{-1}\left(\frac{X_{tot}}{R}\right) = \tan^{-1}\left(\frac{70\ \Omega}{3.3\ \text{k}\Omega}\right) = 1.22°$$

I leads V_s by 1.22°.

(c) At $f = 5$ kHz,

$$X_C = \frac{1}{2\pi(5\ \text{kHz})(0.02\ \mu\text{F})} = 1.59\ \text{k}\Omega$$

$$X_L = 2\pi(5\ \text{kHz})(100\ \text{mH}) = 3.14\ \text{k}\Omega$$

The circuit is now predominantly inductive because $X_L > X_C$. The impedance and phase angle are

$$Z = \sqrt{(3.3\ \text{k}\Omega)^2 + (3.14\ \text{k}\Omega - 1.59\ \text{k}\Omega)^2} = 3.65\ \text{k}\Omega$$

$$\theta = \tan^{-1}\left(\frac{X_{tot}}{R}\right) = \tan^{-1}\left(\frac{1.55\ \text{k}\Omega}{3.3\ \text{k}\Omega}\right) = 25.2°$$

I lags V_s by 25.2°.

Notice how the circuit changed from capacitive to inductive as the frequency increased. The phase condition changed from the current leading to the current lagging. It is interesting to note that both the impedance and the phase angle decreased to a minimum and then began increasing again as the frequency went up.

EXAMPLE C.24-13

Find the voltages across each element in Figure C.24-13, and draw a complete voltage phasor diagram. Also find the voltage across L and C combined.

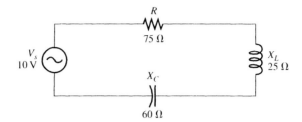

FIGURE C.24-13

Solution   First find the total impedance.

$$Z = \sqrt{R^2 + (X_L - X_C)^2} = \sqrt{(75\ \Omega)^2 + (25\ \Omega - 60\ \Omega)^2}$$
$$= \sqrt{(75\ \Omega)^2 + (35\ \Omega)^2} = 82.8\ \Omega$$

Apply Ohm's law to find the current.

$$I = \frac{V_s}{Z} = \frac{10\ \text{V}}{82.8\ \Omega} = 121\ \text{mA}$$

Now apply Ohm's law to find the voltages across R, L, and C.

$$V_R = IR = (121\ \text{mA})(75\ \Omega) = 9.08\ \text{V}$$
$$V_L = IX_L = (121\ \text{mA})(25\ \Omega) = 3.03\ \text{V}$$
$$V_C = IX_C = (121\ \text{mA})(60\ \Omega) = 7.26\ \text{V}$$

The voltage across L and C combined is

$$V_{LC} = V_C - V_L = 7.26\ \text{V} - 3.03\ \text{V} = 4.23\ \text{V}$$

The circuit phase angle is

$$\theta = \tan^{-1}\left(\frac{X_{tot}}{R}\right) = \tan^{-1}\left(\frac{|X_L - X_C|}{R}\right) = \tan^{-1}\left(\frac{35\ \Omega}{75\ \Omega}\right) = 25°$$

Since the circuit is capacitive ($X_C > X_L$), the current leads the source voltage by 25°.

The phasor diagram is shown in Figure C.24-14. Notice that V_L is leading V_R by 90°, and V_C is lagging V_R by 90°. Also, there is a 180° phase difference between V_L and V_C. If the current phasor were shown, it would be at the same angle as V_R.

FIGURE C.24-14

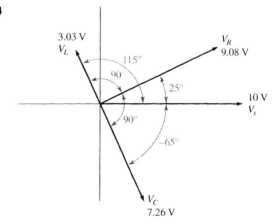

True/False

1. In a series *RC* circuit, the phase angle on the impedance is always 90 degrees.
2. In a parallel *RC* circuit, the equivalent impedance goes down as frequency increases.

Multiple Choice

3. What is the impedance of a series circuit with a 10 ohm resistor and 5 ohm capacitor?
   a. 10 ohms.
   b. 11.18 ohms.
   c. 15 ohms.
4. What is the phase angle of the impedance in Question 3?
   a. −26 degrees.
   b. −45 degrees.
   c. −90 degrees.
5. What is the impedance of a parallel circuit containing a 10 ohm resistor and 40 ohm inductor?
   a. 8 ohms.
   b. 9.7 ohms.
   c. 10 ohms.

C.25 Fabricate and demonstrate AC *RC*, *RL*, and *RLC* circuits

This material is covered in Sections C.9, C.12, C.19, and C.22.

C.26 Troubleshoot and repair AC *RC*, *RL*, and *RLC* circuits

INTRODUCTION

In this section, the effects that typical component failures or degradation have on the response of basic *RC* circuits are considered.

Effect of an Open Resistor   It is easy to see how an open resistor affects the operation of a basic series *RC* circuit, as shown in Figure C.26-1. Obviously, there is no path for current, so the capacitor voltage remains at zero; thus, the total voltage, V_s, appears across the open resistor.

FIGURE C.26-1
Effect of an open resistor.

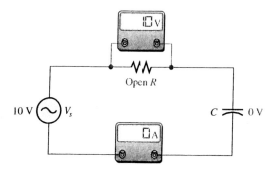

Effect of an Open Capacitor   When the capacitor is open, there is no current; thus, the resistor voltage remains at zero. The total source voltage is across the open capacitor, as shown in Figure C.26-2.

Effect of a Shorted Capacitor   When a capacitor shorts out, the voltage across it is zero, the current equals V_s/R, and the total voltage appears across the resistor, as shown in Figure C.26-3.

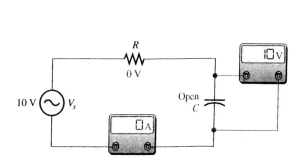

FIGURE C.26-2
Effect of an open capacitor.

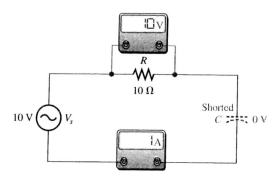

FIGURE C.26-3
Effect of a shorted capacitor.

Effect of a Leaky Capacitor When a capacitor exhibits a high leakage current, the leakage resistance effectively appears in parallel with the capacitor, as shown in Figure C.26-4(a). When the leakage resistance is comparable in value to the circuit resistance, R, the circuit response is drastically affected. The circuit, looking from the capacitor toward the source, can be thevenized, as shown in Figure C.26-4(b). The Thevenin equivalent resistance is R in parallel with R_{leak} (the source appears as a short), and the Thevenin equivalent voltage is determined by the voltage-divider action of R and R_{leak}.

$$R_{th} = R \parallel R_{leak}$$

$$R_{th} = \frac{RR_{leak}}{R + R_{leak}}$$

$$V_{th} = \frac{R_{leak}V_{in}}{R + R_{leak}}$$

Obviously, the voltage to which the capacitor will charge is reduced since $V_{th} < V_{in}$. Also, the circuit time constant is reduced, and the current is increased. The Thevenin equivalent circuit is shown in Figure C.26-4(c).

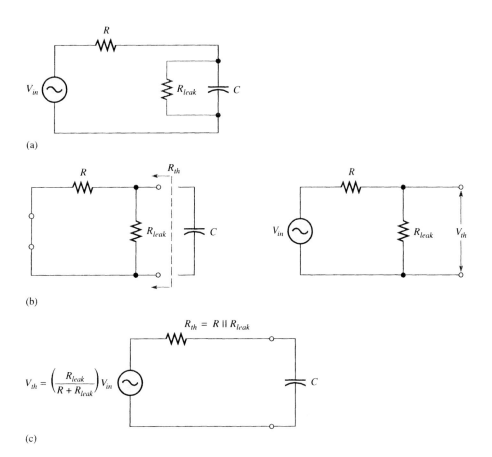

FIGURE C.26-4
Effect of a leaky capacitor.

EXAMPLE C.26-1

Assume that the capacitor in Figure C.26-5 is degraded to a point where its leakage resistance is 10 kΩ. Determine the phase shift from input to output and the output voltage under the degraded condition.

FIGURE C.26-5

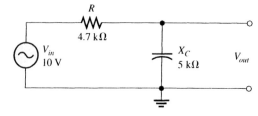

Solution The effective circuit resistance is

$$R_{th} = \frac{RR_{leak}}{R + R_{leak}} = \frac{(4.7 \text{ k}\Omega)(10 \text{ k}\Omega)}{14.7 \text{ k}\Omega} = 3.2 \text{ k}\Omega$$

The phase lag is

$$\phi = 90° - \tan^{-1}\left(\frac{X_C}{R_{th}}\right) = 90° - \tan^{-1}\left(\frac{5 \text{ k}\Omega}{3.2 \text{ k}\Omega}\right) = 32.6°$$

To determine the output voltage, first calculate the Thevenin equivalent voltage.

$$V_{th} = \left(\frac{R_{leak}}{R + R_{leak}}\right)V_{in} = \left(\frac{10 \text{ k}\Omega}{14.7 \text{ k}\Omega}\right)10 \text{ V} = 6.80 \text{ V}$$

$$V_{out} = \left(\frac{X_C}{\sqrt{R_{th}^2 + X_C^2}}\right)V_{th} = \left(\frac{5 \text{ k}\Omega}{\sqrt{(3.2 \text{ k}\Omega)^2 + (5 \text{ k}\Omega)^2}}\right)6.8 \text{ V} = 5.73 \text{ V}$$

Effect of an Open Inductor The most common failure mode for inductors occurs when the winding opens as a result of excessive current or a mechanical contact failure. It is easy to see how an open coil affects the operation of a basic series *RL* circuit, as shown in Figure C.26-6. Obviously, there is no current path; therefore, the resistor voltage is zero, and the total applied voltage appears across the inductor.

Effect of an Open Resistor When the resistor is open, there is no current, and the inductor voltage is zero. The total input voltage is across the open resistor, as shown in Figure C.26-7.

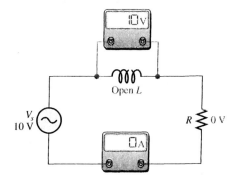

FIGURE C.26-6
Effect of an open coil.

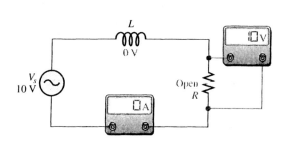

FIGURE C.26-7
Effect of an open resistor.

Open Components in Parallel Circuits In a parallel *RL* circuit, an open resistor or inductor will cause the total current to decrease because the total impedance will increase. Obviously, the branch with the open component will have zero current. Figure C.26-8 illustrates these conditions.

Effect of an Inductor with Shorted Windings It is possible for some of the windings of coils to short together as a result of damaged insulation. This failure mode is much less likely than the open coil. Shorted windings result in a reduction in inductance because the inductance of a coil is proportional to the square of the number of turns.

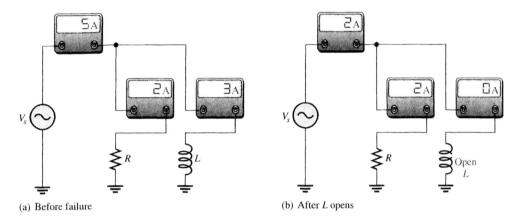

(a) Before failure   (b) After *L* opens

FIGURE C.26-8
Effect of an open component in a parallel circuit with V_s constant.

C.27 Understand principles and operations of AC frequency selective filter circuits

INTRODUCTION

One useful application of reactive circuits is their ability to act as *filters*. For example, a low-pass filter passes signals with low frequency but blocks signals with high frequencies. In this section, we will examine how to develop several different frequency-selective filters.

The *RC* Circuit as a Filter

Filters are frequency-selective circuits that permit signals of certain frequencies to pass from the input to the output while blocking all others. That is, all frequencies but the selected ones are filtered out.

Series *RC* circuits exhibit a frequency-selective characteristic and therefore act as basic filters. There are two types. The first one we examine, called a **low-pass filter,** is realized by taking the output across the capacitor, just as in a lag network. The second type, called a **high-pass filter,** is implemented by taking the output across the resistor, as in a lead network.

Low-Pass Filter   You have seen what happens to the phase angle and the output voltage in the lag network. In terms of its filtering action, we are interested primarily in how the magnitude of the output voltage varies with frequency.

To illustrate low-pass filter action, Figure C.27-1 shows a specific series of measurements in which the frequency starts at zero (DC) and is increased in increments up to 20 kHz. At each value of frequency, the output voltage is measured. As you can see, the capacitive reactance decreases as frequency goes up, thus dropping less voltage across the capacitor while the input voltage is held at a constant 10 V throughout each step. Table C.27-1 summarizes the variation of the circuit parameters with frequency.

TABLE C.27-1

f	X_C (Ω)	Z_{tot} (Ω)	I (mA)	V_{out} (V)
0	∞	∞	0	10
1 kHz	159	187.8	53.25	8.5
10 kHz	15.9	101.3	98.7	1.57
20 kHz	7.96	100.3	99.7	0.79

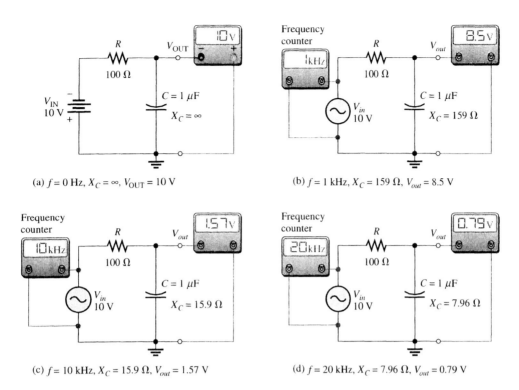

(a) $f = 0$ Hz, $X_C = \infty$, $V_{OUT} = 10$ V

(b) $f = 1$ kHz, $X_C = 159$ Ω, $V_{out} = 8.5$ V

(c) $f = 10$ kHz, $X_C = 15.9$ Ω, $V_{out} = 1.57$ V

(d) $f = 20$ kHz, $X_C = 7.96$ Ω, $V_{out} = 0.79$ V

FIGURE C.27-1
Example of low-pass filter action. As frequency increases, V_{out} decreases.

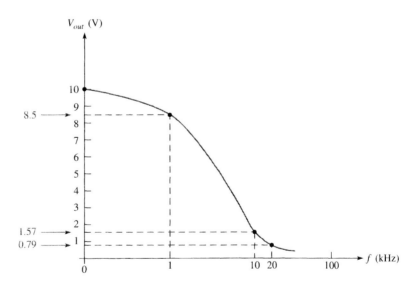

FIGURE C.27-2
Frequency response curve for the low-pass filter in Figure C.27-1.

The **frequency response** of the low-pass filter in Figure C.27-1 is shown in Figure C.27-2, where the measured values are plotted on a graph of V_{out} versus frequency, and a smooth curve is drawn connecting the points. This graph, called a *response curve,* shows that the output voltage is greater at the lower frequencies and decreases as the frequency increases.

High-Pass Filter   To illustrate high-pass filter action, Figure C.27-3 shows a series of specific measurements. Again, the frequency starts at zero (DC) and is increased in increments

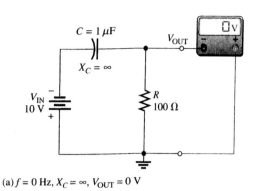

(a) $f = 0$ Hz, $X_C = \infty$, $V_{OUT} = 0$ V

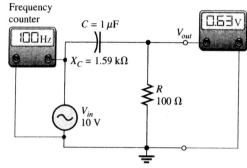

(b) $f = 100$ Hz, $X_C = 1.59$ kΩ, $V_{out} = 0.63$ V

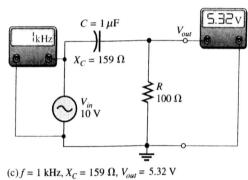

(c) $f = 1$ kHz, $X_C = 159$ Ω, $V_{out} = 5.32$ V

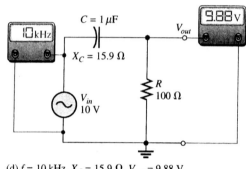

(d) $f = 10$ kHz, $X_C = 15.9$ Ω, $V_{out} = 9.88$ V

FIGURE C.27-3
Example of high-pass filter action. As frequency increases, V_{out} increases.

TABLE C.27-2

f	X_C (Ω)	Z_{tot} (Ω)	I (mA)	V_{out} (V)
0	∞	∞	0	0
100 Hz	1590	1593	6.3	0.63
1 kHz	159	187.8	53.2	5.32
10 kHz	15.9	101.3	98.7	9.88

up to 10 kHz. As you can see, the capacitive reactance decreases as the frequency goes up, thus causing more of the total input voltage to be dropped across the resistor. Table C.27-2 summarizes the variation of circuit parameters with frequency.

In Figure C.27-4, the measured values for the high-pass filter shown in Figure C.27-3 have been plotted to produce a response curve for this circuit. As you can see, the output voltage is greater at the higher frequencies and decreases as the frequency is reduced.

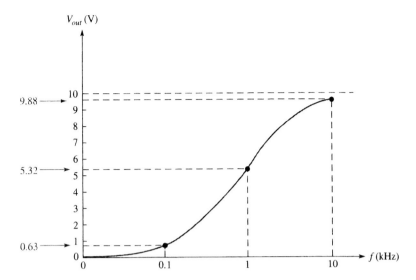

FIGURE C.27-4
Frequency response curve for the high-pass filter in Figure C.27-3.

The Cutoff Frequency and the Bandwidth of a Filter   The frequency at which the capacitive reactance equals the resistance in a low-pass or high-pass *RC* filter is called the **cutoff frequency** and is designated f_c. This condition is expressed as $1/(2\pi f_c C) = R$. Solving for f_c results in the following formula:

$$f_c = \frac{1}{2\pi RC}$$

At f_c, the output voltage of the filter is 70.7% of its maximum value. It is standard practice to consider the cutoff frequency as the limit of a filter's performance in terms of passing or rejecting frequencies. For example, in a high-pass filter, all frequencies above f_c are considered to be passed by the filter, and all those below f_c are considered to be rejected. The reverse is true for a low-pass filter.

The range of frequencies that is considered to be passed by a filter is called the **bandwidth.** Figure C.27-5 illustrates the bandwidth and the cutoff frequency for a low-pass filter.

FIGURE C.27-5
Normalized general response curve of a low-pass filter showing the cutoff frequency and the bandwidth.

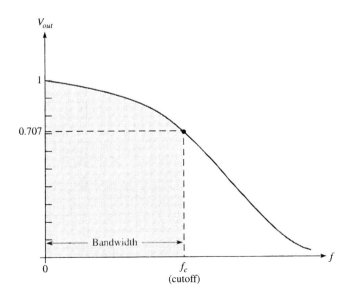

The *RL* Circuit as a Filter

Like *RC* circuits, series *RL* circuits also exhibit a frequency-selective characteristic and therefore act as basic filters.

Low-Pass Filter You have seen what happens to the phase angle and the output voltage in the lag network. In terms of its filtering action, the change in the amplitude of the output voltage with frequency is the primary consideration.

To illustrate low-pass filter action, Figure C.27-6 shows a specific series of measurements in which the frequency starts at zero (DC) and is increased in increments up to 20 kHz. At each value of frequency, the output voltage is measured. As you can see, the inductive reactance increases as frequency goes up, thus causing less voltage to be dropped across the resistor while the input voltage is held at a constant 10 V throughout each step. The response curve for these particular values would appear the same as the response curve for the *RC* low-pass filter.

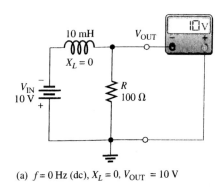

(a) $f = 0$ Hz (dc), $X_L = 0$, $V_{OUT} = 10$ V

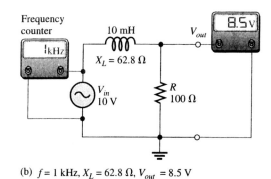

(b) $f = 1$ kHz, $X_L = 62.8 \, \Omega$, $V_{out} = 8.5$ V

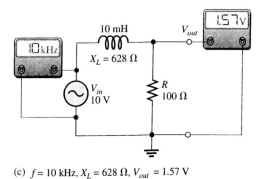

(c) $f = 10$ kHz, $X_L = 628 \, \Omega$, $V_{out} = 1.57$ V

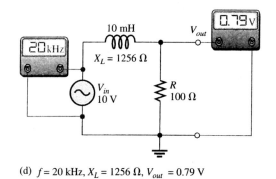

(d) $f = 20$ kHz, $X_L = 1256 \, \Omega$, $V_{out} = 0.79$ V

FIGURE C.27-6
Example of low-pass filter action. Winding resistance has been neglected. As the input frequency increases, the output voltage decreases.

High-Pass Filter To illustrate *RL* high-pass filter action, Figure C.27-7 shows a series of specific measurements. The frequency starts at zero (DC) and is increased in increments up to 10 kHz. As you can see, the inductive reactance increases as the frequency goes up, thus causing more voltage to be dropped across the inductor. Again, when the values are plotted, the response curve is the same as the one for the *RC* high-pass filter.

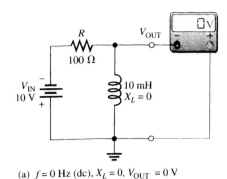

(a) $f = 0$ Hz (dc), $X_L = 0$, $V_{OUT} = 0$ V

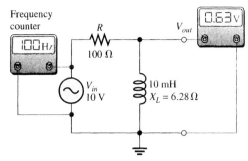

(b) $f = 100$ Hz, $X_L = 6.28$ Ω, $V_{out} = 0.63$ V

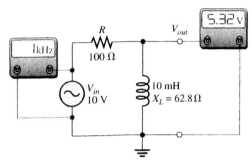

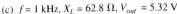

(c) $f = 1$ kHz, $X_L = 62.8$ Ω, $V_{out} = 5.32$ V

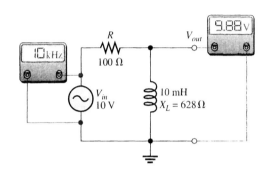

(d) $f = 10$ kHz, $X_L = 628$ Ω, $V_{out} = 9.88$ V

FIGURE C.27-7
Example of high-pass filter action. Winding resistance has been neglected. As the input frequency increases, the output voltage increases.

The Cutoff Frequency of an RL Filter The frequency at which the inductive reactance equals the resistance in a low-pass or high-pass *RL* filter is called the *cutoff frequency* and is designated f_c. This condition is expressed as $2\pi f_c L = R$. Solving for f_c results in the following formula:

$$f_c = \frac{R}{2\pi L}$$

As with the *RC* filter, the output voltage is 70.7% of its maximum value at f_c. In a high-pass filter, all frequencies above f_c are considered to be passed by the filter, and all those below f_c are considered to be rejected. The reverse, of course, is true for a low-pass filter.

SERIES RESONANT FILTERS

The Band-Pass Filter

A basic series resonant **band-pass filter** is shown in Figure C.27-8. Notice that the series *LC* portion is placed between the input and the output and that the output is taken across the resistor.

FIGURE C.27-8
A basic series resonant band-pass filter.

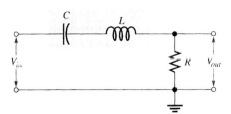

A band-pass filter allows signals at the resonant frequency and at frequencies within a certain band (or range) extending below and above the resonant value to pass from input to output without a significant reduction in amplitude. Signals at frequencies lying outside this specified band (called the **passband**) are reduced in amplitude to below a certain level and are considered to be rejected by the filter.

The filtering action is the result of the impedance characteristic of the filter. The impedance is minimum at resonance and has increasingly higher values below and above the resonant frequency. At very low frequencies, the impedance is very high and tends to block the current. As the frequency increases, the impedance drops, allowing more current and thus more voltage across the output resistor. At the resonant frequency, the impedance is very low and equal to the winding resistance of the coil. At this point there is maximum current and the resulting output voltage is maximum. As the frequency goes above resonance, the impedance again increases, causing the current and the resulting output voltage to drop. Figure C.27-9 illustrates the general frequency response of a series resonant band-pass filter.

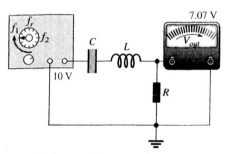

(a) As the frequency increases to f_1, V_{out} increases to 7.07 V.

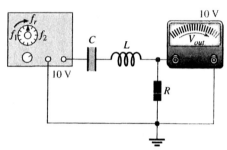

(b) As the frequency increases from f_1 to f_r, V_{out} increases from 7.07 V to 10 V.

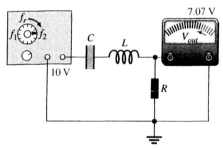

(c) As the frequency increases from f_r to f_2, V_{out} decreases from 10 V to 7.07 V.

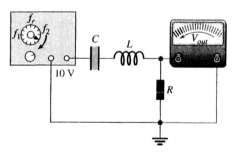

(d) As the frequency increases above f_2, V_{out} decreases below 7.07 V.

FIGURE C.27-9
Example of the frequency response of a series resonant band-pass filter with the input voltage at a constant 10 V rms. The winding resistance of the coil is neglected.

Bandwidth of the Passband

The **bandwidth** (*BW*) of a band-pass filter is the range of frequencies for which the current (or output voltage) is equal to or greater than 70.7% of its value at the resonant frequency. Figure C.27-10 shows the bandwidth on the response curve for the band-pass filter.

The frequencies at which the output of a filter is 70.7% of its maximum are the **cut-off frequencies.** Notice in Figure C.27-10 that frequency f_1, which is below f_r, is the frequency at which I (or V_{out}) is 70.7% of the resonant value (I_{max}); f_1 is commonly called the

FIGURE C.27-10
Generalized response curve of a series resonant band-pass filter.

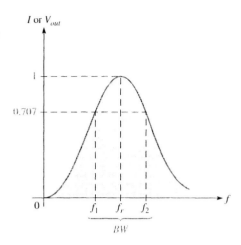

lower cutoff frequency. At frequency f_2, above f_r, the current (or V_{out}) is again 70.7% of its maximum; f_2 is called the *upper cutoff frequency.* Other names for f_1 and f_2 are -3 dB frequencies, *critical frequencies, band frequencies,* and *half-power frequencies.*

The formula for calculating the bandwidth is as follows:

$$BW = f_2 - f_1$$

The unit of bandwidth is the hertz (Hz), the same as for frequency.

The Band-Stop Filter

The basic series resonant **band-stop filter** is shown in Figure C.27-11. Notice that the output voltage is taken across the *LC* portion of the circuit. This filter is still a series *RLC* circuit, just as the band-pass filter is. The difference is that in this case, the output voltage is taken across the combination of *L* and *C* rather than across *R*.

The band-stop filter rejects signals with frequencies between the upper and lower cutoff frequencies and passes those signals with frequencies below and above the cutoff values, as shown in the response curve of Figure C.27-12. The range of frequencies between the lower and upper cutoff points is called the **stopband.** This type of filter is also referred to as a *band-elimination filter, band-reject filter,* or a *notch filter.*

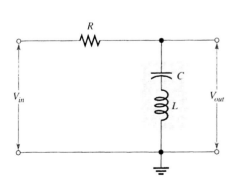

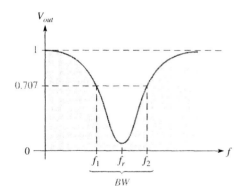

FIGURE C.27-11
A basic series resonant band-stop filter.

FIGURE C.27-12
Generalized response curve for a band-stop filter.

At very low frequencies, the *LC* combination appears as a near open due to the high X_C, thus allowing most of the input voltage to pass through to the output. As the frequency increases, the impedance of the *LC* combination decreases until, at resonance, it is zero (ideally). Thus, the input signal is shorted to ground, and there is very little output voltage. As the frequency goes above its resonant value, the *LC* impedance increases, allowing an increasing amount of voltage to be dropped across it. The general frequency response of a series resonant band-stop filter is illustrated in Figure C.27-13.

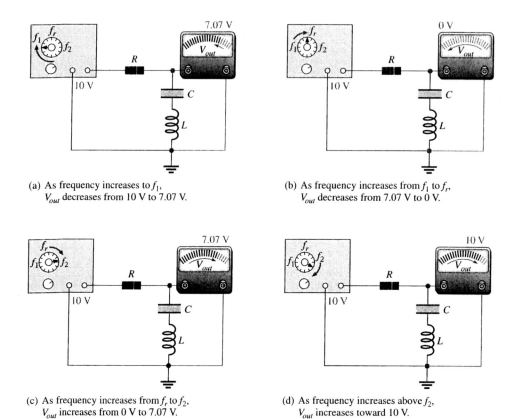

(a) As frequency increases to f_1, V_{out} decreases from 10 V to 7.07 V.

(b) As frequency increases from f_1 to f_r, V_{out} decreases from 7.07 V to 0 V.

(c) As frequency increases from f_r to f_2, V_{out} increases from 0 V to 7.07 V.

(d) As frequency increases above f_2, V_{out} increases toward 10 V.

FIGURE C.27-13
Example of the frequency response of a series resonant band-stop filter with V_{in} at a constant 10 V rms. The winding resistance is neglected.

Characteristics of the Band-Stop Filter

All the characteristics that have been discussed in relation to the band-pass filter (current response, impedance characteristic, bandwidth, selectivity, and *Q*) apply equally to the band-stop filter, with the exception that the response curve of the output voltage is opposite. For the band-pass filter, V_{out} is maximum at resonance. For the band-stop filter, V_{out} is minimum at resonance.

REVIEW QUESTIONS

True/False

1. An *RC* circuit can be used as a low-pass or as a high-pass filter.
2. An *RL* circuit may only be used as a low-pass filter.

Multiple Choice

3. A low-pass filter uses a 1 K ohm resistor and .22 μF capacitor. What is its corner frequency?

   a. 723 Hz.

   b. 1000 Hz.

   c. 4545 Hz.

4. A series *RLC* circuit can be used as a

   a. Low-pass or high-pass filter.

   b. Band-pass filter.

   c. Band-stop filter.

5. The higher the *Q*, the

   a. Lower the bandwidth.

   b. Higher the bandwidth.

   c. Higher the resonant frequency.

C.28 Fabricate and demonstrate AC frequency selective filter circuits

INTRODUCTION

Using actual components (or simulation software such as *Electronics Workbench*), set up and examine the circuit in Figure C.28-1.

Vary the input frequency until you locate the corner frequency. Explain how you found the corner frequency. What is the phase shift at the corner? Measure all component voltages and the circuit current. Lower the frequency to one-tenth the corner frequency and repeat all measurements. Raise the frequency to ten times the corner and repeat all measurements. Explain why the circuit acts like a low-pass filter.

Repeat all tests for the high-pass filter shown in Figure C.28-2.

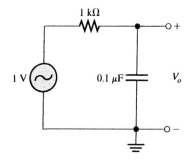
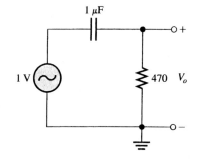

FIGURE C.28-1
Low-pass filter.

FIGURE C.28-2
High-pass filter.

C.29 Troubleshoot and repair AC frequency selective filter circuits

INTRODUCTION

One important point to keep in mind while troubleshooting a filter is the effect a load resistor has on the circuit. For example, loading the low-pass filter causes the capacitor to see the Thevenin equivalent of the load and filter resistance, which is *lower* than the original filter resistance. This causes the corner frequency to *increase*. In addition, a voltage divider action between the two resistors lowers the amount of signal available for filtering, which in turn lowers the gain. The same is true for a high-pass filter, except there is no voltage divider action.

If the load resistance is significantly (at least ten times) greater than the filter resistance, this problem is eliminated.

C.30 Understand principles and operations of AC polyphase circuits

INTRODUCTION

Polyphase (or three-phase) circuits utilize balanced sources and loads. A balanced source has each source set to the same voltage and equal phase shifts between each source (typically 120°). A balanced load contains the same impedance in each phase. In this section we will examine the operation of balanced, polyphase circuits.

EXAMPLE C.30-1

The instantaneous position of a certain Y-connected AC generator is shown in Figure C.30-1. If each phase voltage has a magnitude of 120 V rms, determine the magnitude of each line voltage, and sketch the phasor diagram.

FIGURE C.30-1

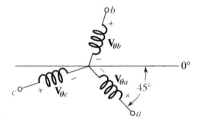

Solution The magnitude of each line voltage is

$$V_L = \sqrt{3}V_\theta = \sqrt{3}(120 \text{ V}) = 208 \text{ V}$$

The phasor diagram for the given instantaneous generator position is shown in Figure C.30-2.

FIGURE C.30-2

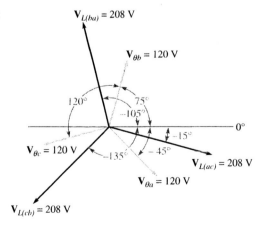

EXAMPLE C.30-2

The three-phase Δ-connected generator represented in Figure C.30-3 is driving a balanced load such that each phase current is 10 A in magnitude. When $\mathbf{I}_{\theta a} = 10\angle 30°$ A, determine the following:
(a) The polar expressions for the other phase currents
(b) The polar expressions for each of the line currents
(c) The complete current phasor diagram

FIGURE C.30-3

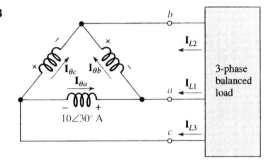

Solution
(a) The phase currents are separated by 120°; therefore,

$$\mathbf{I}_{\theta b} = 10\angle(30° + 120°) = 10\angle 150° \text{ A}$$
$$\mathbf{I}_{\theta c} = 10\angle(30° - 120°) = 10\angle -90° \text{ A}$$

(b) The line currents are separated from the nearest phase current by 30°; therefore,

$$\mathbf{I}_{L1} = \sqrt{3}I_{\theta a}\angle(30° - 30°) = 17.3\angle 0° \text{ A}$$
$$\mathbf{I}_{L2} = \sqrt{3}I_{\theta b}\angle(150° - 30°) = 17.3\angle 120° \text{ A}$$
$$\mathbf{I}_{L3} = \sqrt{3}I_{\theta c}\angle(-90° - 30°) = 17.3\angle -120° \text{ A}$$

(c) The phasor diagram is shown in Figure C.30-4.

FIGURE C.30-4

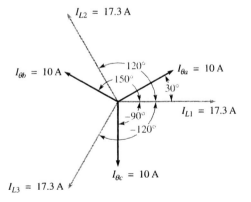

EXAMPLE C.30-3

In the Y-Y system of Figure C.30-5, determine the following:
(a) Each load current   (b) Each line current   (c) Each phase current
(d) Neutral current   (e) Each load voltage

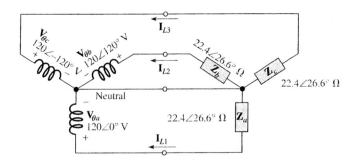

FIGURE C.30-5

Solution This system has a balanced load.

(a) $\mathbf{Z}_a = \mathbf{Z}_b = \mathbf{Z}_c = 22.4\angle 26.6°\ \Omega$

$$\mathbf{I}_{Za} = \frac{\mathbf{V}_{\theta a}}{\mathbf{Z}_a} = \frac{120\angle 0°\ \text{V}}{22.4\angle 26.6°\ \Omega} = 5.36\angle -26.6°\ \text{A}$$

$$\mathbf{I}_{Zb} = \frac{\mathbf{V}_{\theta b}}{\mathbf{Z}_b} = \frac{120\angle 120°\ \text{V}}{22.4\angle 26.6°\ \Omega} = 5.36\angle 93.4°\ \text{A}$$

$$\mathbf{I}_{Zc} = \frac{\mathbf{V}_{\theta c}}{\mathbf{Z}_c} = \frac{120\angle -120°\ \text{V}}{22.4\angle 26.6°\ \Omega} = 5.36\angle -147°\ \text{A}$$

(b) $\mathbf{I}_{L1} = 5.36\angle -26.6°\ \text{A}$

$\mathbf{I}_{L2} = 5.36\angle 93.4°\ \text{A}$

$\mathbf{I}_{L3} = 5.36\angle -147°\ \text{A}$

(c) $\mathbf{I}_{\theta a} = 5.36\angle -26.6°\ \text{A}$

$\mathbf{I}_{\theta b} = 5.36\angle 93.4°\ \text{A}$

$\mathbf{I}_{\theta c} = 5.36\angle -147°\ \text{A}$

(d) $\mathbf{I}_{\text{neut}} = \mathbf{I}_{Za} + \mathbf{I}_{Zb} + \mathbf{I}_{Zc}$

$\quad = 5.36\angle -26.6°\ \text{A} + 5.36\angle 93.4°\ \text{A} + 5.36\angle -147°\ \text{A}$

$\quad = (4.80\ \text{A} - j2.40\ \text{A}) + (-0.33\ \text{A} + j5.35\ \text{A}) + (-4.47\ \text{A} - j2.95\ \text{A}) = 0\ \text{A}$

If the load impedances were not equal (balanced load), the neutral current would have a nonzero value.

(e) The load voltages are equal to the corresponding source phase voltages.

$$\mathbf{V}_{Za} = 120\angle 0°\ \text{V}$$
$$\mathbf{V}_{Zb} = 120\angle 120°\ \text{V}$$
$$\mathbf{V}_{Zc} = 120\angle -120°\ \text{V}$$

EXAMPLE C.30-4

Determine the load voltages and load currents in Figure C.30-6, and show their relationship in a phasor diagram.

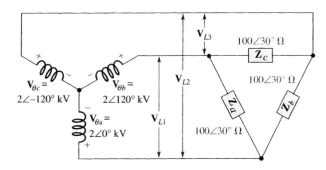

FIGURE C.30-6

Solution

$$\mathbf{V}_{Za} = \mathbf{V}_{L1} = 2\sqrt{3}\angle 150° \text{ kV} = 3.46\angle 150° \text{ kV}$$
$$\mathbf{V}_{Zb} = \mathbf{V}_{L2} = 2\sqrt{3}\angle 30° \text{ kV} = 3.46\angle 30° \text{ kV}$$
$$\mathbf{V}_{Zc} = \mathbf{V}_{L3} = 2\sqrt{3}\angle -90° \text{ kV} = 3.46\angle -90° \text{ kV}$$

The load currents are

$$\mathbf{I}_{Za} = \frac{\mathbf{V}_{Za}}{\mathbf{Z}_a} = \frac{3.46\angle 150° \text{ kV}}{100\angle 30° \text{ }\Omega} = 34.6\angle 120° \text{ A}$$

$$\mathbf{I}_{Zb} = \frac{\mathbf{V}_{Zb}}{\mathbf{Z}_b} = \frac{3.46\angle 30° \text{ kV}}{100\angle 30° \text{ }\Omega} = 34.6\angle 0° \text{ A}$$

$$\mathbf{I}_{Zc} = \frac{\mathbf{V}_{Zc}}{\mathbf{Z}_c} = \frac{3.46\angle -90° \text{ kV}}{100\angle 30° \text{ }\Omega} = 34.6\angle -120° \text{ A}$$

The phasor diagram is shown in Figure C.30-7.

FIGURE C.30-7

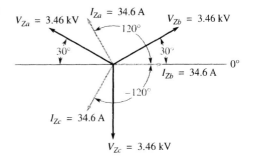

Determine the currents and voltages in the balanced load and the magnitude of the line voltages in Figure C.30-8.

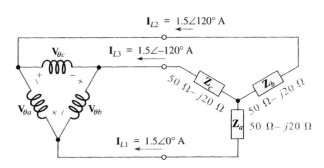

FIGURE C.30-8

Solution   The load currents equal the specified line currents.

$$\mathbf{I}_{Za} = \mathbf{I}_{L1} = 1.5\angle 0° \text{ A}$$
$$\mathbf{I}_{Zb} = \mathbf{I}_{L2} = 1.5\angle 120° \text{ A}$$
$$\mathbf{I}_{Zc} = \mathbf{I}_{L3} = 1.5\angle -120° \text{ A}$$

The load voltages are

$$
\begin{aligned}
\mathbf{V}_{Za} &= \mathbf{I}_{Za}\mathbf{Z}_a \\
&= (1.5\angle 0° \text{ A})(50 \ \Omega - j20 \ \Omega) \\
&= (1.5\angle 0° \text{ A})(53.9\angle -21.8° \ \Omega) = 80.9\angle -21.8° \text{ V}
\end{aligned}
$$

$$
\begin{aligned}
\mathbf{V}_{Zb} &= \mathbf{I}_{Zb}\mathbf{Z}_b \\
&= (1.5\angle 120° \text{ A})(53.9\angle -21.8° \ \Omega) = 80.9\angle 98.2° \text{ V}
\end{aligned}
$$

$$
\begin{aligned}
\mathbf{V}_{Zc} &= \mathbf{I}_{Zc}\mathbf{Z}_c \\
&= (1.5\angle -120° \text{ A})(53.9\angle -21.8° \ \Omega) = 80.9\angle -142° \text{ V}
\end{aligned}
$$

The magnitude of the line voltages is

$$V_L = V_\theta = \sqrt{3}V_Z = \sqrt{3}(80.9 \text{ V}) = 140 \text{ V}$$

EXAMPLE C.30-6

Determine the magnitude of the load currents and the line currents in Figure C.30-9.

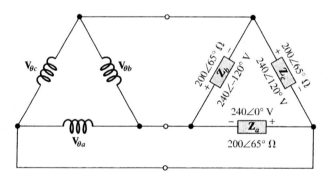

FIGURE C.30-9

Solution $V_{Za} = V_{Zb} = V_{Zc} = 240 \text{ V}$

The magnitude of the load currents is

$$I_{Za} = I_{Zb} = I_{Zc} = \frac{V_{Za}}{Z_a} = \frac{240 \text{ V}}{200 \ \Omega} = 1.20 \text{ A}$$

The magnitude of the line currents is

$$I_L = \sqrt{3}I_Z = \sqrt{3}(1.20 \text{ A}) = 2.08 \text{ A}$$

EXAMPLE C.30-7

In a certain Δ-connected balanced load, the line voltages are 250 V and the impedances are $50\angle 30° \ \Omega$. Determine the total load power.

Solution   In a Δ-connected system, $V_Z = V_L$ and $I_L = \sqrt{3}I_Z$. The load current magnitudes are

$$I_Z = \frac{V_Z}{Z} = \frac{250 \text{ V}}{50 \ \Omega} = 5 \text{ A}$$

and

$$I_L = \sqrt{3}I_Z = \sqrt{3}(5 \text{ A}) = 8.66 \text{ A}$$

The power factor is

$$\cos \theta = \cos 30° = 0.866$$

The total load power is

$$P_{L(tot)} = \sqrt{3}V_L I_L \cos \theta = \sqrt{3}(250 \text{ V})(8.66 \text{ A})(0.866) = 3.25 \text{ kW}$$

REVIEW QUESTIONS

True/False

1. In a balanced three-phase system, the angle between each phase is 120 degrees.
2. Y, Δ, and Z are three ways to connect three-phase sources and loads.

Multiple Choice

3. A balanced Y-connected source with each phase voltage equal to 100 V has a line voltage of
   a. 70.7 V.
   b. 100 V.
   c. 173 V.
4. Which load uses the most power from the same source, Y or Δ?
   a. Y-connected load.
   b. Δ-connected load.
   c. Both loads use the same power.
5. How many ways can a Y or Δ source connect to a Y or Δ load?
   a. 2.
   b. 3.
   c. 4.

C.31 Understand principles and operations of AC phase-locked loop circuits

INTRODUCTION

The phase-locked loop (PLL) is an electronic feedback circuit consisting of a phase detector, a low-pass filter, and a voltage-controlled oscillator (VCO). It is capable of locking onto or synchronizing with an incoming signal. When the phase changes, indicating that the incoming frequency is changing, the phase detector's output voltage (error voltage) increases or decreases just enough to keep the oscillator frequency the same as the incoming frequency. Phase-locked loops are used in a wide variety of communication system applications, including TV receivers, FM demodulation, modems, telemetry, and tone decoders.

Basic Operation

Using the basic block diagram of a phase-locked loop (PLL) in Figure C.31-1 as a reference, the general operation is as follows. When there is no input signal, the error voltage is zero and the frequency, f_o, of the voltage-controlled oscillator (VCO) is called the *free-running* or *center frequency*. When an input signal is applied, the phase detector compares the phase and frequency of the input signal with the VCO frequency and produces error voltage, V_e. This error voltage is proportional to the phase and frequency difference of the incoming frequency and the VCO frequency. The error voltage contains components that are the sum and difference of the two compared frequencies. The low-pass filter passes only the difference frequency, V_d, which is the lower of the two components. This signal is amplified and fed back to the VCO as a control voltage, V_{CONT}. The control voltage forces the VCO frequency to change in a direction that reduces the difference between the incoming frequency, f_i, and the VCO frequency, f_o. When f_i and f_o are sufficiently close in value, the feedback action of the PLL causes the VCO to lock onto the incoming signal. Once the VCO locks, its frequency is the same as the input frequency with a slight difference in phase. This phase difference, ϕ, is necessary to keep the PLL in the lock condition.

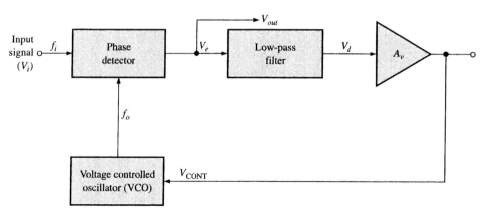

FIGURE C.31-1
Basic phase-locked loop block diagram.

Lock Range   Once the PLL is locked, it can track frequency changes in the incoming signal. The range of frequencies over which the PLL can maintain lock with the incoming signal is called the *lock* or *tracking range*. The lock range is usually expressed as a percentage of the VCO frequency.

Capture Range   The range of frequencies over which the PLL can acquire lock with an input signal is called the *capture range*. The parameter is normally expressed in percentage of f_o, also.

Sum and Difference Frequencies   The phase detector operates as a multiplier circuit to produce the sum and difference of the input frequency, f_i, and the VCO frequency, f_o. This action can best be described mathematically as follows. Recall from AC circuit theory that a sinusoidal voltage can be expressed as

$$v = V_p \sin 2\pi f t$$

where V_p is the peak value, f is the frequency, and t is the time. Using this basic expression, the input signal voltage, v_i, and the VCO voltage, v_o, can be written as

$$v_i = V_{ip} \sin 2\pi f_i t$$
$$v_o = V_{op} \sin 2\pi f_o t$$

When these two signals are multiplied in the phase detector, we get a product at the output as follows:

$$V_{out} = V_{ip} V_{op} (\sin 2\pi f_i t)(\sin 2\pi f_o t)$$

Applying the trigonometric identity,

$$(\sin A)(\sin B) = \frac{1}{2}[\cos(A - B) - \cos(A + B)]$$

to the preceding equation for V_{out}, we get

$$V_{out} = \frac{V_{ip} V_{op}}{2}[\cos(2\pi f_i t - 2\pi f_o t) - \cos(2\pi f_i t + 2\pi f_o t)]$$

$$V_{out} = \frac{V_{ip} V_{op}}{2} \cos 2\pi(f_i - f_o)t - \frac{V_{ip} V_{op}}{2} \cos 2\pi(f_i + f_o)t$$

You can see that V_{out} of the phase detector consists of a difference frequency component $(f_i - f_o)$ and a sum frequency component $(f_i + f_o)$. This concept is illustrated in Figure C.31-2 with frequency spectrum graphs. Each vertical line represents a specific signal frequency, and the height is its amplitude, A.

FIGURE C.31-2
Frequency spectrum of the phase detector.

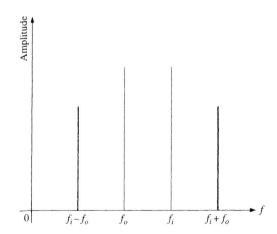

EXAMPLE C.31-1

A 10 kHz signal f_i and an 8 kHz signal f_o are applied to a phase detector. Determine the sum and difference frequencies.

Solution

$$f_i + f_o = 10 \text{ kHz} + 8 \text{ kHz} = 18 \text{ kHz}$$
$$f_i - f_o = 10 \text{ kHz} - 8 \text{ kHz} = 2 \text{ kHz}$$

When the PLL Is in Lock

When the PLL is in a lock condition, the VCO frequency equals the input frequency ($f_o = f_i$). Thus, the difference frequency is $f_o - f_i = 0$. A zero frequency indicates a DC component. The low-pass filter removes the sum frequency ($f_i + f_o$) and passes the DC (0-frequency) component, which is amplified and fed back to the VCO. When the PLL is in lock, the difference frequency component is always DC and is always passed by the filter, so the lock range is independent of the bandwidth of the low-pass filter. This is illustrated in Figure C.31-3.

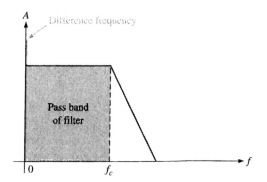

FIGURE C.31-3
When the PLL is in lock, the difference frequency is zero and passes through the filter. The lock range is independent of the filter bandwidth.

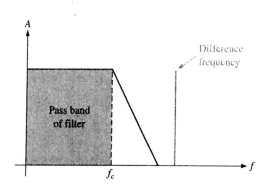

FIGURE C.31-4
When the PLL is not in lock, the difference frequency can be outside the pass band.

When the PLL Is Not in Lock

In the case where the PLL has not yet locked onto the incoming frequency, the phase detector still produces sum and difference frequencies. In this case, however, the difference frequency may lie outside the pass band of the low-pass filter and will not be fed back to the VCO, as illustrated in Figure C.31-4. The VCO will remain at its center frequency (free-running) as long as this condition exists.

As the input frequency approaches the VCO frequency, the difference component produced by the phase detector decreases and eventually falls into the pass band of the filter and drives the VCO toward the incoming frequency. When the VCO reaches the incoming frequency, the PLL locks on it.

REVIEW QUESTIONS

True/False

1. A phase-locked loop uses a high-pass filter to lock in on the center frequency.

2. A phase difference is required to keep the PLL locked in.

Multiple Choice

3. When locked, the PLL's difference frequency is

   a. Zero.

   b. The same as F_{in}.

   c. The same as F_{out}.

4. Input frequencies of 12 KHz and 8 KHz cause the phase detector to generate a

   a. 4 KHz output.

   b. 20 KHz output.

   c. Both a and b.

5. VCO stands for

   a. Variable capacitance output.

   b. Voltage coupled oscillator.

   c. Voltage-controlled oscillator.

C.32 Troubleshoot and repair AC phase-locked loop circuits

INTRODUCTION

Typically, phase-locked loop circuits come prepackaged in an integrated circuit. A few components are required to complete the circuit (R and C for the low-pass filter section). Assuming all component values are correct and the wiring has been checked, the PLL should operate as planned. One quick check of a nonworking circuit is to swap PLL IC's with a working circuit (to see if the problem moves).

Discrete Solid State Devices

D.01   Demonstrate an understanding of the properties of semiconductor materials

D.02   Demonstrate an understanding of *pn* junctions

D.03   Demonstrate an understanding of bipolar transistors

D.04   Demonstrate an understanding of field effect transistors (FETs/MOSFETs)

D.05   Demonstrate an understanding of special diodes and transistors

D.06   Understand principles and operations of diode circuits

D.07   Fabricate and demonstrate diode circuits

D.08   Troubleshoot and repair diode circuits

D.09   Understand principles and operations of optoelectronic circuits (gate isolators, interrupt sensors, infrared sensors, etc.)

D.10   Fabricate and demonstrate optoelectronic circuits (gate isolators, interrupt sensors, infrared sensors, etc.)

D.11   Troubleshoot and repair optoelectronic circuits (gate isolators, interrupt sensors, infrared sensors, etc.)

D.12   Understand principles and operations of single-stage amplifiers

D.13   Fabricate and demonstrate single-stage amplifiers

D.14   Troubleshoot and repair single-stage amplifiers

D.15   Understand principles and operations of thyristor circuitry (SCR, triac, diac, etc.)

D.16   Fabricate and demonstrate thyristor circuitry (SCR, triac, diac, etc.)

D.17   Troubleshoot and repair thyristor circuitry (SCR, triac, diac, etc.)

D.01 Demonstrate an understanding of the properties of semiconductor materials

INTRODUCTION

The study of semiconductors requires a good understanding of atoms and their interactions with each other. In this section we examine the principles behind *n*- and *p*-type semiconductors.

Silicon and Germanium Atoms

Two types of semiconductive materials are **silicon** and **germanium.** Both the silicon and the germanium atoms have four valence electrons. These atoms differ in that silicon has 14 protons in its nucleus and germanium has 32. Figure D.01-1 shows a representation of the atomic structure for both materials.

FIGURE D.01-1
Diagrams of the silicon and germanium atoms.

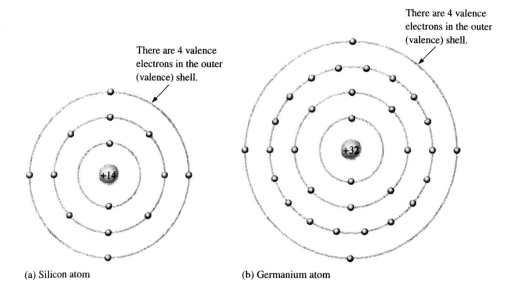

There are 4 valence electrons in the outer (valence) shell.

There are 4 valence electrons in the outer (valence) shell.

(a) Silicon atom   (b) Germanium atom

The valence electrons in germanium are in the fourth shell while the ones in silicon are in the third shell, closer to the nucleus. This means that the germanium valence electrons are at higher energy levels than those in silicon and, therefore, require a smaller amount of additional energy to escape from the atom. This property makes germanium more unstable at high temperatures, and this is a basic reason why silicon is the most widely used semiconductive material.

ATOMIC BONDING

Figure D.01-2 shows how each silicon atom positions itself with four adjacent atoms to form a silicon **crystal.** A silicon atom with its four valence electrons shares an electron with each of its four neighbors. This effectively creates eight valence electrons for each atom and produces a state of chemical stability. Also, this sharing of valence electrons produces the **covalent** bonds that hold the atoms together; each shared electron is attracted equally by two adjacent atoms which share it. Covalent bonding in an intrinsic silicon crystal is shown in Figure D.01-3. An **intrinsic** crystal is one that has no impurities. Covalent bonding for germanium is similar because it also has four valence electrons.

FIGURE D.01-2
Covalent bonds in silicon.

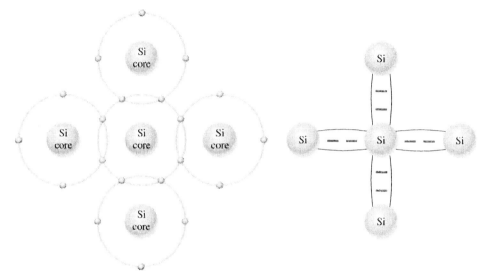

(a) The center atom shares an electron with each of the four surrounding atoms creating a covalent bond with each. The surrounding atoms are in turn bonded to other atoms, and so on.

(b) Bonding diagram. The negative signs represent the shared valence electrons.

FIGURE D.01-3
Covalent bonds form a crystal structure.

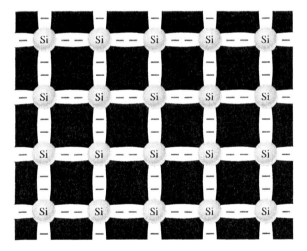

CONDUCTION IN SEMICONDUCTORS

An energy band diagram is shown in Figure D.01-4 for an unexcited silicon atom (no external energy). This condition occurs only at absolute zero temperature.

Conduction Electrons and Holes

An intrinsic (pure) silicon crystal at room temperature derives heat (thermal) energy from the surrounding air, causing some valence electrons to gain sufficient energy to jump the gap from the valence band into the conduction band, becoming free electrons. This

FIGURE D.01-4
Energy band diagram for unexcited silicon atom. There are no electrons in the conduction band.

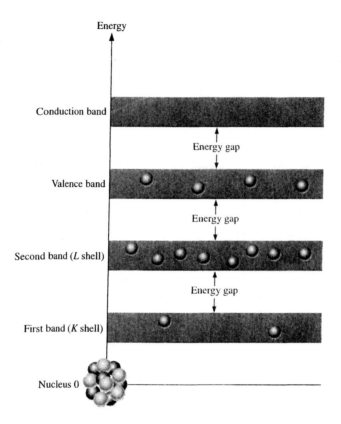

FIGURE D.01-5
Creation of an electron-hole pair in an excited silicon atom. An electron in the conduction band is a free electron.

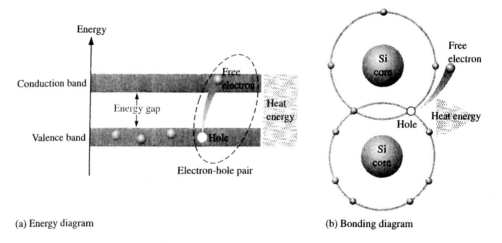

(a) Energy diagram   (b) Bonding diagram

situation is illustrated in the energy diagram of Figure D.01-5(a) and in the bonding diagram of Figure D.01-5(b).

When an electron jumps to the conduction band, a vacancy is left in the valence band. This vacancy is called a **hole.** For every electron raised to the conduction band by external energy, there is one hole left in the valence band, creating what is called an *electron-hole pair.* **Recombination** occurs when a conduction-band electron loses energy and falls back into a hole in the valence band.

To summarize, a piece of intrinsic silicon at room temperature has, at any instant, a number of conduction-band (free) electrons that are unattached to any atom and are essentially drifting randomly throughout the material. Also, an equal number of holes are

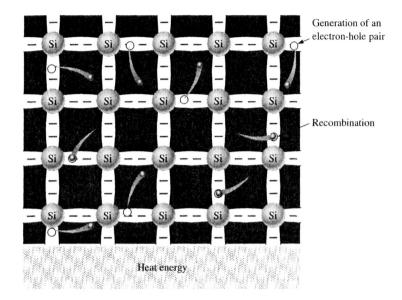

FIGURE D.01-6
Electron-hole pairs in a silicon crystal. Free electrons are being generated continuously while some recombine with holes.

Generation of an electron-hole pair

Recombination

Heat energy

created in the valence band when these electrons jump into the conduction band. This is illustrated in Figure D.01-6.

Electron and Hole Current

When a voltage is applied across a piece of silicon, as shown in Figure D.01-7, the thermally generated free electrons in the conduction band, which are free to move randomly in the crystal structure, are now easily attracted toward the positive end. This movement of free electrons is one type of current in a semiconductor and is called *electron current.*

FIGURE D.01-7
Free electron current in intrinsic silicon is produced by the movement of thermally generated free electrons.

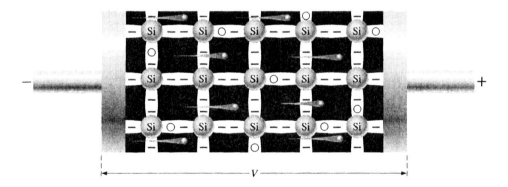

Another type of current occurs at the valence level, where the holes created by the free electrons exist. Electrons remaining in the valence band are still attached to their atoms and are not free to move randomly in the crystal structure. However, a valence electron can "fall" into a nearby hole, with little change in its energy level, thus leaving another hole where it came from. The hole has effectively, although not physically, moved from one place to another in the crystal structure, as illustrated in Figure D.01-8. This current is called *hole current.*

Semiconductors, Conductors, and Insulators

In an **intrinsic** (pure) **semiconductor,** there are relatively few free electrons, so neither silicon nor germanium is very useful in its intrinsic state. Pure semiconductors are neither insulators nor good conductors because current in a material depends directly on the number of free electrons.

FIGURE D.01-8
Hole current in intrinsic silicon.

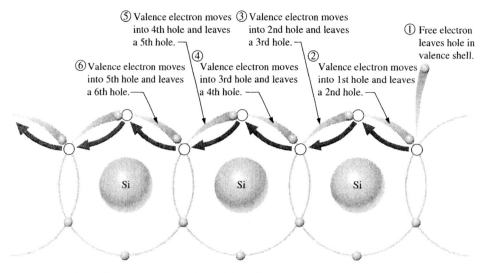

As a valence electron moves left to right to fill a hole while leaving another hole behind, the hole has effectively moved from right to left.

A comparison of the energy bands in Figure D.01-9 for the three types of materials shows the essential differences among them regarding conduction. The energy gap for an insulator is so wide that hardly any electrons acquire enough energy to jump into the conduction band. The valence band and the conduction band in a conductor (such as copper) overlap so that there are always many conduction electrons, even without the application of external energy. A semiconductor, as Figure D.01-9(b) shows, has an energy gap that is much narrower than that in an insulator.

FIGURE D.01-9
Energy diagrams for the three types of materials.

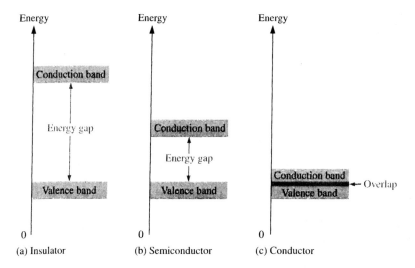

n-TYPE AND *p*-TYPE SEMICONDUCTORS

Doping

The conductivities of silicon and germanium can be drastically increased and controlled by the addition of impurities to the intrinsic (pure) semiconductor. This process, called **doping,** increases the number of current carriers (electrons or holes). The two categories of impurities are *n-type* and *p-type.*

n-Type Semiconductor

To increase the number of conduction-band electrons in intrinsic silicon, pentavalent impurity atoms are added. These are atoms with five valence electrons, such as arsenic (As), phosphorus (P), and antimony (Sb).

As illustrated in Figure D.01-10, each pentavalent atom (antimony, in this case) forms covalent bonds with four adjacent silicon atoms. Four of the antimony atom's valence electrons are used to form the covalent bonds, leaving one extra electron. This extra electron becomes a conduction electron because it is not attached to any atom. The number of conduction electrons can be controlled by the number of impurity atoms added to the silicon.

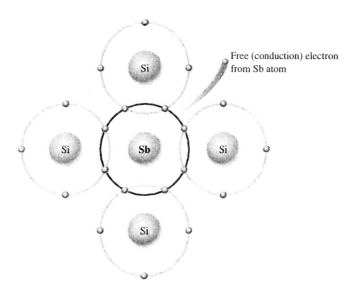

Free (conduction) electron
from Sb atom

FIGURE D.01-10
Pentavalent impurity atom in a silicon crystal. An antimony (Sb) impurity atom is shown in the center. The extra electron from the Sb atom becomes a free electron.

Majority and Minority Carriers Since most of the current carriers are electrons, silicon (or germanium) doped in this way is an *n*-type semiconductor (the *n* stands for the negative charge on an electron). The electrons are called the *majority carriers* in *n*-type material. Although the majority of current carriers in *n*-type material are electrons, there are some holes. Holes in an *n*-type material are called *minority carriers.*

p-Type Semiconductor

To increase the number of holes in intrinsic silicon, trivalent impurity atoms are added. These are atoms with three valence electrons, such as aluminum (Al), boron (B), and gallium (Ga).

As illustrated in Figure D.01-11, each trivalent atom (boron, in this case) forms covalent bonds with four adjacent silicon atoms. All three of the boron atom's valence electrons are used in the covalent bonds; and, since four electrons are required, a hole is formed with each trivalent atom. The number of holes can be controlled by the amount of trivalent impurity added to the silicon.

FIGURE D.01-11
*Trivalent impurity atom
in a silicon crystal. A
boron (B) impurity atom
is shown in the center.*

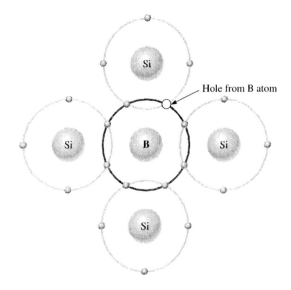

Hole from B atom

Majority and Minority Carriers   Since most of the current carriers are holes, silicon (or germanium) doped in this way is a *p*-type semiconductor because holes can be thought of as positive charges. The holes are the majority carriers in *p*-type material. Although the majority of current carriers in *p*-type material are holes, there are some electrons. Electrons in *p*-type material are the minority carriers.

REVIEW QUESTIONS

True/False

1. Current is carried by free electrons.
2. A semiconductor contains pure silicon.

Multiple Choice

3. To make an *n*-type semiconductor, dope with
   a. Trivalent atoms.
   b. Pentavalent atoms.
   c. Neutral atoms.
4. When an electron recombines, it
   a. Loses energy.
   b. Gains energy.
   c. Stays at the same energy level.
5. Electrons are the majority carriers in a(n)
   a. *n*-type semiconductor.
   b. *p*-type semiconductor.
   c. The energy gap.

D.02 Demonstrate an understanding of *pn* junctions

INTRODUCTION

If you take a block of silicon and dope half of it with a trivalent impurity and the other half with a pentavalent impurity, a boundary called the *pn* junction is formed between the resulting *p*-type and *n*-type portions. The *pn* junction is the feature that allows diodes, transistors, and other devices to work. This section and the next one will provide a basis for the discussion of the diode.

Formation of the Depletion Region

A *pn* **junction** is illustrated in Figure D.02-1. The *n* region has many conduction electrons, and the *p* region has many holes. With no external voltage, the conduction electrons in the *n* region are aimlessly drifting in all directions. At the instant of junction formation, some of the electrons near the junction drift across into the *p* region and recombine with holes near the junction as shown in part (a).

For each electron that crosses the junction and recombines with a hole, a pentavalent atom is left with a net positive charge in the *n* region near the junction, making it a positive ion. Also, when the electron recombines with a hole in the *p* region, a trivalent atom acquires net negative charge, making it a negative ion.

As a result of this recombination process, a large number of positive and negative ions builds up near the *pn* junction. As this buildup occurs, the electrons in the *n* region must overcome both the attraction of the positive ions and the repulsion of the negative

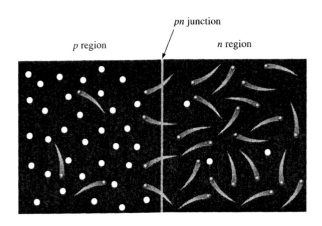

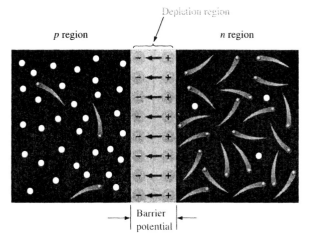

(a) At the instant of junction formation, free electrons in the *n* region near the *pn* junction begin to diffuse across the junction and fall into holes near the junction in the *p* region.

(b) For every electron that diffuses across the junction and combines with a hole, a positive charge is left in the *n* region and a negative charge is created in the *p* region, forming a barrier potential. This action continues until the voltage of the barrier repels further diffusion.

FIGURE D.02-1
Formation of the depletion region in a pn junction.

ions in order to migrate into the *p* region. Thus, as the ion layers build up, the area on both sides of the junction becomes essentially depleted of any conduction electrons or holes and is known as the *depletion region.* This condition is illustrated in Figure D.02-1(b). When an equilibrium condition is reached, the depletion region has widened to a point where no more electrons can cross the *pn* junction.

The existence of the positive and negative ions on opposite sides of the junction creates a **barrier potential** across the depletion region, as indicated in Figure D.02-1(b). The barrier potential, V_B, is the amount of energy required to move electrons through the electric field. At 25°C, it is approximately 0.7 V for silicon and 0.3 V for germanium. As the junction temperature increases, the barrier potential decreases, and vice versa.

Energy Diagram of the *pn* Junction

Now, let's look at the operation of the *pn* junction in terms of its energy level. First consider the *pn* junction at the instant of its formation. The energy bands of the trivalent impurity atoms in the *p*-type material are at a slightly higher level than those of the pentavalent impurity atom in the *n*-type material, as shown in the graph of Figure D.02-2(a). They are higher because the core attraction for the valence electrons (+3) in the trivalent atom is less than the core attraction for the valence electrons (+5) in the pentavalent atom. Thus, the trivalent valence electrons are in a slightly higher orbit and, thus, at a higher energy level.

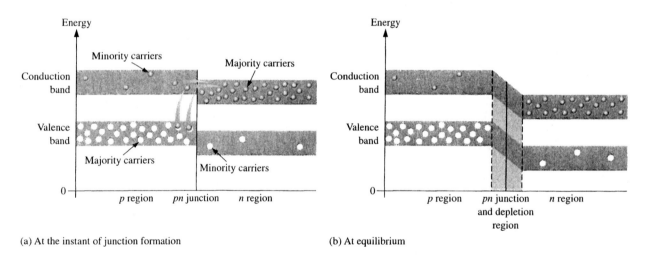

(a) At the instant of junction formation   (b) At equilibrium

FIGURE D.02-2
Energy diagrams illustrating the formation of the pn junction and depletion region.

Notice in Figure D.02-2(a) that there is some overlap of the conduction bands in the *p* and *n* regions and also some overlap of the valence bands in the *p* and *n* regions. This overlap permits the electrons of higher energy near the top of the *n*-region conduction band to begin diffusing across the junction into the lower part of the *p*-region conduction band. As soon as an electron diffuses across the junction, it recombines with a hole in the valence band. As diffusion continues, the depletion region begins to form. Also, the energy bands in the *n* region "shift" down as the electrons of higher energy are lost to diffusion. When the top of the *n*-region conduction band reaches the same level as the bottom of the *p*-region conduction band, diffusion ceases and the equilibrium condition is reached. This condition is shown in terms of energy levels in Figure D.02-2(b). There is an energy gradient across the depletion region rather than an abrupt change in energy level.

BIASING THE *pn* JUNCTION

Forward Bias

Forward bias is the condition that permits current across a *pn* junction. The term **bias** in electronics normally refers to a fixed dc voltage that sets the operating conditions for a semiconductive device. Figure D.02-3 shows a DC voltage connected in a direction to forward-bias the junction. Notice that the negative **terminal** of the battery is connected to the *n* region, and the positive terminal is connected to the *p* region.

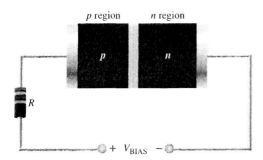

FIGURE D.02-3
Forward-bias connection. The resistor limits the forward current in order to prevent damage to the semiconductor.

A discussion of the basic operation of forward bias follows: The negative terminal of the battery pushes the conduction-band electrons in the *n* region toward the junction, while the positive terminal pushes the holes in the *p* region also toward the junction. Recall that like charges repel each other.

When it overcomes the barrier potential, the external voltage source provides the *n*-region electrons with enough energy to penetrate the depletion region and cross the junction, where they combine with the *p*-region holes. As electrons leave the *n* region, more flow in from the negative terminal of the battery. Thus, current through the *n* region is formed by the movement of conduction electrons (majority carriers) toward the junction.

Once the conduction electrons enter the *p* region and combine with holes, they become valence electrons. Then they move as valence electrons from hole to hole toward the positive connection of the battery. The movement of these valence electrons is the same as the movement of holes in the opposite direction. Thus, current in the *p* region is formed by the movement of holes (majority carriers) toward the junction. Figure D.02-4 illustrates current in a forward-biased *pn* junction.

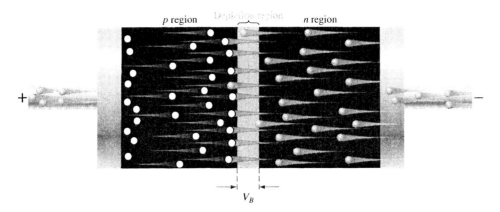

FIGURE D.02-4
Current in a forward-biased pn junction.

The Effect of the Barrier Potential on Forward Bias The effect of the barrier potential in the depletion region is to oppose forward bias. This is because the negative ions near the junction in the p region tend to prevent electrons from crossing the junction into the p region. You can think of the barrier potential effect as simulating a small battery connected in a direction to oppose the forward-bias voltage, as shown in Figure D.02-5. The resistances R_p and R_n represent the dynamic resistances of the p and n materials.

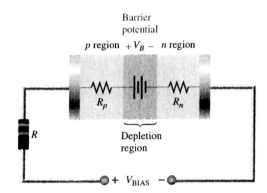

FIGURE D.02-5
Barrier potential and dynamic resistance equivalent for a pn junction.

The external bias voltage must overcome the effect of the barrier potential before the *pn* junction conducts, as illustrated in Figure D.02-6. Conduction occurs at approximately 0.7 V for silicon and 0.3 V for germanium. Once the *pn* junction is conducting in the forward direction, the voltage drop across it remains at approximately the barrier potential and changes very little with changes in forward current (I_F), as illustrated in Figure D.02-6.

Energy Diagram for Forward Bias When a *pn* junction is forward-biased, the *n*-region conduction band is raised to a higher energy level that overlaps with the *p*-region conduction band. Then large numbers of free electrons have enough energy to climb the "energy hill" and enter the *p* region where they combine with holes in the valence band. Forward bias is illustrated by the energy diagram in Figure D.02-7.

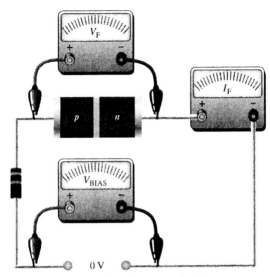

(a) No bias voltage. *PN* junction is at equilibrium.

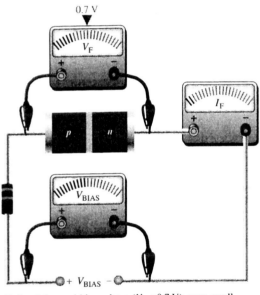

(b) Small forward-bias voltage ($V_F < 0.7$ V), very small forward current.

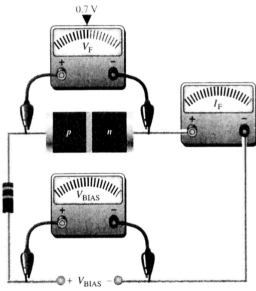

(c) Forward voltage reaches and remains at 0.7 V. Forward current continues to increase as the bias voltage is increased.

FIGURE D.02-6
Illustration of pn junction operation under forward-bias conditions.

FIGURE D.02-7
Energy diagram for forward bias, showing recombination in the p region as conduction electrons move across the junction.

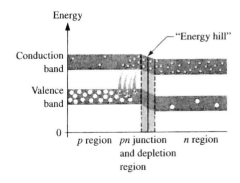

Reverse Bias

Reverse bias is the condition that prevents current across the *pn* junction. Figure D.02-8 shows a DC voltage source connected to reverse-bias the diode. Notice that the negative terminal of the battery is connected to the *p* region, and the positive terminal is connected to the *n* region.

A discussion of the basic operation for reverse bias follows: The negative terminal of the battery attracts holes in the *p* region away from the *pn* junction, while the positive terminal also attracts electrons away from the junction. As electrons and holes move away from the junction, the depletion region widens; more positive ions are created in the *n* region, and more negative ions are created in the *p* region, as shown in Figure D.02-9(a). The initial flow of majority carriers away from the junction is called *transient current* and lasts only for a very short time upon application of reverse bias.

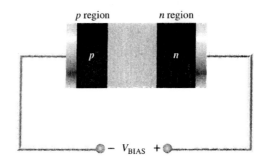

FIGURE D.02-8
Reverse-bias connection.

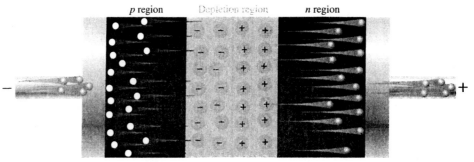

(a) There is transient current as depletion region widens.

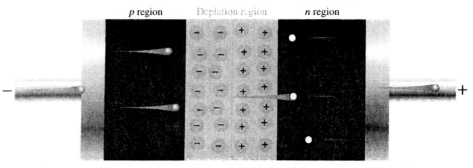

(b) Majority current ceases when barrier potential equals bias voltage. There is an extremely small reverse current due to minority carriers.

FIGURE D.02-9
Illustration of reverse bias.

The depletion region widens until the potential difference across it equals the external bias voltage. At this point, the holes and electrons stop moving away from the junction, and majority current ceases, as indicated in Figure D.02-9(b).

When the diode is reverse-biased, the depletion region effectively acts as an insulator between the layers of oppositely charged ions, forming an effective capacitance. Since the depletion region widens with increased reverse-biased voltage, the capacitance decreases, and vice versa. This internal capacitance is called the *depletion-region capacitance.*

Reverse Current   As you have learned, majority current quickly becomes zero when reverse bias is applied. There is, however, a very small current produced by minority carriers during reverse bias. Germanium, as a rule, has a greater reverse current than silicon. This current is typically in the µA or nA range. A relatively small number of thermally produced electron-hole pairs exist in the depletion region. Under the influence of the external voltage, some electrons manage to diffuse across the *pn* junction before recombination. This process establishes a small minority carrier current throughout the material.

The reverse current is dependent primarily on the junction temperature and not on the amount of reverse-biased voltage. A temperature increase causes an increase in reverse current.

Energy Diagram for Reverse Bias   When a *pn* junction is reverse-biased, the *n*-region conduction band remains at an energy level that prevents the free electrons from crossing into the *p* region. There are a few free minority electrons in the *p*-region conduction band that can easily flow down the "energy hill" into the *n* region where they combine with minority holes in the valence band. This reverse current is extremely small compared to the current in forward bias and can normally be considered negligible. Figure D.02-10 illustrates reverse bias.

FIGURE D.02-10
Energy diagram for reverse bias.

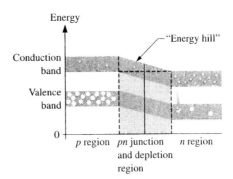

Reverse Breakdown

If the external reverse-bias voltage is increased to a large enough value, **reverse breakdown** occurs. The following describes what happens: Assume that one minority conduction-band electron acquires enough energy from the external source to accelerate it toward the positive end of the *pn* junction. During its travel, it collides with an atom and imparts enough energy to knock a valence electron into the conduction band. There are now two conduction-band electrons. Each will collide with an atom, knocking two more valence electrons into the conduction band. There are now four conduction band electrons which, in turn, knock four more into the conduction band. This rapid multiplication of conduction-band electrons, known as an *avalanche effect,* results in a rapid buildup of reverse current.

A single *pn* junction device is called a **diode.** Most diodes normally are not operated in reverse breakdown and can be damaged if they are. However, a particular type of diode known as a zener diode is specially designed for reverse-breakdown operation.

True/False

1. The voltage drop across a silicon *pn* junction is 0.7 V.
2. The *pn* junction contains a depletion region.

Multiple Choice

3. A *pn* junction is designed to conduct current when
   a. Forward biased.
   b. Reverse biased.
   c. Exposed to magnetic fields.
4. The depletion region contains a certain amount of
   a. Capacitance.
   b. Inductance.
   c. Holes.
5. If a *pn* junction is reverse biased with too much voltage,
   a. Avalanche breakdown may occur.
   b. The junction will simply remain open.
   c. Reverse current oscillation will occur.

D.03 Demonstrate an understanding of bipolar transistors

INTRODUCTION

The basic structure of the bipolar junction transistor, BJT, determines its operating characteristics. In this section, you will see how semiconductive materials are joined to form a transistor, and you will learn the standard transistor symbols. Also, you will see how important DC bias is to the operation of transistors in terms of setting up proper currents and voltages in a transistor circuit. Finally, two important parameters, α_{DC} and β_{DC}, are introduced.

The **bipolar junction transistor** (BJT) is constructed with three doped semiconducting regions separated by two *pn* junctions. The three regions are called **emitter, base,** and **collector.** The two types of bipolar transistors are shown in Figure D.03-1. One type consists of two *n* regions separated by a *p* region (*npn*), and the other consists of two *p* regions separated by an *n* region (*pnp*).

FIGURE D.03-1
Basic construction of bipolar transistors.

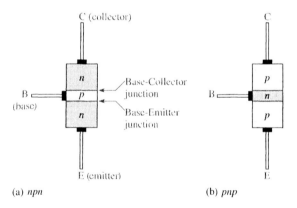

(a) *npn*   (b) *pnp*

The *pn* junction joining the base region and the emitter region is called the *base-emitter junction.* The junction joining the base region and the collector region is called the *base-collector* junction, as indicated in Figure D.03-1(a). A wire lead connects to each of the three regions, as shown. These leads are labeled E, B, and C for emitter, base, and collector, respectively. The base material is lightly doped and very narrow compared to the heavily doped emitter and collector materials. The reason for this is discussed in the next section.

Figure D.03-2 shows the schematic symbols for the *npn* and *pnp* bipolar transistors. The term **bipolar** refers to the use of both holes and electrons as carriers in the transistor structure.

FIGURE D.03-2
Transistor symbols.

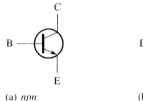

(a) *npn*   (b) *pnp*

Transistor Biasing

In order for the **transistor** to operate properly as an amplifier, the two *pn* junctions must be correctly biased with external DC voltages. We will use the *npn* transistor to illustrate transistor biasing. The operation of the *pnp* is the same as for the *npn* except that the roles of the electrons and holes, the bias voltage polarities, and the current directions are all reversed. Figure D.03-3 shows the proper bias arrangement for both *npn* and *pnp* transistors. Notice that in both cases the base-emitter (BE) junction is forward-biased and the base-collector (BC) junction is reverse-biased. This is called *forward-reverse bias.*

FIGURE D.03-3
Forward-reverse bias of a bipolar transistor.

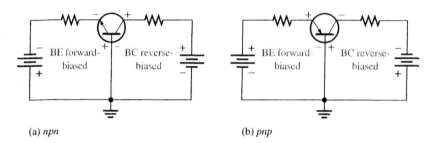

(a) *npn*   (b) *pnp*

To illustrate transistor action, let's examine what happens inside the *npn* transistor when the junctions are forward-reverse biased. The forward bias from base to emitter narrows the BE depletion region, and the reverse bias from base to collector widens the BC depletion region, as depicted in Figure D.03-4. The heavily doped *n*-type emitter region is teeming with conduction-band (free) electrons that easily diffuse through the forward-biased BE junction into the *p*-type base region where they become minority carriers, just as in a forward-biased diode. The base region is lightly doped and very thin so that it has a limited number of holes. Thus, only a small percentage of all the electrons flowing through the BE junction can combine with the available holes in the base. These relatively few recombined electrons flow out of the base lead as valence electrons, forming the small base electron current, as shown in Figure D.03-4.

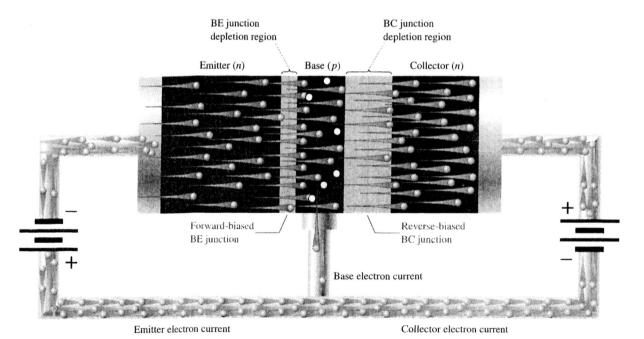

FIGURE D.03-4
Current flow in a properly biased transistor.

Most of the electrons flowing from the emitter into the thin, lightly doped base region do not recombine but diffuse into the BC depletion region. Once in this region they are pulled through the reverse-biased BC junction by the electric field set up by the force of attraction between the positive and negative ions. Actually, you can think of the electrons as being pulled across the reverse-biased BC junction by the attraction of the collector supply voltage. The electrons now move through the collector region, out through the collector lead, and into the positive terminal of the collector voltage source. This forms the collector electron current, as shown in Figure D.03-4.

Transistor Currents

The directions of current in an *npn* and a *pnp* transistor are as shown in Figure D.03-5. An examination of these diagrams shows that the emitter current is the sum of the collector and base currents, expressed as follows:

$$I_E = I_C + I_B$$

As mentioned before, I_B is very small compared to I_E or I_C. The capital-letter subscripts indicate DC values.

These direct currents (emitter, base, and collector) are also related by two parameters: the DC **alpha** (α_{DC}), which is the ratio I_C/I_E and the DC **beta** (β_{DC}), which is the ratio I_C/I_B. β_{DC} is the direct current gain and is usually designated as h_{FE} on the transistor data sheets.

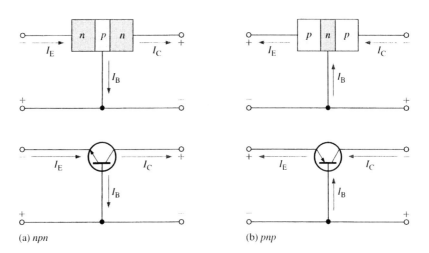

(a) *npn*　　　　　　　(b) *pnp*

FIGURE D.03-5
Transistor currents.

EXAMPLE D.03-1

Find I_B, I_C, I_E, V_B, and V_C in Figure D.03-6, where β_{DC} is 50.

FIGURE D.03-6

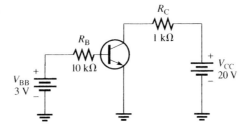

Solution   Since the emitter is at ground, $V_B = 0.7$ V. The drop across R_B is $V_{BB} - V_B$, so I_B is calculated as follows:

$$I_B = \frac{V_{BB} - V_B}{R_B} = \frac{3 \text{ V} - 0.7 \text{ V}}{10 \text{ k}\Omega} = 230 \text{ }\mu\text{A}$$

Now you can find I_C, I_E, and V_C.

$$I_C = \beta_{DC}I_B = 50(230 \text{ }\mu\text{A}) = 11.5 \text{ mA}$$
$$I_E = I_C + I_B = 11.5 \text{ mA} + 230 \text{ }\mu\text{A} = 11.73 \text{ mA}$$
$$V_C = V_{CC} - I_C R_C = 20 \text{ V} - (11.5 \text{ mA})(1 \text{ k}\Omega) = 8.5 \text{ V}$$

EXAMPLE D.03-2

Determine V_B, V_E, V_C, V_{CE}, I_B, I_E, and I_C in Figure D.03-7.

FIGURE D.03-7

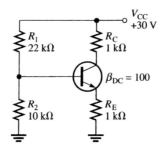

Solution   The input resistance at the base is

$$R_{IN} \cong \beta_{DC}R_E = 100(1 \text{ k}\Omega) = 100 \text{ k}\Omega$$

Since R_{IN} is ten times greater than R_2, you can neglect it, and the base voltage is approximately

$$V_B \cong \left(\frac{R_2}{R_1 + R_2}\right)V_{CC} = \left(\frac{10 \text{ k}\Omega}{32 \text{ k}\Omega}\right)30 \text{ V} = 9.38 \text{ V}$$

Therefore, $V_E = V_B - 0.7 \text{ V} = 8.68 \text{ V}$.

Now that you know V_E, you can find I_E by Ohm's law.

$$I_E = \frac{V_E}{R_E} = \frac{8.68 \text{ V}}{1 \text{ k}\Omega} = 8.68 \text{ mA}$$

Since α_{DC} is so close to 1 for most transistors, it is a good approximation to assume that $I_C \cong I_E$. Thus,

$$I_C \cong 8.68 \text{ mA}$$

Use $I_C = \beta_{DC}I_B$ and solve for I_B.

$$I_B = \frac{I_C}{\beta_{DC}} = \frac{8.68 \text{ mA}}{100} = 86.8 \text{ }\mu\text{A}$$

Since you know I_C, you can find V_C.

$$V_C = V_{CC} - I_C R_C = 30 \text{ V} - (8.68 \text{ mA})(1 \text{ k}\Omega) = 30 \text{ V} - 8.68 \text{ V} = 21.3 \text{ V}$$

Since V_{CE} is the collector-to-emitter voltage, it is the difference of V_C and V_E.

$$V_{CE} = V_C - V_E = 21.3 \text{ V} - 8.68 \text{ V} = 12.6 \text{ V}$$

Collector Characteristic Curves

Using a circuit like that shown in Figure D.03-8(a), you can generate a set of collector characteristic curves that show how the collector current, I_C, varies with the collector-to-emitter voltage, V_{CE}, for specified values of base current, I_B. Notice in the circuit diagram that both V_{BB} and V_{CC} are variable sources of voltage.

Assume that V_{BB} is set to produce a certain value of I_B, and V_{CC} is zero. For this condition, both the base-emitter junction and the base-collector junction are forward-biased because the base is at approximately 0.7 V while the emitter and the collector are at 0 V.

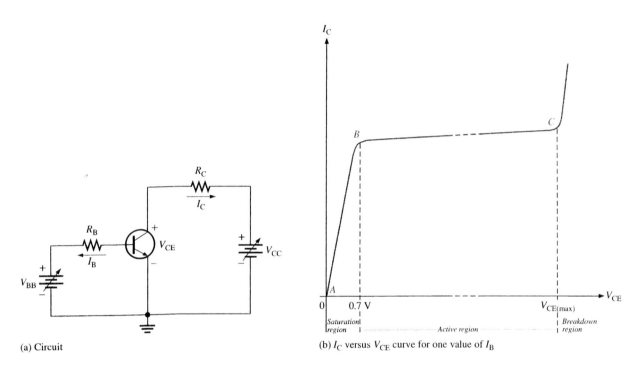

(a) Circuit

(b) I_C versus V_{CE} curve for one value of I_B

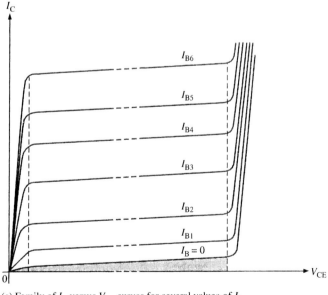

(c) Family of I_C versus V_{CE} curves for several values of I_B
($I_{B1} < I_{B2} < I_{B3,}$ etc.)

FIGURE D.03-8
Collector characteristic curves.

The base current is through the base-emitter junction because of the low impedance path to ground and, therefore, I_C is zero. When both junctions are forward-biased, the transistor is in the **saturation** region of its operation.

As V_{CC} is increased, V_{CE} increases gradually as the collector current increases. This is indicated by the portion of the characteristic curve between points A and B in Figure D.03-8(b). I_C increases as V_{CC} is increased because V_{CE} remains less than 0.7 V due to the forward-biased base-collector junction.

When V_{CE} exceeds 0.7 V, the base-collector junction becomes reverse-biased and the transistor goes into the *active* or **linear** region of its operation. Once the base-collector junction is reverse-biased, I_C levels off and remains essentially constant for a given value of I_B as V_{CE} continues to increase. Actually, I_C increases very slightly as V_{CE} increases due to widening of the base-collector depletion region. This results in fewer holes for recombination in the base region which effectively causes a slight increase in β_{DC}. This is shown by the portion of the characteristic curve between points B and C in Figure D.03-8(b). For this portion of the characteristic curve, the value of I_C is determined only by the relationship expressed as $I_C = \beta_{DC}I_B$.

When V_{CE} reaches a sufficiently high voltage, the reverse-biased base-collector junction goes into breakdown; and the collector current increases rapidly as indicated by the part of the curve to the right of point C in Figure D.03-8(b). A transistor should never be operated in this breakdown region.

A family of collector characteristic curves is produced when I_C versus V_{CE} is plotted for several values of I_B, as illustrated in Figure D.03-8(c). When $I_B = 0$, the transistor is in the **cutoff** region although there is a very small collector leakage current as indicated. The amount of collector leakage current for $I_B = 0$ is exaggerated on the graph for purposes of illustration.

Cutoff

As previously mentioned, when $I_B = 0$, the transistor is in the cutoff region of its operation. This is shown in Figure D.03-9 with the base lead open, resulting in a base current of zero. Under this condition, there is a very small amount of collector leakage current, I_{CEO}, due mainly to thermally produced carriers. Because I_{CEO} is extremely small, it will usually be neglected in circuit analysis so that $V_{CE} = V_{CC}$. In cutoff, both the base-emitter and the base-collector junctions are reverse-biased.

FIGURE D.03-9
Collector leakage current (I_{CEO}) in cutoff.

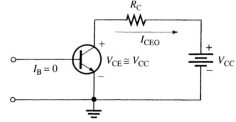

Saturation

When the base current in Figure D.03-10 is increased, the collector current also increases ($I_C = \beta_{DC}I_B$) and V_{CE} decreases as a result of more drop across the collector resistor ($V_{CE} = V_{CC} - I_C R_C$). This is illustrated in Figure D.03-10. When V_{CE} reaches its saturation value, $V_{CE(sat)}$, the base-collector junction becomes forward-biased and I_C can increase no further even with a continued increase in I_B. At the point of saturation, the relation $I_C = \beta_{DC}I_B$ is no longer valid. $V_{CE(sat)}$ for a transistor occurs somewhere below

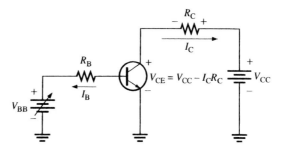

FIGURE D.03-10

As I_B increases due to increasing V_{BB}, I_C also increases and V_{CE} decreases due to the increased voltage drop across R_C.

the knee of the collector curves, and it is usually only a few tenths of a volt for silicon transistors.

DC Load Line

Cutoff and saturation can be illustrated in relation to the collector characteristic curves by the use of a load line. Figure D.03-11 shows a DC load line drawn on a family of curves connecting the cutoff point and the saturation point. The bottom of the load line is at ideal cutoff where $I_C = 0$ and $V_{CE} = V_{CC}$. The top of the load line is at saturation where $I_C = I_{C(sat)}$ and $V_{CE} = V_{CE(sat)}$. In between cutoff and saturation along the load line is the active region of the transistor's operation. Load line operation is discussed more later.

FIGURE D.03-11

DC load line on a family of collector characteristic curves illustrating the cutoff and saturation conditions.

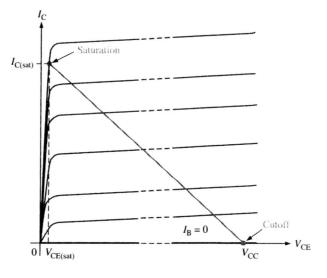

Transistor Data Sheet

A partial data sheet for the 2N3903 and 2N3904 *npn* transistors is shown in Figure D.03-12. Notice that the maximum collector-emitter voltage (V_{CEO}) is 40 V. The "O" in the subscript indicates that the voltage is measured from collector (C) to emitter (E) with the base open (O). In the text, we use $V_{CE(max)}$ for purposes of clarity. Also notice that the maximum collector current is 200 mA.

The β_{DC} (h_{FE}) is specified for several values of I_C and, as you can see, h_{FE} varies with I_C as we previously discussed.

The collector-emitter saturation voltage, $V_{CE(sat)}$, is 0.2 V maximum for $I_{C(sat)} = 10$ mA and increases with the current.

Maximum Ratings

Rating	Symbol	Value	Unit
Collector-Emitter voltage	V_{CEO}	40	V dc
Collector-Base voltage	V_{CBO}	60	V dc
Emitter-Base voltage	V_{EBO}	6.0	V dc
Collector current — continuous	I_C	200	mA dc
Total device dissipation @ $T_A = 25°C$ Derate above 25°C	P_D	625 5.0	mW mW/C°
Total device dissipation @ $T_C = 25°C$ Derate above 25°C	P_D	1.5 12	Watts mW/C°
Operating and storage junction Temperature range	T_J, T_{stg}	−55 to +150	°C

Thermal Characteristics

Characteristic	Symbol	Max	Unit
Thermal resistance, junction to case	$R_{\theta JC}$	83.3	C°/W
Thermal resistance, junction to ambient	$R_{\theta JA}$	200	C°/W

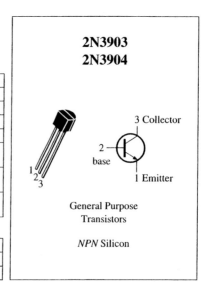

2N3903
2N3904

General Purpose
Transistors

NPN Silicon

Electrical Characteristics ($T_A = 25°C$ unless otherwise noted.)

Characteristic		Symbol	Min	Max	Unit
OFF Characteristics					
Collector-Emitter breakdown voltage ($I_C = 1.0$ mA dc, $I_B = 0$)		$V_{(BR)CEO}$	40	–	V dc
Collector-Base breakdown voltage ($I_C = 10$ μA dc, $I_E = 0$)		$V_{(BR)CBO}$	60	–	V dc
Emitter-Base breakdown voltage ($I_E = 10$ μA dc, $I_C = 0$)		$V_{(BR)EBO}$	6.0	–	V dc
Base cutoff current ($V_{CE} = 30$ V dc, $V_{EB} = 3.0$ V dc)		I_{BL}	–	50	nA dc
Collector cutoff current ($V_{CE} = 30$ V dc, $V_{EB} = 3.0$ V dc)		I_{CEX}	–	50	nA dc
ON Characteristics					
DC current gain ($I_C = 0.1$ mA dc, $V_{CE} = 1.0$ V dc)	2N3903 2N3904	h_{FE}	20 40	– –	–
($I_C = 1.0$ mA dc, $V_{CE} = 1.0$ V dc)	2N3903 2N3904		35 70	– –	
($I_C = 10$ mA dc, $V_{CE} = 1.0$ V dc)	2N3903 2N3904		50 100	150 300	
($I_C = 50$ mA dc, $V_{CE} = 1.0$ V dc)	2N3903 2N3904		30 60	– –	
($I_C = 100$ mA dc, $V_{CE} = 1.0$ V dc)	2N3903 2N3904		15 30	– –	
Collector-Emitter saturation voltage ($I_C = 10$ mA dc, $I_B = 1.0$ mA dc) ($I_C = 50$ mA dc, $I_B = 5.0$ mA dc)		$V_{CE(sat)}$	– –	0.2 0.3	V dc
Base-Emitter saturation voltage ($I_C = 10$ mA dc, $I_B = 1.0$ mA dc) ($I_C = 50$ mA dc, $I_B = 5.0$ mA dc)		$V_{BE(sat)}$	0.65 –	0.85 0.95	V dc

FIGURE D.03-12
Partial transistor data sheet.

REVIEW QUESTIONS

True/False

1. The base current controls the collector current.

2. Bipolar transistors must be biased properly before they can be used.

Multiple Choice

3. The base-emitter junction must be

    a. Forward-biased to turn transistor on.

    b. Reverse-biased to turn transistor on.

    c. Rapidly forward- and reverse-biased to enable current gain.

4. Three modes of transistor operation are

    a. Cutoff, amplification, and normal.

    b. Cutoff, saturation, and normal.

    c. Cutoff, saturation, and active.

5. In a power transistor, the metal case is typically the

    a. Base.

    b. Collector.

    c. Emitter.

D.04 Demonstrate an understanding of field effect transistors (FETs/MOSFETs)

INTRODUCTION

Recall that the bipolar junction transistor (BJT) is a current-controlled device; that is, the base current controls the amount of collector current. The *field-effect transistor (FET)* is different; it is a voltage-controlled device in which the voltage at the gate terminal controls the amount of current through the device. Also, compared to the BJT, the FET has a very high input resistance, which makes it superior in certain applications.

The **junction field-effect transistor (JFET)** is a type of FET that operates with a reverse-biased junction to control current in the channel. Depending on their structure, JFETs fall into either of two categories: *n* channel or *p* channel. Figure D.04-1(a) shows the basic structure of an *n* channel JFET. Wire leads are connected to each end of the *n* channel; the **drain** is at the upper end and the **source** is at the lower end. Two *p*-type regions are diffused in the *n*-type material to form a channel, and both *p*-type regions are connected to the **gate** lead. In the remaining structure diagrams, the interconnection of both *p*-type regions is omitted for simplicity, with a connection to only one shown. A *p*-channel JFET is shown in Figure D.04-1(b).

FIGURE D.04-1
Basic structure of the two types of JFET.

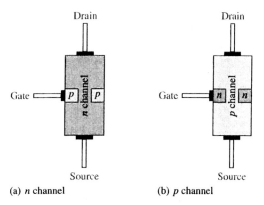

(a) *n* channel   (b) *p* channel

Basic Operation

To illustrate the operation of a JFET, bias voltages are shown applied to an *n*-channel device in Figure D.04-2(a). V_{DD} provides a drain-to-source voltage and supplies current from drain to source. V_{GG} sets the reverse-biased voltage between the gate and the source, as shown.

The JFET is always operated with the gate-to-source pn junction reverse-biased. Reverse-biasing of the gate-source junction with a negative gate voltage produces a depletion region in the *n* channel and thus increases its resistance. The channel width can be controlled by varying the gate voltage; thereby the amount of drain current, I_D, can also be controlled. This concept is illustrated in Figure D.04-2(b) and (c).

The white areas surrounding the *p* material of the gate represent the depletion region created by the reverse bias. This depletion region is wider toward the drain end of the channel because the reverse-biased voltage between the gate and the drain is greater than that between the gate and the source.

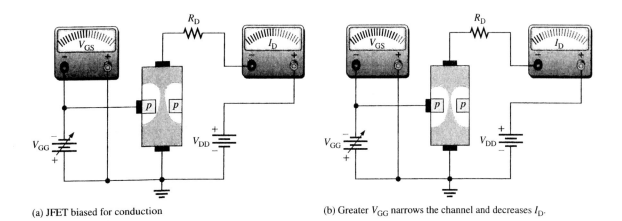

(a) JFET biased for conduction

(b) Greater V_{GG} narrows the channel and decreases I_D.

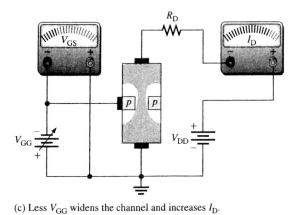

(c) Less V_{GG} widens the channel and increases I_D.

FIGURE D.04-2
Effects of V_{GG} on channel width and drain current ($V_{GG} = V_{GS}$).

FIGURE D.04-3
JFET schematic symbols.

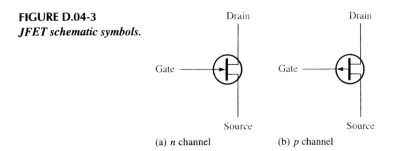

(a) *n* channel   (b) *p* channel

JFET Symbols

The schematic symbols for both *n*-channel and *p*-channel JFETs are shown in Figure D.04-3. Notice that the arrow on the gate points "in" for *n* channel and "out" for *p* channel.

JFET CHARACTERISTICS

First let's consider the case where the gate-to-source voltage is 0 ($V_{GS} = 0$ V). This voltage is produced by shorting the gate to the source, as in Figure D.04-4(a) where both are grounded. As V_{DD} (and thus V_{DS}) is increased from 0 V, I_D will increase proportionally, as shown in the graph of Figure D.04-4(b) between points A and B. In this region, the channel

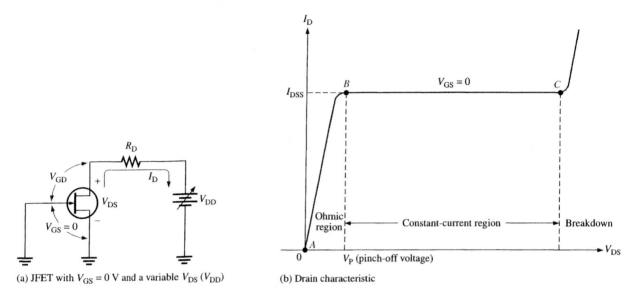

(a) JFET with $V_{GS} = 0$ V and a variable V_{DS} (V_{DD})

(b) Drain characteristic

FIGURE D.04-4
The drain characteristic curve of a JFET for $V_{GS} = 0$ V, showing pinch-off.

resistance is essentially constant because the depletion region is not large enough to have significant effect. This region is called the *ohmic region* because V_{DS} and I_D are related by Ohm's law.

At point B in Figure D.04-4(b), the curve levels off and I_D becomes essentially constant. As V_{DS} increases from point B to point C, the reverse-bias voltage from gate to drain (V_{GD}) produces a depletion region large enough to offset the increase in V_{DS}, thus keeping I_D relatively constant.

Pinch-Off Voltage

For $V_{GS} = 0$ V, the value of V_{DS} at which I_D becomes essentially constant (point B on the curve in Figure D.04-4(b)) is the **pinch-off voltage,** V_P. For a given JFET, V_P has a fixed value. As you can see, a continued increase in V_{DS} above the pinch-off voltage produces an almost constant drain current. This value of drain current is I_{DSS} (*D*rain to *S*ource current with gate *S*horted) and is always specified on JFET data sheets. I_{DSS} is the *maximum* drain current that a specific JFET can produce regardless of the external circuit, and it is always specified for the condition, $V_{GS} = 0$ V.

Continuing along the graph in Figure D.04-4(b), breakdown occurs at point C when I_D begins to increase very rapidly with any further increase in V_{DS}. Breakdown can result in irreversible damage to the device, so JFETs are always operated below breakdown and within the constant-current region (between points B and C on the graph).

V_{GS} Controls I_D

Let's connect a bias voltage, V_{GG}, from gate to source as shown in Figure D.04-5(a). As V_{GS} is set to increasingly more negative values by adjusting V_{GG}, a family of drain characteristic curves is produced as shown in Figure D.04-5(b). Notice that I_D decreases as the magnitude of V_{GS} is increased to larger negative values. Also notice that, for each increase in V_{GS}, the JFET reaches pinch-off (where constant current begins) at values of V_{DS} less than V_P. So, the amount of drain current is controlled by V_{GS}.

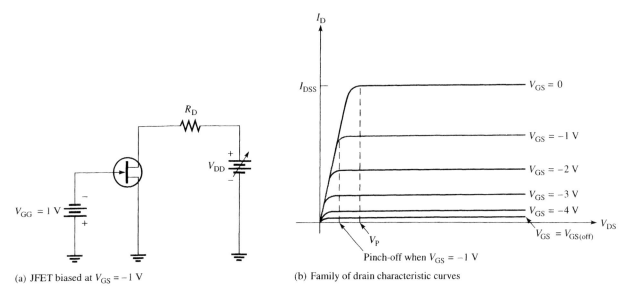

(a) JFET biased at $V_{GS} = -1$ V

(b) Family of drain characteristic curves

FIGURE D.04-5
Pinch-off occurs at a lower V_{DS} as V_{GS} is increased to more negative values.

Cutoff Voltage

The value of V_{GS} that makes I_D approximately zero is the cutoff voltage, $V_{GS(off)}$. The JFET must be operated between $V_{GS} = 0$ V and $V_{GS(off)}$. For this range of gate-to-source voltages, I_D will vary from a maximum of I_{DSS} to a minimum of almost zero.

As you have seen, for an *n*-channel JFET, the more negative V_{GS} is, the smaller I_D becomes in the constant-current region. When V_{GS} has a sufficiently large negative value, I_D is reduced to zero. This cutoff effect is caused by the widening of the depletion region to a point where it completely closes the channel as shown in Figure D.04-6.

The basic operation of a *p*-channel JFET is the same as for an *n*-channel device except that it requires a negative V_{DD} and a positive V_{GS}.

FIGURE D.04-6
JFET at cutoff.

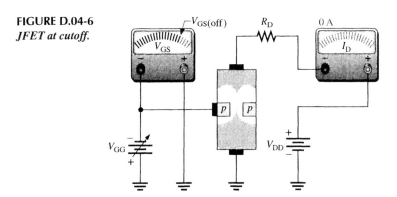

Comparison of Pinch-Off and Cutoff

As you have seen, there is definitely a difference between pinch-off and cutoff. There is also a connection. V_P is the value of V_{DS} at which the drain current becomes constant and is always measured at $V_{GS} = 0$ V. However, pinch-off occurs for V_{DS} values less than V_P when V_{GS} is nonzero. So, although V_P is a constant, the minimum value of V_{DS} at which I_D becomes constant varies with V_{GS}.

$V_{GS(off)}$ and V_P are always equal in magnitude but opposite in sign. A data sheet usually will give either $V_{GS(off)}$ or V_P, but not both. However, when you know one, you have the other. For example, if $V_{GS(off)} = -5$ V, then $V_P = +5$ V.

EXAMPLE D.04-1

For the JFET in Figure D.04-7, $V_{GS(off)} = -3.5$ V and $I_{DSS} = 6$ mA. Determine the minimum value of V_{DD} required to put the device in the constant-current region of operation.

FIGURE D.04-7

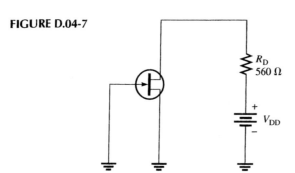

Solution   Since $V_{GS(off)} = -3.5$ V, $V_P = 3.5$ V. The minimum value of V_{DS} for the JFET to be in its constant-current region is

$$V_{DS} = V_P = 3.5 \text{ V}$$

In the constant-current region with $V_{GS} = 0$ V,

$$I_D = I_{DSS} = 6 \text{ mA}$$

The drop across the drain resistor is

$$V_{R_D} = (6 \text{ mA})(560 \text{ } \Omega) = 3.36 \text{ V}$$

Applying Kirchhoff's law around the drain circuit gives

$$V_{DD} = V_{DS} + V_{R_D} = 3.5 \text{ V} + 3.36 \text{ V} = 6.86 \text{ V}$$

This is the value of V_{DD} to make $V_{DS} = V_P$ and put the device in the constant-current region.

JFET Input Resistance and Capacitance

A JFET operates with its gate-source junction reverse-biased. Therefore, the input resistance at the gate is very high. This high input resistance is one advantage of the JFET over the bipolar transistor. (Recall that a bipolar transistor operates with a forward-biased base-emitter junction.)

JFET data sheets often specify the input resistance by giving a value for the gate reverse current, I_{GSS}, at a certain gate-to-source voltage. The input resistance can then be determined using the following equation. The vertical lines indicate an absolute value (no sign).

$$R_{IN} = \left| \frac{V_{GS}}{I_{GSS}} \right|$$

For example, the 2N3970 data sheet lists a maximum I_{GSS} of 250 pA for $V_{GS} = -20$ V at 25°C. I_{GSS} increases with temperature, so the input resistance decreases.

The input capacitance C_{iss} of a JFET is considerably greater than that of a bipolar transistor because the JFET operates with a reverse-biased pn junction. Recall that a reverse-biased pn junction acts as a capacitor whose capacitance depends on the amount of reverse voltage. For example, the 2N3970 has a maximum C_{iss} of 25 pF for $V_{GS} = 0$ V.

THE METAL-OXIDE SEMICONDUCTOR FET (MOSFET)

The metal-oxide semiconductor field-effect transistor (MOSFET) is the second category of field-effect transistor. The MOSFET differs from the JFET in that it has no *pn* junction structure; instead, the gate of the MOSFET is insulated from the channel by a silicon dioxide (SiO_2) layer. The two basic types of MOSFETs are depletion (D) and enhancement (E).

Depletion MOSFET (D-MOSFET)

One type of **MOSFET** is the depletion MOSFET (D-MOSFET), and Figure D.04-8 illustrates its basic structure. The drain and source are diffused into the substrate material and then connected by a narrow channel adjacent to the insulated gate. Both *n*-channel and *p*-channel devices are shown in the figure. We will use the *n*-channel device to describe the basic operation. The *p*-channel operation is the same, except the voltage polarities are opposite those of the *n*-channel device.

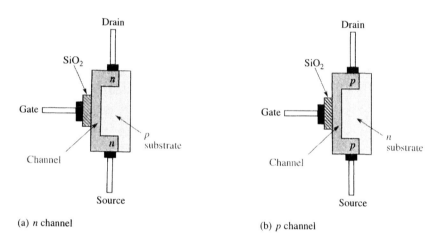

(a) *n* channel (b) *p* channel

FIGURE D.04-8
Basic structure of D-MOSFETs.

The D-MOSFET can be operated in either of two modes—the depletion mode or the enhancement mode—and is sometimes called a *depletion/enhancement MOSFET.* Since the gate is insulated from the channel, either a positive or a negative gate voltage can be applied. The *n*-channel MOSFET operates in the depletion mode when a negative gate-to-source voltage is applied and in the enhancement mode when a positive gate-to-source voltage is applied.

Depletion Mode   Visualize the gate as one plate of a parallel plate capacitor and the channel as the other plate. The silicon dioxide insulating layer is the dielectric. With a negative gate voltage, the negative charges on the gate repel conduction electrons from the channel, leaving positive ions in their place. Thereby, the *n*-channel is depleted of some of its electrons, thus decreasing the channel conductivity. The greater the negative voltage on the gate, the greater the depletion of *n*-channel electrons. At a sufficiently negative gate-to-source voltage, $V_{GS(off)}$, the channel is totally depleted and the drain current is zero. This depletion mode is illustrated in Figure D.04-9(a).

Like the *n*-channel JFET, the *n*-channel D-MOSFET conducts drain current for gate-to-source voltages between $V_{GS(off)}$ and 0 V. In addition, the D-MOSFET conducts for values of V_{GS} above 0 V.

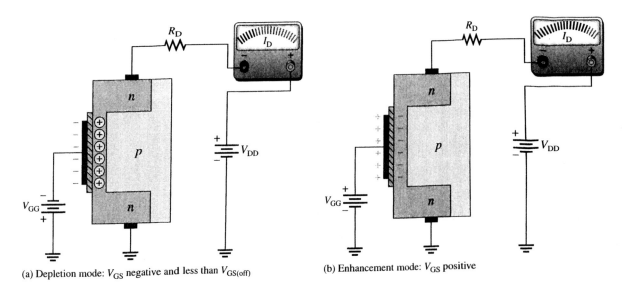

(a) Depletion mode: V_{GS} negative and less than $V_{GS(off)}$

(b) Enhancement mode: V_{GS} positive

FIGURE D.04-9
Operation of n-channel D-MOSFET.

Enhancement Mode   With a positive gate voltage, more conduction electrons are attracted into the channel, thus increasing (enhancing) the channel conductivity, as illustrated in Figure D.04-9(b).

D-MOSFET Symbols   The schematic symbols for both the *n*-channel and the *p*-channel depletion/enhancement MOSFETs are shown in Figure D.04-10. The substrate, indicated by the arrow, is normally (but not always) connected internally to the source. An inward substrate arrow is for *n*-channel, and an outward arrow is for *p*-channel.

FIGURE D.04-10
D-MOSFET schematic symbols.

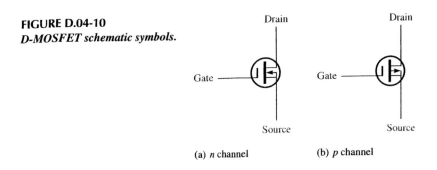

(a) *n* channel

(b) *p* channel

Enhancement MOSFET (E-MOSFET)

This type of MOSFET operates *only* in the enhancement mode and has no depletion mode. It differs in construction from the D-MOSFET in that it has no structural channel. Notice in Figure D.04-11(a) that the substrate extends completely to the SiO_2 layer.

For an *n*-channel device, a positive gate voltage above a threshold value, $V_{GS(th)}$, *induces* a channel by creating a thin layer of negative charges in the substrate region adjacent to the SiO_2 layer, as shown in Figure D.04-11(b). The conductivity of the channel is enhanced by increasing the gate-to-source voltage, thus pulling more electrons into the channel. For any gate voltage below the threshold value, there is no channel.

EXAMPLE D.04-2

Find V_{DS} and V_{GS} in Figure D.04-14, given that $I_D = 5$ mA.

FIGURE D.04-14

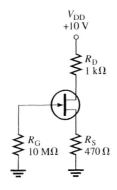

Solution $V_S = I_D R_S = (5 \text{ mA})(470 \ \Omega) = 2.35 \text{ V}$

$V_D = V_{DD} - I_D R_D = 10 \text{ V} - (5 \text{ mA})(1 \text{ k}\Omega) = 10 \text{ V} - 5 \text{ V} = 5 \text{ V}$

Therefore,

$$V_{DS} = V_D - V_S = 5 \text{ V} - 2.35 \text{ V} = 2.65 \text{ V}$$

Since $V_G = 0$ V,

$$V_{GS} = V_G - V_S = 0 \text{ V} - 2.35 \text{ V} = -2.35 \text{ V}$$

D-MOSFET Bias

Recall that depletion/enhancement MOSFETs can be operated with either positive or negative values of V_{GS}. A simple bias method is to set $V_{GS} = 0$ V so that an AC signal at the gate varies the gate-to-source voltage above and below this bias point. A MOSFET with zero bias is shown in Figure D.04-15.

Since $V_{GS} = 0$ V, $I_D = I_{DSS}$ as indicated. The drain-to-source voltage is expressed as follows:

$$V_{DS} = V_{DD} - I_{DSS} R_D$$

FIGURE D.04-15
A zero-biased D-MOSFET.

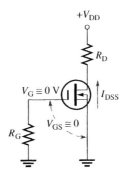

EXAMPLE D.04-3

Determine the drain current in the circuit of Figure D.04-16. The MOSFET data sheet gives $V_{GS(off)} = -8$ V and $I_{DSS} = 12$ mA.

FIGURE D.04-16

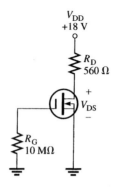

Solution   Since $I_D = I_{DSS} = 12$ mA, the drain-to-source voltage is calculated as follows:

$$V_{DS} = V_{DD} - I_{DSS}R_D = 18 \text{ V} - (12 \text{ mA})(560 \text{ }\Omega) = 11.3 \text{ V}$$

E-MOSFET Bias

Recall that enhancement-only MOSFETs must have a V_{GS} greater than the threshold value, $V_{GS(th)}$. Figure D.04-17 shows two ways to bias an E-MOSFET where an *n*-channel device is used for illustration. In either bias arrangement, the purpose is to make the gate voltage more positive than the source by an amount exceeding $V_{GS(th)}$.

FIGURE D.04-17
E-MOSFET biasing arrangements.

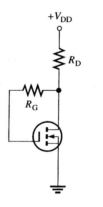

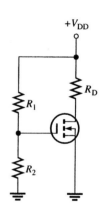

(a) Drain-feedback bias   (b) Voltage-divider bias

In the drain feedback bias circuit in Figure D.04-17(a), there is negligible gate current and, therefore, no voltage drop across R_G. As a result, $V_{GS} = V_{DS}$.

Equations for the voltage-divider bias in Figure D.04-17(b) are as follows:

$$V_{GS} = \left(\frac{R_2}{R_1 + R_2}\right)V_{DD}$$

$$V_{DS} = V_{DD} - I_D R_D$$

MOSFET Data Sheet

Figure D.04-18 shows a partial data sheet for a 2N3796/2N3797 MOSFET. What are the significant parameters listed?

Maximum Ratings

Rating	Symbol	Value	Unit
Drain-Source voltage 2N3796 2N3797	V_{DS}	25 20	V dc
Gate-Source voltage	V_{GS}	±10	V dc
Drain current	I_D	20	mA dc
Total device dissipation @ $T_A = 25°C$ Derate above 25°C	P_D	200 1.14	mW mW/C°
Junction temperature range	T_J	+175	°C
Storage channel temperature range	T_{stg}	–65 to +200	°C

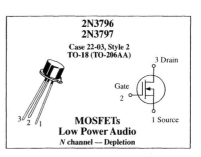

2N3796
2N3797

Case 22-03, Style 2
TO-18 (TO-206AA)

3 Drain

Gate
2

1 Source

3 2 1

MOSFETs
Low Power Audio
N channel — Depletion

Electrical Characteristics ($T_A = 25°C$ unless otherwise noted.)

Characteristic		Symbol	Min	Typ	Max	Unit
OFF Characteristics						
Drain-Source breakdown voltage ($V_{GS} = -4.0$ V, $I_D = 5.0$ µA) ($V_{GS} = -7.0$ V, $I_D = 5.0$ µA)	2N3796 2N3797	$V_{(BR)DSX}$	25 20	30 25	– –	V dc
Gate reverse current ($V_{GS} = -10$ V, $V_{DS} = 0$) ($V_{GS} = -10$ V, $V_{DS} = 0, T_A = 150°C$)		I_{GSS}	– –	– –	1.0 200	pA dc
Gate-Source cutoff voltage ($I_D = 0.5$ µA, $V_{DS} = 10$ V) ($I_D = 2.0$ µA, $V_{DS} = 10$ V)	2N3796 2N3797	$V_{GS(off)}$	– –	–3.0 –5.0	–4.0 –7.0	V dc
Drain-Gate reverse current ($V_{DG} = 10$ V, $I_S = 0$)		I_{DGO}	–	–	1.0	pA dc
ON Characteristics						
Zero-Gate-Voltage drain current ($V_{DS} = 10$ V, $V_{GS} = 0$)	2N3796 2N3797	I_{DSS}	0.5 2.0	1.5 2.9	3.0 6.0	mA dc
On-State drain current ($V_{DS} = 10$ V, $V_{GS} = +3.5$ V)	2N3796 2N3797	$I_{D(on)}$	7.0 9.0	8.3 14	14 18	mA dc
Small-Signal Characteristics						
Forward-transfer admittance ($V_{DS} = 10$ V, $V_{GS} = 0, f = 1.0$ kHz)	2N3796 2N3797	$\|y_{fs}\|$	900 1500	1200 2300	1800 3000	µmhos or µS
($V_{DS} = 10$ V, $V_{GS} = 0, f = 1.0$ MHz)	2N3796 2N3797		900 1500	– –	– –	
Output admittance ($V_{DS} = 10$ V, $V_{GS} = 0, f = 1.0$ kHz)	2N3796 2N3797	$\|y_{os}\|$	– –	12 27	25 60	µmhos or µS
Input capacitance ($V_{DS} = 10$ V, $V_{GS} = 0, f = 1.0$ MHz)	2N3796 2N3797	C_{iss}	– –	5.0 6.0	7.0 8.0	pF
Reverse transfer capacitance ($V_{DS} = 10$ V, $V_{GS} = 0, f = 1.0$ MHz)		C_{rss}	–	0.5	0.8	pF
Functional Characteristics						
Noise figure ($V_{DS} = 10$ V, $V_{GS} = 0, f = 1.0$ kHz, $R_S = 3$ megohms)		NF	–	3.8	–	dB

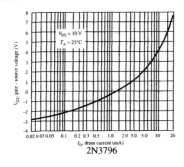

2N3796

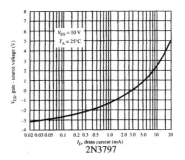

2N3797

FIGURE D.04-18
Partial data sheet for the 2N3797 D-MOSFET.

True/False

1. A JFET is a current-controlled device.

2. The GS terminals of a JFET must be forward-biased for proper operation.

Multiple Choice

3. The three terminals on a JFET are the
   a. Drain, supply, and grid.
   b. Drain, sink, and gate.
   c. Drain, source, and gate.

4. The region of a JFET curve that indicates a constant channel resistance is the
   a. Channel region.
   b. I-V region.
   c. Ohmic region.

5. The MOSFET differs from the JFET in that it
   a. Has a lower input resistance.
   b. Only works for positive supply voltages.
   c. Has no *pn* junction.

D.05 Demonstrate an understanding of special diodes and transistors

INTRODUCTION

In this section we will examine the operation of several special components, including zener and light-emitting diodes, UJTs, PUTs, and phototransistors.

ZENER DIODES

Figure D.05-1 shows the schematic symbol for a zener diode. The **zener diode** is a silicon *pn* junction device that differs from the rectifier diode in that it is designed for operation in the reverse breakdown region. The breakdown voltage of a zener diode is set by carefully controlling the doping level during the manufacturing process. From the discussion of the diode characteristic curve in the last chapter, recall that when a diode reaches reverse breakdown, its voltage remains almost constant even though the current may change drastically. This volt-ampere characteristic is shown again in Figure D.05-2.

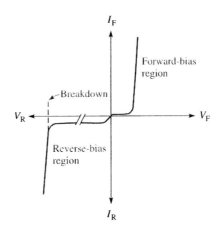

FIGURE D.05-1
Zener diode symbol.

FIGURE D.05-2
General diode characteristic.

Zener Breakdown

Two types of reverse breakdown in a zener diode are *avalanche* and *zener.* The avalanche breakdown also occurs in rectifier diodes at a sufficiently high reverse voltage. Zener breakdown occurs in a zener diode at low reverse voltages. A zener diode is heavily doped to reduce the breakdown voltage, causing a very narrow depletion region. As a result, an intense electric field exists within the depletion region. Near the breakdown voltage (V_Z), the field is intense enough to pull electrons from their valence bands and create current.

Zener diodes with breakdown voltages of less than approximately 5 V operate predominantly in zener breakdown. Those with breakdown voltages greater than approximately 5 V operate predominantly in avalanche breakdown. Both types, however, are

269

called *zener diodes.* Zeners with breakdown voltages of 1.8 V to 200 V are commercially available.

Breakdown Characteristics   Figure D.05-3 shows the reverse portion of the characteristic curve of a zener diode. Notice that as the reverse voltage (V_R) is increased, the reverse current (I_R) remains extremely small up to the "knee" of the curve. At this point, the breakdown effect begins; the zener resistance (R_Z) begins to decrease as the current (I_Z) increases rapidly. From the bottom of the knee, the breakdown voltage (V_Z) remains essentially constant. This regulating ability is the key feature of the zener diode: *A zener diode maintains an essentially constant voltage across its terminals over a specified range of reverse current values.*

FIGURE D.05-3
Reverse characteristic of a zener diode. V_Z is usually specified at the zener test current, I_{ZT}, and is designated V_{ZT}.

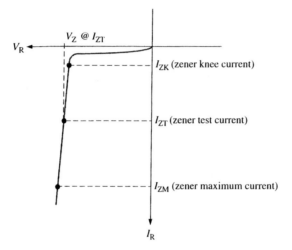

A minimum value of reverse current, I_{ZK}, must be maintained in order to keep the diode in regulation. You can see on the curve that when the reverse current is reduced below the knee of the curve, the voltage changes drastically and regulation is lost. Also, there is a maximum current, I_{ZM}, above which the diode may be damaged.

Thus, basically, the zener diode maintains a nearly constant voltage across its terminals for values of reverse current ranging from I_{ZK} to I_{ZM}. A nominal zener test voltage, V_{ZT}, is usually specified on a data sheet at a value of reverse current called the *zener test current, I_{ZT}.*

Zener Equivalent Circuit

Figure D.05-4(a) shows the ideal model of a zener diode in reverse breakdown. It acts simply as a battery having a value equal to the zener voltage. Figure D.05-4(b) represents the practical equivalent of a zener, where the zener resistance (R_Z) is included. The zener resistance is actually an AC resistance because it is dependent on the ratio of a change in voltage to a change in current and can be different for different portions of the characteristic curve. Since the voltage curve is not ideally vertical, a change in reverse current (ΔI_Z) produces a small change in zener voltage (ΔV_Z), as illustrated in Figure D.05-4(c).

The ratio of ΔV_Z to ΔI_Z is the zener resistance, expressed as follows:

$$R_Z = \frac{\Delta V_Z}{\Delta I_Z}$$

Normally, R_Z is specified at I_{ZT}, the zener test current. In most cases, this value of R_Z is approximately constant over the full range of reverse-current values.

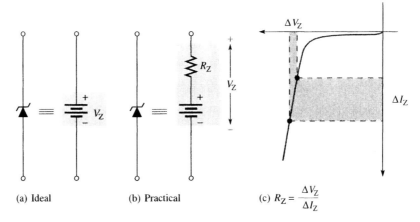

(a) Ideal   (b) Practical   (c) $R_Z = \dfrac{\Delta V_Z}{\Delta I_Z}$

FIGURE D.05-4
Zener equivalent circuits.

EXAMPLE D.05-1

A certain zener diode has a resistance of 5 Ω. The data sheet gives $V_{ZT} = 6.8$ V at $I_{ZT} = 20$ mA, $I_{ZK} = 1$ mA, and $I_{ZM} = 50$ mA. What is the voltage across its terminals when the current is 30 mA? What is the voltage when $I = 10$ mA?

Solution   Figure D.05-5 represents the diode. The 30 mA current is a 10 mA increase (ΔI_Z) above $I_{ZT} = 20$ mA.

$$\Delta I_Z = +10 \text{ mA}$$
$$\Delta V_Z = \Delta I_Z R_Z = (10 \text{ mA})(5 \ \Omega) = +50 \text{ mV}$$

FIGURE D.05-5

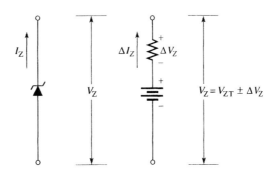

The change in voltage due to the increase in current above the I_{ZT} value causes the zener terminal voltage to increase. The zener voltage for $I_Z = 30$ mA is

$$V_Z = 6.8 \text{ V} + \Delta V_Z = 6.8 \text{ V} + 50 \text{ mV} = 6.85 \text{ V}$$

The 10 mA current is a 10 mA decrease below $I_{ZT} = 20$ mA.

$$\Delta I_Z = -10 \text{ mA}$$
$$\Delta V_Z = \Delta I_Z R_Z = (-10 \text{ mA})(5 \ \Omega) = -50 \text{ mV}$$

The change in voltage due to the decrease in current below I_{ZT} causes the zener terminal voltage to decrease. The zener voltage for $I_Z = 10$ mA is

$$V_Z = 6.8 \text{ V} - \Delta V_Z = 6.8 \text{ V} - 50 \text{ mV} = 6.75 \text{ V}$$

Varactor Diodes

A **varactor** is basically a reverse-biased *pn* junction that utilizes the inherent capacitance of the depletion region. The depletion region, created by the reverse bias, acts as a capacitor dielectric because of its nonconductive characteristics. The *p* and *n* regions are conductive and act as the capacitor plates, as illustrated in Figure D.05-6.

When the reverse-bias voltage increases, the depletion region widens, effectively increasing the dielectric thickness (*d*) and thus decreasing the capacitance. When the reverse-bias voltage decreases, the depletion region narrows, thus increasing the capacitance. This action is shown in Figure D.05-7, parts (a) and (b). A general curve of capacitance versus voltage is shown in Figure D.05-7(c).

FIGURE D.05-6
The reverse-biased varactor diode acts as a variable capacitor.

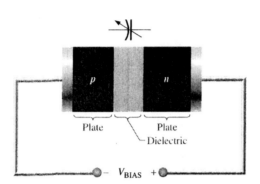

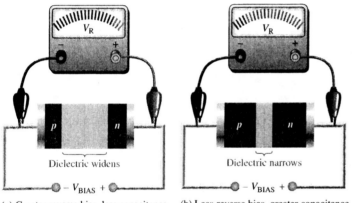

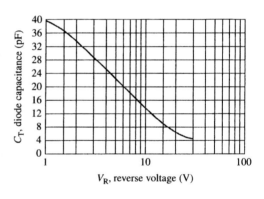

(a) Greater reverse bias, less capacitance   (b) Less reverse bias, greater capacitance   (c) Graph of diode capacitance versus reverse voltage

FIGURE D.05-7
Varactor diode capacitance varies with reverse voltage.

Recall that capacitance is determined by the plate area (*A*), dielectric constant (ε), and dielectric thickness (*d*), as expressed in the following formula:

$$C = \frac{A\epsilon}{d}$$

In a varactor diode, the capacitance parameters are controlled by the method of doping in the depletion region and the size and geometry of the diode's construction. Varactor capacitances typically range from a few picofarads to a few hundred picofarads.

Figure D.05-8(a) shows a common symbol for a varactor, and part (b) shows a simplified equivalent circuit. R_S is the reverse series resistance, and C_V is the variable capacitance.

FIGURE D.05-8
Varactor diode.

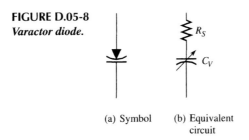

(a) Symbol (b) Equivalent
circuit

LEDs AND PHOTODIODES

The Light-Emitting Diode (LED)

The basic operation of the **light-emitting diode (LED)** is as follows: When the device is forward-biased, electrons cross the *pn* junction from the *n*-type material and recombine with holes in the *p*-type material. Recall that these free electrons are in the conduction band and at a higher energy level than the holes in the valence band. When recombination takes place, the recombining electrons release energy in the form of heat and light. A large exposed surface area on one layer of the semiconductor permits the photons to be emitted as visible light. Figure D.05-9 illustrates this process, which is called *electroluminescence*.

FIGURE D.05-9
Electroluminescence in an LED.

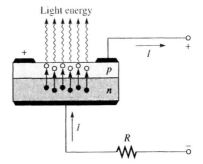

  The semiconductive materials used in LEDs are gallium arsenide (GaAs), gallium arsenide phosphide (GaAsP), and gallium phosphide (GaP). Silicon and germanium are not used because they are essentially heat-producing materials and are very poor at producing light. GaAs LEDs emit infrared (IR) radiation, GaAsP produces either red or yellow visible light, and GaP emits red or green visible light. The symbol for an LED is shown in Figure D.05-10.
  The LED emits light in response to a sufficient forward current, as shown in Figure D.05-11(a). The amount of light output is directly proportional to the forward current, as indicated in Figure D.05-11(b). Typical LEDs are shown in part (c).

FIGURE D.05-10
Symbol for an LED.

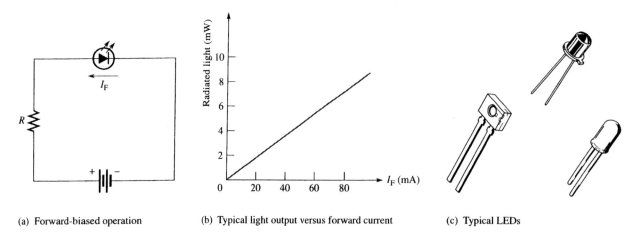

(a) Forward-biased operation   (b) Typical light output versus forward current   (c) Typical LEDs

FIGURE D.05-11
Light-emitting diodes (LEDs).

Applications   LEDs are commonly used for indicator lamps and readout displays on a wide variety of instruments, ranging from consumer appliances to scientific apparatus. A common type of display device using LEDs is the 7-segment display. Combinations of the segments form the ten decimal digits. Also, IR-emitting diodes are used in optical coupling applications, often in conjunction with fiber optics.

The Photodiode

The **photodiode** is a *pn* junction device that operates in reverse bias, as shown in Figure D.05-12(a). Note the schematic symbol for the photodiode. The photodiode has a small transparent window that allows light to strike the *pn* junction. Typical photodiodes are shown in Figure D.05-12(b), and an alternate symbol is shown in part (c).

FIGURE D.05-12
Photodiode.

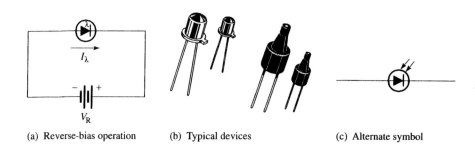

(a) Reverse-bias operation   (b) Typical devices   (c) Alternate symbol

Recall that when reverse-biased, a rectifier diode has a very small reverse leakage current. The same is true for the photodiode. The reverse-biased current is produced by thermally generated electron hole pairs in the depletion region, which are swept across the junction by the electric field created by the reverse voltage. In a rectifier diode, the reverse current increases with temperature due to an increase in the number of electron hole pairs.

In a photodiode, the reverse current increases with the light intensity at the exposed *pn* junction. When there is no incident light, the reverse current (I_λ) is almost negligible and is called the *dark current*. An increase in the amount of light intensity, expressed as irradiance (mW/cm^2), produces an increase in the reverse current, as shown by the graph in Figure D.05-13(a).

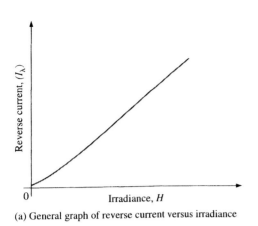

(a) General graph of reverse current versus irradiance

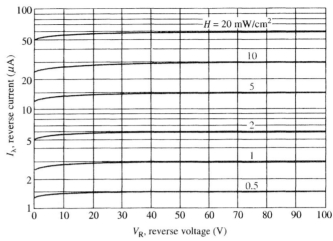

(b) Example of a graph of reverse current versus reverse voltage for several values of irradiance, H

FIGURE D.05-13
Typical photodiode characteristics.

From the graph in Figure D.05-13(b), the reverse current for this particular device is approximately 1.4 μA at a reverse-bias voltage of 10 V. Therefore, the reverse resistance of the device with an irradiance of 0.5 mW/cm^2 is

$$R_R = \frac{V_R}{I_\lambda} = \frac{10 \text{ V}}{1.4 \text{ } \mu\text{A}} = 7.14 \text{ M}\Omega$$

At 20 mW/cm^2, the current is approximately 55 μA at $V_R = 10$ V. The reverse resistance under this condition is

$$R_R = \frac{V_R}{I_\lambda} = \frac{10 \text{ V}}{55 \text{ } \mu\text{A}} = 182 \text{ k}\Omega$$

These calculations show that the photodiode can be used as a variable-resistance device controlled by light intensity.

UNIJUNCTION TRANSISTORS (UJTs)

Figure D.05-14(a) shows the construction of a **unijunction transistor (UJT)**. The base contacts are made to the *n*-type bar. The emitter lead is connected to the *p* region. The UJT schematic symbol is shown in Figure D.05-14(b). Do not confuse this symbol with that of a JFET; *the difference is that the arrow is at an angle for the UJT.*

FIGURE D.05-14
Unijunction transistor (UJT).

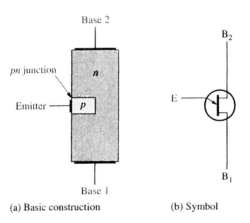

(a) Basic construction (b) Symbol

UJT Operation

In normal UJT operation, base 2 (B_2) and the emitter are biased positive with respect to base 1 (B_1). Figure D.05-15 will illustrate the operation. This equivalent circuit represents the internal UJT characteristics. The total resistance between the two bases, R_{BB}, is the resistance of the *n*-type material. R_{B1} is the resistance between point *k* and base 1. R_{B2} is the resistance between point *k* and base 2. The sum of these two resistances makes up the total resistance, R_{BB}. The diode represents the *pn* junction between the emitter and the *n*-type material.

FIGURE D.05-15
Equivalent circuit for a UJT.

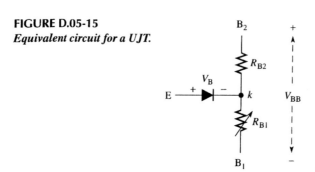

The ratio R_{B1}/R_{BB} is designated η (the Greek letter *eta*) and is defined as the *intrinsic standoff ratio.* It takes an emitter voltage of $V_B + \eta V_{BB}$ to turn the UJT on. This voltage is called the *peak voltage.* Once the device is on, resistance R_{B1} drops in value. Thus, as emitter current increases, emitter voltage decreases because of the decrease in R_{B1}. This characteristic is the negative resistance characteristic of the UJT.

As the emitter voltage decreases, it reaches a value called the *valley voltage.* At this point, the *pn* junction is no longer forward-biased, and the UJT turns off.

THE PROGRAMMABLE UNIJUNCTION TRANSISTOR (PUT)

The structure of the **programmable unijunction transistor (PUT)** is similar to that of an SCR (four-layer) except that the gate is brought out as shown in Figure D.05-16. Notice that the gate is connected to the *n* region adjacent to the anode. This *pn* junction controls the *on* and *off* states of the device. The gate is always biased positive with respect to the cathode. When the anode voltage exceeds the gate voltage by approximately 0.7 V, the *pn* junction is forward-biased and the PUT turns on. The PUT stays on until the anode voltage falls back below this level, then the PUT turns off.

FIGURE D.05-16
The programmable unijunction transistor (PUT).

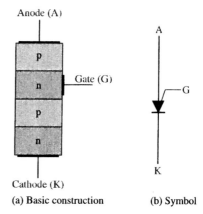

(a) Basic construction   (b) Symbol

FIGURE D.05-17
PUT biasing.

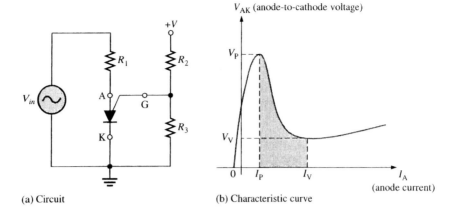

(a) Circuit   (b) Characteristic curve

Setting the Trigger Voltage

The gate can be biased to a desired voltage with an external voltage divider, as shown in Figure D.05-17(a), so that when the anode voltage exceeds this "programmed" level, the PUT turns on.

THE PHOTOTRANSISTOR

The relationship between the collector current and the light-generated base current in a phototransistor is

$$I_C = \beta_{DC} I_\lambda$$

The schematic symbol and some typical phototransistors are shown in Figure D.05-18. Since the actual photogeneration of base current occurs in the collector-base region, the larger the physical area of this region, the more base current is generated. Thus, a typical phototransistor is designed to offer a large area to the incident light, as the simplified structure diagram in Figure D.05-19 illustrates.

FIGURE D.05-18
Phototransistor.

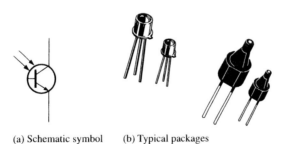

(a) Schematic symbol   (b) Typical packages

FIGURE D.05-19
Typical phototransistor chip structure.

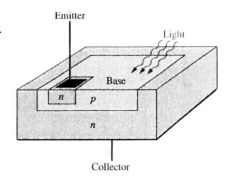

A **phototransistor** can be either a two-lead or a three-lead device. In the three-lead configuration, the base lead is brought out so that the device can be used as a conventional bipolar transistor with or without the additional light-sensitivity feature. In the two-lead configuration, the base is not electrically available, and the device can be used only with light as the input. In many applications, the phototransistor is used in the two-lead version. Figure D.05-20 shows a phototransistor with a biasing circuit and typical collector characteristic curves. Notice that each individual curve on the graph corresponds to a certain value of light intensity (in this case, the units are mW/cm^2) and that the collector current increases with light intensity.

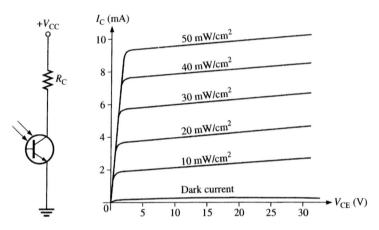

FIGURE D.05-20
Phototransistor bias circuit and typical collector characteristic curves.

FIGURE D.05-21
Typical phototransistor spectral response.

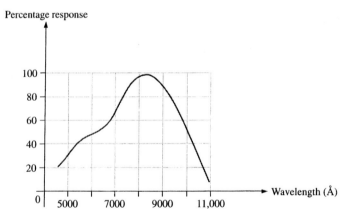

Phototransistors are not sensitive to all light but only to light within a certain range of wavelengths. They are most sensitive to particular wavelengths, as shown by the peak of the spectral response curve in Figure D.05-21.

Photodarlington

The photodarlington consists of a phototransistor connected in a darlington arrangement with a conventional transistor, as shown in Figure D.05-22. Because of the higher current gain, this device has a much higher collector current and exhibits a greater light sensitivity than does a regular phototransistor.

FIGURE D.05-22
Photodarlington.

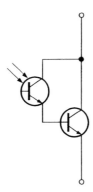

REVIEW QUESTIONS

True/False

1. A zener diode is designed to be operated reverse-biased.

2. A varactor diode has a constant capacitor over a wide reverse-bias range.

Multiple Choice

3. A diode that gives off light when forward-biased is the
    a. Photodiode.
    b. Light-emitting diode.
    c. Zener diode.

4. A UJT has a certain
    a. Feedback factor.
    b. Noise margin.
    c. Intrinsic standoff ratio.

5. In a phototransistor, if the light intensity increases, the collector current
    a. Decreases.
    b. Increases.
    c. Stays the same.

D.06 Understand principles and operations of diode circuits

INTRODUCTION

In this exercise we examine several diode applications: rectifier circuits, clippers and clampers, and voltage multipliers.

HALF-WAVE RECTIFIERS

The Basic DC Power Supply

The DC **power supply** converts the standard 110 V, 60 Hz AC available at wall outlets into a constant DC voltage. It is one of the most common electronic circuits that you will find. The DC voltage produced by a power supply is used to power all types of electronic circuits, such as television receivers, stereo systems, VCRs, CD players, and laboratory equipment.

A basic block diagram for a power supply is shown in Figure D.06-1. The **rectifier** can be either a half-wave rectifier or a full-wave rectifier. The rectifier converts the AC input voltage to a pulsating DC voltage, which is half-wave rectified as shown. The **filter** eliminates the fluctuations in the rectified voltage and produces a relatively smooth DC voltage. The **regulator** is a circuit that maintains a constant DC voltage for variations in the input line voltage or in the load. Regulators vary from a single device to more complex circuits. The load block is usually a circuit for which the power supply is producing the DC voltage and load current.

FIGURE D.06-1
Block diagram of a DC power supply.

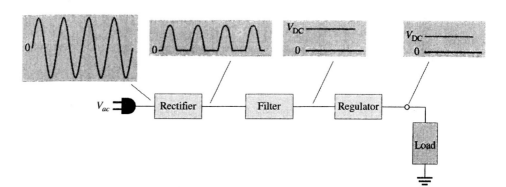

The Half-Wave Rectifier

Figure D.06-2 illustrates the process called *half-wave rectification.* In part (a), a diode is connected to an AC source and a load resistor R_L, forming a **half-wave rectifier.** Let's examine what happens during one cycle of the input voltage using the ideal model for the diode. When the sinusoidal input voltage (V_{in}) goes positive, the diode is forward-biased and conducts current through the load resistor, as shown in part (b). The current produces an output voltage across the load R_L, which has the same shape as the positive half-cycle of the input voltage.

FIGURE D.06-2
*Half-wave rectifier
operation. The diode is
considered to be ideal.*

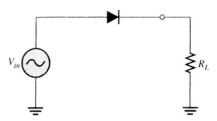

(a) Half-wave rectifier circuit

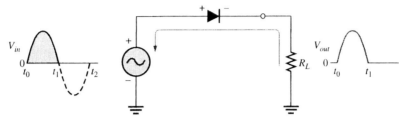

(b) Operation during positive alternation of the input voltage

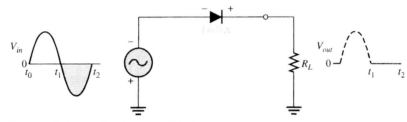

(c) Operation during negative alternation of the input voltage

(d) Half-wave output voltage for three input cycles

When the input voltage goes negative during the second half of its cycle, the diode is reverse-biased. There is no current, so the voltage across the load resistor is 0 V, as shown in Figure D.06-2(c). The net result is that only the positive half-cycles of the AC input voltage appear across the load. Since the output does not change polarity, it is a *pulsating DC voltage* with a frequency of 60 Hz as shown in part (d).

Average Value of the Half-Wave Output Voltage   The average value of the half-wave rectified output voltage is the value you would measure on a DC voltmeter. It is determined by finding the area under the curve over a full cycle, as illustrated in Figure D.06-3, and then dividing by the period, *T*.

$$V_{AVG} = \frac{V_p}{\pi}$$

V_p is the peak value of the voltage.

FIGURE D.06-3
*Average value of half-
wave rectified signal.*

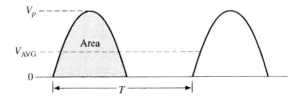

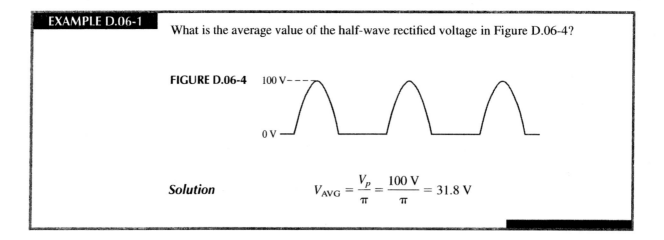

EXAMPLE D.06-1

What is the average value of the half-wave rectified voltage in Figure D.06-4?

FIGURE D.06-4   100 V－－－

0 V

Solution
$$V_{\text{AVG}} = \frac{V_p}{\pi} = \frac{100 \text{ V}}{\pi} = 31.8 \text{ V}$$

Effect of the Barrier Potential on the Half-Wave Rectifier Output

In the previous discussion, the diode was considered ideal. When the practical diode model is used and the barrier potential is taken into account, this is what happens. During the positive half-cycle, the input voltage must overcome the barrier potential before the diode becomes forward-biased. For a silicon diode this results in a half-wave output with a peak value that is 0.7 V less than the peak value of the input, as shown in Figure D.06-5. For silicon, the expression for the peak output voltage is

$$V_{p(out)} = V_{p(in)} - 0.7 \text{ V}$$

In working with diode circuits, it is usually practical to neglect the effect of barrier potential when the peak value of the applied voltage is much greater than the barrier potential (at least 10 V, as a rule of thumb). We will always use the practical model of a silicon diode, taking the barrier potential into account unless stated otherwise.

FIGURE D.06-5
Effect of barrier potential on half-wave rectified output voltage (silicon diode shown).

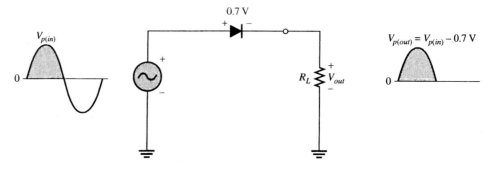

Peak Inverse Voltage (PIV)

The maximum value of reverse voltage, designated as *peak inverse voltage* (PIV), occurs at the peak of the negative alternation of the input cycle when the diode is reverse-biased. This condition is illustrated in Figure D.06-6. The PIV equals the peak value of the input voltage, and the diode must be capable of withstanding this amount of repetitive reverse voltage.

$$\text{PIV} = V_{p(in)}$$

FIGURE D.06-6
The PIV occurs at the peak of the half-cycle when the diode is reverse-biased. In this circuit, the PIV occurs at the peak of the negative half-cycle.

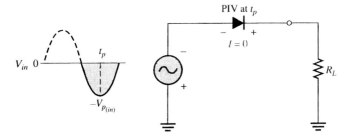

FIGURE D.06-7
Half-wave rectifier with transformer-coupled input voltage.

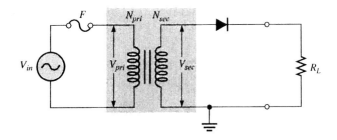

Half-Wave Rectifier with Transformer-Coupled Input Voltage

A transformer is often used to couple the AC input voltage from the source to the rectifier circuit, as shown in Figure D.06-7. Transformer coupling provides two advantages. First, it allows the source voltage to be stepped up or stepped down as needed. Second, the AC power source is electrically isolated from the rectifier circuit, thus reducing the shock hazard.

From basic AC circuits recall that the secondary voltage of a transformer equals the turns ratio (N_{sec}/N_{pri}) times the primary voltage.

$$V_{sec} = \left(\frac{N_{sec}}{N_{pri}}\right)V_{pri}$$

If $N_{sec} > N_{pri}$, the secondary voltage is greater than the primary voltage. If $N_{sec} < N_{pri}$, the secondary voltage is less than the primary voltage. If $N_{sec} = N_{pri}$, then $V_{sec} = V_{pri}$.

EXAMPLE D.06-2

Determine the peak value of the output voltage for Figure D.06-8.

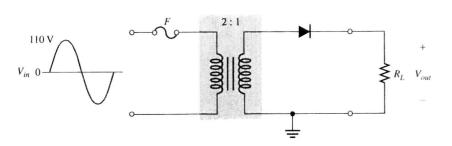

FIGURE D.06-8

Solution $V_{p(pri)} = V_{p(in)} = 110 \text{ V}$

The secondary peak voltage is

$$V_{p(sec)} = \left(\frac{N_{sec}}{N_{pri}}\right)V_{p(pri)} = 0.5(110 \text{ V}) = 55 \text{ V}$$

The peak rectified output voltage is

$$V_{p(out)} = V_{p(sec)} - 0.7 \text{ V} = 55 \text{ V} - 0.7 \text{ V} = 54.3 \text{ V}$$

The difference between full-wave and half-wave rectification is that a **full-wave rectifier** allows unidirectional (one-way) current to the load during the entire 360° of the input cycle, and the half-wave rectifier allows this only during one half-cycle. The result of full-wave rectification is an output voltage with a frequency twice the input that pulsates every half-cycle of the input, as shown in Figure D.06-9.

Since the number of positive alternations that make up the full-wave rectified voltage is twice that of the half-wave voltage for the same time interval, the average value for a full-wave rectified sinusoidal voltage is twice that of the half-wave, as shown in the following formula:

$$V_{AVG} = \frac{2V_p}{\pi}$$

FIGURE D.06-9
Full-wave rectification.

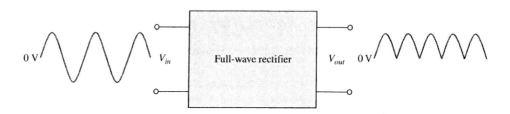

EXAMPLE D.06-3

Find the average value of the full-wave rectified voltage in Figure D.06-10.

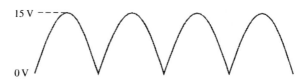

FIGURE D.06-10

Solution

$$V_{AVG} = \frac{2V_p}{\pi} = \frac{2(15 \text{ V})}{\pi} = 9.55 \text{ V}$$

The Full-Wave Center-Tapped Rectifier

The full-wave **center-tapped rectifier** uses two diodes connected to the secondary of a center-tapped transformer, as shown in Figure D.06-11. The input voltage is coupled through the transformer to the center-tapped secondary. Half of the total secondary voltage appears between the center tap and each end of the secondary winding as shown.

For a positive half-cycle of the input voltage, the polarities of the secondary voltages are as shown in Figure D.06-12(a). This condition forward-biases the upper diode D_1 and

FIGURE D.06-11
A full-wave center-tapped rectifier.

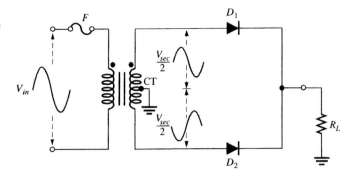

FIGURE D.06-12
Basic operation of a full-wave center-tapped rectifier. Note that the current through the load resistor is in the same direction during the entire input cycle.

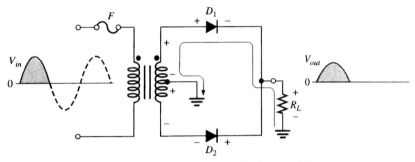

(a) During positive half-cycles, D_1 is forward-biased and D_2 is reverse-biased.

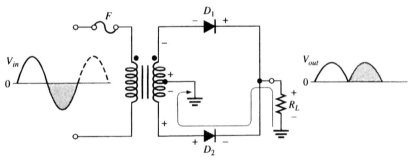

(b) During negative half-cycles, D_2 is forward-biased and D_1 is reverse-biased.

reverse-biases the lower diode D_2. The current path is through D_1 and the load resistor R_L, as indicated. For a negative half-cycle of the input voltage, the voltage polarities on the secondary are as shown in Figure D.06-12(b). This condition reverse-biases D_1 and forward-biases D_2. The current path is through D_2 and R_L, as indicated. Because the output current during both the positive and negative portions of the input cycle is in the same direction through the load, the output voltage developed across the load resistor is a full-wave rectified DC voltage.

Effect of the Turns Ratio on the Output Voltage   If the transformer's turns ratio is 1, the peak value of the rectified output voltage equals half the peak value of the primary input voltage less the barrier potential as illustrated in Figure D.06-13. Incidentally, we will begin referring to the forward voltage due to the barrier potential as the **diode drop.** This is because half of the primary voltage appears across each half of the secondary winding ($V_{p(sec)} = V_{p(pri)}$).

In order to obtain an output voltage with a peak equal to the input peak (less the diode drop), a step-up transformer with a turns ratio of 2 must be used, as shown in Figure D.06-14. In this case, the total secondary voltage V_{sec} is twice the primary voltage ($2V_{pri}$), so the voltage across each half of the secondary is equal to V_{pri}.

FIGURE D.06-13
Full-wave center-tapped rectifier with a transformer turns ratio of 1. $V_{p(pri)}$ is the peak value of the primary voltage.

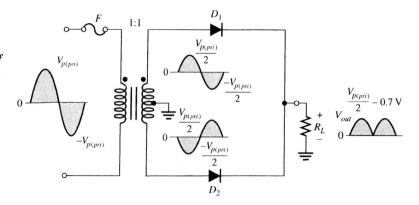

FIGURE D.06-14
Full-wave center-tapped rectifier with a transformer turns ratio of 2.

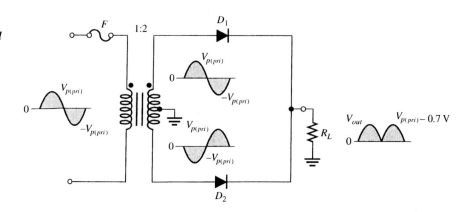

In any case, the output voltage of a full-wave center-tapped rectifier is always one-half of the total secondary voltage less the diode drop, no matter what the turns ratio.

$$V_{out} = \frac{V_{sec}}{2} - 0.7 \text{ V}$$

Peak Inverse Voltage   Each diode in the full-wave rectifier is alternately forward-biased and then reverse-biased. The maximum reverse voltage that each diode must withstand is the peak secondary voltage $V_{p(sec)}$. This is shown in Figure D.06-15.

When the total secondary voltage has the polarity shown, the maximum anode voltage of D_1 is $+V_{p(sec)}/2$ and the maximum anode voltage of D_2 is $-V_{p(sec)}/2$. Since D_1 is forward-biased, its cathode is at the same voltage as its anode minus the diode drop; this is

FIGURE D.06-15
Diode reverse voltage (D_2 shown reverse-biased). The PIV is twice the peak value of the output voltage plus a diode drop.

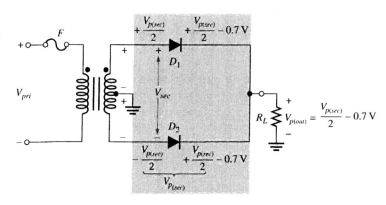

also the voltage on the cathode of D_2. Applying Kirchhoff's voltage law around the loop, the peak inverse voltage across D_2 is

$$\text{PIV} = \left(\frac{V_{p(sec)}}{2} - 0.7 \text{ V}\right) - \left(\frac{-V_{p(sec)}}{2}\right) = \frac{V_{p(sec)}}{2} + \frac{V_{p(sec)}}{2} - 0.7 \text{ V} = V_{p(sec)} - 0.7 \text{ V}$$

Since

$$V_{p(out)} = \frac{V_{p(sec)}}{2} - 0.7 \text{ V}$$

the peak inverse voltage across either diode in the full-wave center-tapped rectifier in terms of $V_{p(out)}$ is

$$\text{PIV} = 2V_{p(out)} + 0.7 \text{ V}$$

EXAMPLE D.06-4

Show the voltage waveforms across each half of the secondary winding and across R_L when a 100 V peak sine wave is applied to the primary winding in Figure D.06-16. Also, what PIV rating must the diodes have?

FIGURE D.06-16

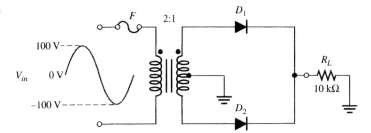

Solution The total peak secondary voltage is

$$V_{p(sec)} = \left(\frac{N_{sec}}{N_{pri}}\right)V_{p(pri)} = (0.5)100 \text{ V} = 50 \text{ V}$$

There is a 25 V peak across each half of the secondary. The output load voltage has a peak value of 25 V, less the 0.7 V drop across the diode. Each diode must have a minimum PIV rating of 50 V (neglecting diode drop). The waveforms are shown in Figure D.06-17.

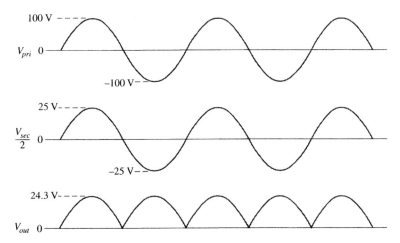

FIGURE D.06-17

The Full-Wave Bridge Rectifier

The full-wave **bridge rectifier** uses four diodes, as shown in Figure D.06-18. When the input cycle is positive as in part (a), diodes D_1 and D_2 are forward-biased and conduct current in the direction shown. A voltage is developed across R_L which looks like the positive half of the input cycle. During this time, diodes D_3 and D_4 are reverse-biased.

When the input cycle is negative as in Figure D.06-18(b), diodes D_3 and D_4 are forward-biased and conduct current in the same direction through R_L as during the positive half-cycle. During the negative half-cycle, D_1 and D_2 are reverse-biased. A full-wave rectified output voltage appears across R_L as a result of this action.

FIGURE D.06-18
Operation of a full-wave bridge rectifier.

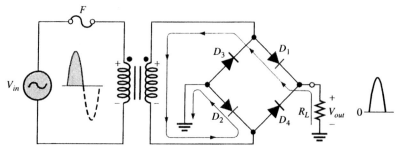

(a) During positive half-cycle of the input, D_1 and D_2 are forward-biased and conduct current. D_3 and D_4 are reverse-biased.

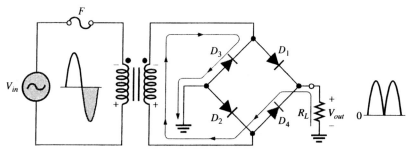

(b) During negative half-cycle of the input, D_3 and D_4 are forward-biased and conduct current. D_1 and D_2 are reverse-biased.

Bridge Output Voltage A bridge rectifier with a transformer-coupled input is shown in Figure D.06-19(a). During the positive half-cycle of the total secondary voltage, diodes D_1 and D_2 are forward-biased. Neglecting the diode drops, the secondary voltage V_{sec} appears across the load resistor. The same is true when D_3 and D_4 are forward-biased during the negative half-cycle.

$$V_{out} = V_{sec}$$

As you can see in Figure D.06-19(b), two diodes are always in series with the load resistor during both the positive and negative half-cycles. If these diode drops are taken into account, the output voltage is

$$V_{p(out)} = V_{p(sec)} - 1.4 \text{ V}$$

Peak Inverse Voltage Let's assume that D_1 and D_2 are forward-biased and examine the reverse voltage across D_3 and D_4. Visualizing D_1 and D_2 as shorts (ideally), as in Figure D.06-20(a), we can see that D_3 and D_4 have a peak inverse voltage equal to the peak secondary voltage. Since the output voltage is *ideally* equal to the secondary voltage,

$$\text{PIV} = V_{p(out)}$$

FIGURE D.06-19
Bridge operation during positive half-cycle of the secondary voltage.

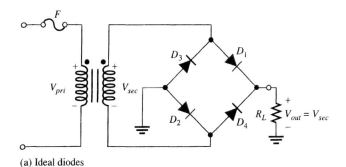

(a) Ideal diodes

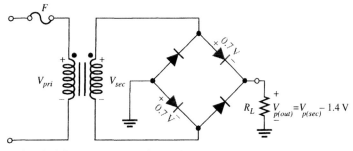

(b) Practical diodes (Diode drops included)

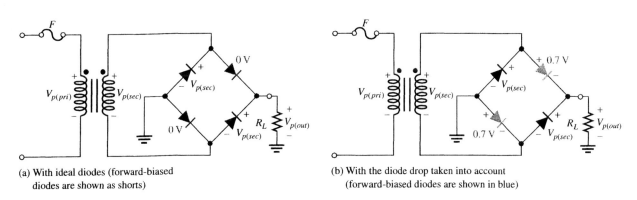

(a) With ideal diodes (forward-biased diodes are shown as shorts)

(b) With the diode drop taken into account (forward-biased diodes are shown in blue)

FIGURE D.06-20
Peak inverse voltage in a bridge rectifier during the positive half-cycle of the secondary voltage.

If the diode drops of the forward-biased diodes are included as shown in Figure D.06-20(b), the peak inverse voltage across each reverse-biased diode in terms of $V_{p(out)}$ is

$$PIV = V_{p(out)} + 0.7 \text{ V}$$

The PIV rating of the bridge diodes is less than that required for the center-tapped configuration. If the diode drop is neglected, the bridge rectifier requires diodes with half the PIV rating of those in a center-tapped rectifier for the *same* output voltage.

EXAMPLE D.06-5

Determine the output voltage for the bridge rectifier in Figure D.06-21. What PIV rating is required for the silicon diodes? The transformer is specified to have a 12 V rms secondary voltage for the standard 110 V across the primary.

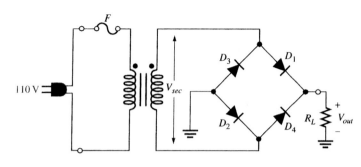

FIGURE D.06-21

Solution The peak output voltage is (taking into account the two diode drops)

$$V_{p(sec)} = 1.414V_{rms} = 1.414(12\ V) \cong 17\ V$$
$$V_{p(out)} = V_{p(sec)} - 1.4\ V = 17\ V - 1.4\ V = 15.6\ V$$

The PIV for each diode is

$$PIV = V_{p(out)} + 0.7\ V = 15.6\ V + 0.7\ V = 16.3\ V$$

DIODE LIMITING AND CLAMPING CIRCUITS

Diode Limiters

Figure D.06-22(a) shows a diode **limiter** circuit **(clipper)** that limits or clips the positive part of the input voltage. As the input voltage goes positive, the diode becomes forward-biased. Since the cathode is at ground potential (0 V), the anode cannot exceed 0.7 V. So point *A* is limited to +0.7 V when the input voltage exceeds this value. When the input voltage goes back below 0.7 V, the diode is reverse-biased and appears as an open. The

FIGURE D.06-22
Example of a diode limiter (clipper).

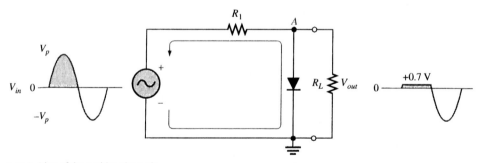

(a) Limiting of the positive alternation

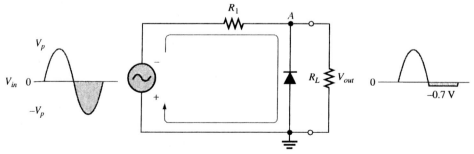

(b) Limiting of the negative alternation

output voltage looks like the negative part of the input voltage, but with a magnitude determined by the voltage divider formed by R_1 and the load resistor, R_L, as follows:

$$V_{out} = \left(\frac{R_L}{R_1 + R_L}\right)V_{in}$$

If R_1 is small compared to R_L, then $V_{out} = V_{in}$.

Turn the diode around, as in Figure D.06-22(b), and the negative part of the input voltage is clipped off. When the diode is forward-biased during the negative part of the input voltage, point A is held at -0.7 V by the diode drop. When the input voltage goes above -0.7 V, the diode is no longer forward-biased; and a voltage appears across R_L proportional to the input voltage.

EXAMPLE D.06-6

What would you expect to see displayed on an oscilloscope connected across R_L in Figure D.06-23?

FIGURE D.06-23

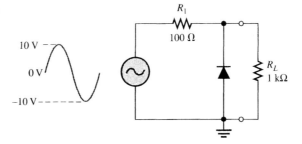

Solution The diode conducts when the input voltage goes below -0.7 V. So, a negative limiter with a peak output voltage can be determined by the following equation:

$$V_{p(out)} = \left(\frac{R_L}{R_1 + R_L}\right)V_{p(in)} = \left(\frac{1\ k\Omega}{1.1\ k\Omega}\right)10\ V = 9.09\ V$$

The scope will display an output waveform as shown in Figure D.06-24.

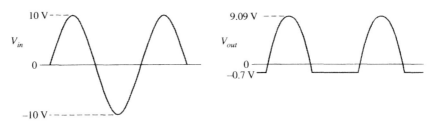

FIGURE D.06-24
Waveforms for Figure D.06-23.

Biased Limiters The level to which an AC voltage is limited can be adjusted by adding a bias voltage, V_{BIAS}, in series with the diode, as shown in Figure D.06-25. The voltage at point A must equal $V_{BIAS} + 0.7$ V before the diode will conduct. Once the diode begins to conduct, the voltage at point A is limited to $V_{BIAS} + 0.7$ V so that all input voltage above this level is clipped off.

FIGURE D.06-25
A positive limiter.

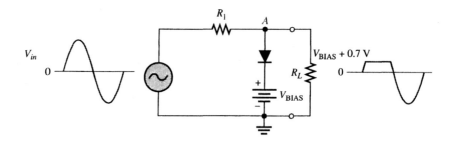

If the bias voltage is varied up or down, the limiting level changes correspondingly, as shown in Figure D.06-26.

To limit a voltage to a specified negative level, the diode and bias voltage must be connected as in Figure D.06-27. In this case, the voltage at point A must go below $-V_{BIAS} - 0.7$ V to forward-bias the diode and initiate limiting action as shown.

By turning the diode around, the positive limiter can be modified to limit the output voltage to the portion of the input voltage waveform above $V_{BIAS} - 0.7$ V, as shown in Figure D.06-28(a). Similarly, the negative limiter can be modified to limit the output voltage to the portion of the input voltage waveform below $-V_{BIAS} + 0.7$ V, as shown in part (b).

FIGURE D.06-26
A positive limiter with variable bias.

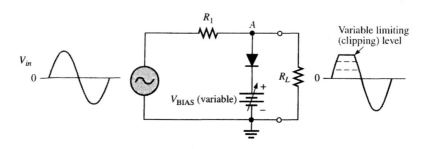

FIGURE D.06-27
A negative limiter.

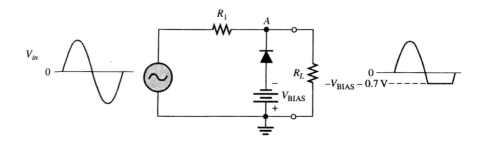

Diode Clampers

A clamper adds a DC level to an AC voltage. **Clampers** are sometimes known as *DC restorers*. Figure D.06-29 shows a diode clamper that inserts a positive DC level. The operation of this circuit can be seen by considering the first negative half-cycle of the input voltage. When the input voltage initially goes negative, the diode is forward-biased, allowing the capacitor to charge to near the peak of the input ($V_{p(in)} - 0.7$ V), as shown in Figure D.06-29(a). Just after the negative peak, the diode is reverse-biased. This is because the cathode is held near $V_{p(in)} - 0.7$ V by the charge on the capacitor. The capacitor can only discharge through the high resistance of R_L. So, from the peak of one negative half-cycle to the next, the capacitor discharges very little. The amount that is discharged, of course, depends on the value of R_L. For good clamping action, the RC time constant should be at least ten times the period of the input frequency.

FIGURE D.06-28

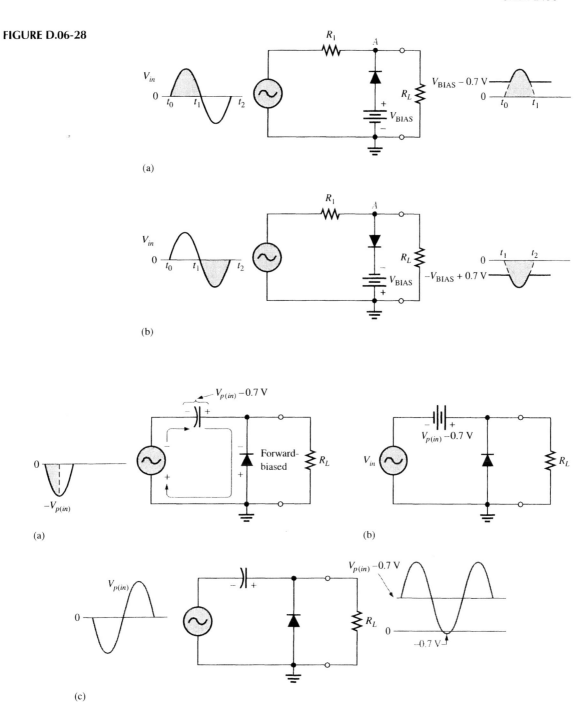

(a)

(b)

(a)

(b)

(c)

FIGURE D.06-29
Positive clamper operation.

The net effect of the clamping action is that the capacitor retains a charge approximately equal to the peak value of the input less the diode drop. The capacitor voltage acts essentially as a battery in series with the input voltage, as shown in Figure D.06-29(b). The DC voltage of the capacitor adds to the input voltage by superposition, as in part (c). If the diode is turned around, a negative DC voltage is added to the output voltage as shown in Figure D.06-30.

FIGURE D.06-30
Negative clamper.

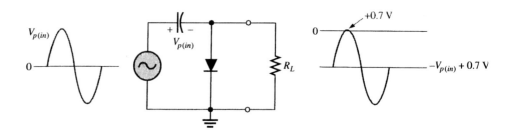

EXAMPLE D.06-7

What is the output voltage that you would expect to observe across R_L in the clamper circuit of Figure D.06-31? Assume that RC is large enough to prevent significant capacitor discharge.

FIGURE D.06-31

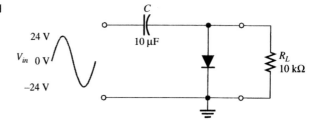

Solution   Ideally, a negative DC value equal to the input peak less the diode drop is inserted by the clamping circuit.

$$V_{DC} \cong -(V_{p(in)} - 0.7 \text{ V}) = -(24 \text{ V} - 0.7 \text{ V}) = -23.3 \text{ V}$$

Actually, the capacitor will discharge slightly between peaks, and, as a result, the output voltage will have an average value of slightly less than that calculated above. The output waveform goes to approximately +0.7 V, as shown in Figure D.06-32.

FIGURE D.06-32
Output waveform for Figure D.06-31.

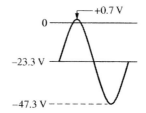

VOLTAGE MULTIPLIERS

Voltage Doubler

Half-Wave Voltage Doubler   A voltage doubler is a **voltage multiplier** with a multiplication factor of two. A half-wave voltage doubler is shown in Figure D.06-33. During the positive half-cycle of the secondary voltage, diode D_1 is forward-biased and D_2 is reverse-biased. Capacitor C_1 is charged to the peak of the secondary voltage (V_p) less the diode drop with the polarity shown in part (a). During the negative half-cycle, diode D_2 is forward-biased and D_1 is reverse-biased, as shown in part (b). Since C_1 can't discharge, the

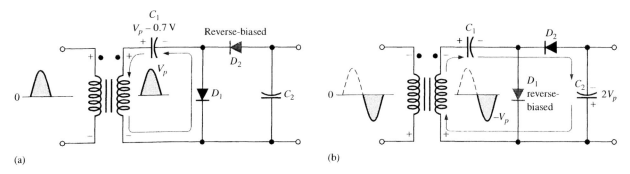

FIGURE D.06-33
Half-wave voltage doubler operation. V_p is the peak secondary voltage.

peak voltage on C_1 adds to the secondary voltage to charge C_2 to approximately $2V_p$. Applying Kirchhoff's law around the loop as shown in part (b),

$$V_{C1} - V_{C2} + V_p = 0$$
$$V_{C2} = V_p + V_{C1}$$

Neglecting the diode drop of D_2, $V_{C1} = V_p$. Therefore,

$$V_{C2} = V_p + V_p = 2V_p$$

Under a no-load condition, C_2 remains charged to approximately $2V_p$. If a load resistance is connected across the output, C_2 discharges slightly through the load on the next positive half-cycle and is again recharged to $2V_p$ on the following negative half-cycle. The resulting output is a half-wave, capacitor-filtered voltage. The peak inverse voltage across each diode is $2V_p$.

Full-Wave Voltage Doubler A full-wave doubler is shown in Figure D.06-34. When the secondary voltage is positive, D_1 is forward-biased and C_1 charges to approximately V_p, as shown in part (a). During the negative half-cycle, D_2 is forward-biased and C_2 charges to approximately V_p, as shown in part (b). The output voltage, $2V_p$, is taken across the two capacitors in series.

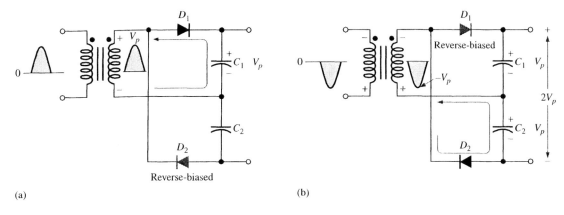

FIGURE D.06-34
Full-wave voltage doubler operation.

REVIEW QUESTIONS

True/False

1. Half-wave and full-wave rectifiers output the same voltage for the same transformer.
2. If one diode fails in a full-wave center-tapped rectifier, the output drops to zero.

Multiple Choice

3. A diode circuit that limits the size of a signal is called a
    a. Clipper.
    b. Clamper.
    c. Doubler.

4. Adding a DC level to an AC signal is done with a
    a. Clipper.
    b. Clamper.
    c. Doubler.

5. How many diodes are required to make a voltage doubler?
    a. One.
    b. Two.
    c. Three.

D.07 Fabricate and demonstrate diode circuits

This material is covered in C.16.

D.08 Troubleshoot and repair diode circuits

INTRODUCTION

In this section we examine several ways faulty diode circuits can be investigated.

Testing a Diode

A multimeter can be used as a fast and simple way to check a diode. As you know, a good diode will show an extremely high resistance (or open) with reverse bias and a very low resistance with forward bias. A defective open diode will show an extremely high resistance (or open) for both forward and reverse bias. A defective shorted or resistive diode will show zero or a low resistance for both forward and reverse bias. An open diode is the most common type of failure.

The DMM Diode Test Position   Many digital multimeters (DMMs) have a diode test position which provides a convenient way to test a diode. A typical DMM, as shown in Figure D.08-1, has a small diode symbol to mark the position of the function switch. When set to *diode test,* the meter provides an internal voltage sufficient to forward bias and reverse bias a diode. This internal voltage may vary among different makes of DMM, but 2.5 V to 3.5 V is a typical range of values. The meter provides a voltage reading or other indication to show the condition of the diode under test.

FIGURE D.08-1
Diode test on a properly functioning diode.

(a) Forward-bias test   (b) Reverse-bias test

When the Diode Is Working In Figure D.08-1(a), the red (positive) lead of the meter is connected to the anode and the black (negative) lead is connected to the cathode to forward bias the diode. If the diode is good, you will get a reading of between 0.5 V and 0.9 V, with 0.7 V being typical for forward bias.

In Figure D.08-1(b), the leads are switched to reverse-bias the diode as shown. If the diode is working properly, you will get a voltage reading based on the meter's internal voltage source. The 2.6 V shown in the figure represents a typical value and indicates that the diode has an extremely high reverse resistance with essentially all of the internal voltage appearing across it.

When the Diode Is Defective When a diode has failed open, you get an open circuit voltage reading (2.6 V is typical) for both the forward-bias and the reverse-bias condition, as illustrated in Figure D.08-2(a) and (b). If a diode is shorted, the meter reads 0 V in both forward- and reverse-bias tests, as indicated in part (c). Sometimes, a failed diode may exhibit a small resistance for both bias conditions rather than a pure short. In this case, the meter will show a small voltage much less than the correct open voltage. For example, a resistive diode may result in a reading of 1.1 V in both directions rather than the correct readings of 0.7 V forward and 2.6 V reverse.

(a) Forward-bias test, open diode

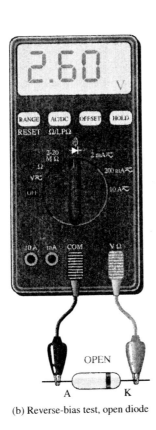

(b) Reverse-bias test, open diode

(c) Forward- and reverse-bias tests for a shorted diode give same 0 V reading. If the diode is resistive, the reading is less than 2.6 V.

FIGURE D.08-2
Testing a defective diode.

TROUBLESHOOTING RECTIFIER CIRCUITS

Several types of failures can occur in power supply rectifiers. Let us examine some possible failures and the effects they would have on a circuit's operation.

Open Diode

A half-wave rectifier with a diode that has opened (a common failure mode) is shown in Figure D.08-3. In this case, you would measure 0 V DC at the rectifier output, as shown.

Now consider the full-wave, center-tapped rectifier in Figure D.08-4. Assume that diode D_1 has failed open. With an oscilloscope connected to the output, as shown in part (a), you would observe the following: You would see a larger-than-normal ripple voltage at a frequency of 60 Hz rather than 120 Hz. Disconnecting the filter capacitor, you would observe a half-wave rectified voltage, as in part (b). Now let's examine the reason for these observations. If diode D_1 is open, there will be current through R_L only during the negative half-cycle of the input signal. During the positive half-cycle, an open path prevents current through R_L. The result is a half-wave voltage, as illustrated.

FIGURE D.08-3
The effect of an open diode in a half-wave rectifier.

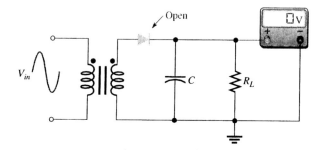

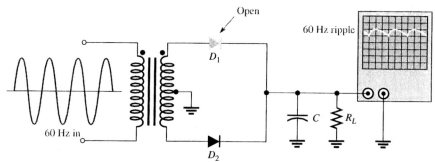

(a) Ripple should be less and have a frequency of 120 Hz.

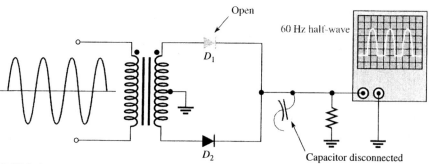

(b) With C removed, output should be a full-wave 120 Hz signal.

FIGURE D.08-4
Symptoms of an open diode in a full-wave, center-tapped rectifier.

With the filter capacitor in the circuit, the half-wave signal will allow it to discharge more than it would with a normal full-wave signal, resulting in a larger ripple voltage. Basically, the same observations would be made for an open failure of diode D_2.

An open diode in a bridge rectifier would create symptoms identical to those just discussed for the center-tapped rectifier. As illustrated in Figure D.08-5, the open diode would prevent current through R_L during half of the input cycle (in this case, the negative half). As a result, there would be a half-wave output and an increased ripple voltage at 60 Hz, as discussed before.

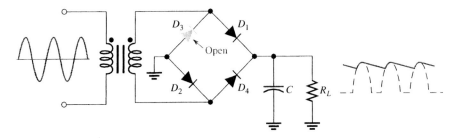

FIGURE D.08-5
Effect of an open diode in a bridge rectifier.

Shorted Diode

A shorted diode is one that has failed such that it has a very low resistance in both directions. If a diode suddenly became shorted, it is likely that a sufficiently high current would exist during one-half of the input cycle such that the fuse in the primary circuit would open. An unfused primary could cause the shorted diode to burn open or the other diode in series with it to burn open. Also, the transformer could be damaged, as illustrated for the case of a bridge rectifier in Figure D.08-6 with D_1 shorted.

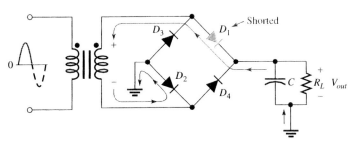

(a) Positive half-cycle: The shorted diode acts as a forward-biased diode, so the load current is normal.

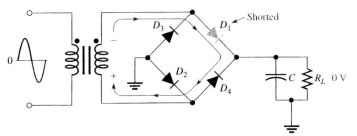

(b) Negative half-cycle: The shorted diode produces a short circuit across the source. As a result D_1, D_4, or the transformer secondary will probably burn open.

FIGURE D.08-6
Effect of a shorted diode in a bridge rectifier.

In part (a) of Figure D.08-6, current is supplied to the load through the shorted diode during the first positive half-cycle, just as though it were forward-biased. During the negative half-cycle, the current is shorted through D_1 and D_4, as shown in part (b). Again, damage to the transformer is possible. Also, it is likely that this excessive current would burn either or both of the diodes open. If only one of the diodes opened, the circuit will operate as a half-wave rectifier. If both diodes opened, there would be no voltage developed across the load. These conditions are illustrated in Figure D.08-7.

FIGURE D.08-7
Effect of open diodes.

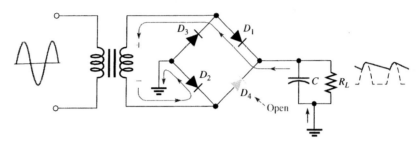

(a) One open diode produces a half-wave output.

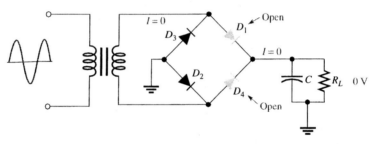

(b) Two open diodes produce a 0 V output.

D.09 Understand principles and operations of optoelectronic circuits (gate isolators, interrupt sensors, infrared sensors, etc.)

INTRODUCTION

Optical couplers are designed to provide complete electrical isolation between an input circuit and an output circuit. The usual purpose of isolation is to provide protection from high-voltage transients, surge voltage, or low-level noise that could possibly result in an erroneous output or damage to the device. Optical couplers also allow interfacing circuits with different voltage levels, different grounds, and so on.

The input circuit of an optical coupler is typically an LED, but the output circuit can take several forms, such as the phototransistor shown in Figure D.09-1(a). When the input voltage forward-biases the LED, light transmitted to the phototransistor turns it on, producing current through the external load, as shown in Figure D.09-1(b). Some typical devices are shown in Figure D.09-1(c).

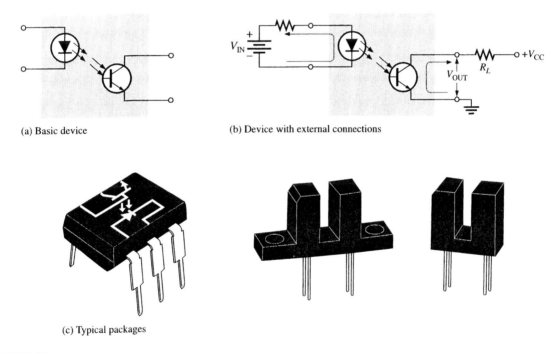

(a) Basic device

(b) Device with external connections

(c) Typical packages

FIGURE D.09-1
Phototransistor couplers.

Several other types of couplers are shown in Figure D.09-2. The darlington transistor coupler in Figure D.09-2(a) can be used when increased output current capability is needed beyond that provided by the phototransistor output. The disadvantage is that the photodarlington has a switching speed less than that of the phototransistor.

An LASCR output coupler is shown in Figure D.09-2(b). This device can be used in applications where, for example, a low-level input voltage is required to latch a high-voltage relay for the purpose of activating some type of electromechanical device.

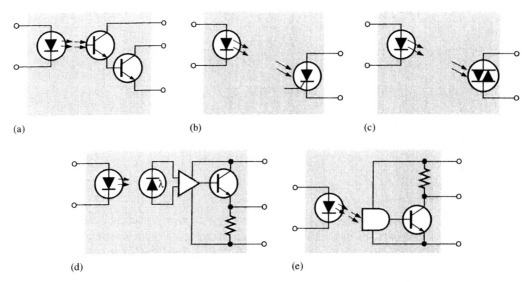

FIGURE D.09-2
Common types of optical-coupling devices.

A phototriac output coupler is illustrated in Figure D.09-2(c). This device is designed for applications requiring isolated triac triggering, such as switching a 110 V AC line from a low-level input.

Figure D.09-2(d) shows an optically isolated AC linear coupler. This device converts an input current variation to an output voltage variation. The output circuit consists of an amplifier with a photodiode across its input terminals. Light variations emitted from the LED are picked up by the photodiode, providing an input signal to the amplifier. The output of the amplifier is buffered with the emitter-follower stage. The optically isolated AC linear coupler can be used for telephone line coupling, peripheral equipment isolation, and audio applications.

Part (e) of Figure D.09-2 shows a digital output coupler. This device consists of a high-speed detector circuit followed by a transistor buffer stage. When there is current through the input LED, the detector is light-activated and turns on the output transistor, so that the collector switches to a low-voltage level. When there is no current through the LED, the output is at the high-voltage level. The digital output coupler can be used in applications requiring compatibility with digital circuits, such as interfacing computer terminals with peripheral devices.

Isolation Voltage   The *isolation voltage* of an optical coupler is the maximum voltage that can exist between the input and output terminals without dielectric breakdown occurring. Typical values are about 7500 V AC peak.

DC Current Transfer Ratio   This parameter is the ratio of the output current to the input current through the LED. It is usually expressed as a percentage. For a phototransistor output, typical values range from 2% to 100%. For a photodarlington output, typical values range from 50% to 500%.

LED Trigger Current   This parameter applies to the LASCR output coupler and the phototriac output device. The *trigger current* is the value of current required to trigger the thyristor output device. Typically, the trigger current is in the mA range.

Transfer Gain   This parameter applies to the optically isolated AC linear coupler. The *transfer gain* is the ratio of output voltage to input current, and a typical value is 200 mV/mA.

Fiber Optics

Fiber optics provides a means for coupling a photo-emitting device to a photodetector via a light-transmitting cable consisting of optical fibers. Typical applications include medical electronics, industrial controls, microprocessor systems, security systems, and communications. Glass fiber cables are used to maximize the optical coupling between optoelectronic devices. Fiber optics is based on the principle of internal reflection.

Every material that conducts light has an **index of refraction, *n*.** A ray of light striking the interface between two materials with different indices of refraction will either be reflected or refracted, depending on the angle at which the light strikes the interface surface, as illustrated in Figure D.09-3. If the incident angle, θ_i, is equal to or greater than a certain value called the *critical angle,* θ_c, the light is reflected. If θ_i is less than the critical angle, the light is refracted.

FIGURE D.09-3
Refraction and reflection of light rays.

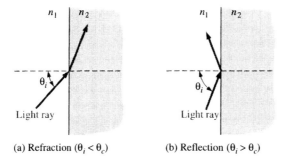

(a) Refraction ($\theta_i < \theta_c$)   (b) Reflection ($\theta_i > \theta_c$)

FIGURE D.09-4
A light ray in optical fiber.

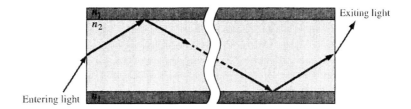

A glass fiber is clad with a layer of glass with a lower index of refraction than the fiber. A ray of light entering the end of the cable will be refracted as shown in Figure D.09-4. If, after refraction, it approaches and hits the glass interface at an angle greater than the critical angle, it is reflected within the fiber as shown. Since, according to a law of physics, the angle of reflection must equal the angle of incidence, the light ray will bounce down the fiber and emerge, refracted, at the other end, as shown in the figure.

THE LIGHT-ACTIVATED SCR (LASCR)

Figure D.09-5 shows a **light-activated silicon-controlled rectifier (LASCR)** schematic symbol and typical packages. The LASCR is most sensitive to light when the gate terminal is open. If necessary, a resistor from the gate to the cathode can be used to reduce the sensitivity. Figure D.09-6 shows an LASCR used to energize a latching relay. The input source turns on the lamp; and the resulting incident light triggers the LASCR. The anode current energizes the relay and closes the contact. Notice that the input source is electrically isolated from the rest of the circuit.

FIGURE D.09-5
Light-activated SCRs.

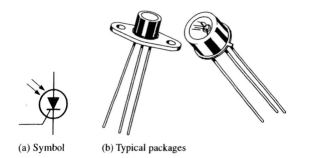

(a) Symbol   (b) Typical packages

FIGURE D.09-6
A light-activated SCR circuit.

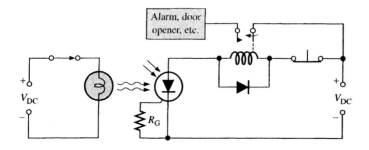

REVIEW QUESTIONS

True/False

1. An optoisolator provides electrical isolation between its input and output.
2. The input-output connection in an optoisolator is a beam of light.

Multiple Choice

3. LASCR stands for
    a. Laser SCR.
    b. Light-activated SCR.
    c. LED-activated SCR.
4. A fiber optic cable relies on
    a. Internal reflection.
    b. External reflection.
    c. Magnetic field propagation.
5. The amount of voltage an optoisolator can withstand between input and output is the
    a. Breakdown voltage.
    b. Breakover voltage.
    c. Isolation voltage.

D.10 Fabricate and demonstrate optoelectronic circuits (gate isolators, interrupt sensors, infrared sensors, etc.)

INTRODUCTION

This section requires an actual breadboard for fabrication and demonstration, since software simulation does not adequately show the properties inherent in optoelectronic devices.

Testing LEDs

Set up the circuit shown in Figure D.10-1. Set the supply voltage (E) to 0 volts. Slowly turn the voltage up. At what voltage (E) does the LED begin to glow? What is the voltage across the LED when it begins to glow?

FIGURE D.10-1
LED biasing circuit.

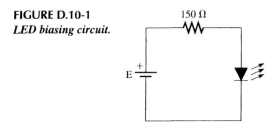

Turn the voltage up until the LED has a sufficient amount of brightness. Record the supply voltage and the LEDs voltage again. Use both sets of measurements to calculate the LED current. How does the LED's current relate to its intensity?

Change from a DC source to an AC source and set the frequency to 1 Hz. Adjust the source voltage to obtain a reasonable amount of brightness when the LED is on. Slowly increase the frequency until the LED appears to be on all the time (even though it is still flashing). What is the frequency that causes this effect?

Testing Optoisolators

Set up the optoisolator circuit shown in Figure D.10-2. Record the output voltage for each setting of E shown in Table D.10-1. Is there any current flow in the output when the input is off? At what input voltage does the output switch from off to on?

FIGURE D.10-2
Optoisolator test circuit.

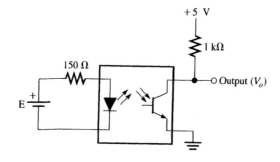

TABLE D.10-1
Test voltages.

E	V_o
0.0	
0.5	
1.0	
1.5	
2.0	
2.5	
3.0	
5.0	

D.11 Troubleshoot and repair optoelectronic circuits (gate isolators, interrupt sensors, infrared sensors, etc.)

INTRODUCTION

Troubleshooting optoelectronic devices can sometimes pose challenges not found in ordinary electronic circuits. For example, how do you identify the problem with a fiber optic circuit? Is it the transmitter (faulty laser diode)? Is it the receiver (faulty photodiode or phototransistor)? Or is there something wrong with the fiber itself? If you do not have access to expensive equipment, such as an optical time domain reflectometer, there may be little you can do to discover the problem.

On the other hand, making simple voltage measurements may be enough to tell you that a LED is not properly biased, or has internally opened or shorted due to stress. Optoelectronic circuits are like all other electronic circuits. They must still obey Ohm's law and all other applicable rules. Data sheets on the device being investigated are especially important as well, since they will state the acceptable range of values during normal operation.

Also, keep in mind that a circuit that once worked failed for a reason. This makes it even more important to verify that the voltages and currents found in the device are acceptable.

D.12 Understand principles and operations of single-stage amplifiers

INTRODUCTION

In this section we examine the basic amplifier configurations for both BJT and FET amplifiers.

COMMON-EMITTER AMPLIFIERS

Figure D.12-1 shows a typical **common-emitter (CE)** amplifier. The one shown has voltage-divider bias, although other types of bias methods are possible. C_1 and C_2 are coupling capacitors used to pass the signal into and out of the amplifier such that the source or load will not affect the DC bias voltages. C_3 is a bypass capacitor that shorts the emitter signal voltage (AC) to ground without disturbing the DC emitter voltage. Because of the bypass capacitor, the emitter is at signal ground (but not DC ground), thus making the circuit a common-emitter amplifier. The purpose of the bypass capacitor is to increase the signal voltage **gain.** (The reason why this increase occurs will be discussed shortly.) Notice that the input signal is applied to the base, and the output signal is taken off the collector. All capacitors are assumed to have a reactance of approximately zero at the signal frequency.

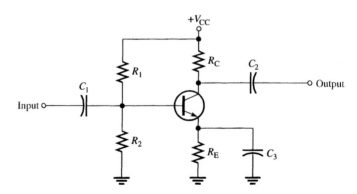

FIGURE D.12-1
Typical common-emitter (CE) amplifier.

A Bypass Capacitor Increases Voltage Gain

The bypass capacitor shorts the signal around the emitter resistor, R_E, in order to increase the voltage gain. To understand why, let's consider the amplifier without the bypass capacitor and see what the voltage gain is. The CE amplifier with the bypass capacitor removed is shown in Figure D.12-2.

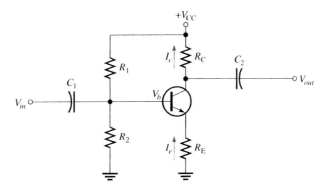

FIGURE D.12-2
CE amplifier with bypass capacitor removed.

As before, lowercase italic subscripts indicate signal (AC) voltages and signal (alternating) currents. The **voltage gain** of the amplifier is V_{out}/V_{in}. The output signal voltage is

$$V_{out} = I_c R_C$$

The signal voltage at the base is approximately equal to

$$V_b \cong V_{in} \cong I_e(r_e + R_E)$$

where r_e is the internal emitter resistance of the transistor. The voltage gain A_v can now be expressed as

$$A_v = \frac{V_{out}}{V_{in}} = \frac{I_c R_C}{I_e(r_e + R_E)}$$

Since $I_c \cong I_e$, the currents cancel and the gain is the ratio of the resistances.

$$A_v = \frac{R_C}{r_e + R_E}$$

Keep in mind that this formula is for the CE configuration without the bypass capacitor. If R_E is much greater than r_e, then $A_v \cong R_C/R_E$.

If the bypass capacitor is connected across R_E, it effectively shorts the signal to ground, leaving only r_e in the emitter. Thus, the voltage gain of the CE amplifier with the bypass capacitor shorting R_E is

$$A_v = \frac{R_C}{r_e}$$

Now, r_e is a very important transistor parameter because it determines the voltage gain of a CE amplifier in conjunction with R_C. A formula for estimating r_e is given without derivation in the following equation:

$$r_e \cong \frac{25 \text{ mV}}{I_E}$$

EXAMPLE D.12-4

Determine the voltage gain of the amplifier in Figure D.12-3 both with and without a bypass capacitor.

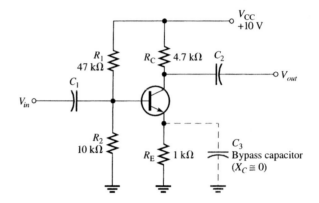

FIGURE D.12-3

Solution   First determine r_e. To do so, you need to find I_E. Thus,

$$V_B \cong \left(\frac{R_2}{R_1 + R_2} \right) V_{CC} = \left(\frac{10 \text{ k}\Omega}{47 \text{ k}\Omega + 10 \text{ k}\Omega} \right) 10 \text{ V} = 1.75 \text{ V}$$

$$V_E = V_B - 0.7 \text{ V} = 1.05 \text{ V}$$

$$I_E = \frac{V_E}{R_E} = \frac{1.05 \text{ V}}{1 \text{ k}\Omega} = 1.05 \text{ mA}$$

$$r_e \cong \frac{25 \text{ mV}}{I_E} \cong \frac{25 \text{ mV}}{1.05 \text{ mA}} = 23.8 \ \Omega$$

The voltage gain without a bypass capacitor is

$$A_v = \frac{R_C}{r_e + R_E} = \frac{4.7 \text{ k}\Omega}{1023.8 \ \Omega} = 4.59$$

The voltage gain with the bypass capacitor installed is

$$A_v = \frac{R_C}{r_e} = \frac{4.7 \text{ k}\Omega}{23.8 \ \Omega} = 197$$

As you can see, the voltage gain is greatly increased by the addition of the bypass capacitor. In terms of decibels (dB), the voltage gain is

$$A_v = 20 \log(197) = 45.9 \text{ dB}$$

Phase Inversion

The output voltage at the collector is 180° out of phase with the input voltage at the base. Therefore, the CE amplifier is characterized by a phase inversion between the input and the output. This inversion is sometimes indicated by a negative voltage gain.

AC Input Resistance

The input resistance "seen" by the signal at the base is derived by having the emitter resistor bypassed to ground.

$$R_{in} = \frac{V_b}{I_b}$$

$$V_b = I_e r_e$$

$$I_e \cong \beta_{ac} I_b$$

$$R_{in} \cong \frac{\beta_{ac} I_b r_e}{I_b}$$

The I_b terms cancel, leaving

$$R_{in} \cong \beta_{ac} r_e$$

Total Input Resistance to a CE Amplifier

Viewed from the base, R_{in} is the AC resistance. The actual resistance seen by the source includes that of bias resistors. We will now develop an expression for the total input resistance. The concept of AC ground was mentioned earlier. At this point it needs some additional explanation because it is important in the development of the formula for total input resistance, $R_{in(tot)}$.

You have already seen that the bypass capacitor effectively makes the emitter appear as ground to the AC signal, because the X_C of the capacitor is nearly zero at the signal frequency. Of course, to a DC signal the capacitor looks like an open and thus does not affect the DC emitter voltage.

In addition to seeing ground through the bypass capacitor, the signal also sees ground through the DC supply voltage source, V_{CC}. It does so because there is zero signal voltage at the V_{CC} terminal. Thus, the $+V_{CC}$ terminal effectively acts as AC ground. As a result, the two bias resistors, R_1 and R_2, appear in parallel to the AC input, because one end of R_2 goes to actual ground and one end of R_1 goes to AC ground (V_{CC} terminal). Also, R_{in} at the base appears in parallel with $R_1 \parallel R_2$. This situation is illustrated in Figure D.12-4.

FIGURE D.12-4
Total input resistance.

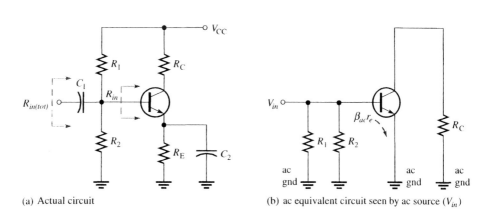

(a) Actual circuit (b) ac equivalent circuit seen by ac source (V_{in})

The expression for the total input resistance to the CE amplifier as seen by the AC source is as follows:

$$R_{in(tot)} = R_1 \parallel R_2 \parallel R_{in}$$

R_C has no effect because of the reverse-biased, base-collector junction.

EXAMPLE D.12-2

Determine the total input resistance seen by the signal source in the CE amplifier in Figure D.12-5. $\beta_{ac} = 150$.

FIGURE D.12-5

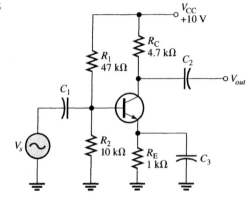

Solution In Example D.12-1, you found r_e for the same circuit. Thus,

$$R_{in} = \beta_{ac}r_e = 150(23.8\ \Omega) = 3.57\ k\Omega$$
$$R_{in(tot)} = R_1 \parallel R_2 \parallel R_{in} = 47\ k\Omega \parallel 10\ k\Omega \parallel 3.57\ k\Omega = 2.49\ k\Omega$$

Current Gain

The signal **current gain** of a CE amplifier is

$$A_i = \frac{I_c}{I_s}$$

where I_s is the source current and is calculated by $V_s/R_{in(tot)}$.

Power Gain

The **power gain** of a CE amplifier is the product of the voltage gain and the current gain.

$$A_p = A_v A_i$$

EXAMPLE D.12-3

Determine the voltage gain, current gain, and power gain for the CE amplifier in Figure D.12-6. $\beta_{ac} = 100$.

FIGURE D.12-6

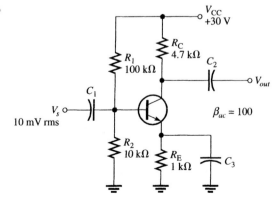

Solution First find r_e. To do so, you must find I_E. Begin by calculating V_B. Since R_{IN} is ten times greater than R_2, it can be neglected. Thus,

$$V_B \cong \left(\frac{R_2}{R_1 + R_2}\right)V_{CC} = \left(\frac{10 \text{ k}\Omega}{110 \text{ k}\Omega}\right)30 \text{ V} = 2.73 \text{ V}$$

$$I_E = \frac{V_E}{R_E} = \frac{V_B - 0.7 \text{ V}}{R_E} = \frac{2.03 \text{ V}}{1 \text{ k}\Omega} = 2.03 \text{ mA}$$

$$r_e \cong \frac{25 \text{ mV}}{I_E} = \frac{25 \text{ mV}}{2.03 \text{ mA}} = 12.3 \text{ }\Omega$$

The AC voltage gain is

$$A_v = \frac{R_C}{r_e} = \frac{4.7 \text{ k}\Omega}{12.3 \text{ }\Omega} = 382$$

Determine the signal current gain by first finding $R_{in(tot)}$ to get I_s.

$$R_{in(tot)} = R_1 \| R_2 \| \beta_{ac}r_e = 100 \text{ k}\Omega \| 10 \text{ k}\Omega \| 1.23 \text{ k}\Omega = 1.08 \text{ k}\Omega$$

$$I_s = \frac{V_s}{R_{in(tot)}} = \frac{10 \text{ mV}}{1.08 \text{ k}\Omega} = 9.26 \text{ }\mu\text{A}$$

Next determine I_c.

$$I_c = \frac{V_{out}}{R_C} = \frac{A_v V_s}{R_C} = \frac{(382)(10 \text{ mV})}{4.7 \text{ k}\Omega} = 813 \text{ }\mu\text{A}$$

$$A_i = \frac{I_c}{I_s} = \frac{813 \text{ }\mu\text{A}}{9.26 \text{ }\mu\text{A}} = 87.8$$

The power gain is

$$A_p = A_v A_i = (382)(87.8) = 33,540$$

The voltage gain and the power gain in decibels are as follows:

$$A_v = 20 \log(382) = 51.6 \text{ dB}$$
$$A_p = 10 \log(33,540) = 45.3 \text{ dB}$$

EXAMPLE D.12-4

Determine the input resistance of the emitter-follower in Figure D.12-7. Also find the voltage gain, current gain, and power gain.

FIGURE D.12-7

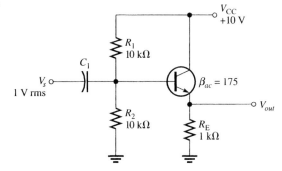

Solution The approximate input resistance viewed from the base is

$$R_{in} \cong \beta_{ac}R_E = (175)(1 \text{ k}\Omega) = 175 \text{ k}\Omega$$

The total input resistance is

$$R_{in(tot)} = R_1 \parallel R_2 \parallel R_{in} = 10 \text{ k}\Omega \parallel 10 \text{ k}\Omega \parallel 175 \text{ k}\Omega = 4.86 \text{ k}\Omega$$

The voltage gain is, neglecting r_e,

$$A_v \cong 1$$

The current gain is

$$A_i = \frac{I_e}{I_s}$$

$$I_e = \frac{V_e}{R_E} = \frac{A_v V_b}{R_E} \cong \frac{1 \text{ V}}{1 \text{ k}\Omega} = 1 \text{ mA}$$

$$I_s = \frac{V_s}{R_{in(tot)}} = \frac{1 \text{ V}}{4.86 \text{ k}\Omega} = 206 \ \mu\text{A}$$

$$A_i = \frac{1 \text{ mA}}{206 \ \mu\text{A}} = 4.85$$

The power gain is

$$A_p \cong A_i = 4.85$$

EXAMPLE D.12-5

Find the input resistance, voltage gain, current gain, and power gain for the CB amplifier in Figure D.12-8.

FIGURE D.12-8

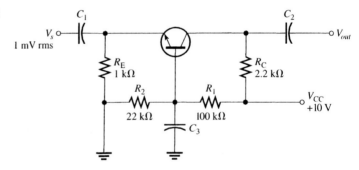

Solution First find I_E so that you can determine r_e. Then $R_{in} = r_e$. Thus,

$$V_B \cong \left(\frac{R_2}{R_1 + R_2}\right) V_{CC} = \left(\frac{22 \text{ k}\Omega}{122 \text{ k}\Omega}\right) 10 \text{ V} = 1.80 \text{ V}$$

$$V_E = V_B - 0.7 \text{ V} = 1.80 \text{ V} - 0.7 \text{ V} = 1.10 \text{ V}$$

$$I_E = \frac{V_E}{R_E} = \frac{1.10 \text{ V}}{1 \text{ k}\Omega} = 1.10 \text{ mA}$$

$$R_{in} = r_e \cong \frac{25 \text{ mV}}{1.10 \text{ mA}} = 22.7 \ \Omega$$

The signal voltage gain is

$$A_v = \frac{R_C}{r_e} = \frac{2.2 \text{ k}\Omega}{22.7 \ \Omega} = 96.9$$

Thus,

$$A_i \cong 1$$
$$A_p \cong 96.9$$

FET AMPLIFIERS

Transconductance of an FET

Recall that in a bipolar transistor, the base current controls the collector current, and the relationship between these two currents is expressed as $I_c = \beta_{ac}I_b$. In an FET, the gate voltage controls the drain current. An important FET parameter is the **transconductance, g_m,** which is defined as

$$g_m = \frac{I_d}{V_{gs}}$$

The transconductance is one factor that determines the voltage gain of an FET amplifier. On data sheets, the transconductance is sometimes called the *forward transadmittance* and is designated y_{fs}.

Common-Source (CS) Amplifiers

A self-biased *n*-channel JFET with an AC source capacitively coupled to the gate is shown in Figure D.12-9. The resistor R_G serves two purposes: (1) It keeps the gate at approximately 0 V DC (because I_{GSS} is extremely small), and (2) its large value (usually several megohms) prevents loading of the AC signal source. The bias voltage is created by the drop across R_S. The bypass capacitor, C_3, keeps the source of the FET effectively at AC ground.

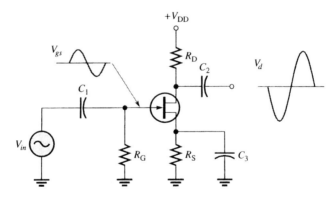

FIGURE D.12-9
JFET common-source amplifier.

The signal voltage causes the gate-to-source voltage to swing above and below its *Q*-point value, causing a swing in drain current. As the drain current increases, the voltage drop across R_D also increases, causing the drain voltage (with respect to ground) to decrease.

The drain current swings above and below its *Q*-point value in phase with the gate-to-source voltage. The drain-to-source voltage swings above and below its *Q*-point value 180° out of phase with the gate-to-source voltage, as illustrated in Figure D.12-9.

D-MOSFET   A zero-biased n-channel D-MOSFET with an AC source capacitively coupled to the gate is shown in Figure D.12-10. The gate is at approximately 0 V DC and the source terminal is at ground, thus making $V_{GS} = 0$ V.

The signal voltage causes V_{gs} to swing above and below its 0 value, producing a swing in I_d. The negative swing in V_{gs} produces the depletion mode, and I_d decreases. The positive swing in V_{gs} produces the enhancement mode, and I_d increases.

E-MOSFET   Figure D.12-11 shows a voltage-divider-biased, n-channel E-MOSFET with an AC signal source capacitively coupled to the gate. The gate is biased with a positive voltage such that $V_{GS} > V_{GS(th)}$, where $V_{GS(th)}$ is the threshold value.

As with the JFET and D-MOSFET, the signal voltage produces a swing in V_{gs} above and below its Q-point value. This swing, in turn, causes a swing in I_d. Operation is entirely in the enhancement mode.

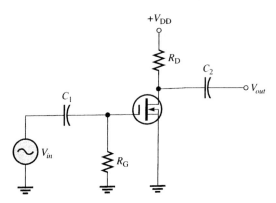

FIGURE D.12-10
Zero-biased D-MOSFET common-source amplifier.

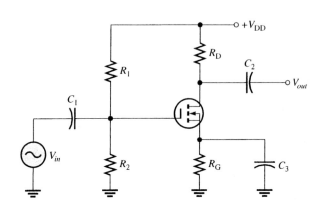

FIGURE D.12-11
Common-source E-MOSFET amplifier with voltage-divider bias.

Voltage Gain   Voltage gain, A_v, of an amplifier always equals V_{out}/V_{in}. In the case of the CS amplifier, V_{in} is equal to V_{gs}, and V_{out} is equal to the signal voltage developed across R_D, which is $I_d R_D$. Thus,

$$A_v = \frac{I_d R_D}{V_{gs}}$$

Since $g_m = I_d/V_{gs}$, the common-source voltage gain is

$$A_v = g_m R_D$$

Input Resistance   Because the input to a CS amplifier is at the gate, the input resistance is extremely high. Ideally, it approaches infinity and can be neglected. As you know, the high input resistance is produced by the reverse-biased pn junction in a JFET and by the insulated gate structure in a MOSFET.

The actual input resistance seen by the signal source is the gate-to-ground resistor R_G in parallel with the FET's input resistance, V_{GS}/I_{GSS}. The reverse leakage current I_{GSS} is typically given on the data sheet for a specific value of V_{GS} so that the input resistance of the device can be calculated.

EXAMPLE D.12-6

(a) What is the total output voltage (DC + AC) of the amplifier in Figure D.12-12? The g_m is 1800 μS, I_D is 2 mA, $V_{GS(off)}$ is -3.5 V, and I_{GSS} is 15 nA.

(b) What is the input resistance seen by the signal source?

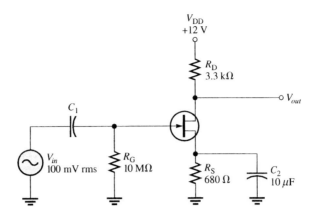

FIGURE D.12-12

Solution

(a) First, find the DC output voltage.

$$V_D = V_{DD} - I_D R_D = 12 \text{ V} - (2 \text{ mA})(3.3 \text{ k}\Omega) = 5.4 \text{ V}$$

Next, find the AC output voltage by using the gain formula.

$$A_v = \frac{V_{out}}{V_{in}} = g_m R_D$$

$$V_{out} = g_m R_D V_{in} = (1800 \ \mu\text{S})(3.3 \text{ k}\Omega)(100 \text{ mV}) = 594 \text{ mV rms}$$

The total output voltage is an AC signal with a peak-to-peak value of 594 mV ×
2.828 = 1.67 V, riding on a DC level of 5.4 V.

(b) The input resistance is determined as follows (since $V_G = 0$ V):

$$V_{GS} = I_D R_S = (2 \text{ mA})(680 \ \Omega) = 1.36 \text{ V}$$

The input resistance at the gate of the JFET is

$$R_{IN(gate)} = \frac{V_{GS}}{I_{GSS}} = \frac{1.36 \text{ V}}{15 \text{ nA}} = 91 \text{ M}\Omega$$

The input resistance seen by the signal source is

$$R_{in} = R_G \parallel R_{IN(gate)} = 10 \text{ M}\Omega \parallel 91 \text{ M}\Omega = 9.0 \text{ M}\Omega$$

EXAMPLE D.12-7

(a) Determine the voltage gain of the amplifier in Figure D.12-13(a) using the data
sheet information in Figure D.12-13(b).

(b) Also determine the input resistance. Assume minimum data sheet values where
available.

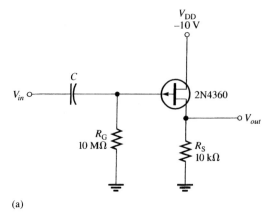

(a)

*ELECTRICAL CHARACTERISTICS (T_A = 25°C unless otherwise noted)

Characteristic	Symbol	Min	Max	Unit
OFF CHARACTERISTICS				
Gate-Source Breakdown Voltage (I_G = 10 μAdc, V_{DS} = 0)	$V_{(BR)GSS}$	20	–	Vdc
Gate-Source Cutoff Voltage (V_{DS} = –10 Vdc, I_D = 1.0 μAdc)	$V_{GS(off)}$	0.7	10	Vdc
Gate Reverse Current (V_{GS} = 15 Vdc, V_{DS} = 0) (V_{GS} = 15 Vdc, V_{DS} = 0, T_A = 65°C)	I_{GSS}	– –	10 0.5	nAdc μAdc
ON CHARACTERISTICS				
Zero-Gate Voltage Drain Current (Note 1) (V_{DS} = –10 Vdc, V_{GS} = 0)	I_{DSS}	3.0	30	mAdc
Gate-Source Breakdown Voltage (V_{DS} = –10 Vdc, I_D = 0.3 mAdc)	V_{GS}	0.4	9.0	Vdc
SMALL SIGNAL CHARACTERISTICS				
Drain-Source "ON" Resistance (V_{GS} = 0, I_D = 0, f = 1.0 kHz)	$r_{ds(on)}$	–	700	Ohms
Forward Transadmittance (Note 1) (V_{DS} = –10 Vdc, V_{GS} = 0, f = 1.0 kHz)	$|y_{fs}|$	2000	8000	μmhos
Forward Transconductance (V_{DS} = –10 Vdc, V_{GS} = 0, f = 1.0 MHz)	$Re(y_{fs})$	1500	–	μmhos
Output Admittance (V_{DS} = –10 Vdc, V_{GS} = 0, f = 1.0 kHz)	$|y_{os}|$	–	100	μmhos
Input Capacitance (V_{DS} = –10 Vdc, V_{GS} = 0, f = 1.0 MHz)	C_{iss}	–	20	pF
Reverse Transfer Capacitance (V_{DS} = –10 Vdc, V_{GS} = 0, f = 1.0 MHz)	C_{rss}	–	5.0	pF
Common-Source Noise Figure (V_{DS} = –10 Vdc, I_D = 1.0 mAdc, R_G = 1.0 Megohm, f = 100 Hz)	NF	–	5.0	dB
Equivalent Short-Circuit Input Noise Voltage (V_{DS} = –10 Vdc, I_D = 1.0 mAdc, f = 100 Hz, BW = 15 Hz)	E_n	–	0.19	μV/$\sqrt{Hz}$

*Indicates JEDEC Registered Data.

Note 1: Pulse Test: Pulse Width ≤ 630 ms, Duty Cycle ≤ 10%.

(b)

FIGURE D.12-13

Solution

(a) From the data sheet, $g_m = y_{fs} = 2000$ μS minimum. The gain is

$$A_v \cong \frac{g_m R_S}{1 + g_m R_S} = \frac{(2000 \ \mu S)(10 \ k\Omega)}{1 + (2000 \ \mu S)(10 \ k\Omega)} = 0.952$$

(b) From the data sheet, $I_{GSS} = 10$ nA at V_{GS} = 15 V. Therefore,

$$R_{IN(gate)} = \frac{15 \ V}{10 \ nA} = 1500 \ M\Omega$$

$$R_{IN} = R_G \parallel R_{IN(gate)} = 10 \ M\Omega \parallel 1500 \ M\Omega = 9.93 \ M\Omega$$

EXAMPLE D.12-8

(a) Determine the voltage gain of the amplifier in Figure D.12-14.
(b) Determine the input resistance.

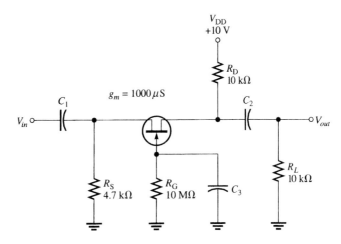

FIGURE D.12-14

Solution

(a) This CG amplifier has a load resistor effectively in parallel with R_D, so the effective drain resistance is $R_D \parallel R_L$ and the gain is

$$A_v = g_m(R_D \parallel R_L) = (1000 \ \mu S)(10 \ k\Omega \parallel 10 \ k\Omega) = 5$$

(b) The input resistance at the source terminal is

$$R_{in(source)} = \frac{1}{g_m} = \frac{1}{1000 \ \mu S} = 1 \ k\Omega$$

The signal source actually sees R_S in parallel with $R_{in(source)}$, so the total input resistance is

$$R_{in(tot)} = R_{in(source)} \parallel R_S = 1 \ k\Omega \parallel 4.7 \ k\Omega = 825 \ \Omega$$

REVIEW QUESTIONS

True/False

1. The bypass capacitor increases the voltage gain of the CE amplifier.
2. Coupling capacitors affect the high-frequency operation of an amplifier.

Multiple Choice

3. Which configuration has a voltage gain close to 1.0?
   a. Common base.
   b. Common collector.
   c. Common emitter.

4. In a class A amplifier the Q point is biased
   a. Near cutoff.
   b. Near the middle of the DC load line.
   c. Near saturation.

5. The internal capacitance of a transistor causes trouble at
   a. Low frequencies.
   b. High frequencies.
   c. No frequency (effect is negligible).

D.13 Fabricate and demonstrate single-stage amplifiers

INTRODUCTION

In this section we examine the use of a software package called *Electronics Workbench* to evaluate the design of a common emitter amplifier. The schematic of the amplifier is entered graphically and simulated in real time. Virtual instruments record the necessary information.

The amplifier is simulated as shown in Figure D.13-1. During operation, a virtual oscilloscope connected to the amplifier shows the following waveforms presented in Figure D.13-2.

Using *EWB,* the amplitude and frequency of the source can be varied to get the total response of the amplifier (or you can use the virtual Bode plotter instrument). DC meters can be connected as well, allowing you to measure the voltages and currents present in the biasing components.

FIGURE D.13-1
CE amplifier design.

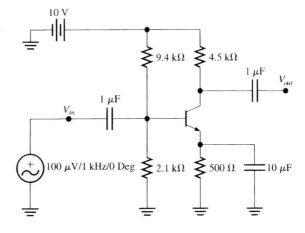

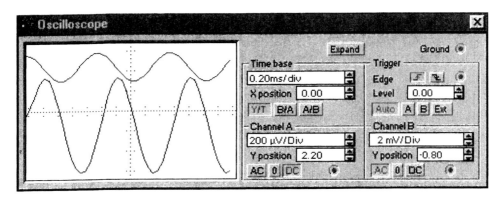

FIGURE D.13-2
EWB oscilloscope display.

D.14 Troubleshoot and repair single-stage amplifiers

INTRODUCTION

In this section we examine troubleshooting procedures for both BJT and FET amplifiers. The two-stage amplifiers used as examples illustrate how a stage may be isolated and tested individually.

TROUBLESHOOTING THE BJT AMPLIFIER

When you are faced with having to troubleshoot a circuit, the first thing you must have is a schematic with the proper DC and signal voltages labeled. You must know what the correct voltages in the circuit should be before you can identify an incorrect voltage. Schematics of some circuits are available with voltages indicated at certain points. If this is not the case, you must use your knowledge of the circuit operation to determine the correct voltages. Figure D.14-1 is the schematic for a two-stage BJT amplifier. The correct voltages are indicated at each point.

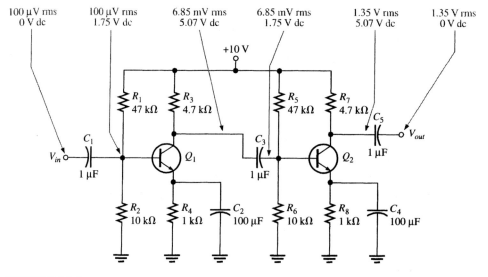

FIGURE D.14-1
Two-stage common-emitter amplifier with correct voltages indicated.

The Complete Troubleshooting Process

Beginning with a malfunctioning or nonfunctioning circuit or system, a typical troubleshooting procedure may go as follows:

1. Identify the symptom(s).

2. Perform a power check.

3. Perform a sensory check.

4. Apply a signal-tracing technique to isolate the fault to a single circuit.

5. Apply fault analysis to isolate the fault further to a single component or group of components.

6. Use replacement or repair to fix the problem.

Applying the Troubleshooting Process to a Two-Stage Amplifier To determine the faulty component in a multistage amplifier, use the general six-step troubleshooting procedure which is illustrated as follows.

Step 1: *Check the input and output voltages.* Assume the measurements indicate that the input signal voltage is correct. However, there is no output signal voltage or the output signal voltage is much less than it should be, as shown by the diagram in Figure D.14-2.

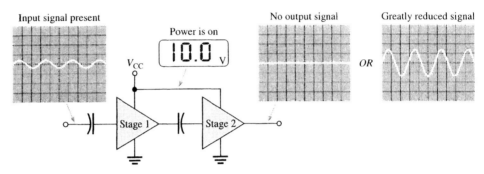

FIGURE D.14-2
Initial check of a faulty two-stage amplifier.

Step 2: *Perform a power check.* Assume the DC supply voltage is correct as indicated in Figure D.14-2.

Step 3: *Perform a sensory check.* Assume there are no detectable signs of a fault.

Step 4: *Apply the midpoint method of signal tracing.* Check the voltages at the output of the first stage. No signal voltage or a reduced signal voltage indicates that the problem is in the first stage. An incorrect DC voltage also indicates a first stage problem. If the signal voltage and the DC voltage are correct at the output of the first stage, the problem is in the second stage. After this check, you have narrowed the problem to one of the two stages. This step is illustrated in Figure D.14-3.

FIGURE D.14-3
Midpoint signal tracing isolates the faulty stage.

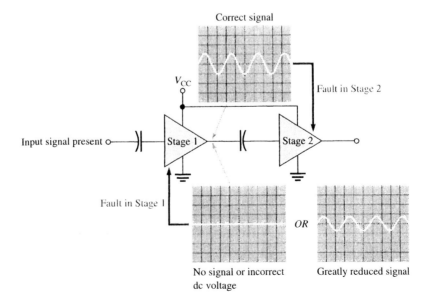

Step 5:   *Apply fault analysis.* Focus on the faulty stage and determine the component failure that can produce the incorrect output.

Symptom: DC voltages incorrect.
Faults: A failure of any resistor or the transistor will produce an incorrect DC bias voltage. A leaky bypass or coupling capacitor will also affect the DC bias voltages. Further measurements in the stage are necessary to isolate the faulty component.

Incorrect AC voltages and the most likely fault(s) are illustrated in Figure D.14-4 as follows:

(a) *Symptom 1:* Signal voltage at output missing; DC voltage correct.
   Symptom 2: Signal voltage at base missing; DC voltage correct.
   Fault: Input coupling capacitor open. This prevents the signal from getting to the base.

(b) *Symptom:* Correct signal at base but no output signal.
   Fault: Transistor base open.

(c) *Symptom:* Signal voltage at output greatly reduced; DC voltage correct.
   Fault: Bypass capacitor open.

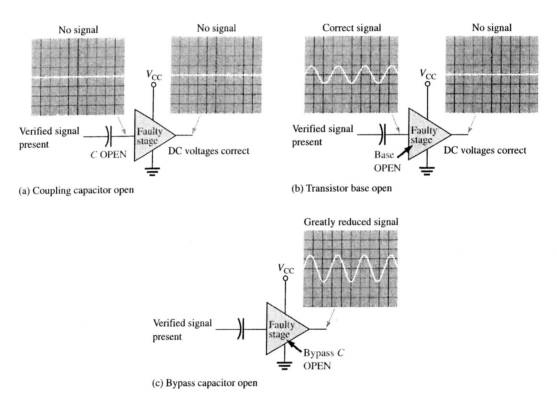

FIGURE D.14-4
Troubleshooting a faulty stage.

Step 6:   *Replace or repair.* With the power turned off, replace the defective component or repair the defective connection. Turn on the power, and check for proper operation.

EXAMPLE D.14-1

The two-stage amplifier in Figure D.14-1 has malfunctioned. Specify the step-by-step troubleshooting procedure for an assumed fault.

Solution   The troubleshooting procedure for a certain fault scenario is as follows:

Step 1:   There is a verified input signal voltage, but no output signal voltage is measured.

Step 2:   There is power to the circuit as indicated by a correct V_{CC} measurement.

Step 3:   Assume there are no visual or other indications of a problem such as a charred resistor, solder splash, wire clipping, broken connection, or extremely hot component.

Step 4:   The signal voltage and the DC voltage at the collector of Q_1 are correct. This means that the problem is in the second stage or in the coupling capacitor C_3 between the stages.

Step 5:   The correct signal voltage and DC bias voltage are measured at the base of Q_2. This eliminates the possibility of a fault in C_3 or the second-stage bias circuit.

   The collector of Q_2 is at 10 V and there is no signal voltage. This measurement, made directly on the transistor collector, indicates that either the collector is shorted to V_{CC} or the transistor is internally open.

   A visual inspection reveals no physical shorts between V_{CC} and the collector connection. It is unlikely that the collector resistor R_7 is shorted, but to verify, turn off the power and use an ohmmeter to check.

   The possibility of a short is eliminated by the ohmmeter check. The other possible faults are (a) transistor Q_2 internally open or (b) emitter resistor or connection open. Use a transistor tester and/or ohmmeter to check each of these possible faults with power off.

Step 6:   Replace the faulty component or repair open connection and retest the circuit for proper operation.

TROUBLESHOOTING FET AMPLIFIERS

Assume that you are given a circuit board pulled from the audio amplifier section of a sound system and told simply that it is not working properly. The first step is to obtain the system schematic and locate this particular circuit on it. The circuit is a two-stage FET amplifier, as shown in Figure D.14-5. The problem is approached in the following sequence.

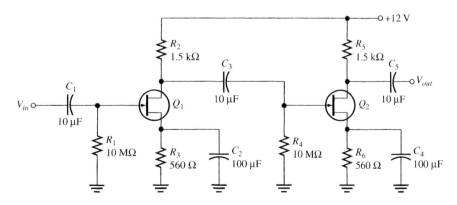

FIGURE D.14-5
A two-stage FET amplifier circuit.

Step 1: Determine what the voltage levels in the circuit should be so that you know what to look for. First, pull a data sheet on the particular transistor (assume both Q_1 and Q_2 are found to be the same type of transistor) and determine the g_m so that you can calculate the voltage gain. Assume that for this particular device, a typical g_m of 5000 μS is specified. Calculate the expected typical voltage gain of each stage (notice they are identical). Because the input resistance is very high, the second stage does not significantly load the first stage, as in a bipolar amplifier. So, the unloaded voltage gain for each stage is

$$A_v = g_m R_2 = (5000 \ \mu S)(1.5 \ k\Omega) = 7.5$$

Since the stages are identical, the typical overall gain should be

$$A'_v = (7.5)(7.5) = 56.3$$

We will ignore DC levels at this time and concentrate on signal tracing.

Step 2: Arrange a test setup to permit connection of an input test signal, a DC supply voltage, and ground to the circuit. The schematic shows that the DC supply voltage must be +12 V. Choose 10 mV rms as an input test signal. This value is arbitrary (although the capability of your signal source is a factor), but it must be small enough that the expected output signal voltage is well below the absolute peak-to-peak limit of 12 V set by the supply voltage and ground (you know that the output voltage swing cannot go higher than 12 V or lower than 0 V). Set the frequency of the sine wave signal source to an arbitrary value in the audio range (say, 10 kHz), since you know this is an audio amplifier. The audio frequency range is generally accepted as 20 Hz to 20 kHz.

Step 3: Check the input signal at the gate of Q_1 and the output signal at the drain of Q_2 with an oscilloscope. The results are shown in Figure D.14-6. The measured output voltage has a peak value of 226 mV. The expected typical peak output voltage is

$$V_{out} = V_{in} A'_v = (14.14 \ mV)(56.3) = 796 \ mV \ peak$$

The output is much less than it should be.

Step 4: Trace the signal from the output back toward the input to determine the fault. Figure D.14-6 shows the oscilloscope displays of the measured signal voltages. The voltage at the gate of Q_2 is 106 mV peak, as expected (14.14 mV × 7.5 = 106 mV). This signal is properly coupled from the drain of Q_1. Therefore, the problem lies in the second stage. From the oscilloscope displays, the gain of Q_2 is much lower than it should be (213 mV/100 mV = 2.13 instead of 7.5).

Step 5: Analyze the possible causes of the observed malfunction. There are three possible reasons the gain is low:

1. Q_2 has a lower transconductance (g_m) than the specified typical value. Check the data sheet to see if the minimum g_m accounts for the lower measured gain.

2. R_5 has a lower value than shown on the schematic.

3. The bypass capacitor C_4 is open.

The only way to check the g_m is by replacing Q_2 with a new transistor of the same type and rechecking the output signal. You can make certain that R_5 is the proper value by removing one end of the resistor from the circuit board and measuring the resistance with an ohmmeter. To avoid having to unsolder a component, the best way to start isolating the fault is by checking the signal voltage at the source of Q_2. If the capacitor is working properly, there will be only a DC voltage at the source. The presence of a signal voltage at the source indicates that C_4 is open. With R_6 unbypassed, the gain expression is $g_m R_d/(1 + g_m R_d)$ rather than simply $g_m R_d$, thus resulting in less gain.

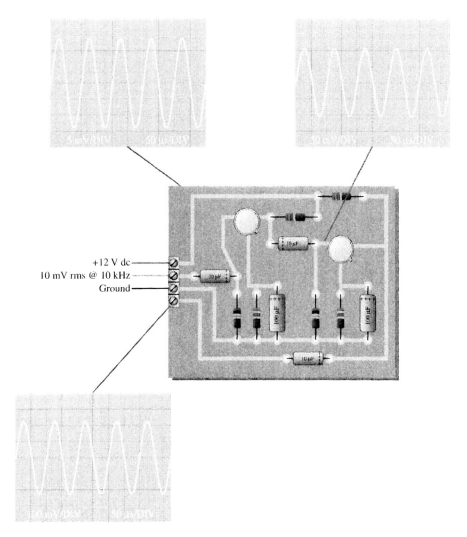

FIGURE D.14-6
Oscilloscope displays of signals in the two-stage FET amplifier.

D.15 Understand principles and operations of thyristor circuitry (SCR, triac, diac, etc.)

INTRODUCTION

SCRs, diacs, and triacs are the semiconductors of choice for applications requiring full DC or AC control of voltage or current. In this section we examine the operation of these important semiconductor devices.

THE SILICON-CONTROLLED RECTIFIER (SCR)

The basic structure of a **silicon-controlled rectifier (SCR)** is shown in Figure D.15-1(a) and the schematic symbol is shown in Figure D.15-1(b). Typical SCR packages are shown in Figure D.15-1(c). Other types of thyristors are found in the same or similar packages.

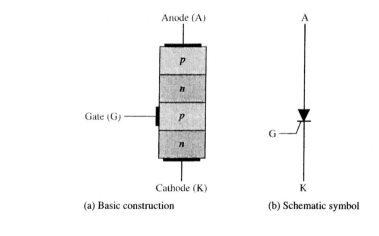

(a) Basic construction   (b) Schematic symbol

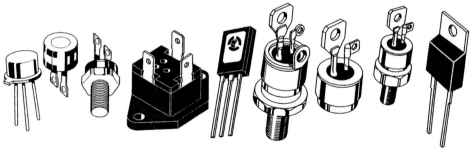

(c) Typical packages

FIGURE D.15-1
The silicon-controlled rectifier (SCR).

SCR Equivalent Circuit

Like the Shockley diode operation, the SCR operation can best be understood by thinking of its internal *pnpn* structure as a two-transistor arrangement, as shown in Figure D.15-2. This structure is like that of the Shockley diode except for the gate connection. The upper *pnp* layers act as a transistor, Q_1, and the lower *npn* layers act as a transistor, Q_2. Again, notice that the two middle layers are "shared."

FIGURE D.15-2
SCR equivalent circuit.

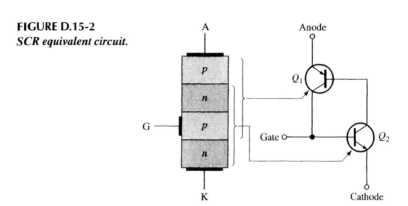

Turning the SCR On

When the gate current, I_G, is zero, as shown in Figure D.15-3(a), the device acts as a Shockley diode in the *off* state. In this state, the very high resistance between the anode and cathode can be approximated by an open switch, as indicated. When a positive pulse of current (**trigger**) is applied to the gate, both transistors turn on (the anode must be more positive than the cathode). This action is shown in Figure D.15-3(b). I_{B2} turns on Q_2, providing a path for I_{B1} out of the Q_2 collector, thus turning on Q_1. The collector current of Q_1 provides additional base current for Q_2 so that Q_2 stays in conduction after the trigger pulse is removed from the gate. By this regenerative action, Q_2 sustains the saturated conduction of Q_1 by providing a path for I_{B1}; in turn, Q_1 sustains the saturated conduction of Q_2 by providing I_{B2}. Thus, the device stays on (latches) once it is triggered on, as shown in Figure D.15-3(c). In this state, the very low resistance between the anode and cathode can be approximated by a closed switch, as indicated.

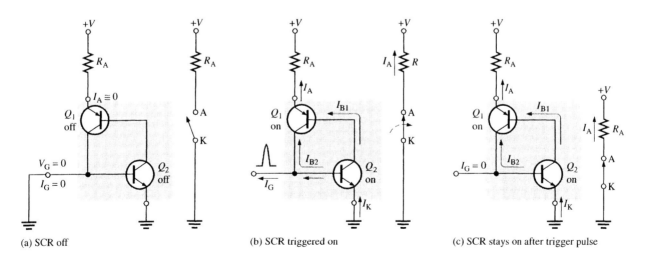

(a) SCR off   (b) SCR triggered on   (c) SCR stays on after trigger pulse

FIGURE D.15-3
The SCR turn-on process with the switch equivalents shown.

Like the Shockley diode, an SCR can also be turned on without gate triggering by increasing anode-to-cathode voltage to a value exceeding the forward-breakover voltage $V_{BR(F)}$, as shown on the characteristic curve in Figure D.15-4(a). The forward-breakover voltage decreases as I_G is increased above 0 V, as shown by the set of curves in Figure D.15-4(b). Eventually, a value of I_G is reached at which the SCR turns on at a very low anode-to-cathode voltage. So, as you can see, the gate current controls the value of forward voltage $V_{BR(F)}$ required for turn-on.

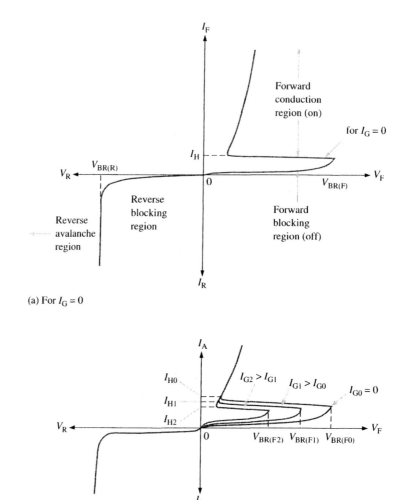

(a) For $I_G = 0$

(b) For various I_G values

FIGURE D.15-4
SCR characteristic curves.

Although anode-to-cathode voltages in excess of $V_{BR(F)}$ will not damage the device if current is limited, this situation should be avoided because the normal control of the SCR is lost. It should normally be triggered on only with a pulse at the gate.

Turning the SCR Off

When the gate returns to 0 V after the trigger pulse is removed, the SCR cannot turn off; it stays in the forward-conduction region. The anode current must drop below the value of the holding current, I_H, in order for turn-off to occur. The holding current is indicated in Figure D.15-4.

There are two basic methods for turning off an SCR: *anode current interruption* and *forced commutation*. The anode current can be interrupted by either a momentary series or parallel switching arrangement, as shown in Figure D.15-5. The series switch in part (a)

FIGURE D.15-5
SCR turn-off by anode current interruption.

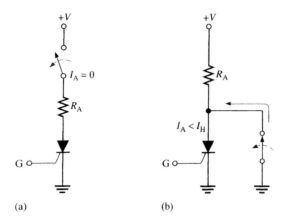

(a)   (b)

simply reduces the anode current to zero and causes the SCR to turn off. The parallel switch in part (b) routes part of the total current away from the SCR, thereby reducing the anode current to a value less than I_H.

The **forced commutation** method basically requires momentarily forcing current through the SCR in the direction opposite to the forward conduction so that the net forward current is reduced below the holding value. The basic circuit, as shown in Figure D.15-6, consists of a switch (normally a transistor switch) and a battery in parallel with the SCR. While the SCR is conducting, the switch is open, as shown in part (a). To turn off the SCR, the switch is closed, placing the battery across the SCR and forcing current through it opposite to the forward current, as shown in part (b). Typically, turn-off times for SCRs range from a few microseconds up to about 30 μs.

FIGURE D.15-6
SCR turn-off by forced commutation.

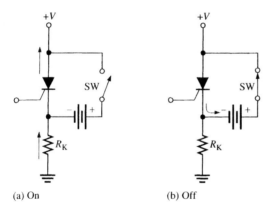

(a) On   (b) Off

SCR Characteristics and Ratings

Several of the most important SCR characteristics and ratings are defined as follows. Use the curve in Figure D.15-4(a) for reference where appropriate.

Forward-breakover voltage, $V_{BR(F)}$ This is the voltage at which the SCR enters the forward-conduction region. The value of $V_{BR(F)}$ is maximum when $I_G = 0$ and is designated $V_{BR(F0)}$. When the gate current is increased, $V_{BR(F)}$ decreases and is designated $V_{BR(F1)}$, $V_{BR(F2)}$, and so on, for increasing steps in gate current (I_{G1}, I_{G2}, and so on).

Holding current, I_H   This is the value of anode current below which the SCR switches from the forward-conduction region to the forward-blocking region. The value increases with decreasing values of I_G and is maximum for $I_G = 0$.

Gate trigger current, I_{GT}   This is the value of gate current necessary to switch the SCR from the forward-blocking region to the forward-conduction region under specified conditions.

Average forward current, $I_{F(avg)}$ This is the maximum continuous anode current (DC) that the device can withstand in the conduction state under specified conditions.

Forward-conduction region   This region corresponds to the *on* condition of the SCR where there is forward current from cathode to anode through the very low resistance (approximate short) of the SCR.

Forward- and reverse-blocking regions   These regions correspond to the *off* condition of the SCR where the forward current from cathode to anode is blocked by the effective open circuit of the SCR.

Reverse-breakdown voltage, $V_{BR(R)}$ This parameter specifies the value of reverse voltage from cathode to anode at which the device breaks into the avalanche region and begins to conduct heavily (the same as in a *pn* junction diode).

THE SILICON-CONTROLLED SWITCH (SCS)

The symbol and terminal identification for the **silicon-controlled switch (SCS)** are shown in Figure D.15-7.

FIGURE D.15-7
The silicon-controlled switch (SCS).

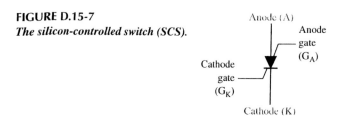

As with the previous four-layer devices, the basic operation of the SCS can be understood by referring to the transistor equivalent, shown in Figure D.15-8. To start, assume that both Q_1 and Q_2 are off, and therefore that the SCS is not conducting. A positive pulse on the cathode gate drives Q_2 into conduction and thus provides a path for Q_1 base current. When Q_1 turns on, its collector current provides base current for Q_2, thus sustaining the *on* state of the device. This regenerative action is the same as in the turn-on process of the SCR and the Shockley diode and is illustrated in Figure D.15-8(a).

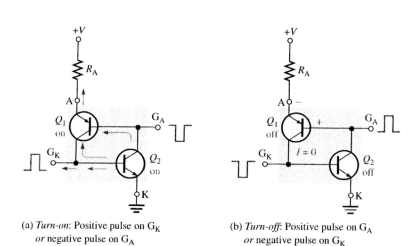

(a) *Turn-on*: Positive pulse on G_K
   or negative pulse on G_A

(b) *Turn-off*: Positive pulse on G_A
   or negative pulse on G_K

FIGURE D.15-8
SCS operation.

The SCS can also be turned on with a negative pulse on the anode gate, as indicated in Figure D.15-8(a). This drives Q_1 into conduction which, in turn, provides base current for Q_2. Once Q_2 is on, it provides a path for Q_1 base current, thus sustaining the *on* state.

To turn the SCS off, a positive pulse is applied to the anode gate. This reverse-biases the base-emitter junction of Q_1 and turns it off. Q_2, in turn, cuts off and the SCS ceases conduction, as shown in Figure D.15-8(b). The device can also be turned off with a negative pulse on the cathode gate, as indicated in part (b). The SCS typically has a faster turn-off time than the SCR.

In addition to the positive pulse on the anode gate or the negative pulse on the cathode gate, there are other methods for turning off an SCS. Figure D.15-9(a) and (b) shows two switching methods to reduce the anode current below the holding value. In each case, the bipolar transistor acts as a switch.

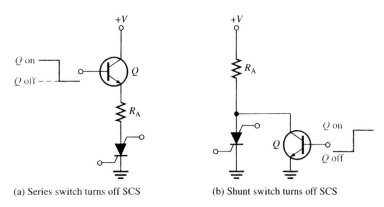

(a) Series switch turns off SCS   (b) Shunt switch turns off SCS

FIGURE D.15-9
Transistor switches reduce I_A below I_H and turn off the SCS.

Applications

The SCS and SCR are used in similar applications. The SCS has the advantage of faster turn-off with pulses on either gate terminal; however, it is more limited in terms of maximum current and voltage ratings. Also, the SCS is sometimes used in digital applications such as counters, registers, and timing circuits.

THE DIAC AND TRIAC

Both the diac and the triac are types of thyristors that can conduct current in both directions (bilateral). The difference between the two devices is that the diac has two terminals, while the triac has a third terminal, which is the gate. The diac functions basically like two parallel Shockley diodes turned in opposite directions. The triac functions basically like two parallel SCRs turned in opposite directions with a common gate terminal.

The Diac

The **diac** basic construction and schematic symbol are shown in Figure D.15-10. Notice that there are two terminals, labelled A_1 and A_2. Conduction occurs in the diac when the breakover voltage is reached with either polarity across the two terminals. The curve in Figure D.15-11 illustrates this characteristic. Once breakover occurs, current is in a direction depending on the polarity of the voltage across the terminals. The device turns off when the current drops below the holding value.

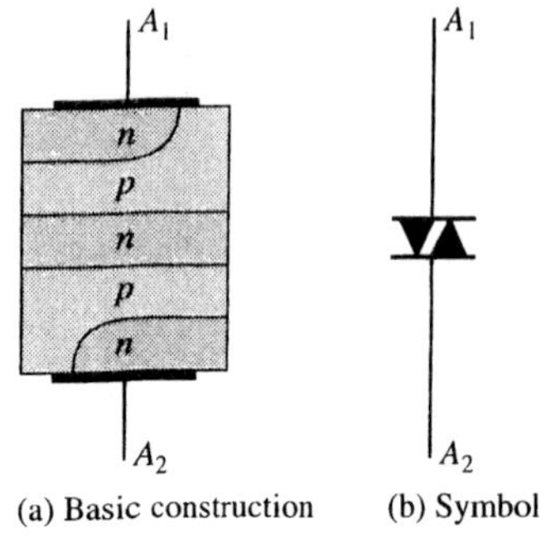

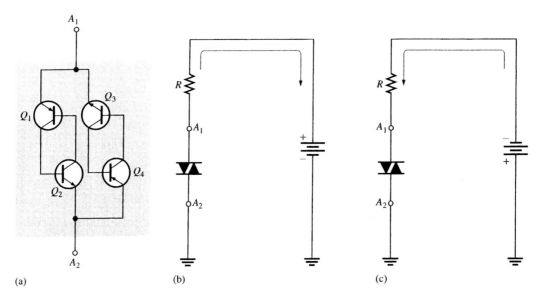

(a) Basic construction   (b) Symbol

FIGURE D.15-10
The diac.

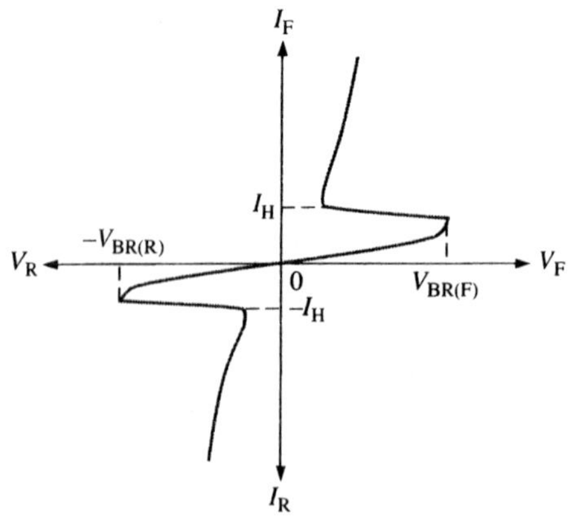

FIGURE D.15-11
Diac characteristic curve.

The equivalent circuit of a diac consists of four transistors arranged as shown in Figure D.15-12(a). When the diac is biased as in Figure D.15-12(b), the *pnpn* structure from A_1 to A_2 provides the four-layer device operation as was described for the Shockley diode. In the equivalent circuit, Q_1 and Q_2 are forward-biased, and Q_3 and Q_4 are reverse-biased. The device operates on the upper right portion of the characteristic curve in Figure D.15-11 under this bias condition. When the diac is biased as shown in Figure D.15-12(c), the *pnpn* structure from A_2 and A_1 is used. In the equivalent circuit, Q_3 and Q_4 are forward-biased, and Q_1 and Q_2 are reverse-biased. Under this bias condition, the device operates on the lower left portion of the characteristic curve, as shown in Figure D.15-11.

(a)   (b)   (c)

FIGURE D.15-12
Diac equivalent circuit and bias conditions.

The Triac

The **triac** is like a diac with a gate terminal. The triac can be turned on by a pulse of gate current and does not require the breakover voltage to initiate conduction, as does the diac. Basically, the triac can be thought of simply as two SCRs connected in parallel and in opposite directions with a common gate terminal. Unlike the SCR, the triac can conduct current in either direction when it is triggered on, depending on the polarity of the voltage across its A_1 and A_2 terminals. Figure D.15-13(a) and (b) shows the basic construction and schematic symbol for the triac. The characteristic curve is shown in Figure D.15-14. Notice that the breakover potential decreases as the gate current increases, just as with the SCR.

As with other four-layer devices, the triac ceases to conduct when the anode current drops below the specified value of the holding current, I_H. The only way to turn off the triac is to reduce the current to a sufficiently low level.

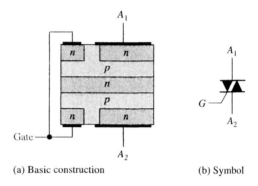

(a) Basic construction   (b) Symbol

FIGURE D.15-13
The triac.

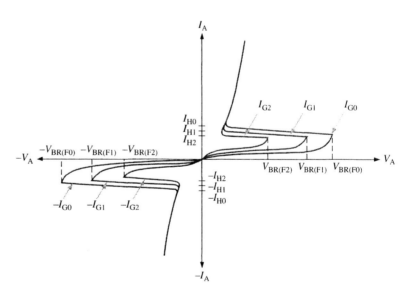

FIGURE D.15-14
Triac characteristic curves.

Figure D.15-15 shows the triac being triggered into both directions of conduction. In part (a), terminal A_1 is biased positive with respect to A_2, so the triac conducts as shown when triggered by a positive pulse at the gate terminal. The transistor equivalent circuit in part (b) shows that Q_1 and Q_2 conduct when a positive trigger pulse is applied. In part (c), terminal A_2 is biased positive with respect to A_1, so the triac conducts as shown. In this case, Q_3 and Q_4 conduct as indicated in part (d) upon application of a positive trigger pulse.

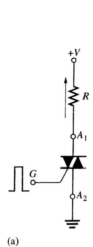

(a)

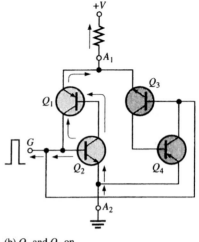

(b) Q_1 and Q_2 on

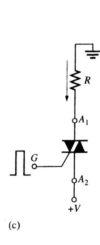

(c)

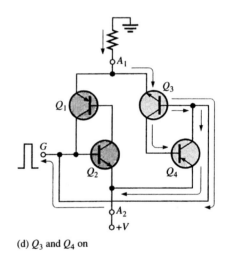

(d) Q_3 and Q_4 on

FIGURE D.15-15
Bilateral operation of a triac.

REVIEW QUESTIONS

True/False

1. Thyristors are designed for AC operation only.

2. The SCR remains on until its current drops below the holding current.

Multiple Choice

3. Forced commutation is used to

   a. Turn an SCR on.

   b. Turn an SCR off.

   c. Trigger an SCS.

4. Exceeding the breakover voltage is the normal way to turn on a(n)

   a. SCR.

   b. Diac.

   c. Triac.

5. Controlled current conduction in both directions is possible with a(n)

   a. SCR.

   b. Diac.

   c. Triac.

D.16 Fabricate and demonstrate thyristor circuitry (SCR, triac, diac, etc.)

INTRODUCTION

Your company is in the process of designing an optical counting and control system for controlling the rate at which objects on an assembly line are moved. In this system, objects passing on an assembly line conveyor belt are counted, and the speed of the conveyor is adjusted according to a predetermined rate. The focus of this application is the conveyor motor speed-control circuit.

Basic Operation of the System

The system controls the speed of the conveyor so that a preset average number of un-equally spaced parts flow past a point on the production line in a specified period of time. This is to allow a proper amount of time for the production line workers to perform certain tasks on each part. A basic diagram of the conveyor speed-control system is shown in Figure D.16-1.

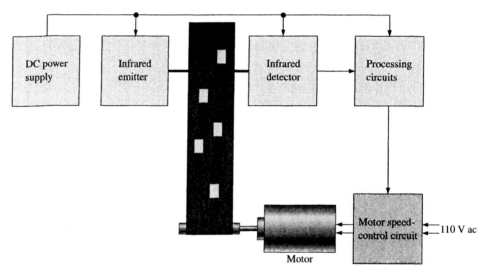

FIGURE D.16-1
Diagram of the conveyor control system.

Each time a part on the moving conveyor belt passes the optical detector and interrupts the infrared light beam, a digital counter in the processing circuits is advanced by one. The count of the passing parts is accumulated over a specified period of time and converted to a proportional voltage by the processing circuits based on the desired number of parts in a specified time interval. The more parts that pass the sensor, the higher the voltage. The proportional voltage is applied to the motor speed-control circuit which, in turn, adjusts the speed of the electric motor that drives the conveyor belt in order to maintain the desired number of parts per unit time.

The Motor Speed-Control Circuit The proportional voltage from the processing circuits is applied to the gate of a PUT on the motor speed-control circuit board. This voltage determines the point in the AC cycle that the SCR is triggered on. For a higher PUT gate voltage, the SCR turns on later in the half-cycle and therefore delivers less average power to the motor causing its speed to decrease. For a lower PUT gate voltage, the SCR turns on earlier in the half-cycle, delivering more average power to the motor and increasing its speed. This process continually adjusts the motor speed to maintain the required number of objects per unit time. The potentiometer is used for calibration of the SCR trigger point.

The Motor Speed-Control Circuit Board

☐ Make sure that the circuit board shown in Figure D.16-2(a) is correctly assembled by comparison with the schematic in part (b). Device pin diagrams are shown.

☐ Label a copy of the board with component and input/output designations in agreement with the schematic.

FIGURE D.16-2
Motor speed-control circuit board.

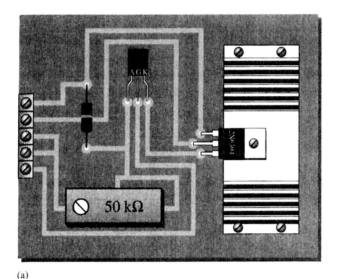 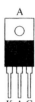

(a)

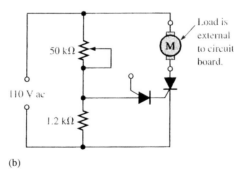

(b)

Analysis of the Motor Speed-Control Circuit

Refer to the schematic in Figure D.16-2(b). A 1 kΩ resistor is connected in place of the motor and 110 V, 60 Hz voltage is applied across the input terminals.

☐ Determine the voltage waveforms at the anode, cathode, and gate of the SCR and the PUT with respect to ground for a PUT gate voltage of 0 V and with the potentiometer set at 25 kΩ.

☐ Determine voltage across the 1 kΩ resistor, for each of the following PUT gate voltages: 0 V, 2 V, 4 V, 6 V, 8 V, and 10 V. The potentiometer is still at 25 kΩ.

Test Procedure

☐ Develop a step-by-step set of instructions on how to check the motor speed-control circuit board for proper operation using the numbered test points indicated in the test bench setup of Figure D.16-3.

☐ Specify voltage values for all the measurements to be made.

☐ Provide a fault analysis for all possible component failures.

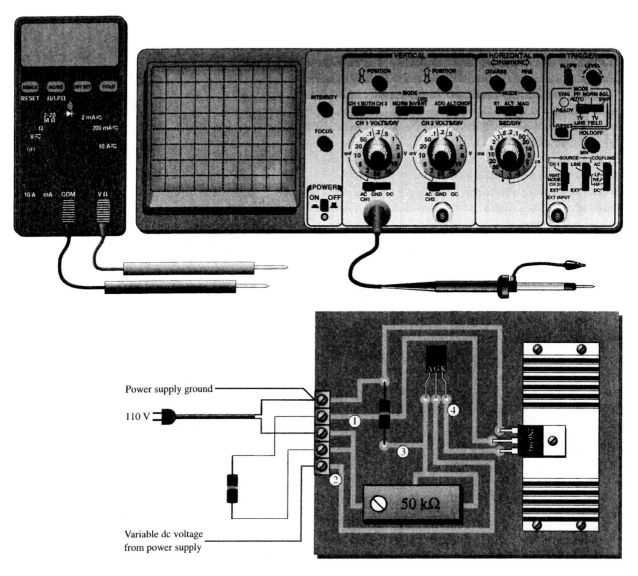

FIGURE D.16-3
Test bench setup for the motor speed-control circuit board.

Troubleshooting

Problems have developed in four boards. Based on the test bench measurements for each board indicated in Figure D.16-4, determine the most likely fault in each case. The circled numbers indicate test point connections to the circuit board. Assume that each board has 110 V AC applied.

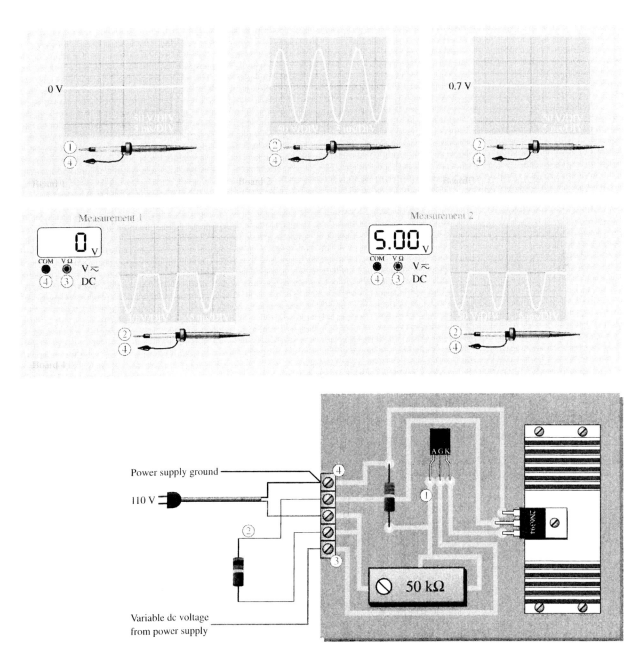

FIGURE D.16-4
Test results for four faulty circuit boards.

D.17 Troubleshoot and repair thyristor circuitry (SCR, diac, triac, etc.)

This material is covered in D.16.

E   Analog Circuits

E.01	Understand principles and operations of multistage amplifiers	**E.16**	Understand principles and operations of regulated and switching power supply circuits
E.02	Fabricate and demonstrate multistage amplifiers	**E.17**	Troubleshoot and repair regulated and switching power supply circuits
E.03	Troubleshoot and repair multistage amplifiers	**E.18**	Understand principles and operations of active filter circuits
E.04	Understand principles and operations of IF circuits	**E.19**	Troubleshoot and repair active filter circuits
E.05	Fabricate and demonstrate IF circuits	**E.20**	Understand principles and operations of sinusoidal and nonsinusoidal oscillator circuits
E.06	Troubleshoot and repair IF circuits		
E.07	Understand principles and operations of linear power supplies and filters	**E.21**	Troubleshoot and repair sinusoidal and nonsinusoidal oscillator circuits
E.08	Fabricate and demonstrate linear power supplies and filters	**E.22**	Understand principles and operations of fiber-optic circuits using photodiodes or lasers
E.09	Troubleshoot and repair linear power supplies and filters	**E.23**	Troubleshoot and repair fiber-optic circuits using photodiodes or lasers
E.10	Understand principles and operations of operational amplifier circuits	**E.24**	Understand principles and operations of RF circuits
E.11	Fabricate and demonstrate operational amplifier circuits	**E.25**	Fabricate and demonstrate RF circuits
		E.26	Troubleshoot and repair RF circuits
E.12	Troubleshoot and repair operational amplifier circuits	**E.27**	Understand principles and operations of signal modulation systems (AM, FM, stereo)
E.13	Understand principles and operations of audio power amplifiers	**E.28**	Troubleshoot and repair signal modulation systems (AM, FM, stereo)
E.14	Fabricate and demonstrate audio power amplifiers	**E.29**	Demonstrate an understanding of motor phase shift control circuits
E.15	Troubleshoot and repair audio power amplifiers	**E.30**	Understand the principles and operations of microwave circuits

E.01 Understand principles and operations of multistage amplifiers

INTRODUCTION

Several amplifiers can be connected in a cascaded arrangement with the output of one, amplifier driving the input of the next. Each amplifier in the cascaded arrangement is known as a stage. The basic purpose of a multistage arrangement is to increase the overall voltage gain.

Multistage Voltage Gain

The overall voltage gain, A'_v, of **cascaded** amplifiers, as shown in Figure E.01-1, is the product of the individual gains.

$$A'_v = A_{v1}A_{v2}A_{v3} \cdots A_{vn}$$

where n is the number of **stages.**

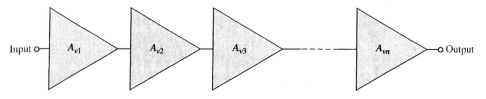

FIGURE E.01-1
Cascaded amplifiers. Each triangular symbol represents a separate amplifier.

Voltage Gain Expressed in Decibels

Amplifier voltage gain is often expressed in **decibels** (dB) as follows:

$$A_{v(dB)} = 20 \log A_v$$

This is particularly useful in **multistage** systems because the overall voltage gain in dB is the *sum* of the individual gains in dB.

$$A'_{v(dB)} = A_{v1(dB)} + A_{v2(dB)} + \cdots + A_{vn(dB)}$$

EXAMPLE E.01-1

A certain cascaded amplifier arrangement has the following voltage gains: $A_{v1} = 10$, $A_{v2} = 15$, and $A_{v3} = 20$. What is the overall voltage gain? Also express each gain in decibels (dB) and determine the total voltage gain in dB.

Solution

$$A'_v = A_{v1}A_{v2}A_{v3} = (10)(15)(20) = 3000$$

$$A_{v1(\text{dB})} = 20 \log 10 = 20.0 \text{ dB}$$

$$A_{v2(\text{dB})} = 20 \log 15 = 23.5 \text{ dB}$$

$$A_{v3(\text{dB})} = 20 \log 20 = 26.0 \text{ dB}$$

$$A'_{v(\text{dB})} = 20 \text{ dB} + 23.5 \text{ dB} + 26.0 \text{ dB} = 69.5 \text{ dB}$$

Multistage Amplifier Analysis

For purposes of illustration, we will use the two-stage capacitively coupled amplifier in Figure E.01-2. Notice that both stages are identical common-emitter amplifiers with the output of the first stage capacitively coupled to the input of the second stage. Capacitive coupling prevents the DC bias of one stage from affecting that of the other but allows the AC signal to pass without attenuation because $X_C \cong 0 \ \Omega$ at the frequency of operation. Notice, also, that the transistors are labeled Q_1 and Q_2.

FIGURE E.01-2
A two-stage common-emitter amplifier.

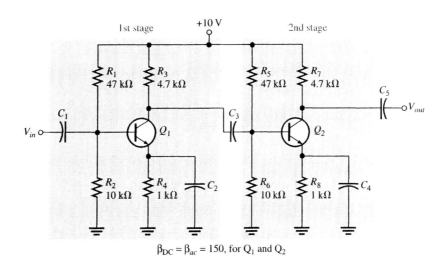

$\beta_{\text{DC}} = \beta_{ac} = 150$, for Q_1 and Q_2

Loading Effects In determining the voltage gain of the first stage, you must consider the loading effect of the second stage. Because the coupling capacitor C_3 effectively appears as a short at the signal frequency, the total input resistance of the second stage presents an AC load to the first stage. Looking from the collector of Q_1, the two biasing resistors in the second stage, R_5 and R_6, appear in parallel with the input resistance at the base of Q_2. In other words, the signal at the collector of Q_1 "sees" R_3 and R_5, R_6 and $R_{in(base2)}$ of the second stage all in parallel to AC ground. Thus, the effective AC collector resistance of Q_1 is the total of all these resistances in parallel, as Figure E.01-3 illustrates. The voltage gain of

FIGURE E.01-3
AC equivalent of first stage in Figure E.01-2, showing loading from second stage.

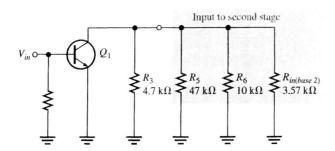

the first stage is reduced by the loading of the second stage, because the effective AC collector resistance of the first stage is less than the actual value of its collector resistor, R_3. Remember that $A_v = R_c/r'_e$.

Voltage Gain of the First Stage   The AC collector resistance of the first stage is

$$R_{c1} = R_3 \parallel R_5 \parallel R_6 \parallel R_{in(base2)}$$

Remember that lowercase italic subscripts denote AC quantities such as for R_c.

You can verify that $I_E = 1.05$ mA, $r'_e = 23.8$ Ω, and $R_{in(base2)} = 3.57$ kΩ. The effective AC collector resistance of the first stage is as follows:

$$R_{c1} = 4.7 \text{ k}\Omega \parallel 47 \text{ k}\Omega \parallel 10 \text{ k}\Omega \parallel 3.57 \text{ k}\Omega = 1.63 \text{ k}\Omega$$

Therefore, the base-to-collector voltage gain of the first stage is

$$A_{v1} = \frac{R_{c1}}{r'_e} = \frac{1.63 \text{ k}\Omega}{23.8 \text{ }\Omega} = 68.5$$

Voltage Gain of the Second Stage   The second stage has no load resistor, so the AC collector resistance is R_7, and the gain is

$$A_{v2} = \frac{R_7}{r'_e} = \frac{4.7 \text{ k}\Omega}{23.8 \text{ }\Omega} = 197$$

Compare this to the gain of the first stage, and notice how much the loading from the second stage reduced the gain.

Overall Voltage Gain   The overall amplifier gain with no load on the output is

$$A'_v = A_{v1}A_{v2} = (68.5)(197) \cong 13{,}495$$

If an input signal of 100 μV, for example, is applied to the first stage and if there is no attenuation in the input base circuit due to the source resistance, an output from the second stage of $(100 \text{ }\mu\text{V})(13{,}495) \cong 1.35$ V will result. The overall voltage gain can be expressed in dB as follows:

$$A'_{v(\text{dB})} = 20 \log (13{,}495) = 82.6 \text{ dB}$$

DC Voltages in the Capacitively Coupled Multistage Amplifier   Since both stages in Figure E.01-2 are identical, the DC voltages for Q_1 and Q_2 are the same. Since $\beta_{DC}R_4 \gg R_2$ and $\beta_{DC}R_8 \gg R_6$, the DC base voltage for Q_1 and Q_2 is

$$V_B \cong \left(\frac{R_2}{R_1 + R_2}\right)10 \text{ V} = \left(\frac{10 \text{ k}\Omega}{57 \text{ k}\Omega}\right)10 \text{ V} = 1.75 \text{ V}$$

The DC emitter and collector voltages are as follows:

$$V_E = V_B - 0.7 \text{ V} = 1.05 \text{ V}$$

$$I_E = \frac{V_E}{R_4} = \frac{1.05 \text{ V}}{1 \text{ k}\Omega} = 1.05 \text{ mA}$$

$$I_C \cong I_E = 1.05 \text{ mA}$$

$$V_C = V_{CC} - I_C R_3 = 10 \text{ V} - (1.05 \text{ mA})(4.7 \text{ k}\Omega) = 5.07 \text{ V}$$

Direct-Coupled Multistage Amplifiers

A basic two-stage, direct-coupled amplifier is shown in Figure E.01-4. Notice that there are no coupling or bypass capacitors in this circuit. The DC collector voltage of the first stage provides the base-bias voltage for the second stage. Because of the direct coupling, this type of amplifier has a better low-frequency response than the capacitively coupled type in which the reactance of coupling and bypass capacitors at very low frequencies may become excessive. The increased reactance at lower frequencies produces signal loss and gain reduction.

FIGURE E.01-4
A basic two-stage direct-coupled amplifier.

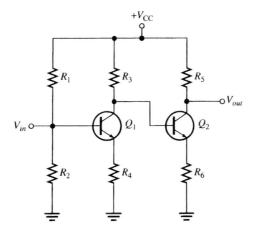

Direct-coupled amplifiers, on the other hand, can be used to amplify low frequencies all the way down to DC (0 Hz) without loss of voltage gain because there are no capacitive reactances in the circuit. The disadvantage of direct-coupled amplifiers is that small changes in the DC bias voltages from temperature effects or power-supply variation are amplified by the succeeding stages, which can result in a significant drift in the DC levels throughout the circuit.

Transformer-Coupled Multistage Amplifiers

A basic transformer-coupled two-stage amplifier is shown in Figure E.01-5. Transformer coupling is often used in high-frequency amplifiers such as those in the RF (radio frequency) and IF (intermediate frequency) sections of radio and TV receivers. At lower frequency ranges such as **audio,** the size of transformers is usually prohibitive. Capacitors are usually connected across the primary windings of the transformers to obtain resonance and increased selectivity for the band of frequencies to be amplified.

FIGURE E.01-5
A basic two-stage transformer-coupled amplifier.

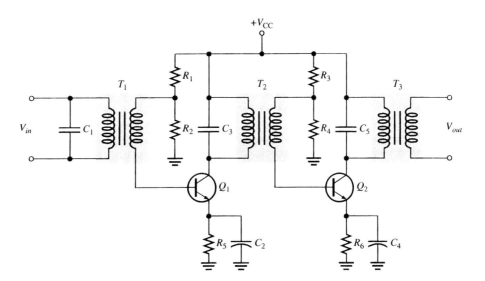

REVIEW QUESTIONS

True/False

1. The gain of a multistage amplifier is the sum of the individual gains.
2. Loading an amplifier reduces its gain.

Multiple Choice

3. Individual amplifier stages may be
   a. Capacitively or directly coupled.
   b. Parallel coupled.
   c. Both a and b.

4. Compared to the capacitively coupled amplifier, the direct coupled amplifier has
   a. Better low-frequency response.
   b. Worse low-frequency response.
   c. The same low-frequency response.

5. Transfer-coupled amplifiers are primarily used for
   a. Low frequencies.
   b. High frequencies.
   c. High voltage gain applications.

E.02 Fabricate and demonstrate multistage amplifiers

INTRODUCTION

Using actual components or simulation software (such as *Electronics Workbench*), set up the cascaded amplifier shown in Figure E.02-1. Apply a small sinusoidal input voltage, say 2 mV p-p, and examine the input and output waveforms. What is the phase relationship? What is the overall voltage gain? If the output is distorted, lower the input voltage and re-measure.

Next, measure the voltage at the collector of Q_1. Use it to calculate the individual gains of each stage. Does the product of the individual gains equal the overall gain?

Finally, connect a 2.2 K ohm load resistor to the output. Determine the new voltage gain.

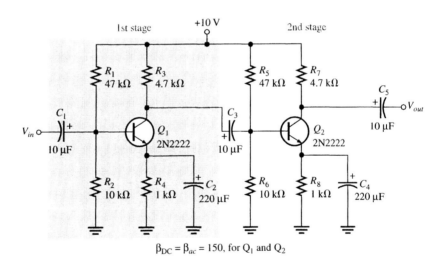

$\beta_{DC} = \beta_{ac} = 150$, for Q_1 and Q_2

FIGURE E.02-1
A two-stage common-emitter amplifier.

E.03 Troubleshoot and repair multistage amplifiers

INTRODUCTION

The proper signal levels and DC voltage levels for the capacitively coupled two-stage amplifier are shown in Figure E.03-1.

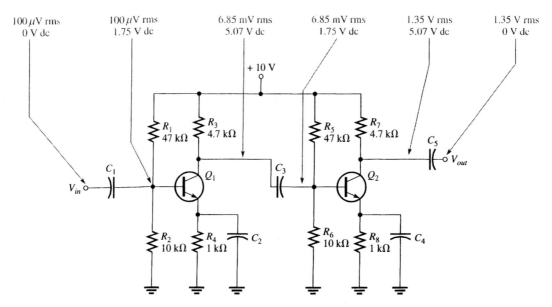

100 μV rms	100 μV rms	6.85 mV rms	6.85 mV rms	1.35 V rms	1.35 V rms
0 V dc	1.75 V dc	5.07 V dc	1.75 V dc	5.07 V dc	0 V dc

FIGURE E.03-1
Two-stage amplifier with proper AC and DC voltage levels indicated.

Troubleshooting Procedure

A basic procedure for troubleshooting called *signal tracing,* which is usually done with an oscilloscope, is illustrated in Figure E.03-2 using the two-stage amplifier as an example. This general procedure can be expanded to any number of stages.

Let's begin by assuming that the final output signal of the amplifier has been determined to be missing. Let's also assume that there is an input signal and that this has been verified. When troubleshooting, you generally start at the point where the signal is missing and work back point-by-point toward the input until a correct voltage is found. The fault then lies somewhere between the point of the first good voltage check and the missing or incorrect voltage point.

Before you begin checking the voltages, it is usually a good idea to visually check the circuit board or assembly for obvious problems such as broken or poor connections, solder splashes, wire clippings, or burned components.

Step 1: Check the DC supply voltage. Often something as simple as a blown fuse or the power switch being in the *off* position may be the problem. The circuit will certainly not work without power.

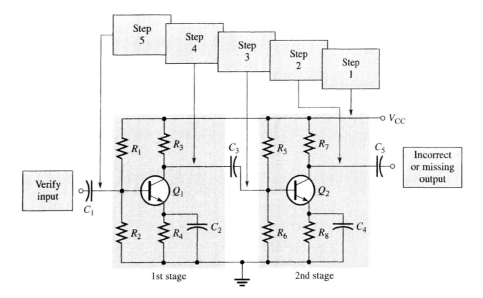

FIGURE E.03-2
Basic troubleshooting procedure for a two-stage amplifier with an incorrect or missing output signal.

Step 2: Check the voltage at the collector of Q_2. If the correct signal is present at this point, the coupling capacitor, C_5, is open. If there is no signal at this point, proceed to Step 3.

Step 3: Check the signal at the base of Q_2. If the signal is present at this point, the fault is in the second stage of the amplifier. First, do an in-circuit check of the transistor. If it is OK, then one of the biasing resistors may be open; if this is the case, the DC voltages will be incorrect. If there is no signal at this point, proceed to Step 4.

Step 4: Check the signal at the collector of Q_1. If the signal is present at this point, the coupling capacitor, C_3, is open. If there is no signal at this point, proceed to Step 5.

Step 5: Check the signal at the base of Q_1. If the signal is present at this point, the fault is in the first stage of the amplifier. First, do an in-circuit check of the transistor. If it is OK, then one of the biasing resistors may be open; if this is the case, the DC voltages will be incorrect. If there is no signal at this point, the coupling capacitor, C_1, is open because we started out by verifying that there is a correct signal at the input.

As another example of isolating a component failure in a circuit, let's use a class A amplifier with the output monitored by an oscilloscope, as shown in Figure E.03-3. As shown, the amplifier has a normal sine wave output when a sinusoidal input signal is applied.

Now, several incorrect output waveforms will be considered and the most likely causes discussed. In Figure E.03-4(a), the scope displays a DC level equal to the DC supply voltage, indicating that the transistor is in cutoff. The two possible causes of this condition are (1) the transistor is open from collector to emitter, or (2) R_4 is open, preventing collector and emitter current.

In Figure E.03-4(b), the scope displays a DC level at the collector approximately equal to the emitter voltage. The two possible causes of this indication are (1) the transistor is shorted from collector to emitter, or (2) R_2 is open, causing the transistor to be biased in saturation. In the second case, a sufficiently large input signal can bring the transistor out of saturation on its negative peaks, resulting in short pulses on the output.

FIGURE E.03-3
Class A amplifier with proper output display.

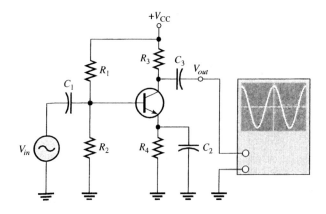

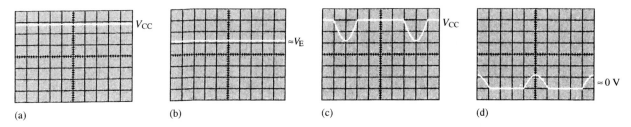

(a)   (b)   (c)   (d)

FIGURE E.03-4
Oscilloscope displays of output voltage for the amplifier in Figure E.03-3, illustrating several types of failures.

In Figure E.03-4(c), the scope displays an output waveform that is clipped at cutoff. Possible causes of this indication are (1) the Q-point has shifted down due to a drastic out-of-tolerance change in a resistor value, or (2) R_1 is open, biasing the transistor in cutoff. In the second case, the input signal is sufficient to bring it out of cutoff for a small portion of the cycle.

In Figure E.03-4(d), the scope displays an output waveform that is clipped at saturation. Again, it is possible that a resistance change has caused a drastic shift in the Q-point up toward saturation, or R_2 is open, causing the transistor to be biased in saturation, and the input signal is bringing it out of saturation for a small portion of the cycle.

E.04 Understand principles and operations of IF circuits

INTRODUCTION

In this exercise we examine the nature and application of IF amplifiers.

IF Amplifiers

IF amplifiers provide the bulk of a receiver's gain (and thus are a major influence on its sensitivity) and selectivity characteristics. An IF amplifier is not a whole lot different from an RF stage except it operates at a fixed frequency. This allows the use of fixed double-tuned inductively coupled circuits to allow for the sharply defined bandpass response characteristic of superheterodyne receivers.

The number of IF stages in any given receiver varies, but from two to four is typical. Some typical IF amplifiers are shown in Figure E.04-1. (Figure E.04-1 continues on the next page.) The circuit at (a) uses the 40673 dual-gate MOSFET while the other two use LICs specially made for IF amplifier applications. Notice the double-tuned LC circuits at the input and output of all three circuits. They are shown within dashed lines to indicate they are one complete assembly. They can be economically purchased for all common IF frequencies and have a variable slug in the transformer core for fine tuning their center frequency. All three of the circuits have provision for the AGC level. Not all receivers utilize AGC to control the gain of mixer and/or RF stages, but they invariably do control the gain of IF stages.

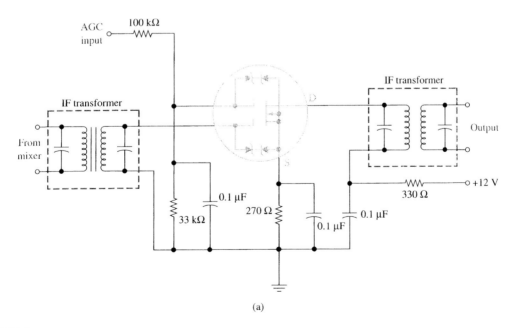

(a)

FIGURE E.04-1
Typical IF amplifiers.

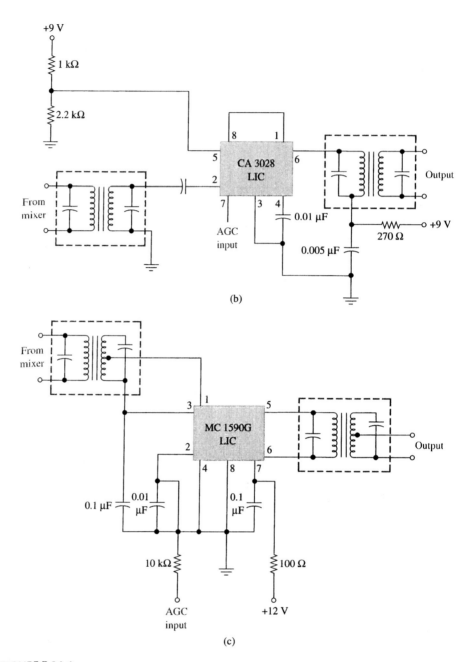

(b)

(c)

FIGURE E.04-1
(Continued)

THE FRONT END AND IF AMPLIFIERS IN A TELEVISION RECEIVER

The front end of a TV receiver is also called the **tuner** and contains the RF amplifier, mixer, and local oscillator. Its output is fed into the first IF amplifier. It is the obvious function of the tuner to select the desired station and to reject all others, but these important functions are also performed:

1. It provides amplification.

2. It prevents the local oscillator signal from being driven into the antenna and thus radiating unwanted interference.

3. It steps the received RF signal down to the frequency required for the IF stages.

4. It provides proper impedance matching between the antenna–feed line combination into the tuner itself. This allows for the largest possible signal into the tuner and thus the largest possible signal-to-noise ratio.

Figure E.04-2 provides a block diagram of a VHF/UHF tuner. The large majority of tuners are synthesized, which allows for the remote control feature that is found on most sets. We will analyze Figure E.04-2 in two steps, starting with the UHF portion inside the dashed box.

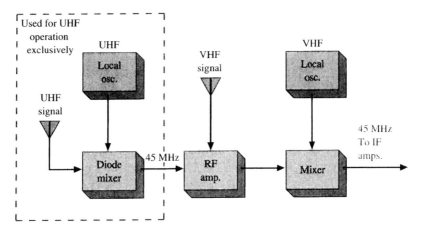

FIGURE E.04-2
VHF/UHF tuner block diagram.

Note that there is no RF amplifier for the UHF front end; the antenna signal goes directly to the mixer. RF amplifiers at these frequencies are expensive and suffer from relatively poor noise performance.

When the tuner is switched to VHF channel 1, power is automatically removed from the VHF local oscillator and applied to the UHF local oscillator. At the same time, the tuned circuits of the RF amplifier and the mixer to its right are switched to 45 MHz, effectively converting them to IF amplifiers for the output of the UHF diode mixer. This compensates for the gain lost because of the missing RF amplifier and brings the UHF output up to equal the level of the VHF output.

Switching the tuner to a VHF channel removes power from the UHF oscillator and sends it to the VHF LO. In this way, only one oscillator operates at a time, thereby preventing the generation of signals that could cause interference. VHF signals are RF amplified, mixed, and sent to the 45 MHz IF.

IF Amplifiers

The IF amplifier section is fed from the mixer output of the tuner. It is often referred to as the video IF even though it is also processing the sound signal. Sets that process the sound and video in the same IF stages are known as **intercarrier systems.** Very early sets used completely separate IF amps for the sound and video signals. The IF stages of intercarrier sets are often referred to as the video IF even though they also handle the sound signal because the sound signal is also processed by another IF stage after it has been extracted from the video signal. From now on the video IF will be referred to simply as the IF.

The major functions of the TV IF stage are the same as in a regular radio receiver: to provide the bulk of the set's selectivity and amplification. The standard IF frequencies are 45.75 MHz for the picture carrier and 41.25 MHz (45.75 MHz minus 4.5 MHz) for the sound carrier. Recall that mixer action causes a reversal in frequency when the IF amplifier accepts the difference between the higher local oscillator frequency and the incoming RF signal.

TABLE E.04-1
IF Signal Frequency Inversion.

Channel 5 76–82 MHz	Transmitted RF Frequency (MHz)	Local Oscillator Frequency (MHz)	IF Frequency (MHz)
Upper channel frequency	82	123	41
Sound carrier	81.75	123	41.25
Picture carrier	77.25	123	45.75
Lower channel frequency	76	123	47

Therefore, the sound carrier that is 4.5 MHz above the picture carrier in the RF signal ends up being 4.5 MHz below it in the mixer output into the first IF stage. The inversion effect of IF frequencies when receiving channel 5 is shown in Table E.04-1. The IF frequencies are always equal to the difference between the local oscillator and RF frequencies.

Stagger Tuning

A major difference between radio and TV IF amplifiers is that most radio receivers require relatively high-Q tuned circuits since the desired bandwidth is often less than 10 kHz. A TV IF amp requires a passband of about 6 MHz because of the wide frequency range necessary for video signals. Hence, the problem here is not how to get a very narrow bandwidth with high-Q components, but instead how to get a wide enough bandwidth but still have relatively sharp falloff at the passband edges. Most TV IF amplifiers solve this problem through the use of **stagger tuning.** Stagger tuning is the technique of cascading a number of tuned circuits with slightly different resonant frequencies, as shown in Figure E.04-3. The response of three separate LC tuned circuits is used to obtain the total resultant passband shown with dashed lines. The use of a lower-Q tuned circuit in the middle helps provide a flatter overall response than would otherwise be possible.

Another interesting point illustrated in Figure E.04-3 is the attenuation given to the video side frequencies right around the picture carrier. This is done to reverse the vestigial-sideband characteristic generated at the transmitter. If the receiver IF response were equal for all the video frequencies, the lower ones (up to 0.75 MHz) would have excessive output because they have both upper- and lower-sideband components.

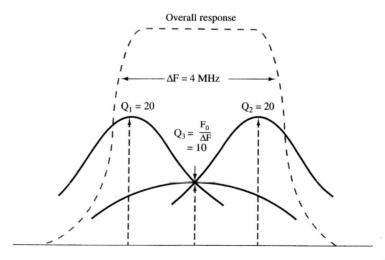

FIGURE E.04-3
Stagger tuning.

IF Amplifier Response

The ideal overall IF response curve in Figure E.04-4 provides some interesting food for thought. The sound IF carrier and its narrow sidebands are amplified at only one-tenth the midband IF gain. This is done to minimize interference effects that the sound would otherwise have on the picture. You may have noticed a TV with normal picture when no audio is present but visual interference in step with the sound output. This is an indication that the sound signal in the IF is not attenuated enough, and it can often be remedied by adjustment of the set's fine-tuning control.

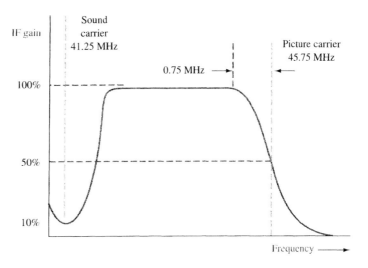

FIGURE E.04-4
Ideal IF response curve.

REVIEW QUESTIONS

True/False

1. IF amplifiers operate over a wide range of frequencies.

2. Typically, more than one IF amplifier stage is used in a circuit.

Multiple Choice

3. In a television receiver, the IF frequency for the picture carrier is
   a. 41 MHz.
   b. 41.25 MHz.
   c. 45.75 MHz.
4. Cascading multiple tuned circuits with slightly different resonant frequencies is called
   a. Staggered resonance.
   b. Stagger tuning.
   c. IF cascading.
5. Compared with radio IF circuits, television IF circuits have
   a. Lower bandwidth.
   b. The same bandwidth.
   c. Higher bandwidth.

E.05 Fabricate and demonstrate IF circuits

INTRODUCTION

Many electronics stores and magazines contain electronic project kits that allow you to build your own circuit. A printed circuit board is typically included so that the only skill required is soldering. The purpose of this exercise is for you to obtain a suitable kit to build and examine. To demonstrate your knowledge of IF circuits, try to find one of the following projects:

☐ AM radio receiver

☐ FM radio receiver

☐ Wireless microphone

☐ Aircraft receiver

☐ Remote control transmitter/receiver

After constructing the kit, use an oscilloscope to view the various signals in the circuit. Attach a spectrum analyzer as well. What are the important frequencies involved?

E.06 Troubleshoot and repair IF circuits

INTRODUCTION

In this section we examine IF troubleshooting through the example of a television receiver.

The TV set is basically an AM receiver for picture information and an FM receiver for sound reception. The basic approach for troubleshooting a TV set is to proceed as if it were two radios. In this section you will learn techniques that will help you identify and isolate faulty sections within the TV receiver. These techniques may be used when troubleshooting any kind of TV set.

Upon completing this section you should be able to

☐ Identify defective stages in a TV receiver

☐ Describe faulty sections within the TV based on viewed symptoms

Looking into the back of a TV set for the first time can be a fearful sight. Inside the TV set are many very complex circuits and hundreds of components. To be an effective service technician, you must have a thorough knowledge and understanding of how a TV works. It becomes quite apparent that service literature is needed when servicing a TV. Make sure you get the service literature pertaining to the model and chassis that you are repairing. The TV setup should be as close as you can get it to normal viewing. Organize your thinking in a logical pattern before ever making any circuit measurements. Observe and classify any abnormality such as sound, raster, video, or color problems. Study the service literature and identify function sections in the TV set. For example, identify the location of the horizontal output section, vertical section, video section, and others. Try to localize the problem to a specific section from the symptoms being observed.

The TV set stands out from other communications receivers in that circuit defects often show up on the screen. From these visual symptoms faulty sections can be singled out before ever opening the back of the TV set. For example, symptoms like no video, no sound, and good raster would lead us to check the IF section of the TV since both picture information and sound information are amplified there. The **raster** refers to CRT illumination by the scan lines when no signal is received and/or being displayed.

Study the problem. Make a decision where to start troubleshooting based on the viewed symptoms. Try to isolate the defective stage from the symptoms on the screen. Is a picture present? Does the picture roll up and down or left to right? Does the picture have snow in it? Is the sound present or not? By observing these signposts you quickly determine the defective section within the TV. Table E.06-1 gives symptoms, causes, and the area in which to look. It is not all-inclusive of the problems that might occur in a TV set.

Consult the manufacturer's service literature for diagnostic charts or other troubleshooting guidelines. Often, when these troubleshooting aids are available, they relate to common failures incurred for that model TV receiver or manufacturing defects that may exist in the set.

Pull the back off the set, and with the power off, look for obvious problems. Look for loose wires or connectors, burned components, broken or burned PCB traces, and cold solder joints. With the power on, listen for unusual sounds like hissing (normally associated with horizontal output transformers that have developed high-voltage leaks), arcing, and

TABLE E.06-1

Symptom	Cause	Stage/Area of Trouble
Set is dead, no sound, no video, no raster	No power to circuits	Check main power supply, start-up circuits, main fuses, and line cord
Set blows fuses	A short circuit exists in main power supply or horizontal output	Check for shorted diodes, shorted regulator transistors, and shorted filter capacitors in main power supply, or shorted horizontal output transistor and shorted horizontal output transformer
Sound normal, no video, no raster	No high voltage	Check the horizontal output circuit/high-voltage section
Normal raster, no video, no sound	Video and sound signal missing	Check antenna, tuner, and IF amplifiers
Raster and video normal	Sound signal missing	Check sound IF amplifiers, detector section, audio amplifiers, and speaker
Raster, video, and sound normal, no color	Color signal missing	Check color killer and color processing circuits
Picture has snow, noise heard in sound	Signal-to-noise ratio high	Check RF amplifier in tuner
Vertical roll (up and down)	Vertical sync missing	Check sync separator, vertical oscillator
Horizontal white line across screen	Vertical output signal missing	Check the vertical output circuit, vertical oscillator, or yoke
Horizontal roll (left and right)	Horizontal sync missing	Check sync separator circuit, horizontal AFC, horizontal oscillator
Vertical white line on screen	Horizontal output signal missing	Check horizontal output circuit, horizontal oscillator, yoke

high-pitched squeals from the horizontal oscillator. Do you smell anything burning? Look for brown areas on the PCB indicating over-heating components.

If the preliminary inspection fails to localize a defective component, continue troubleshooting by doing voltage and resistance measurements on the suspected stage in the TV set. Compare the results with specified values from the service literature. Use the oscilloscope to check for proper waveforms in the defective section and associated sections. Schematic diagrams usually give pictures indicating what the correct waveform looks like for all the common signals in the set. Base your troubleshooting on an organized approach and not a disorganized one. Having an organized strategy will save you valuable time and enhance your troubleshooting.

E.07 Understand principles and operations of linear power supplies and filters

INTRODUCTION

Because of their ability to conduct current in one direction and block current in the other direction, diodes are used in circuits called rectifiers that convert AC voltage into DC voltage. Rectifiers are found in all DC power supplies that operate from an AC voltage source. Power supplies are an essential part of all electronic systems from the simplest to the most complex.

The Basic DC Power Supply

The DC **power supply** converts the standard 110 V, 60 Hz AC available at wall outlets into a constant DC voltage. It is one of the most common electronic circuits that you will find. The DC voltage produced by a power supply is used to power all types of electronic circuits, such as television receivers, stereo systems, VCRs, CD players, and laboratory equipment.

A basic block diagram for a power supply is shown in Figure E.07-1. The **rectifier** can be either a half-wave rectifier or a full-wave rectifier. The rectifier converts the AC input voltage to a pulsating DC voltage, which is half-wave rectified as shown. The **filter** eliminates the fluctuations in the rectified voltage and produces a relatively smooth DC voltage. The **regulator** is a circuit that maintains a constant DC voltage for variations in the input line voltage or in the load. Regulators vary from a single device to more complex circuits. The load block is usually a circuit for which the power supply is producing the DC voltage and load current.

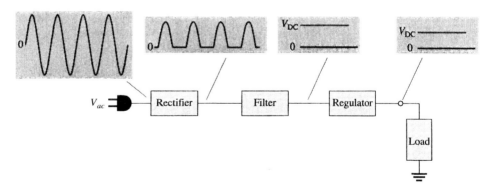

FIGURE E.07-1
Block diagram of a DC power supply.

The Half-Wave Rectifier

The most basic type of rectifier is the **half-wave rectifier.** Figure E.07-2 illustrates the process called *half-wave rectification.* In part (a), an AC source is connected to a load resistor through a diode. Let's examine what happens during one cycle of the input voltage using the ideal model for the diode. When the sinusoidal input voltage goes positive, the diode is forward-biased and conducts current to the load resistor, as shown in part (b). The

363

FIGURE E.07-2
Operation of a half-wave rectifier.

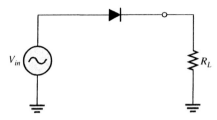

(a) Half-wave rectifier circuit

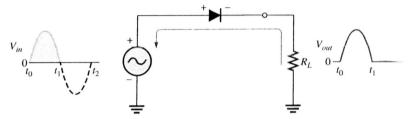

(b) Operation during positive alternation of the input voltage

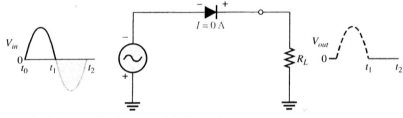

(c) Operation during negative alternation of the input voltage

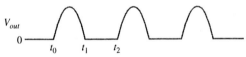

(d) Half-wave output voltage for three input cycles

current produces a voltage across the load which has the same shape as the positive half-cycle of the input voltage. When the input voltage goes negative during the second half of its cycle, the diode is reversed-biased. There is no current, so the voltage across the load resistor is zero, as shown in part (c). The net result is that only the positive half-cycles of the AC input voltage appear across the load. Since the output does not change polarity, it is a pulsating DC voltage, as shown in part (d).

Average Value of the Half-Wave Output Voltage The average value of a half-wave output voltage is the value that would be indicated by a DC voltmeter. It can be calculated with the following equation where $V_{p(out)}$ is the peak value of the half-wave output voltage:

$$V_{AVG} = \frac{V_{p(out)}}{\pi}$$

Figure E.07-3 shows the half-wave voltage with its average value indicated by the red dashed line.

FIGURE E.07-3
Average value of the half-wave rectified signal.

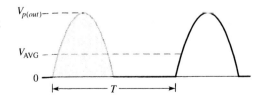

EXAMPLE E.07-1

What is the average (DC) value of the half-wave rectified output voltage waveform in Figure E.07-4?

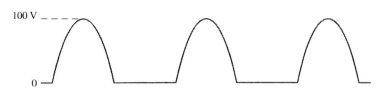

FIGURE E.07-4

Solution

$$V_{AVG} = \frac{V_{p(out)}}{\pi} = \frac{100\text{ V}}{\pi} = 31.8\text{ V}$$

FULL-WAVE RECTIFIERS

The difference between full-wave and half-wave rectification is that a **full-wave rectifier** allows unidirectional current to the load during the entire input cycle, and the half-wave rectifier allows this only during one half of the cycle. The result of full-wave rectification is a DC output voltage that pulsates every half-cycle of the input, as shown in Figure E.07-5.

The average value for a full-wave rectified output voltage is twice that of the half-wave, expressed as follows:

$$V_{AVG} = \frac{2V_{p(out)}}{\pi}$$

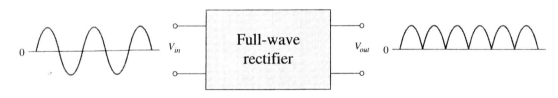

FIGURE E.07-5
Full-wave rectification.

EXAMPLE E.07-2

Find the average value of the full-wave rectified output voltage in Figure E.07-6.

FIGURE E.07-6

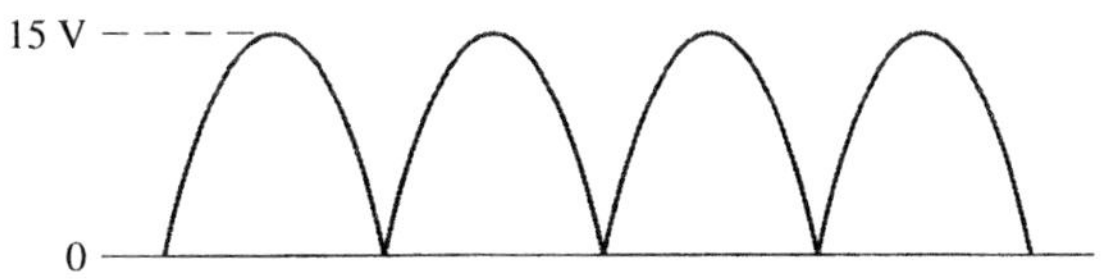

Solution

$$V_{AVG} = \frac{V_{p(out)}}{\pi} = \frac{2(15\text{ V})}{\pi} = 9.55\text{ V}$$

Center-Tapped Full-Wave Rectifier

The center-tapped (CT) full-wave rectifier circuit uses two diodes connected to the secondary of a center-tapped transformer, as shown in Figure E.07-7. The input signal is coupled through the transformer to the secondary. Half of the secondary voltage appears between the **center tap** and each end of the secondary winding as shown.

For a positive half-cycle of the input voltage, the polarities of the secondary voltages are as shown in Figure E.07-8(a). This condition forward-biases the upper diode D_1 and reverse-biases the lower diode D_2. The current path is through D_1 and the load resistor, as indicated. For a negative half-cycle of the input voltage, the voltage polarities on the secondary are as shown in Figure E.07-8(b). This condition reverse-biases D_1 and forward-biases D_2. The current path is through D_2 and the load resistor, as indicated. Because the current during both the positive and the negative portions of the input cycle is in the same direction through the load, the output voltage developed across the load is a full-wave rectified DC voltage.

FIGURE E.07-7
A center-tapped full-wave rectifier.

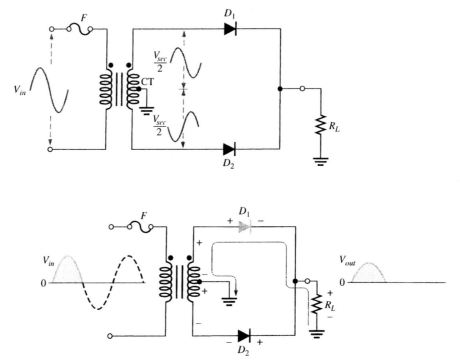

(a) During positive half-cycles, D_1 is forward-biased and D_2 is reverse-biased.

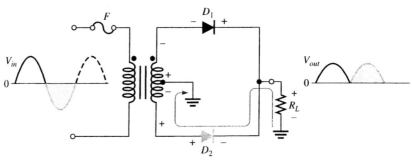

(b) During negative half-cycles, D_2 is forward-biased and D_1 is reverse-biased.

FIGURE E.07-8
Basic operation of a center-tapped full-wave rectifier. Note that the current through the load resistor is in the same direction during the entire input cycle.

Full-Wave Bridge Rectifier

The full-wave bridge rectifier uses four diodes, as shown in Figure E.07-9. When the input cycle is positive as in part (a), diodes D_1 and D_2 are forward-biased and conduct current in the direction shown. A voltage is developed across R_L which looks like the positive half of the input cycle. During this time, diodes D_3 and D_4 are reverse-biased.

When the input cycle is negative, as in Figure E.07-9(b), diodes D_3 and D_4 are forward-biased and conduct current in the same direction through R_L as during the positive half-cycle. During the negative half-cycle, D_1 and D_2 are reverse-biased. A full-wave rectified output voltage appears across R_L as a result of this action.

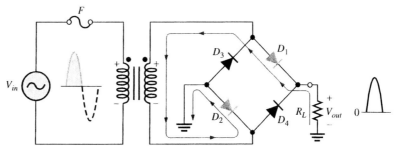

(a) During positive half-cycle of the input, D_1 and D_2 are forward-biased and conduct current. D_3 and D_4 are reverse-biased.

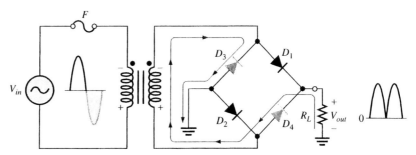

(b) During negative half-cycle of the input, D_3 and D_4 are forward-biased and conduct current. D_1 and D_2 are reverse-biased.

FIGURE E.07-9
Operation of full-wave bridge rectifier.

Bridge Output Voltage   As you can see in Figure E.07-9, two diodes are always in series with the load resistor during both the positive and the negative half-cycles. When these diode drops (barrier potentials) are taken into account, the output voltage is a full-wave rectified voltage with a peak value equal to the peak secondary voltage less the two diode drops.

$$V_{p(out)} = V_{p(sec)} - 1.4 \text{ V}$$

EXAMPLE E.07-3

(a) Determine the peak output voltage for the bridge rectifier in Figure E.07-10.
(b) What minimum PIV rating is required for the silicon diodes?

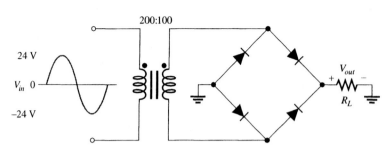

FIGURE E.07-10

Solution

(a) The peak output voltage is

$$V_{p(out)} = V_{p(sec)} - 1.4\ \text{V} = \left(\frac{N_{sec}}{N_{pri}}\right)V_{p(in)} - 1.4\ \text{V} = \left(\frac{100}{200}\right)24\ \text{V} - 1.4\ \text{V}$$

$$= 12\ \text{V} - 1.4\ \text{V} = 10.6\ \text{V}$$

(b) The PIV for each diode is

$$\text{PIV} = V_{p(out)} + 0.7\ \text{V} = 10.6\ \text{V} + 0.7\ \text{V} = 11.3\ \text{V}$$

POWER SUPPLY FILTERS

In most power supply applications, the standard 60 Hz AC power line voltage must be converted to a sufficiently constant DC voltage. The 60 Hz pulsating DC output of a half-wave rectifier or the 120 Hz pulsating output of a full-wave rectifier must be filtered to reduce the large voltage variations. Figure E.07-11 illustrates the filtering concept showing a smooth DC output voltage. The full-wave rectifier voltage is applied to the input of the filter, and, ideally, a constant DC level appears on the output.

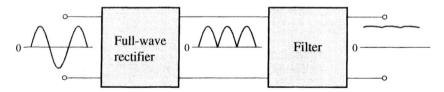

FIGURE E.07-11
Basic block diagram of a power supply that converts 60 Hz AC to DC.

Capacitor Filter

A half-wave rectifier with a capacitor filter is shown in Figure E.07-12. We will use the half-wave rectifier to illustrate the filtering principle; then we will expand the concept to the full-wave rectifier.

During the positive first quarter-cycle of the input, the diode is forward-biased, allowing the capacitor to charge to within a diode drop of the input peak, as illustrated in Figure E.07-12(a). When the input begins to decrease below its peak, as shown in part (b), the capacitor retains its charge and the diode becomes reverse-biased. During the remaining part of the cycle, the capacitor can discharge only through the load resistance at a rate determined by the R_LC time constant. The larger the time constant, the less the capacitor will discharge.

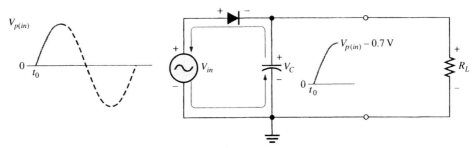

(a) Initial charging of capacitor (diode is forward-biased) happens only once when power is turned on.

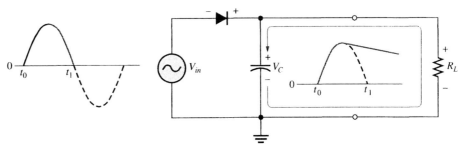

(b) Discharging through R_L after peak of positive alternation (diode is reverse-biased)

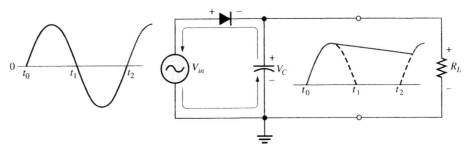

(c) Charging back to peak of input (diode is forward-biased)

FIGURE E.07-12
Operation of a half-wave rectifier with a capacitor filter.

Because the capacitor charges to a peak value of $V_{p(in)} - 0.7$ V, the peak inverse voltage of the diode in this application is

$$PIV = 2V_{p(in)} - 0.7 \text{ V}$$

During the first quarter of the next cycle, as illustrated in Figure E.07-12(c), the diode again will become forward-biased when the input voltage exceeds the capacitor voltage by approximately a diode drop.

Ripple Voltage As you have seen, the capacitor quickly charges at the beginning of a cycle and slowly discharges after the positive peak (when the diode is reverse-biased). The variation in the output voltage due to the charging and discharging is called the **ripple voltage.** The smaller the ripple, the better the filtering action, as illustrated in Figure E.07-13.

For a given input frequency, the output frequency of a full-wave rectifier is twice that of a half-wave rectifier. As a result, a full-wave rectifier is easier to filter. When filtered, the full-wave rectified voltage has a smaller ripple than does a half-wave signal for the same load resistance and capacitor values. A smaller ripple occurs because the capacitor

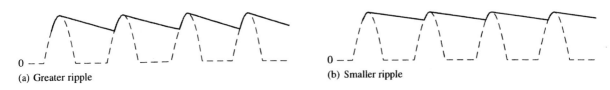

(a) Greater ripple (b) Smaller ripple

FIGURE E.07-13
Half-wave ripple voltage (solid line).

discharges less during the shorter interval between full-wave pulses, as shown in Figure E.07-14. A good rule of thumb for effective filtering is to make $R_L C \geq 10T$, where T is the period of the rectified voltage.

The ripple factor (r) is an indication of the effectiveness of the filter and is defined as the ratio of the ripple voltage (V_r) to the DC (average) value of the filter output voltage (V_{DC}). These parameters are illustrated in Figure E.07-15.

$$r = \left(\frac{V_r}{V_{DC}}\right)100\%$$

The lower the ripple factor, the better the filter. The ripple factor can be decreased by increasing the value of the filter capacitor.

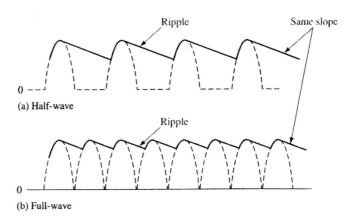

(a) Half-wave

(b) Full-wave

FIGURE E.07-14
Comparison of ripple voltages for half-wave and full-wave signals with same filter and same input frequency.

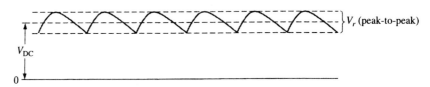

FIGURE E.07-15
V_r and V_{DC} determine the ripple factor.

Surge Current in the Capacitor Filter   Before the switch in Figure E.07-16(a) is closed, the filter capacitor is uncharged. At the instant the switch is closed, voltage is connected to the bridge and the capacitor appears as a short, as shown. An initial surge of current is produced through the two forward-biased diodes. The worst-case situation occurs when the switch is closed at a peak of the input voltage and a maximum surge current $I_{surge(max)}$ is produced, as illustrated in the figure.

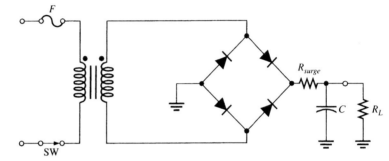

(a) Maximum surge current occurs when switch is closed at peak of input cycle.

(b) A series resistor (R_{surge}) limits the surge current.

FIGURE E.07-16
Surge current in a capacitor filter.

It is possible that the surge current could destroy the diodes, and for this reason a surge-limiting resistor R_{surge} is sometimes connected, as shown in Figure E.07-16(b). The value of this resistor must be small compared to R_L. Also, the diodes must have a forward current rating such that they can withstand the momentary surge of current.

A general rule-of-thumb for estimating the value of the surge-limiting resistor is

$$R_{surge} > \frac{V_{p(sec)} - 1.4\ V}{I_{FSM}}$$

where I_{FSM} is the specified nonrepetitive maximum forward surge current for the diode.

REVIEW QUESTIONS

True/False

1. The rectifier converts AC into pulsating DC.
2. The output voltage of a rectifier increases when a filter capacitor is added.

Multiple Choice

3. The peak input voltage to a half-wave rectifier is 17 V. What is the average output voltage?

   a. 0.7 V.

   b. 5.18 V.

   c. 10.36 V.

4. The peak input voltage to a full-wave bridge rectifier is 17 V. What is the average output voltage?

   a. 1.4 V.

   b. 4.96 V.

   c. 9.93 V.

5. Increasing the size of the filter capacitor

   a. Reduces the amount of ripple voltage.

   b. Increases the amount of ripple voltage.

   c. Has no effect on the ripple voltage.

E.08 Fabricate and demonstrate linear power supplies and filters

INTRODUCTION

Using actual components or simulation software (such as *Electronics Workbench*), set up and examine each of the filtered power supplies shown in Figure E.08-1.

For each supply, measure the average DC output voltage. Then connect a 10 μF/ 25 V electrolytic capacitor and remeasure the DC output voltage. Has it gone up? What is the peak-to-peak ripple on the output? Change the load resistor to 4.7 K ohms and remeasure the DC output and ripple. Repeat for 470 ohms.

FIGURE E.08-1
Test circuits.

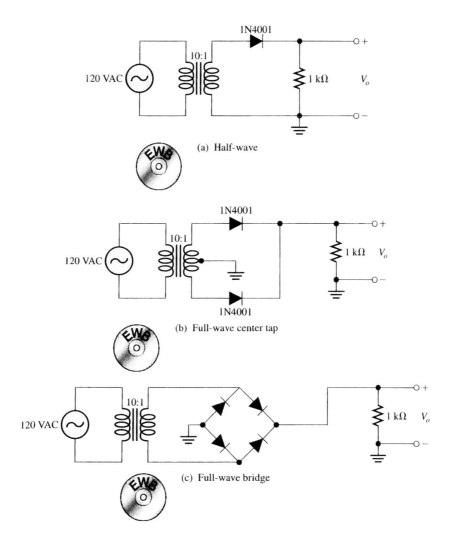

(a) Half-wave

(b) Full-wave center tap

(c) Full-wave bridge

E.09 Troubleshoot and repair linear power supplies and filters

INTRODUCTION

Several types of failures can occur in power supply rectifiers. In this section, we will examine some possible failures and the effects they would have on a circuit's operation.

Open Diode

A half-wave rectifier with a diode that has opened (a common failure mode) is shown in Figure E.09-1. In this case, you would measure 0 V DC at the rectifier output, as shown.

Now consider the full-wave, center-tapped rectifier in Figure E.09-2. Assume that diode D_1 has failed open. With an oscilloscope connected to the output, as shown in part

FIGURE E.09-1
The effect of an open diode in a half-wave rectifier.

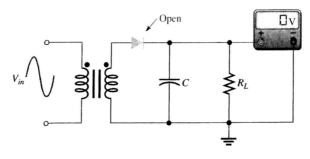

FIGURE E.09-2
Symptoms of an open diode in a full-wave, center-tapped rectifier.

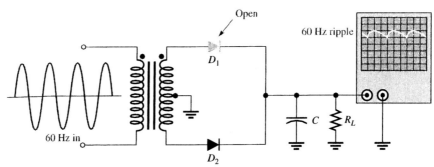

(a) Ripple should be less and have a frequency of 120 Hz.

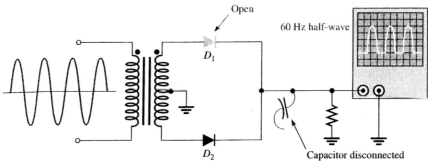

(b) With *C* removed, output should be a full-wave 120 Hz signal.

(a), you would observe the following: You would see a larger-than-normal ripple voltage at a frequency of 60 Hz rather than 120 Hz. Disconnecting the filter capacitor, you would observe a half-wave rectified voltage, as in part (b). Now let's examine the reason for these observations. If diode D_1 is open, there will be current through R_L only during the negative half-cycle of the input signal. During the positive half-cycle, an open path prevents current through R_L. The result is a half-wave voltage, as illustrated.

With the filter capacitor in the circuit, the half-wave signal will allow it to discharge more than it would with a normal full-wave signal, resulting in a larger ripple voltage. Basically, the same observations would be made for an open failure of diode D_2.

An open diode in a bridge rectifier would create symptoms identical to those just discussed for the center-tapped rectifier. As illustrated in Figure E.09-3, the open diode would prevent current through R_L during half of the input cycle (in this case, the negative half). As a result, there would be a half-wave output and an increased ripple voltage at 60 Hz, as discussed before.

FIGURE E.09-3
Effect of an open diode in a bridge rectifier.

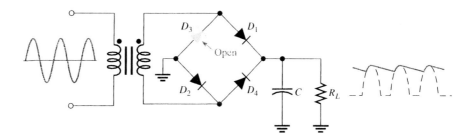

Shorted Diode

A shorted diode is one that has failed such that it has a very low resistance in both directions. If a diode suddenly became shorted, it is likely that a sufficiently high current would exist during one-half of the input cycle such that the fuse in the primary circuit would open. An unfused primary could cause the shorted diode to burn open or the other diode in series with it to burn open. Also, the transformer could be damaged, as illustrated for the case of a bridge rectifier in Figure E.09-4 with D_1 shorted.

FIGURE E.09-4
Effect of a shorted diode in a bridge rectifier.

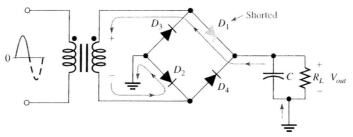

(a) Positive half-cycle: The shorted diode acts as a forward-biased diode, so the load current is normal.

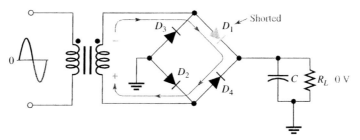

(b) Negative half-cycle: The shorted diode produces a short circuit across the source. As a result D_1, D_4, or the transformer secondary will probably burn open.

In part (a) of Figure E.09-4, current is supplied to the load through the shorted diode during the first positive half-cycle, just as though it were forward-biased. During the negative half-cycle, the current is shorted through D_1 and D_4, as shown in part (b). Again, damage to the transformer is possible. Also, it is likely that this excessive current would burn either or both of the diodes open. If only one of the diodes opened, the circuit will operate as a half-wave rectifier. If both diodes opened, there would be no voltage developed across the load. These conditions are illustrated in Figure E.09-5.

FIGURE E.09-5
Effect of open diodes.

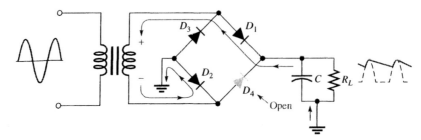

(a) One open diode produces a half-wave output.

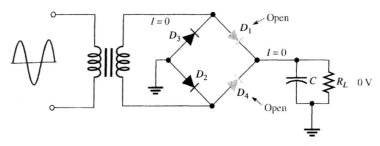

(b) Two open diodes produce a 0 V output.

Open Filter Capacitor

An open filter capacitor would result in a full-wave rectified voltage on the output because the filtering action of the capacitor is lost.

Shorted or Leaky Filter Capacitor

A shorted capacitor would most likely cause some or all of the diodes in a full-wave rectifier to open due to excessive current. In any event, there would be no DC voltage on the output.

A leaky capacitor can be represented by a leakage resistance in parallel with the capacitor, as shown in Figure E.09-6(a). The effect of the leakage resistance is to reduce the discharging time constant, causing an increase in ripple voltage on the output, as shown in Figure E.09-6(b).

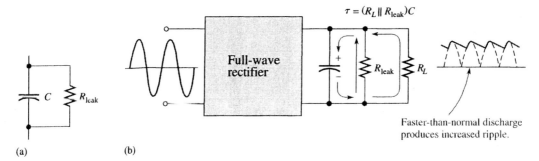

(a)   (b)

FIGURE E.09-6
Effect of a leaky filter capacitor on the output voltage of a full-wave rectifier.

E.10 Understand principles and operations of operational amplifier circuits

INTRODUCTION

Symbol and Terminals

The standard op-amp symbol is shown in Figure E.10-1(a). It has two input terminals, the inverting input (−) and the noninverting input (+), and one output terminal. The typical **operational amplifier** requires two DC supply voltages, one positive and the other negative, as shown in Figure E.10-1(b). Usually these DC voltage terminals are left off the schematic symbol for simplicity, but they are always understood to be there. Typical integrated circuit packages are shown in part (c).

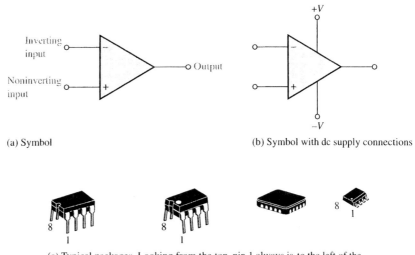

(a) Symbol

(b) Symbol with dc supply connections

(c) Typical packages. Looking from the top, pin 1 always is to the left of the notch or dot on the DIP and SO packages.

FIGURE E.10-1
Op-amp symbols and packages.

The Ideal Op-Amp

To illustrate what an op-amp is, we will consider its *ideal* characteristics. A practical op-amp, of course, falls short of these ideal standards, but it is much easier to understand and analyze the device from an ideal point of view.

First, the ideal op-amp has infinite voltage gain and infinite bandwidth. Also, it has an infinite input impedance (open), so that it does not draw any power from the driving source. Finally, it has a 0 (zero) output impedance. These characteristics are illustrated in Figure E.10-2. The input voltage V_{in} appears between the two input terminals, and the output voltage is $A_v V_{in}$, as indicated by the symbol for internal voltage source. The concept of infinite input impedance is a particularly valuable analysis tool for the various op-amp configurations.

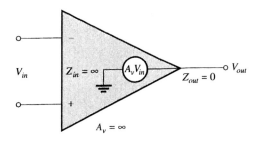

FIGURE E.10-2
Ideal op-amp representation.

FIGURE E.10-3
Practical op-amp representation.

The Practical Op-Amp

Although modern IC op-amps approach parameter values that can be treated as ideal in many cases, the ideal device cannot be made.

Any device has limitations, and the IC op-amp is no exception. Op-amps have both voltage and current limitations. Peak-to-peak output voltage, for example, is usually limited to slightly less than the two supply voltages. Output current is also limited by internal restrictions such as power dissipation and component ratings.

Characteristics of a practical op-amp are high voltage gain, high input impedance, low output impedance, and wide bandwidth, as illustrated in Figure E.10-3.

A Simple Op-Amp Arrangement

Figure E.10-4 shows two diff-amp stages and an emitter-follower connected to form a simple op-amp. The first stage can be used with a single-ended or a differential input. The differential outputs of the first stage feed into the differential inputs of the second stage. The

FIGURE E.10-4
Simplified internal circuitry of a basic op-amp.

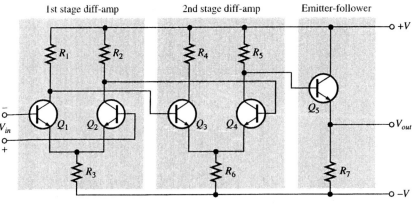

(a) Circuit

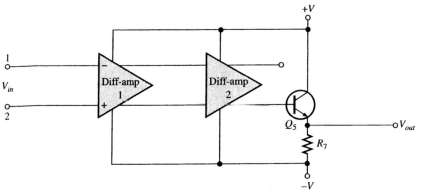

(b) Block diagram

output of the second stage is single-ended to drive an emitter-follower to achieve a relatively low output impedance. Both differential stages together provide a high voltage gain and a high CMRR.

OP-AMP DATA SHEET PARAMETERS

In this section, several important op-amp parameters are defined. These are the input offset voltage, the input offset voltage drift with temperature, the input bias current, the input impedance, the input offset current, the output impedance, the common-mode range, the open-loop voltage gain, the common-mode rejection ratio, the slew rate, and the frequency response. Also four popular IC op-amps are compared in terms of these parameters.

Input Offset Voltage

The ideal op-amp produces zero volts out for zero volts in. In a practical op-amp, however, a small DC voltage appears at the output when no differential input voltage is applied. Its primary cause is a slight mismatch of the base-emitter voltages of the differential input stage, as illustrated in Figure E.10-5(a).

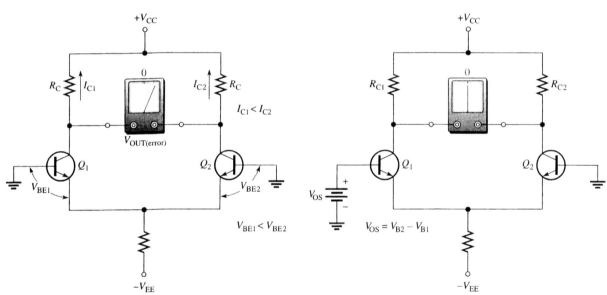

(a) A V_{BE} mismatch causes a small output error voltage.

(b) The input offset voltage is the difference in the voltage between the inputs that is necessary to eliminate the output error voltage (makes $V_{OUT} = 0$)

FIGURE E.10-5
Input offset voltage, V_{OS}.

The output voltage of the differential input stage is expressed as

$$V_{OUT} = I_{C2}R_C - I_{C1}R_C$$

A small difference in the base-emitter voltages of Q_1 and Q_2 causes a small difference in the collector currents. This results in a nonzero value of V_{OUT}. (The collector resistors are equal.)

As specified on an op-amp data sheet, the *input offset voltage, V_{OS},* is the differential DC voltage required between the inputs to force the differential output to zero volts, as demonstrated in Figure E.10-5(b). Typical values of input offset voltage are in the range of 2 mV or less. In the ideal case, it is 0 V.

Input Offset Voltage Drift with Temperature

The *input offset voltage drift* is a parameter related to V_{OS} that specifies how much change occurs in the input offset voltage for each degree change in temperature. Typical values range anywhere from about 5 μV per degree Celsius to about 50 μV per degree Celsius. Usually, an op-amp with a higher nominal value of input offset voltage exhibits a higher drift.

Input Bias Current

You have seen that the input terminals of a diff-amp are the transistor bases and, therefore, the input currents are the base currents.

The *input bias current* is the direct current required by the inputs of the amplifier to properly operate the first stage. By definition, the input bias current is the average of both input currents and is calculated as follows:

$$I_{BIAS} = \frac{I_1 + I_2}{2}$$

The concept of input bias current is illustrated in Figure E.10-6.

FIGURE E.10-6
Input bias current is the average of the two op-amp input currents.

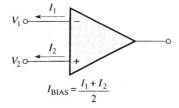

Input Impedance

Two basic ways of specifying the input impedance of an op-amp are the differential and the common mode. The *differential input impedance* is the total resistance between the inverting and the noninverting inputs, as illustrated in Figure E.10-7(a). Differential impedance is measured by determining the change in bias current for a given change in differential input voltage. The *common-mode input impedance* is measured from the inputs to ground and is depicted in Figure E.10-7(b).

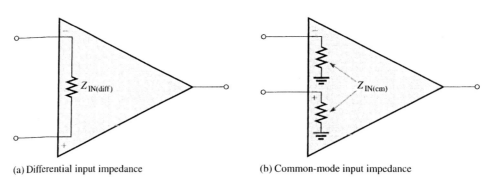

(a) Differential input impedance

(b) Common-mode input impedance

FIGURE E.10-7
Op-amp input impedance.

Input Offset Current

Ideally, the two input bias currents are equal, and thus their difference is zero. In a practical op-amp, however, the bias currents are not exactly equal.

The *input offset current* is the difference of the input bias currents, expressed as

$$I_{OS} = |I_1 - I_2|$$

Actual magnitudes of offset current are usually at least an order of magnitude (ten times) less than the bias current. In many applications, the offset current can be neglected. However, high-gain, high-input impedance amplifiers should have as little I_{OS} as possible, because the difference in currents through large input resistances develops a substantial offset voltage, as shown in Figure E.10-8.

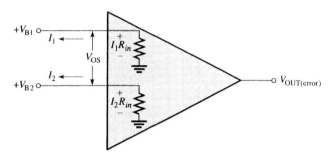

FIGURE E.10-8
Effect of input offset current.

The offset voltage developed by the input offset current is

$$V_{OS} = I_1 R_{in} - I_2 R_{in} = (I_1 - I_2)R_{in}$$

$$V_{OS} = I_{OS}R_{in}$$

The error created by I_{OS} is amplified by the gain A_v of the op-amp and appears in the output as

$$V_{OUT(error)} = A_v I_{OS} R_{in}$$

The change in offset current with temperature is often an important consideration. Values of temperature coefficient in the range of 0.5 nA per degree Celsius are common.

Output Impedance

The *output impedance* is the resistance viewed from the output terminal of the op-amp, as indicated in Figure E.10-9.

FIGURE E.10-9
Op-amp output impedance.

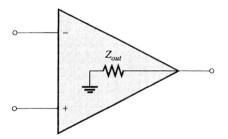

Common-Mode Range

All op-amps have limitations on the range of voltages over which they will operate. The *common-mode range* is the range of input voltages which, when applied to both inputs, will not cause clipping or other output distortion. Many op-amps have common-mode ranges of ± 10 V with DC supply voltages of ± 15 V.

Open-Loop Voltage Gain, A_{ol}

The *open-loop voltage gain* is the gain of the op-amp without any external feedback from output to input. A good op-amp has a very high **open-loop gain;** 50,000 to 200,000 is typical.

Common-Mode Rejection Ratio

The *common-mode rejection ratio (CMRR)*, as discussed in conjunction with the diff-amp, is a measure of an op-amp's ability to reject common-mode signals. An infinite value of CMRR means that the output is zero when the same signal is applied to both inputs (common mode).

An infinite CMRR is never achieved in practice, but a good op-amp does have a very high value of CMRR. As previously discussed, common-mode signals are undesired interference voltages such as 60 Hz power-supply ripple and noise voltages due to pickup of radiated energy. A high CMRR enables the op-amp to virtually eliminate these interference signals from the output.

The accepted definition of CMRR for an op-amp is the open-loop gain (A_{ol}) divided by the common-mode gain.

$$\text{CMRR} = \frac{A_{ol}}{A_{cm}}$$

It is commonly expressed in decibels as follows:

$$\text{CMRR} = 20 \log\left(\frac{A_{ol}}{A_{cm}}\right)$$

EXAMPLE E.10-1

A certain op-amp has an open-loop gain of 100,000 and a common-mode gain of 0.25. Determine the CMRR and express it in decibels.

Solution
$$\text{CMRR} = \frac{A_{ol}}{A_{cm}} = \frac{100{,}000}{0.25} = 400{,}000$$

$$\text{CMRR} = 20 \log(400{,}000) = 112 \text{ dB}$$

Slew Rate

The maximum rate of change of the output voltage in response to a step input voltage is the *slew rate* of an op-amp. The slew rate is dependent upon the frequency response of the amplifier stages within the op-amp.

Slew rate is measured with an op-amp connected as shown in Figure E.10-10(a). This particular op-amp connection is a unity-gain, noninverting configuration that will be discussed later. It gives a worst-case (slowest) slew rate. As shown in part (b), a pulse is applied to the input, and the ideal output voltage is measured. The width of the input pulse

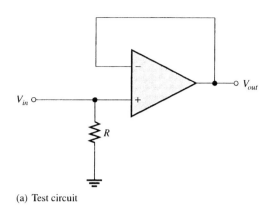

(a) Test circuit

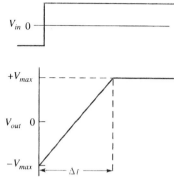

(b) Step input voltage and the resulting output voltage

FIGURE E.10-10
Measurement of slew rate.

must be sufficient to allow the output to "slew" from its lower limit to its upper limit, as shown. As you can see, a certain time interval, Δt, is required for the output voltage to go from its lower limit $-V_{max}$ to its upper limit $+V_{max}$, once the input step is applied. The slew rate is expressed as

$$\text{Slew rate} = \frac{\Delta V_{out}}{\Delta t}$$

where $\Delta V_{out} = +V_{max} - (-V_{max})$. The unit of slew rate is volts per microsecond (V/μs).

EXAMPLE E.10-2

The output voltage of a certain op-amp appears as shown in Figure E.10-11 in response to a step input. Determine the slew rate.

FIGURE E.10-11 V_{out} (V)

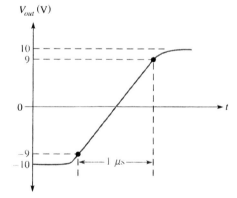

Solution   The output goes from the lower to the upper limit in 1 μs. Since this response is not ideal, the limits are taken at the 90% points, as indicated. Thus, the upper limit is +9 V and the lower limit is −9 V.
   The slew rate is

$$\frac{\Delta V}{\Delta t} = \frac{+9\ \text{V} - (-9\ \text{V})}{1\ \mu\text{s}} = 18\ \text{V/}\mu\text{s}$$

Frequency Response

The internal amplifier stages that make up an op-amp have voltage gains limited by junction capacitances. Although the diff-amps used in op-amps are somewhat different from the basic amplifiers discussed earlier, the same principles apply. An op-amp has no internal coupling capacitors, however; therefore, the low frequency response extends down to DC.

Comparison of Op-Amp Parameters

Table E.10-1 provides a comparison of values of some of the parameters just described for four popular IC op-amps. Any values not listed were not given on the manufacturer's data sheet. All values are typical at 25°C.

TABLE E.10-1

| Parameter | Op-Amp Type | | | |
	741C	LM101A	LM108	LM118
Input offset voltage	1 mV	1 mV	0.7 mV	2 mV
Input bias current	80 nA	120 nA	0.8 nA	120 nA
Input offset current	20 nA	40 nA	0.05 nA	6 nA
Input impedance	2 MΩ	800 kΩ	70 MΩ	3 MΩ
Output impedance	75 Ω	—	—	—
Open-loop gain	200,000	160,000	300,000	200,000
Slew rate	0.5 V/μs	—	—	70 V/μs
CMRR	90 dB	90 dB	100 dB	100 dB

Other Features

Most available op-amps have two very important features: short-circuit protection and no latch-up. Short-circuit protection keeps the circuit from being damaged if the output becomes shorted, and the no-latch-up feature prevents the op-amp from hanging up in one output state (high or low voltage level) under certain input conditions.

OP-AMP CONFIGURATIONS WITH NEGATIVE FEEDBACK

In this section, we will discuss several basic ways in which an op-amp can be connected using negative feedback to stabilize the gain and increase frequency response. The extremely high open-loop gain of an op-amp creates an unstable situation because a small noise voltage on the input can be amplified to a point where the amplifier is driven out of its linear region. Also, unwanted oscillations can occur. In addition, the open-loop gain parameter of an op-amp can vary greatly from one device to the next. Negative feedback takes a portion of the output and applies it back out-of-phase with the input, creating an effective reduction in gain. This closed-loop gain is usually much less than the open-loop gain and independent of it.

Noninverting Amplifier

An op-amp connected in a **closed-loop** configuration as a noninverting amplifier with a controlled amount of voltage gain is shown in Figure E.10-12.

The input signal is applied to the noninverting input. The output is applied back to the inverting input through the feedback circuit formed by R_i and R_f, creating **negative feedback** as follows.

FIGURE E.10-12
*Noninverting amplifier
showing differential
input $V_{in} - V_f$.*

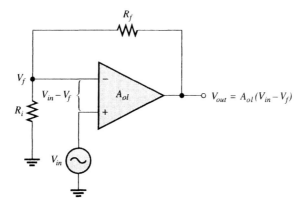

R_i and R_f form a voltage-divider circuit that reduces the output V_{out} and connects the reduced voltage V_f to the inverting input. The feedback voltage is expressed as

$$V_f = \left(\frac{R_i}{R_i + R_f} \right) V_{out}$$

The difference of the input voltage V_{in} and the feedback voltage V_f is the differential input to the op-amp, as shown in Figure E.10-12. This differential voltage is amplified by the open-loop gain of the op-amp (A_{ol}) and produces an output voltage expressed as

$$V_{out} = A_{ol}(V_{in} - V_f)$$

Letting $R_i/(R_i + R_f) = B$ and then substituting BV_{out} for V_f, we get the following algebraic steps:

$$V_{out} = A_{ol}(V_{in} - BV_{out})$$
$$V_{out} = A_{ol}V_{in} - A_{ol}BV_{out}$$
$$V_{out} + A_{ol}BV_{out} = A_{ol}V_{in}$$
$$V_{out}(1 + A_{ol}B) = A_{ol}V_{in}$$

Since the total voltage gain of the amplifier in Figure E.10-12 is V_{out}/V_{in}, it can be expressed as

$$\frac{V_{out}}{V_{in}} = \frac{A_{ol}}{1 + A_{ol}B}$$

The product $A_{ol}B$ is usually much greater than 1, so the equation simplifies to

$$\frac{V_{out}}{V_{in}} \cong \frac{A_{ol}}{A_{ol}B}$$

Since the closed-loop gain of the noninverting (NI) amplifier is

$$A_{cl(NI)} = \frac{V_{out}}{V_{in}}$$

then

$$A_{cl(NI)} \cong \frac{1}{B}$$

or

$$A_{cl(NI)} = 1 + \frac{R_f}{R_i}$$

This shows that the **closed-loop gain,** $A_{cl(NI)}$, of the noninverting amplifier is the reciprocal of the attenuation (B) of the feedback circuit (voltage divider). It is interesting to note that the closed-loop gain is not at all dependent on the op-amp's open-loop gain under the condition $A_{ol}B \gg 1$. The closed-loop gain can be set by selecting values of R_i and R_f.

EXAMPLE E.10-3

Determine the gain of the amplifier in Figure E.10-13. The open-loop voltage gain is 100,000.

FIGURE E.10-13

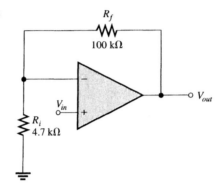

Solution This is a noninverting op-amp configuration. Therefore, the closed-loop gain is

$$A_{cl(NI)} = 1 + \frac{R_f}{R_i} = 1 + \frac{100 \text{ k}\Omega}{4.7 \text{ k}\Omega} = 22.3$$

Voltage-Follower

The voltage-follower configuration is a special case of the noninverting amplifier in which all of the output voltage is fed back to the inverting input, as shown in Figure E.10-14. As you can see, the straight feedback connection has a voltage gain of approximately one. The closed-loop voltage gain of a noninverting amplifier is $1/B$, as previously derived. Since $B = 1$, the closed-loop gain of the voltage-follower (VF) is

$$\boxed{A_{cl(VF)} = 1}$$

FIGURE E.10-14
Op-amp voltage-follower.

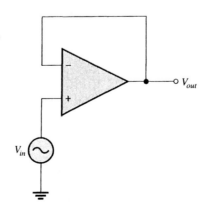

The most important features of the voltage-follower configuration are its very high input impedance and its very low output impedance. These features make it a nearly ideal buffer amplifier for interfacing high-impedance sources and low-impedance loads.

Inverting Amplifier

An op-amp connected as an inverting amplifier with a controlled amount of voltage gain is shown in Figure E.10-15. The input signal is applied through a series input resistor R_i to the inverting input. Also, the output is fed back through R_f to the same input. The noninverting input is grounded.

FIGURE E.10-15
Inverting amplifier.

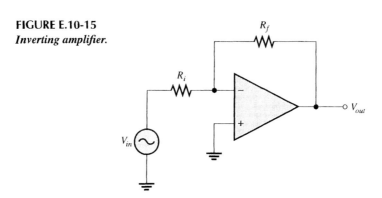

At this point, the ideal op-amp parameters mentioned earlier are useful in simplifying the analysis of this circuit. In particular, the concept of infinite input impedance is of great value. An infinite input impedance implies zero current to the inverting input. If there is zero current through the input impedance, then there must be no voltage drop between the inverting and noninverting inputs. That is, the voltage at the inverting (−) input is zero because the other input (+) is grounded. This zero voltage at the inverting input terminal is referred to as *virtual ground* and is illustrated in Figure E.10-16(a).

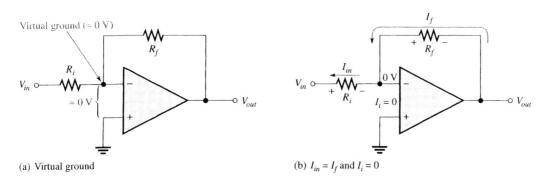

(a) Virtual ground

(b) $I_{in} = I_f$ and $I_i = 0$

FIGURE E.10-16
Virtual ground concept and development of closed-loop voltage gain for the inverting amplifier.

Since there is no current into the inverting input, the current through R_i and the current through R_f are equal, as shown in Figure E.10-16(b).

$$I_{in} = I_f$$

The voltage across R_i equals V_{in} because of virtual ground on the other side of the resistor. Therefore,

$$I_{in} = \frac{V_{in}}{R_i}$$

Also, the voltage across R_f equals $-V_{out}$ because of the virtual ground, and therefore

$$I_f = \frac{-V_{out}}{R_f}$$

Since $I_f = I_{in}$,

$$\frac{-V_{out}}{R_f} = \frac{V_{in}}{R_i}$$

Rearranging the terms,

$$\frac{V_{out}}{V_{in}} = -\frac{R_f}{R_i}$$

Of course, V_{out}/V_{in} is the overall gain of the inverting (I) amplifier.

$$A_{cl(I)} = -\frac{R_f}{R_i}$$

This shows that the closed-loop voltage gain $A_{cl(I)}$ of the inverting amplifier is the ratio of the feedback resistance R_f to the resistance R_i. The *closed-loop gain is independent of the op-amp's internal open-loop gain.* Thus, the negative feedback stabilizes the voltage gain. The negative sign indicates inversion.

EXAMPLE E.10-4

Given the op-amp configuration in Figure E.10-17, determine the value of R_f required to produce a closed-loop voltage gain with an absolute value of 100.

FIGURE E.10-17

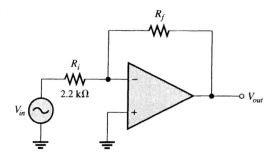

Solution Knowing that $R_i = 2.2 \text{ k}\Omega$ and $|A_{cl(I)}| = 100$, calculate R_f as follows:

$$|A_{cl(I)}| = \frac{R_f}{R_i}$$

$$R_f = |A_{cl(I)}|R_i = (100)(2.2 \text{ k}\Omega) = 220 \text{ k}\Omega$$

REVIEW QUESTIONS

True/False

1. A practical op-amp is close in operation to an ideal amplifier.

2. The front-end of an op-amp is a differential amplifier.

Multiple Choice

3. Which is not an op-amp parameter?

    a. Slew rate.

    b. Differential bias.

    c. Input offset voltage.

4. A noninverting amplifier with R_f equal to 10 k ohms and R_i equal to 1 k ohms has a gain of

    a. 10.

    b. 11.

    c. -10.

5. An inverting amplifier with R_f equal to 10 k ohms and R_i equal to 1 k ohms has a gain of

    a. 10.

    b. 11.

    c. -10.

E.11 Fabricate and demonstrate operational amplifier circuits

INTRODUCTION

Using actual components or simulation software (such as *Electronics Workbench*), set up and examine each operational amplifier circuit shown in Figure E.11-1. Note that in each case, power must be applied to the op-amp ($\pm$ 12 V, for example).

For each circuit, apply a 500 mV peak-to-peak, 1 KHz square wave, triangle wave, and sine wave. View the output waveform and determine the gain and phase shift of the amplifiers. Let R_i equal 1 K ohms and R_f equal 10 K ohms.

Apply a 10 V peak-to-peak square wave to the buffer and slowly increase the applied frequency until the effects of slew rate become noticeable. What is the frequency where this occurs? Where does this condition become noticeable in the inverting amplifier?

FIGURE E.11-1
Test circuits.

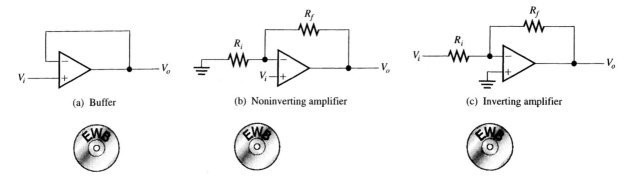

(a) Buffer   (b) Noninverting amplifier   (c) Inverting amplifier

E.12 Troubleshoot and repair operational amplifier circuits

INTRODUCTION

As a technician, you will no doubt encounter situations in which an op-amp or its associated circuitry has malfunctioned. The op-amp is a complex integrated circuit with many types of internal failures possible. However, since you cannot troubleshoot the op-amp internally, you treat it as a single device with only a few connections to it. If it fails, you replace it just as you would a resistor, capacitor, or transistor.

In the basic op-amp configurations, there are only a few external components that can fail. These are the feedback resistor, the input resistor, and the potentiometer used for offset voltage compensation. Also, of course, the op-amp itself can fail or there can be faulty contacts in the circuit. Let's examine the three basic configurations for possible faults and the associated symptoms.

Input Offset Voltage Compensation

Most integrated circuit op-amps provide a means of compensating for offset voltage. This is usually done by connecting an external potentiometer to designated pins on the IC package, as illustrated in Figure E.12-1(a) and (b) for a 741 op-amp. The two terminals are labelled *offset null*. With no input, the potentiometer is simply adjusted until the output voltage reads 0, as shown in Figure E.12-1(c).

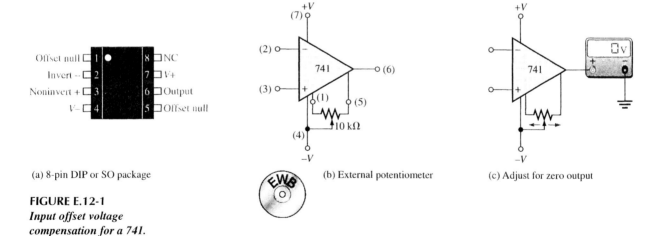

(a) 8-pin DIP or SO package

(b) External potentiometer

(c) Adjust for zero output

FIGURE E.12-1
Input offset voltage compensation for a 741.

Faults in the Noninverting Amplifier

The first thing to do when you suspect a faulty circuit is to check for the proper supply voltage and ground. Having done that, several other possible faults are as follows.

Open Feedback Resistor If the feedback resistor, R_f, in Figure E.12-2 opens, the op-amp is operating with its very high open-loop gain, which causes the input signal to drive the device into nonlinear operation and results in a severely clipped output signal as shown in part (a).

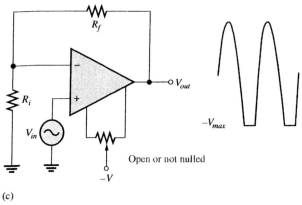

FIGURE E.12-2
Faults in the noninverting amplifier.

Open Input Resistor   In this case, you still have a closed-loop configuration. But, since R_i is open and effectively equal to infinity (∞), the closed-loop gain is

$$A_{cl(\text{NI})} = 1 + \frac{R_f}{R_i} = 1 + \frac{R_f}{\infty} = 1 + 0 = 1$$

This shows that the amplifier acts like a voltage-follower. You would observe an output signal that is the same as the input, as indicated in Figure E.12-2(b).

Open or Incorrectly Adjusted Offset Null Potentiometer   In this situation, the output offset voltage will cause the output signal to begin clipping on only one peak as the input signal is increased to a sufficient amplitude. This is indicated in Figure E.12-2(c).

Faulty Op-Amp   As mentioned, many things can happen to an op-amp. In general, an internal failure will result in a loss or distortion of the output signal. The best approach is to first make sure that there are no external failures or faulty conditions. If everything else is good, then the op-amp must be bad.

Faults in the Voltage-Follower

The voltage-follower is a special case of the noninverting amplifier. Except for a bad op-amp, a bad external connection, or a problem with the offset null potentiometer, about the only thing that can happen in a voltage-follower circuit is an open feedback loop. This would have the same effect as an open feedback resistor as previously discussed.

Faults in the Inverting Amplifier

Open Feedback Resistor   If R_f opens as indicated in Figure E.12-3(a), the input signal still feeds through the input resistor and is amplified by the high open-loop gain of the op-amp. This forces the device to be driven into nonlinear operation, and you will see an output something like that shown. This is the same result as in the noninverting configuration.

Open Input Resistor   This prevents the input signal from getting to the op-amp input, so there will be no output signal, as indicated in Figure E.12-3(b).

Failures in the op-amp itself or the offset null potentiometer have the same effects as previously discussed for the noninverting amplifier.

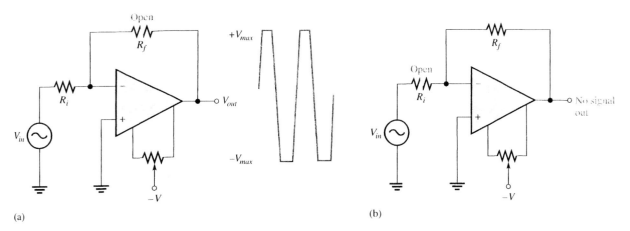

(a)

(b)

FIGURE E.12-3
Faults in the inverting amplifier.

E.13 Understand principles and operations of audio power amplifiers

INTRODUCTION

When a common-emitter, common-collector, or common-base amplifier is biased so that it operates in the linear region for the full 360° of the input cycle, it is a class A amplifier. In this mode of operation, the amplifier does not go into cutoff or saturation; therefore, the output voltage waveform has the same shape as the input waveform. A class A amplifier can be either inverting or noninverting.

The operation of a class A **large-signal** amplifier is illustrated in Figure E.13-1, where the output waveform is an amplified replica of the input and may be either in phase or 180° out of phase with the input.

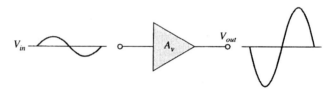

FIGURE E.13-1
Basic class A amplifier operation (inverting).

Centered Q-Point for Maximum Output Signal

When the Q-point is at the center of the AC load line (midway between saturation and cutoff), a maximum class A signal can be obtained. This is graphically illustrated in the AC load line in Figure E.13-2(a). Ideally, the collector current can vary from its Q-point value, I_{CQ}, up to its saturation value, $I_{c(sat)}$, and down to its cutoff value of zero. This operation is indicated in Figure E.13-2(b).

As you can see in Figure E.13-2(b), the peak value of the collector current equals I_{CQ}, and the peak value of the collector-to-emitter voltage equals V_{CEQ}. This is the largest signal achievable from a class A amplifier. When the input signal is too large, the amplifier is driven into cutoff and saturation, as illustrated in Figure E.13-3.

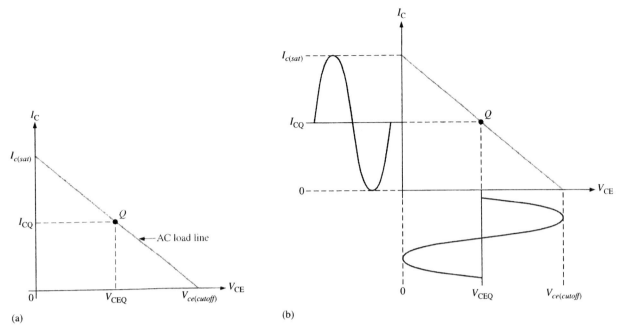

FIGURE E.13-2
For maximum class A operation, the Q-point is centered on the AC load line.

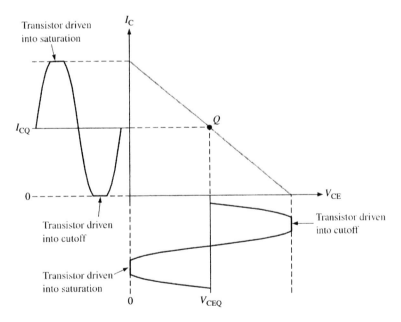

FIGURE E.13-3
Waveforms are clipped at cutoff and saturation because the amplifier is overdriven (too large an input signal).

A Noncentered Q-Point Limits the Output

Q-Point Closer to Cutoff   If the Q-point is not centered on the AC load line, V_{ce} is limited to less than the possible maximum. Figure E.13-4(a) shows an AC load line with the Q-point moved away from center toward cutoff. V_{ce} is limited by cutoff in this case. The collector current can swing only down to near zero and an equal amount above I_{CQ}. The

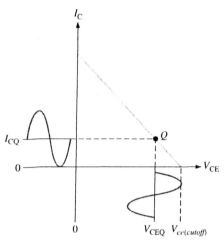
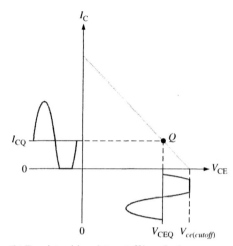

(a) Amplitude of V_{ce} and I_c limited by cutoff

(b) Transistor driven into cutoff by a further increase in input amplitude

FIGURE E.13-4
Q-point closer to cutoff.

collector-to-emitter voltage can swing only up to its cutoff value and an equal amount below V_{CEQ}. If the amplifier is driven any further than this by an increase in the input signal, it will go into cutoff, as shown in Figure E.13-4(b), and the waveforms will be clipped off on one peak.

Q-Point Closer to Saturation   Figure E.13-5(a) shows an AC load line with the Q-point moved away from center toward saturation. In this case, V_{ce} is limited by saturation. The collector current can swing only up to near saturation and an equal amount below I_{CQ}. The collector-to-emitter voltage can swing only down to near its saturation value and an equal amount above V_{CEQ}. If the amplifier is driven any further than this by an increase in the input signal, it will go into saturation, as shown in Figure E.13-5(b).

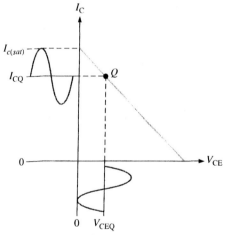
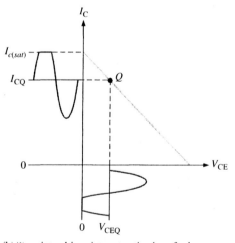

(a) Amplitude of V_{ce} and I_c limited by saturation

(b) Transistor driven into saturation by a further increase in input amplitude

FIGURE E.13-5
Q-point closer to saturation.

Large-Signal Load Line Operation

Recall that an amplifier such as that shown in Figure E.13-6 can be represented in terms of either its DC or its AC equivalent.

DC Load Line Using the DC equivalent circuit in Figure E.13-7(a), you can determine the DC load line as follows: $I_{C(sat)}$ occurs when $V_{CE} \cong 0$, so

$$I_{C(sat)} \cong \frac{V_{CC}}{R_C + R_E}$$

$V_{CE(cutoff)}$ occurs when $I_C \cong 0$, so

$$V_{CE(cutoff)} \cong V_{CC}$$

The DC load line is shown in Figure E.13-7(b).

FIGURE E.13-6
Common-emitter amplifier with signal source.

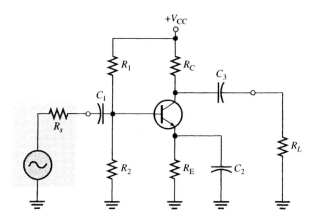

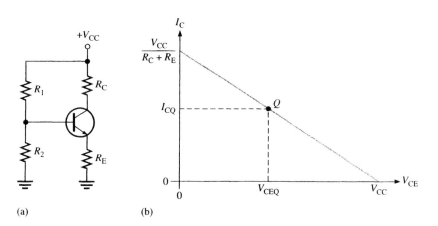

(a)   (b)

FIGURE E.13-7
DC equivalent circuit and DC load line for the amplifier in Figure E.13-6.

AC Load Line From the AC viewpoint, the circuit in Figure E.13-6 looks different than it does from the DC viewpoint. The collector resistance is different because R_L is in parallel with R_C due to the coupling capacitor C_3, and the emitter resistance is zero due to the by-pass capacitor C_2; therefore, the AC load line is different from the DC load line. How much collector current can there be under AC conditions before saturation occurs? To answer this question, refer to the AC equivalent circuit and AC load line in Figure E.13-8.

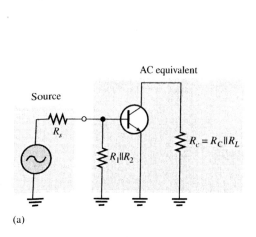
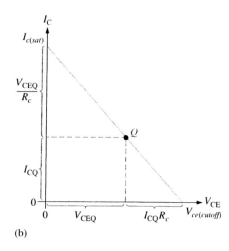

(a)   (b)

FIGURE E.13-8
AC equivalent circuit and AC load line for the amplifier in Figure E.13-6.

Remember that a lowercase italic subscript indicates an AC quantity and an uppercase non-italic subscript indicates a DC quantity. For example, R_c is the AC collector resistance and R_C is the DC collector resistance. I_{CQ} and V_{CEQ} are the DC Q-point coordinates. Going from the Q-point to the saturation point, the collector-to-emitter voltage swings from V_{CEQ} to near 0; that is, $\Delta V_{CE} = V_{CEQ}$. The change in collector current going from the Q-point to saturation is therefore

$$\Delta I_C = \frac{V_{CEQ}}{R_c}$$

where $R_c = R_C \parallel R_L$ is the AC collector resistance. The AC collector current at saturation is

$$I_{c(sat)} = I_{CQ} + \Delta I_C$$

Thus,

$$I_{c(sat)} = I_{CQ} + \frac{V_{CEQ}}{R_c}$$

Going from the Q-point to the cutoff point, the collector current swings from I_{CQ} to near 0; that is, $\Delta I_C = I_{CQ}$. The change in collector-to-emitter voltage going from the Q-point to cutoff is therefore

$$\Delta V_{CE} = (\Delta I_C)R_c = I_{CQ}R_c$$

The cutoff value of AC collector-to-emitter voltage is

$$V_{ce(cutoff)} = V_{CEQ} + I_{CQ}R_c$$

These results are shown on the AC load line of Figure E.13-9. The corresponding DC load line is shown for comparison.

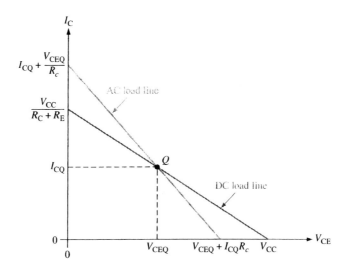

FIGURE E.13-9
DC and AC load lines.

Voltage Gain

The voltage gain of a class A large-signal amplifier is determined in the same way as for a small-signal amplifier with the exception that the formula $r'_e \cong 25$ mV$/I_E$ is not valid for the large-signal amplifier. This is because the signal swings over a large portion of the transconductance curve. Since $r'_e = \Delta V_{BE}/\Delta I_C$, the value is different for large-signal operation than it is for small-signal conditions because of the nonlinearity of the curve.

The large-signal AC emitter resistance, r'_e, can be determined graphically from the transconductance curve, as shown in Figure E.13-10, using the relationship:

$$r'_e = \frac{\Delta V_{BE}}{\Delta I_C}$$

The voltage-gain formula for a common-emitter, large-signal amplifier with R_E completely bypassed is

$$A_v = \frac{R_c}{r'_e}$$

FIGURE E.13-10
Determination of r'_e from the transconductance curve.

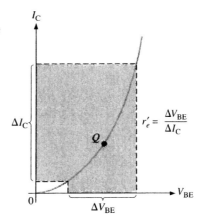

EXAMPLE E.13-1

Find the large-signal voltage gain of the amplifier in Figure E.13-11. Assume that r'_e has been found to be 8 Ω from graphical data.

FIGURE E.13-11

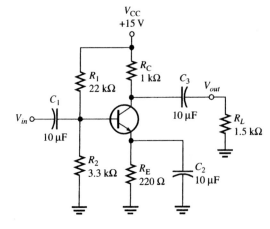

Solution

$$R_c = \frac{(R_C)(R_L)}{R_C + R_L} = \frac{(1\ \text{k}\Omega)(1.5\ \text{k}\Omega)}{2.5\ \text{k}\Omega} = 600\ \Omega$$

$$A_v = \frac{R_c}{r'_e} = \frac{600\ \Omega}{8\ \Omega} = 75$$

Distortion

When the collector current swings over a large portion of the transconductance curve, distortion can occur on the negative half-cycle because of the greater nonlinearity on the lower end of the curve, as shown in Figure E.13-12. Distortion can be sufficiently reduced by keeping the collector current on the more linear part of the curve (at higher values of I_{CQ} and V_{BEQ}). Increasing the base bias voltage will result in more collector current and an increase in V_{BE} due to more voltage drop across r'_e.

FIGURE E.13-12
Example of distortion.

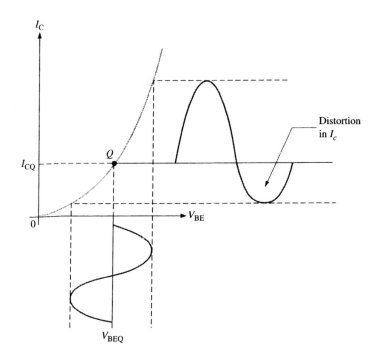

Power Gain

The main purpose of a large-signal amplifier is to achieve power gain. If we assume that the large-signal current gain, A_i, is approximately equal to β_{DC}, then the power gain, A_p, for a common-emitter amplifier is

$$A_p = A_i A_v = \beta_{DC} A_v$$

$$A_p = \beta_{DC}\left(\frac{R_c}{r'_e}\right)$$

Quiescent Power

The power dissipation of a transistor with no signal input is the product of its Q-point current and voltage.

$$P_{DQ} = I_{CQ} V_{CEQ}$$

The quiescent power is the maximum power that the class A transistor must handle; therefore, its power rating should exceed this value.

Output Power

In general, for any Q-point location on the AC load line, the output power of a common-emitter amplifier is the product of the rms collector current and the rms collector-to-emitter voltage.

$$P_{out} = V_{ce} I_c$$

Let's now consider the output power for three cases of Q-point location.

Q-Point Closer to Saturation   When the Q-point is closer to saturation, the maximum collector-to-emitter voltage swing is V_{CEQ}, and the maximum collector current swing is V_{CEQ}/R_c, as shown in Figure E.13-13(a). The AC output power is, therefore,

$$P_{out} = \left(\frac{0.707 V_{CEQ}}{R_c}\right) 0.707 V_{CEQ}$$

$$P_{out} = \frac{0.5 V_{CEQ}^2}{R_c}$$

where $R_c = R_C \parallel R_L$.

Q-Point Closer to Cutoff   When the Q-point is closer to cutoff, the maximum collector current swing is I_{CQ}, and the collector-to-emitter voltage swing is $I_{CQ} R_c$, as shown in Figure E.13-13(b). The AC output power is, therefore,

$$P_{out} = (0.707 I_{CQ})(0.707 I_{CQ} R_c)$$

$$P_{out} = 0.5 I_{CQ}^2 R_c$$

Q-Point Centered   When the Q-point is centered, the maximum collector current swing is I_{CQ}, and the maximum collector-to-emitter voltage swing is V_{CEQ}, as shown in Figure E.13-13(c). The AC output power is, therefore,

$$P_{out} = (0.707 V_{CEQ})(0.707 I_{CQ})$$

$$P_{out} = 0.5 V_{CEQ} I_{CQ}$$

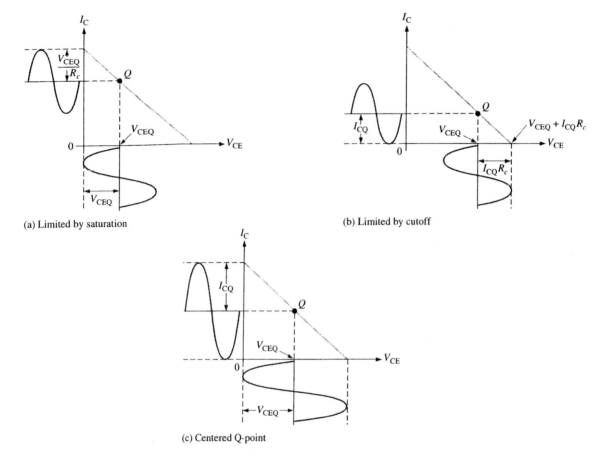

FIGURE E.13-13
AC load line operation showing limitations of output voltage swings.

This is the maximum AC output power from a class A amplifier under signal conditions. Notice that it is one-half the quiescent power dissipation.

Efficiency

Efficiency (η) of an amplifier is the ratio of AC output power to DC input power. The DC input power is the DC supply voltage times the current drawn from the supply.

$$P_{DC} = V_{CC}I_{CC}$$

The average supply current I_{CC} equals I_{CQ}, and the supply voltage V_{CC} is twice V_{CEQ} when the Q-point is centered. The maximum efficiency is therefore

$$\eta_{max} = \frac{P_{out}}{P_{DC}} = \frac{0.5V_{CEQ}I_{CQ}}{V_{CC}I_{CC}} = \frac{0.5V_{CEQ}I_{CQ}}{2V_{CEQ}I_{CQ}} = \frac{0.5}{2}$$

$$\boxed{\eta_{max} = 0.25}$$

Thus, 0.25 or 25% is the highest possible efficiency available from a class A amplifier and is approached only when the Q-point is at the center of the AC load line.

Determine the following values for the amplifier in Figure E.13-14 when operated with the maximum possible output signal:
(a) Minimum transistor power rating
(b) AC output power
(c) Efficiency

FIGURE E.13-14

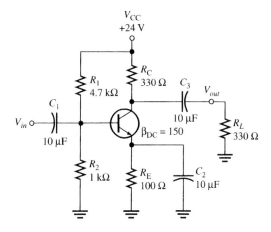

Solution   First determine the DC values. Neglect $R_{\text{IN(base)}}$.

$$V_B \cong \left(\frac{R_2}{R_1 + R_2}\right)V_{CC} = \left(\frac{1\ k\Omega}{5.7\ k\Omega}\right)24\ V = 4.2\ V$$

Then

$$V_E = V_B - V_{BE} = 4.2\ V - 0.7\ V = 3.5\ V$$

$$I_E = \frac{V_E}{R_E} = \frac{3.5\ V}{100\ \Omega} = 35\ mA$$

Therefore, the collector current at the Q-point is

$$I_{CQ} \cong 35\ mA$$

and

$$V_C = V_{CC} - I_{CQ}R_C = 24\ V - (35\ mA)(330\ \Omega) = 12.5\ V$$
$$V_{CEQ} = V_C - V_E = 12.5\ V - 3.5\ V = 9.0\ V$$

(a) The transistor power rating must be greater than

$$P_D = V_{CEQ}I_{CQ} = (9.0\ V)(35\ mA) = 315\ mW$$

(b) To make a calculation of AC output power under a *maximum* signal condition, you must know the location of the Q-point relative to the center. This will tell you whether I_{CQ} or V_{CEQ} is the limiting factor if the Q-point is not centered. The AC load line values are as follows:

$$R_c = R_C \parallel R_L = 330\ \Omega \parallel 330\ \Omega = 165\ \Omega$$

$$I_{c(sat)} = I_{CQ} + \frac{V_{CEQ}}{R_c} = 35\ mA + \frac{9.0\ V}{165\ \Omega} = 89.5\ mA$$

and

$$V_{ce(cutoff)} = V_{CEQ} + I_{CQ}R_c = 9.0\ V + (35\ mA)(165\ \Omega) = 14.8\ V$$

A *centered* Q-point is at

$$I_{CQ} = \frac{89.5 \text{ mA}}{2} = 44.8 \text{ mA}$$

and

$$V_{CEQ} = \frac{14.8 \text{ V}}{2} = 7.37 \text{ V}$$

These Q-point values are shown on the AC load line in Figure E.13-15. The *actual* Q-point for this amplifier is closer to cutoff, as shown in the figure. Therefore, the maximum collector current swing is I_{CQ}, and the AC output power is

$$P_{out} = 0.5I_{CQ}^2 R_c = 0.5(35 \text{ mA})^2(165 \text{ } \Omega) = 101 \text{ mW}$$

FIGURE E.13-15

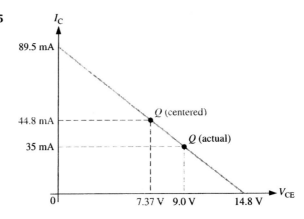

(c) The efficiency is

$$\eta = \frac{P_{out}}{P_{DC}} = \frac{P_{out}}{V_{CC}I_{CC}} = \frac{P_{out}}{V_{CC}I_{CQ}} = \frac{101 \text{ mW}}{(24 \text{ V})(35 \text{ mA})} = 0.12$$

This efficiency of 12% is considerably less than the maximum possible efficiency (25%) because the actual Q-point is not centered.

Maximum Load Power

The maximum power to the load in a class A amplifier occurs when the Q-point is centered, as you can see in Figure E.13-13(c). The maximum peak load voltage equals V_{CEQ}, assuming a negligible voltage drop across the output coupling capacitor.

$$P_L = \frac{V_L^2}{R_L} = \frac{(0.707V_{CEQ})^2}{R_L}$$

$$P_L = \frac{0.5V_{CEQ}^2}{R_L}$$

Sometimes the load power is defined as the output power.

EXAMPLE E.13-3

Determine the maximum load power for the amplifier in Example E.13-2.

Solution   When the Q-point is centered, $V_{CEQ} = 7.37$ V.

$$P_{L(max)} = \frac{0.5V_{CEQ}^2}{R_L} = \frac{0.5(7.37 \text{ V})^2}{330 \text{ }\Omega} = 82.3 \text{ mW}$$

CLASS B AND CLASS AB PUSH-PULL AMPLIFIERS

When an amplifier is biased at cutoff such that it operates in the linear region for 180° of the input cycle and is in cutoff for 180°, it is a class B amplifier. Class AB amplifiers are biased to conduct for slightly more than 180°. The primary advantage of a class B or class AB amplifier over a class A amplifier is that they are more efficient; you can get more output power for a given amount of input power. A disadvantage of class B or class AB is that it is more difficult to implement the circuit in order to get a linear reproduction of the input waveform. As you will see in this section, the term push-pull refers to a common type of class B or class AB amplifier circuit in which the input wave shape is reproduced at the output.

Class B Operation

The class B operation is illustrated in Figure E.13-16, where the output waveform is shown relative to the input.

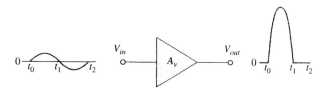

FIGURE E.13-16
Basic class B amplifier operation (noninverting).

The Q-Point Is at Cutoff   The class B amplifier is biased at the cutoff point so that $I_{CQ} = 0$ and $V_{CEQ} = V_{CE(cutoff)}$. It is brought out of cutoff and operates in its linear region when the input signal drives it into conduction. This is illustrated in Figure E.13-17 with an emitter-follower circuit. Obviously, the output is not a replica of the input. Therefore, a two-transistor configuration, known as a **push-pull** amplifier, is necessary to get a sufficiently good reproduction of the input waveform.

Push-Pull Class B Operation   Figure E.13-18 shows one type of push-pull class B amplifier using two emitter-followers. This is called a complementary amplifier because one emitter-follower uses an *npn* transistor and the other a matched *pnp;* the transistors conduct on *opposite* alternations of the input cycle. A matched **complementary pair** of transistors have identical characteristics, except one is an *npn* and the other a *pnp*. The 2N3904 and 2N3906 are examples. Notice that there is no DC base bias voltage ($V_B = 0$); thus, only the signal voltage drives the transistors into conduction. Q_1 conducts during the positive half of the input cycle, and Q_2 conducts during the negative half.

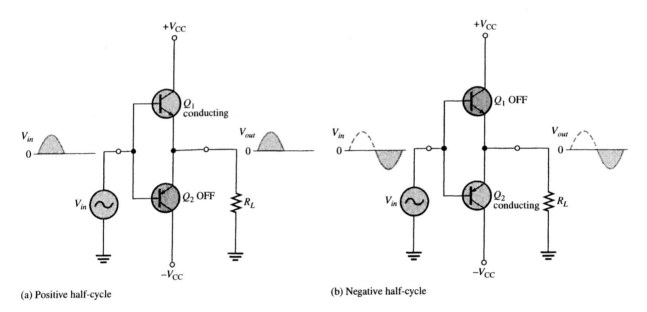

FIGURE E.13-17
Common-collector class B amplifier.

(a) Positive half-cycle

(b) Negative half-cycle

FIGURE E.13-18
Class B push-pull AC operation.

Crossover Distortion   When the DC base voltage is zero, both transistors are off and the input signal voltage must exceed V_{BE} before a transistor conducts. Because of this, there is a time interval between the positive and negative alternations of the input when neither transistor is conducting, as shown in Figure E.13-19. The resulting distortion in the output waveform is quite common and is called **crossover distortion.**

Class AB Operation

To eliminate crossover distortion, both transistors in the push-pull arrangement must be biased slightly above cutoff when there is no signal. This variation of the class B push-pull amplifier is designated as class AB. Class AB biasing can be done with a voltage-divider arrangement, as shown in Figure E.13-20(a). It is, however, difficult to maintain a stable bias point with this circuit due to changes in V_{BE} over temperature changes. (The requirement for dual-polarity power supplies is eliminated when R_L is capacitively coupled.) A

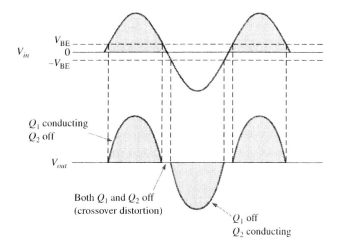

FIGURE E.13-19
Illustration of crossover distortion in a class B push-pull amplifier. The transistors conduct only during portions of the input indicated by the shaded areas.

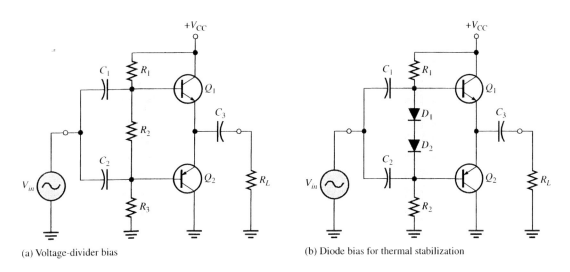

(a) Voltage-divider bias   (b) Diode bias for thermal stabilization

FIGURE E.13-20
Biasing the push-pull amplifier for class AB operation to eliminate crossover distortion.

more suitable arrangement is shown in Figure E.13-20(b). When the diode characteristics of D_1 and D_2 are closely matched to the transconductance characteristics of the transistors, a stable bias over temperature can be maintained. This stabilization can also be accomplished by using the base-emitter junctions of two additional matched transistors instead of D_1 and D_2. Although technically incorrect, class AB amplifiers are often referred to as class B in common practice.

The DC equivalent circuit of the push-pull amplifier is shown in Figure E.13-21. Resistors R_1 and R_2 are of equal value; therefore, the voltage at point A between the two diodes is $V_{CC}/2$. Assuming that both diodes and both transistors are identical, the drop across D_1 equals the V_{BE} of Q_1, and the drop across D_2 equals the V_{BE} of Q_2. As a result, the voltage at the emitters is also $V_{CC}/2$ and, therefore, $V_{CEQ1} = V_{CEQ2} = V_{CC}/2$, as indicated. Because both transistors are biased near cutoff, $I_{CQ} \cong 0$.

FIGURE E.13-21
DC equivalent of push-pull amplifier.

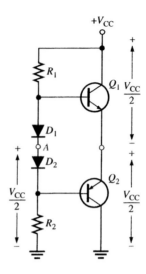

REVIEW QUESTIONS

True/False

1. A class A amplifier is biased near cutoff.

2. The DC and AC load lines of an amplifier are the same.

Multiple Choice

3. Class A amplifiers are biased to operate
   a. 180 degrees linear, 180 degrees cutoff.
   b. 180 degrees linear, 180 degrees saturated.
   c. 360 degrees linear.

4. Class B amplifiers are biased to operate
   a. 180 degrees linear, 180 degrees cutoff.
   b. 180 degrees linear, 180 degrees saturated.
   c. 360 degrees linear.

5. Push-pull amplifiers are biased to operate
   a. 180 degrees linear, 180 degrees cutoff.
   b. 180 degrees linear, 180 degrees saturated.
   c. 360 degrees linear.

E.14 Fabricate and demonstrate audio power amplifiers

INTRODUCTION

Using actual components or a software simulation package (such as *Electronics Workbench*), set up the audio power amplifier shown in Figure E.14-1 and determine the voltage gain, power gain, and efficiency.

FIGURE E.14-1
Audio power amplifier

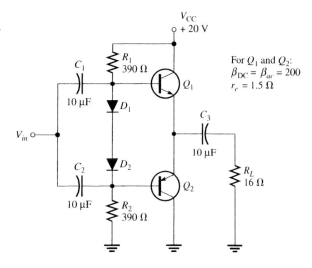

E.15 Troubleshoot and repair audio power amplifiers

INTRODUCTION

In this section, an example of isolating a component failure in a circuit is presented. We will use a class A amplifier with the output voltage monitored by an oscilloscope. Several incorrect output waveforms will be examined and the most likely faults will be discussed.

As shown in Figure E.15-1, the class A amplifier should have a normal sine wave output when a sinusoidal input signal is applied. Now we will consider several incorrect output waveforms and the most likely causes in each case. In Figure E.15-2(a), the scope displays a DC level equal to the DC supply voltage, indicating that the transistor is in cutoff. The two most likely causes of this condition are (1) the transistor has an open *pn* junction, or (2) R_4 is open, preventing collector and emitter current.

FIGURE E.15-1
Class A amplifier with proper output voltage swing.

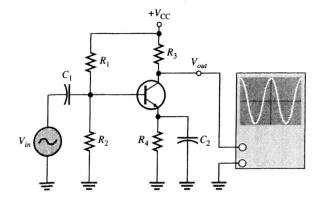

In Figure E.15-2(b), the scope displays a DC level at the collector approximately equal to the DC emitter voltage. The two probable causes of this indication are (1) the transistor is shorted from collector to emitter, or (2) R_2 is open, causing the transistor to be biased in saturation. In the second case, a sufficiently large input signal can bring the transistor out of saturation on its negative peaks, resulting in short pulses on the output.

In Figure E.15-2(c), the scope displays an output waveform that indicates the transistor is cut off except during a small portion of the input cycle. Possible causes of this indication are (1) the Q-point has shifted down due to a drastic out-of-tolerance change in a resistor value, or (2) R_1 is open, biasing the transistor in cutoff. The display shows that the input signal is sufficient to bring it out of cutoff for a small portion of the cycle.

In Figure E.15-2(d), the scope displays an output waveform that indicates the transistor is saturated except during a small portion of the input cycle. Again, it is possible that a resistance change has caused a drastic shift in the Q-point up toward saturation, or R_2 is open, causing the transistor to be biased in saturation, and the input signal is bringing it out of saturation for a small portion of the cycle.

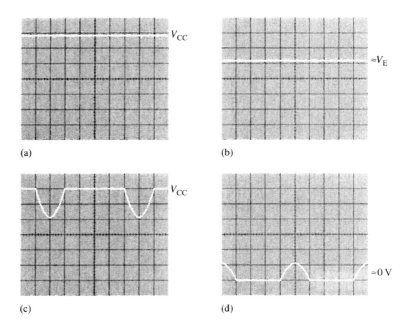

FIGURE E.15-2
Oscilloscope displays of output voltage for the amplifier in Figure E.15-1, illustrating several types of failures.

E.16 Understand principles and operations of regulated and switching power supply circuits

INTRODUCTION

In this section we examine the operation of voltage and current regulation under varying load conditions.

EXAMPLE E.16-1

A certain voltage regulator has a 12 V output when there is no load ($I_L = 0$). When there is a full-load current of 10 mA, the output voltage is 11.9 V. Express the voltage regulation as a percentage change from no-load to full-load and also as a percentage change for each mA change in load current.

Solution   The no-load output voltage is

$$V_{NL} = 12 \text{ V}$$

The full-load output voltage is

$$V_{FL} = 11.9 \text{ V}$$

The load regulation is

$$\text{load regulation} = \left(\frac{V_{NL} - V_{FL}}{V_{FL}}\right)100\% = \left(\frac{12 \text{ V} - 11.9 \text{ V}}{11.9 \text{ V}}\right)100\% = 0.840\%$$

The load regulation can also be expressed as

$$\text{load regulation} = \frac{0.840\%}{10 \text{ mA}} = 0.084\%/\text{mA}$$

where the change in load current from no-load to full-load is 10 mA.

BASIC SERIES REGULATORS

The two fundamental classes of voltage regulators are linear regulators and switching regulators. Both of these are available in integrated circuit form. There are two basic types of linear regulator. One is the series regulator and the other is the shunt regulator. In this section, we will look at the series regulator. The shunt and switching regulators are covered next.

A simple representation of a series type of linear regulator is shown in Figure E.16-1(a), and the basic components are shown in the block diagram in Figure E.16-1(b). The control element is in series with the load between input and output. The output sample circuit senses a change in the output voltage. The error detector compares the sample voltage with a reference voltage and causes the control element to compensate in order to maintain a constant output voltage.

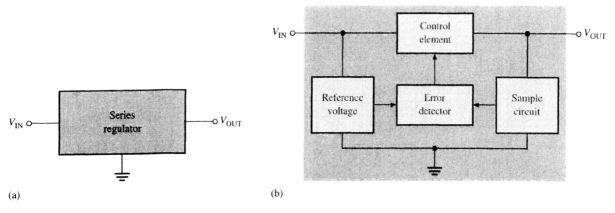

FIGURE E.16-1
Simple series voltage regulator block diagram.

Regulating Action

A basic op-amp series regulator circuit is shown in Figure E.16-2. The resistive voltage divider formed by R_2 and R_3 senses any change in the output voltage. When the output tries to decrease because of a decrease in V_{IN} or because of an increase in I_L (through a load resistor connected to V_{out}), a proportional voltage decrease is applied to the op-amp's inverting input by the voltage divider. Since the zener diode (D_1) holds the other op-amp input at a nearly constant reference voltage V_{REF}, a small difference voltage (error voltage) is developed across the op-amp's inputs. This difference voltage is amplified, and the op-amp's output voltage, V_B, increases. This increase is applied to the base of Q_1, causing the emitter voltage V_{OUT} to increase until the voltage to the inverting input again equals the reference (zener) voltage. This action offsets the attempted decrease in output voltage, thus keeping it nearly constant. The power transistor, Q_1 is usually used with a heat sink because it must handle all of the load current.

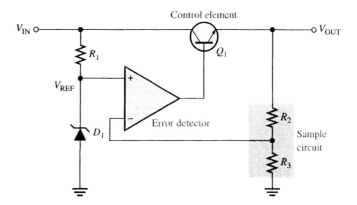

FIGURE E.16-2
Basic op-amp series regulator.

The opposite action occurs when the output tries to increase. The op-amp in the series regulator is actually connected as a noninverting amplifier where the reference voltage V_{REF} is the input at the noninverting terminal, and the R_2/R_3 voltage divider forms the negative feedback circuit. The closed-loop voltage gain is

$$A_{cl} = 1 + \frac{R_2}{R_3}$$

Therefore, the regulated output voltage (neglecting the base-emitter voltage of Q_1) is

$$V_{OUT} \cong \left(1 + \frac{R_2}{R_3}\right)V_{REF}$$

From this analysis, you can see that the output voltage is determined by the zener voltage and the resistors R_2 and R_3. It is relatively independent of the input voltage, and therefore, regulation is achieved (as long as the input voltage and load current are within specified limits).

EXAMPLE E.16-2

Determine the output voltage for the regulator in Figure E.16-3.

FIGURE E.16-3

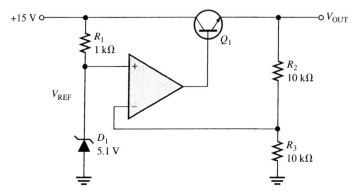

Solution V_{REF} = 5.1 V, the zener voltage. The regulated output voltage is therefore

$$V_{OUT} = \left(1 + \frac{R_2}{R_3}\right)V_{REF} = \left(1 + \frac{10\ k\Omega}{10\ k\Omega}\right)5.1\ V = (2)5.1\ V = 10.2\ V$$

Short-Circuit or Overload Protection

If an excessive amount of load current is drawn, the series-pass transistor can be quickly damaged or destroyed. Most regulators employ some type of excess current protection in the form of a current-limiting mechanism. Figure E.16-4 shows one method of current limiting to prevent overloads called *constant-current limiting*. The current-limiting circuit consists of transistor Q_2 and resistor R_4.

The load current through R_4 creates a voltage from base to emitter of Q_2. When I_L reaches a predetermined maximum value, the voltage drop across R_4 is sufficient to forward-bias the base-emitter junction of Q_2, thus causing it to conduct. Enough Q_1 base current is diverted from the collector of Q_2 so that I_L is limited to its maximum value $I_{L(max)}$. Since the base-to-emitter voltage of Q_2 cannot exceed about 0.7 V for a silicon transistor, the voltage across R_4 is held to this value, and the load current is limited to

$$I_{L(max)} = \frac{0.7\ V}{R_4}$$

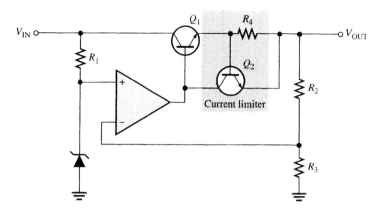

FIGURE E.16-4
Series regulator with constant-current limiting.

EXAMPLE E.16-3

Determine the maximum current that the regulator in Figure E.16-5 can provide to a load.

FIGURE E.16-5

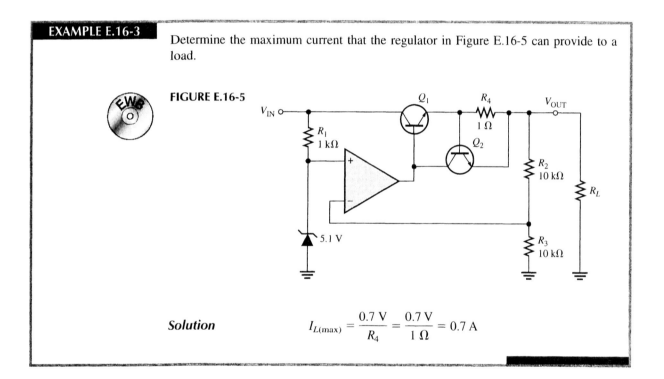

Solution

$$I_{L(max)} = \frac{0.7 \text{ V}}{R_4} = \frac{0.7 \text{ V}}{1 \, \Omega} = 0.7 \text{ A}$$

BASIC SHUNT REGULATORS

The second basic type of linear voltage regulator is the shunt regulator. As you have learned, the control element in the series regulator is the series-pass transistor. In the shunt regulator, the control element is a transistor in parallel (shunt) with the load.

A simple representation of a shunt type of linear regulator is shown in Figure E.16-6(a), and the basic components are shown in the block diagram in part (b).

In the basic shunt regulator, the control element is a transistor Q_1 in parallel with the load, as shown in Figure E.16-7. A resistor, R_1 is in series with the load. The operation of the circuit is similar to that of the series regulator, except that regulation is achieved by controlling the current through the parallel transistor Q_1.

When the output voltage tries to decrease due to a change in input voltage or load current, the attempted decrease is sensed by R_3 and R_4 and applied to the op-amp's noninverting

FIGURE E.16-6
Simple shunt regulator
block diagrams.

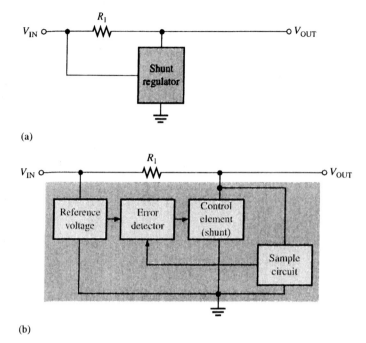

(a)

(b)

FIGURE E.16-7
Basic op-amp shunt
regulator with load
resistor.

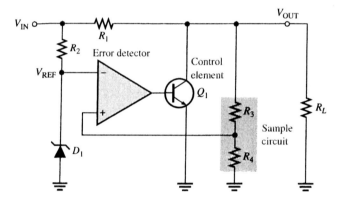

input. The resulting difference voltage reduces the op-amp's output (V_B), driving Q_1 less, thus reducing its collector current (shunt current) and increasing its effective collector-to-emitter resistance r'_{CE}. Since r'_{CE} acts as a voltage divider with R_1, this action offsets the attempted decrease in V_{OUT} and maintains it at an almost constant level.

The opposite action occurs when the output tries to increase. With I_L and V_{OUT} constant, a change in the input voltage produces a change in shunt current (I_S) as follows (Δ means "a change in").

$$\Delta I_S = \frac{\Delta V_{IN}}{R_1}$$

With a constant V_{IN} and V_{OUT}, a change in load current causes an opposite change in shunt current.

$$\Delta I_S = -\Delta I_L$$

This formula says that if I_L increases, I_S decreases, and vice versa.

The shunt regulator is less efficient than the series type but offers inherent short-circuit protection. If the output is shorted ($V_{OUT} = 0$), the load current is limited by the series resistor R_1 to a maximum value as follows ($I_S = 0$).

$$I_{L(\text{max})} = \frac{V_{IN}}{R_1}$$

EXAMPLE E.16-4

In Figure E.16-8, what power rating must R_1 have if the maximum input voltage is 12.5 V?

FIGURE E.16-8

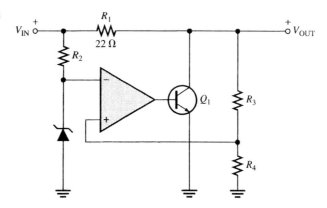

Solution   The worst-case power dissipation in R_1 occurs when the output is short-circuited and $V_{OUT} = 0$. When $V_{IN} = 12.5$ V, the voltage dropped across R_1 is

$$V_{R1} = V_{IN} - V_{OUT} = 12.5 \text{ V}$$

The power dissipation in R_1 is

$$P_{R1} = \frac{V_{R1}^2}{R_1} = \frac{(12.5 \text{ V})^2}{22 \ \Omega} = 7.10 \text{ W}$$

Therefore, a resistor of at least 10 W should be used.

BASIC SWITCHING REGULATORS

The two types of linear regulators, series and shunt, have control elements (transistors) that are conducting all the time, with the amount of conduction varied as demanded by changes in the output voltage or current. The switching regulator is different; the control element operates as a switch. A greater efficiency can be realized with this type of voltage regulator than with the linear types because the transistor is not always conducting. Therefore, switching regulators can provide greater load currents at low voltage than linear regulators because the control transistor doesn't dissipate as much power. Three basic configurations of switching regulators are step-down, step-up, and inverting.

Step-Down Configuration

In the step-down configuration, the output voltage is always less than the input voltage. A basic step-down switching regulator is shown in Figure E.16-9(a), and its simplified equivalent is shown in part (b). Transistor Q_1 is used to switch the input voltage at a duty cycle

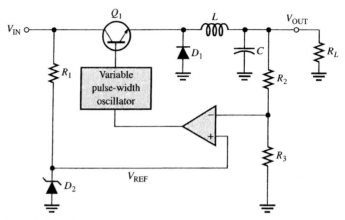

(a) Typical circuit

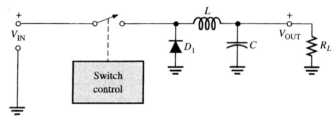

(b) Simplified equivalent circuit

FIGURE E.16-9
Basic step-down switching regulator.

that is based on the regulator's load requirement. The *LC* filter is then used to average the switched voltage. Since Q_1 is either *on* (saturated) or *off,* the power lost in the control element is relatively small. Therefore, the switching regulator is useful primarily in higher power applications or in applications where efficiency is of utmost concern.

The on and off intervals of Q_1 are shown in the waveform of Figure E.16-10(a). The capacitor charges during the on-time (t_{on}) and discharges during the off-time (t_{off}). When the on-time is increased relative to the off-time, the capacitor charges more, thus increasing the output voltage, as indicated in Figure E.16-10(b). When the on-time is decreased relative to the off-time, the capacitor discharges more, thus decreasing the output voltage, as in Figure E.16-10(c). Therefore, by adjusting the duty cycle $t_{on}/(t_{on} + t_{off})$ of Q_1, the output voltage can be varied. The inductor further smooths the fluctuations of the output voltage caused by the charging and discharging action.

The output voltage is expressed as

$$V_{OUT} = \left(\frac{t_{on}}{T}\right)V_{IN}$$

T is the period of the on-off cycle of Q_1 and is related to the frequency by $T = 1/f$. The period is the sum of the on-time and the off-time.

$$T = t_{on} + t_{off}$$

The ratio t_{on}/T is called the *duty cycle.*

The regulating action is as follows. When V_{OUT} tries to decrease, the on-time of Q_1 is increased, causing an additional charge on *C* to offset the attempted decrease. When V_{OUT} tries to increase, the on-time of Q_1 is decreased, causing the capacitor to discharge enough to offset the attempted increase.

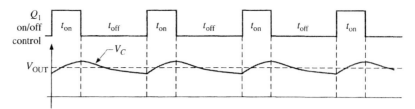

(a) V_{OUT} depends on the duty cycle.

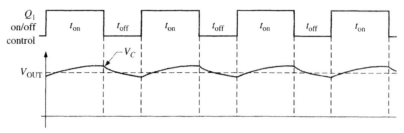

(b) Increase the duty cycle and V_{OUT} increases.

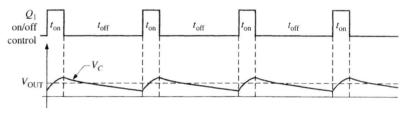

(c) Decrease the duty cycle and V_{OUT} decreases.

FIGURE E.16-10
Switching regulator waveforms. The V_C waveform is for no inductive filtering to illustrate the charge and discharge action. L and C smooth V_C to a nearly constant level, as indicated by the dashed line for V_{OUT}.

Step-Up Configuration

A basic step-up type of switching regulator is shown in Figure E.16-11.

When Q_1 turns on, voltage across L increases instantaneously to $V_{IN} - V_{CE(sat)}$, and the inductor's magnetic field expands quickly. During the on-time (t_{on}) of Q_1, V_L decreases from its initial maximum, as shown. The longer Q_1 is on, the smaller V_L becomes. When Q_1 turns off, the inductor's magnetic field collapses; and its polarity reverses so that its voltage adds to V_{IN}, thus producing an output voltage greater than the input. During the off-time (t_{off}) of Q_1, the diode is forward-biased, allowing the capacitor to charge. The

FIGURE E.16-11
Basic step-up switching regulator.

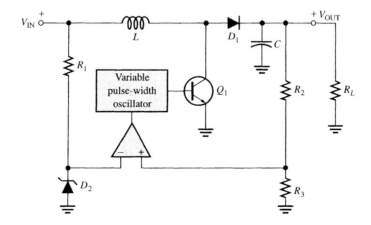

variations in the output voltage due to the charging and discharging action are sufficiently smoothed by the filtering action of L and C.

The shorter the on-time of Q_1, the greater the inductor voltage is, and thus the greater the output voltage is (greater V_L adds to V_{IN}). The longer the on-time of Q_1, the smaller are the inductor voltage and the output voltage (small V_L adds to V_{IN}). When V_{OUT} tries to decrease because of increasing load or decreasing input voltage, t_{on} decreases and the attempted decrease in V_{OUT} is offset. When V_{OUT} tries to increase, t_{on} increases and the attempted increase in V_{OUT} is offset. As you can see, the output voltage is inversely related to the duty cycle of Q_1 and can be expressed as follows.

$$V_{OUT} = \left(\frac{T}{t_{on}}\right)V_{IN}$$

where $T = t_{on} + t_{off}$.

Voltage-Inverter Configuration

A third type of switching regulator produces an output voltage that is opposite in polarity to the input. A basic diagram is shown in Figure E.16-12.

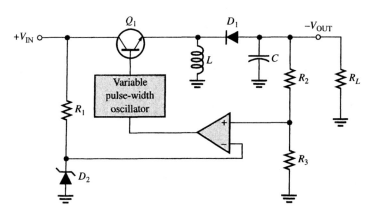

FIGURE E.16-12
Basic inverting switching regulator.

When Q_1 turns on, the inductor voltage jumps to $V_{IN} - V_{CE(sat)}$ and the magnetic field rapidly expands. While Q_1 is on, the diode is reverse-biased and the inductor voltage decreases from its initial maximum. When Q_1 turns off, the magnetic field collapses and the inductor's polarity reverses. This forward-biases the diode, charges C, and produces a negative output voltage, as indicated. The repetitive on-off action of Q_1 produces a repetitive charging and discharging that is smoothed by the LC filter action.

As with the step-up regulator, the less time Q_1 is on, the greater the output voltage is, and vice versa. Switching regulator efficiencies can be greater than 90 percent.

REVIEW QUESTIONS

True/False

1. There are two types of regulators: series and parallel.

2. There are two types of regulators: pass and shunt.

Multiple Choice

3. How is the output sampled in a simple series regulator?

    a. With a pass transistor.

    b. With a reference zener.

    c. With a voltage divider.

4. Switching regulators use

    a. An "always on" pass element.

    b. A variable pulse-width oscillator.

    c. Both a and b.

5. Switching regulators are capable of

    a. Step-up and step-down operation.

    b. Inverting operation.

    c. Both a and b.

E.17 Troubleshoot and repair regulated and switching power supply circuits

INTRODUCTION

In this section we examine the analysis and troubleshooting of a dual-voltage regulated power supply.

This power supply utilizes a full-wave bridge **rectifier** with both the positive and negative rectified voltages taken off the bridge at the appropriate points and filtered by electrolytic capacitors. Integrated circuit voltage regulators (7812 and 7912) provide regulation for the positive and negative voltages.

Step 1: Relate the PC Board to a Schematic

Develop a schematic for the power supply in Figure E.17-1. Add any missing labels and include the IC pin numbers. The rectifier diodes are 1N4001s, the filter capacitors C_1 and C_2 are 100 μF, and the transformer has a turns ratio of 5:1. Determine the backside PC board connections as you develop the schematic.

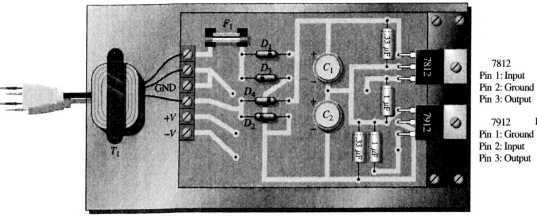

FIGURE E.17-1
Dual-voltage supply circuit board.

Step 2: Analyze the Power Supply Circuits

1. Determine the approximate voltage with respect to ground at each of the four "corners" of the bridge. State whether each voltage is AC or DC.
2. Calculate the peak inverse voltage of the rectifier diodes.
3. Show the voltage waveform across D_1 for a full cycle of the AC input.

Step 3: Troubleshoot the Power Supply

State the probable cause or causes for the following:

1. Both positive and negative output voltages are zero.
2. Positive output voltage is zero and the negative output voltage is -12 V.
3. Negative output voltage is zero and the positive output voltage is $+12$ V.
4. Radical voltage fluctuations on output of positive regulator.

Indicate the voltages you should measure at the four corners of the diode bridge for the following faults:

1. Diode D_1 open.
2. Capacitor C_2 open.

E.18 Understand principles and operations of active filter circuits

INTRODUCTION

Filters are usually categorized by the manner in which the output voltage varies with the frequency of the input voltage. The categories of active filters that we will examine in this section are low-pass, high-pass, band-pass, and band-stop.

Low-Pass Active Filters

Figure E.18-1 shows a basic **active filter** and its response curve. Notice that the input circuit is a single low-pass RC circuit, and unity gain is provided by the op-amp with a negative feedback loop. Simply stated, this is a voltage-follower with an RC filter between the input signal and the noninverting input.

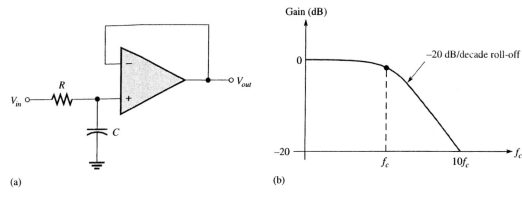

(a)   (b)

FIGURE E.18-1
Single-pole, active low-pass filter and response curve.

The voltage at the noninverting input, $V+$, is as follows:

$$V+ = \left(\frac{X_C}{\sqrt{R^2 + X_C^2}}\right)V_{in}$$

Since the gain of the op-amp is 1, the output voltage is equal to $V+$.

$$V_{out} = \left(\frac{X_C}{\sqrt{R^2 + X_C^2}}\right)V_{in}$$

A filter with one RC circuit that produces a -20 dB/decade **roll-off** beginning at f_c is said to be a *single-pole* or *first-order filter.* The term "-20 dB/decade" means that the voltage gain decreases by ten times (-20 dB) when the frequency increases by ten times (**decade**).

Low-Pass Two-Pole Filters   There are several types of active filters and they can have varying numbers of **poles,** but we will use a two-pole filter to illustrate. Figure E.18-2(a)

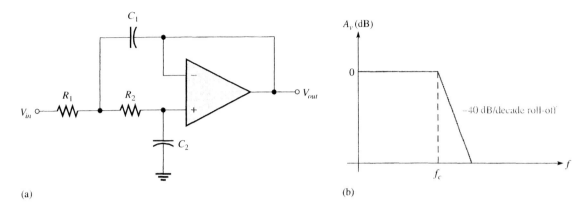

FIGURE E.18-2
Two-pole, active low-pass filter and its ideal response curve.

shows a two-pole (second-order) low-pass filter. Since each RC circuit in a filter is considered to be one-pole, the two-pole filter uses two RC circuits to produce a roll-off rate of -40 dB/decade, as indicated in Figure E.18-2(b). The active filter in Figure E.18-2 has unity gain up to near f_c because the op-amp is connected as a voltage-follower.

One of the RC circuits in Figure E.18-2(a) is formed by R_1 and C_1, and the other by R_2 and C_2. The critical frequency of this filter can be calculated using the following formula:

$$f_c = \frac{1}{2\pi\sqrt{R_1 R_2 C_1 C_2}}$$

Figure E.18-3(a) shows an example of a two-pole low-pass filter with values chosen to produce a response with a critical frequency of 1 kHz. Note that $C_1 = 2C_2$ and $R_1 = R_2$, because these relationships result in a gain of 0.707 (-3 dB) at f_c. For critical frequencies other than 1 kHz, the capacitance values can be scaled inversely with the frequency. For example, as shown in Figure E.18-3(b) and (c), to get a 2 kHz filter, halve the values of C_1 and C_2; for a 500 Hz filter, double the values.

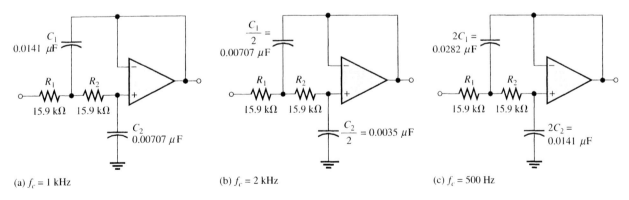

(a) $f_c = 1$ kHz   (b) $f_c = 2$ kHz   (c) $f_c = 500$ Hz

FIGURE E.18-3
Examples of low-pass filters (two-pole).

EXAMPLE E.18-1

Calculate the capacitance values required to produce a 3 kHz critical frequency in the low-pass filter of Figure E.18-4.

FIGURE E.18-4

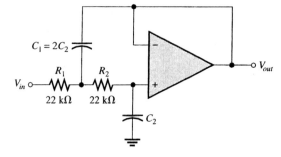

$C_1 = 2C_2$

R_1 R_2

V_{in}

22 kΩ   22 kΩ

C_2

V_{out}

Solution   The resistor values have already been set at 22 kΩ each. Since these differ from the 1 kHz reference filter, you cannot use the scaling method to get the capacitance values. Initially,

$$f_c = \frac{1}{2\pi\sqrt{R_1 R_2 C_1 C_2}}$$

Then square both sides.

$$f_c^2 = \frac{1}{4\pi^2 R_1 R_2 C_1 C_2}$$

Since $C_1 = 2C_2$ and $R_1 = R_2 = R$,

$$f_c^2 = \frac{1}{4\pi^2 R^2 (2C_2^2)}$$

Solve for C_2, and then determine C_1.

$$C_2^2 = \frac{1}{8\pi^2 R^2 f_c^2}$$

$$C_2 = \frac{1}{\sqrt{2}\ 2\pi R f_c} = \frac{0.707}{2\pi R f_c} = \frac{0.707}{2\pi(22\ k\Omega)(3\ kHz)} = 0.0017\ \mu F$$

$$C_1 = 2C_2 = 2(0.0017\ \mu F) = 0.0034\ \mu F$$

High-Pass Active Filters

In Figure E.18-5(a), a high-pass active filter with a 20 dB/decade roll-off is shown. Notice that the input circuit is a single high-pass RC circuit and that unity gain is provided by the op-amp with negative feedback. The response curve is shown in Figure E.18-5(b).

Ideally, a high-pass filter passes all frequencies above f_c without limit, as indicated in Figure E.18-6(a). In practice, of course, such is not the case. All op-amps inherently have internal RC circuits that limit the amplifier's response at high frequencies. Such is the case with the active high-pass filter. There is an upper frequency limit to its response, which essentially makes this type of filter a wide band-pass filter rather than a true high-pass filter, as indicated in Figure E.18-6(b). In many applications, the internal high-frequency cutoff is so much greater than the filter's f_c that the internal high-frequency cutoff can be neglected.

The voltage at the noninverting input is as follows:

$$V+ = \left(\frac{R}{\sqrt{R^2 + X_C^2}}\right)V_{in}$$

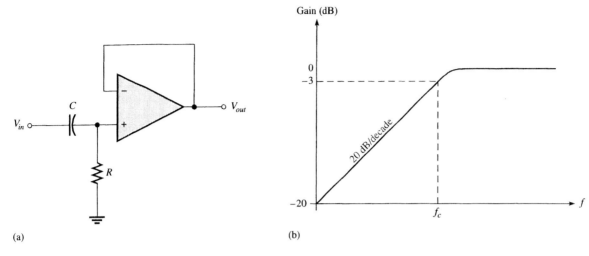

FIGURE E.18-5
Single-pole, active high-pass filter and response curve.

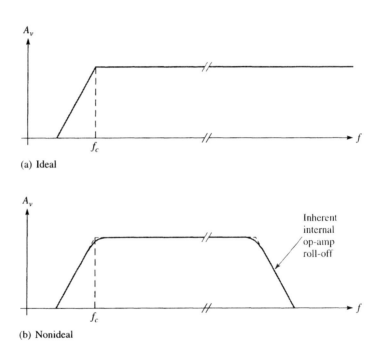

FIGURE E.18-6
High-pass filter response.

Since the op-amp is connected as a voltage-follower with unity gain, the output voltage is the same as $V+$.

$$V_{out} = \left(\frac{R}{\sqrt{R^2 + X_C^2}}\right)V_{in}$$

If the internal critical frequencies of the op-amp are assumed to be much greater than the desired f_c of the filter, the gain will roll off at 20 dB/decade as shown in Figure E.18-6(b). This is a single-pole filter because it has one RC circuit.

High-Pass Two-Pole Filters Figure E.18-7 shows a two-pole active high-pass filter. Notice that it is identical to the corresponding low-pass type, except for the positions of the

FIGURE E.18-7
Two-pole, active high-pass filter.

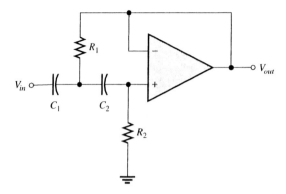

resistors and capacitors. This filter has a roll-off rate of 40 dB/decade below f_c, and the critical frequency is the same as for the low-pass filter.

Figure E.18-8 shows a two-pole high-pass filter with values chosen to produce a response with a critical frequency of 1 kHz. Note that $R_2 = 2R_1$ and $C_1 = C_2$ because these relationships result in a gain of 0.707 (-3 dB) at f_c. For frequencies other than 1 kHz, the resistance values can be scaled inversely, as was done with the capacitors in the low-pass case.

FIGURE E.18-8
Two-pole, high-pass filter ($f_c = 1$ kHz).

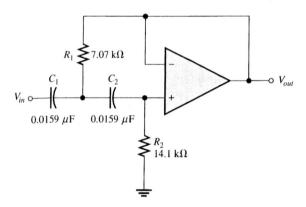

EXAMPLE E.18-2

For the filter of Figure E.18-9, calculate the resistance values required to produce a critical frequency of 5.5 kHz.

FIGURE E.18-9

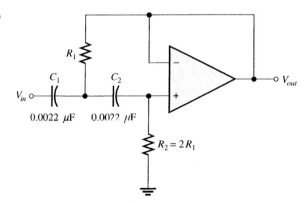

Solution The capacitor values have been preselected to be 0.0022 μF each. Since these differ from the 1 kHz reference filter, you cannot use the scaling method to get the resistor values. Start with:

$$f_c = \frac{1}{2\pi\sqrt{R_1 R_2 C_1 C_2}}$$

$$f_c^2 = \frac{1}{4\pi^2 R_1 R_2 C_1 C_2}$$

Since $R_2 = 2R_1$ and $C_1 = C_2 = C$,

$$f_c^2 = \frac{1}{4\pi^2 (2R_1^2) C^2}$$

Solve for R_1, and then determine R_2.

$$R_1^2 = \frac{1}{8\pi^2 C^2 f_c^2}$$

$$R_1 = \frac{1}{2\sqrt{2}\pi C f_c} = \frac{0.707}{2\pi C f_c} = \frac{0.707}{2\pi (0.0022\ \mu F)(5.5\ kHz)} = 9.3\ k\Omega$$

$$R_2 = 2R_1 = 2(9.3\ k\Omega) = 18.6\ k\Omega$$

Band-Pass Filter Using a High-Pass/Low-Pass Combination

One way to implement a band-pass filter is to use a cascaded arrangement of a high-pass filter followed by a low-pass filter, as shown in Figure E.18-10(a). Each of the filters shown is a two-pole configuration so that the roll-off rates of the response curve are ± 40 dB/decade, as indicated in the composite response curve of part (b). The critical frequency of each filter is chosen so that the response curves overlap, as indicated. The critical frequency of the high-pass filter is lower than that of the low-pass filter.

The lower frequency, f_{c1}, of the passband is set by the critical frequency of the high-pass filter. The upper frequency, f_{c2}, of the passband is the critical frequency of the low-pass filter. Ideally, the center frequency, f_r, of the passband is the geometric average of f_{c1} and f_{c2}. The following formulas express the three frequencies of the band-pass filter in Figure E.18-10:

$$f_{c1} = \frac{1}{2\pi\sqrt{R_1 R_2 C_1 C_2}}$$

$$f_{c2} = \frac{1}{2\pi\sqrt{R_3 R_4 C_3 C_4}}$$

$$f_r = \sqrt{f_{c1} f_{c2}}$$

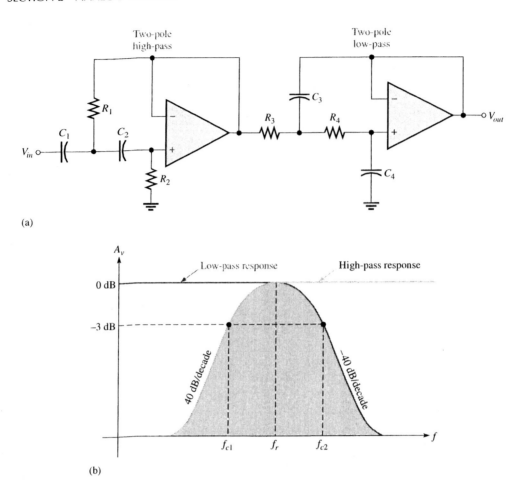

(a)

(b)

FIGURE E.18-10

Band-pass filter formed by combining two-pole, high-pass filter with two-pole, low-pass filter.
(The order in which the filters are cascaded does not matter.)

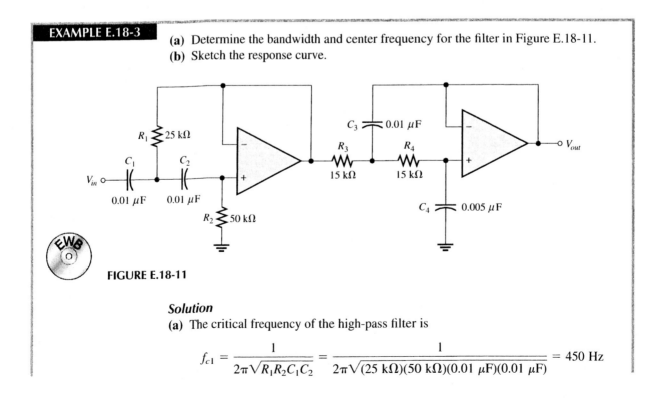

EXAMPLE E.18-3

(a) Determine the bandwidth and center frequency for the filter in Figure E.18-11.
(b) Sketch the response curve.

FIGURE E.18-11

Solution
(a) The critical frequency of the high-pass filter is

$$f_{c1} = \frac{1}{2\pi\sqrt{R_1 R_2 C_1 C_2}} = \frac{1}{2\pi\sqrt{(25 \text{ k}\Omega)(50 \text{ k}\Omega)(0.01 \ \mu\text{F})(0.01 \ \mu\text{F})}} = 450 \text{ Hz}$$

The critical frequency of the low-pass filter is

$$f_{c2} = \frac{1}{2\pi\sqrt{R_3R_4C_3C_4}} = \frac{1}{2\pi\sqrt{(15\ k\Omega)(15\ k\Omega)(0.01\ \mu F)(0.005\ \mu F)}} = 1.5\ kHz$$

$$BW = f_{c2} - f_{c1} = 1.5\ kHz - 450\ Hz = 1.05\ kHz$$

$$f_r = \sqrt{f_{c1}f_{c2}} = \sqrt{(1.5\ kHz)(450\ Hz)} = 822\ Hz$$

(b) The response curve is shown in Figure E.18-12.

FIGURE E.18-12

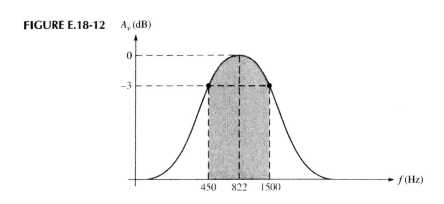

ACTIVE BAND-STOP FILTERS

Band-stop filters reject a specified band of frequencies and pass all others. The response is opposite to that of a band-pass filter.

Multiple-Feedback Band-Stop Filter

Figure E.18-13 shows a multiple-feedback band-stop filter. Notice that this configuration is similar to the band-pass version except that R_3 has been moved and R_4 has been added.

State-Variable Band-Stop Filter

Summing the low-pass and the high-pass responses creates a band-stop response as shown in Figure E.18-14 on the next page. One important application of this filter is minimizing the 60 Hz "hum" in audio systems by setting the center frequency to 60 Hz.

FIGURE E.18-13
Multiple-feedback band-stop filter.

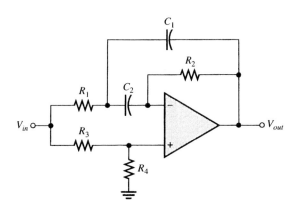

FIGURE E.18-14
State-variable band-stop filter.

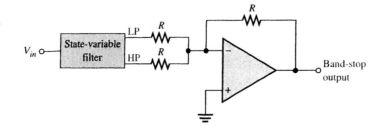

EXAMPLE E.18-4 Verify that the band-stop filter in Figure E.18-15 has a center frequency of 60 Hz, and optimize it for a Q of 30.

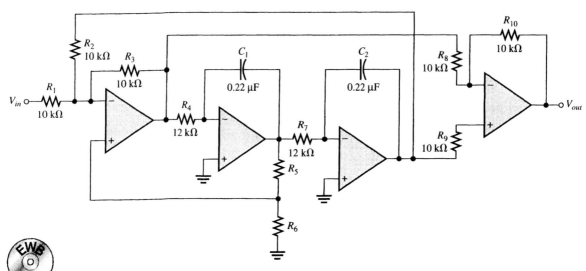

FIGURE E.18-15

Solution f_0 equals the f_c of the integrator stages.

$$f_0 = \frac{1}{2\pi R_4 C_1} = \frac{1}{2\pi R_7 C_2} = \frac{1}{2\pi(12 \text{ k}\Omega)(0.22 \text{ }\mu\text{F})} = 60 \text{ Hz}$$

You can obtain a $Q = 30$ by choosing R_6 and then calculating R_5.

$$Q = \frac{1}{3}\left(\frac{R_5}{R_6} + 1\right)$$

$$R_5 = (3Q - 1)R_6$$

Choose $R_6 = 1 \text{ k}\Omega$. Then

$$R_5 = [3(30) - 1]1 \text{ k}\Omega = 89 \text{ k}\Omega$$

REVIEW QUESTIONS

True/False

1. An active filter is a filter that has gain.

2. A two-pole filter has a rolloff of -40 dB/decade.

Multiple Choice

3. What type of filter has a higher gain at 10 kHz than at 1 kHz?

   a. Low pass.

   b. High pass.

   c. Band-pass.

4. Band-pass filters can be made using

   a. Two low-pass filters.

   b. Two high-pass filters.

   c. A low-pass and a high-pass filter.

5. A band-stop filter

   a. Accepts frequencies within its bandwidth.

   b. Rejects frequencies within its bandwidth.

   c. Is an ideal low-pass or high-pass filter.

E.19 Troubleshoot and repair active filter circuits

INTRODUCTION

In this section, we discuss two methods of determining a filter's response by measurement—discrete point measurement and swept frequency measurement. Being able to make correct measurements is the first requirement of a proper troubleshooting experience.

Discrete Point Measurement

Figure E.19-1 shows an arrangement for taking filter output voltage measurements at discrete values of input frequency using common laboratory instruments. The general procedure is as follows:

1. Set the amplitude of the sine wave generator to a desired voltage level.
2. Set the frequency of the sine wave generator to a value well below the expected critical frequency of the filter under test. For a low-pass filter, set the frequency as near as possible to 0 Hz. For a band-pass filter, set the frequency well below the expected lower critical frequency.
3. Increase the frequency in predetermined steps sufficient to allow enough data points for an accurate response curve.
4. Maintain a constant input voltage amplitude while varying the frequency.
5. Record the output voltage at each value of frequency.
6. After recording a sufficient number of points, plot a graph of output voltage versus frequency.

If the frequencies to be measured exceed the response of the DMM, an oscilloscope may have to be used instead.

FIGURE E.19-1

Test setup for discrete point measurement of the filter response. (Readings are arbitrary and for display only.)

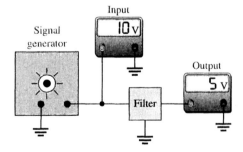

Swept Frequency Measurement

The swept frequency method requires more elaborate test equipment than does the discrete point method, but it is much more efficient and can result in a more accurate response curve. A general test setup is shown in Figure E.19-2 using a swept frequency generator and a spectrum analyzer.

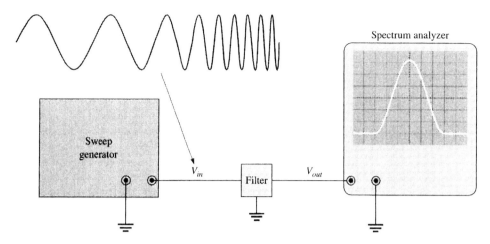

FIGURE E.19-2
Test setup for swept frequency measurement of the filter response.

The swept frequency generator produces a constant amplitude output signal whose frequency increases linearly between two preset limits, as indicated in Figure E.19-2. The spectrum analyzer is essentially an elaborate oscilloscope that can be calibrated for a desired *frequency span/division* rather than for the usual *time/division* setting. Therefore, as the input frequency to the filter sweeps through a preselected range, the response curve is traced out on the screen of the spectrum analyzer. Test equipment called network analyzers perform these types of tests and many others. A sample analyzer is shown in Figure E.19-3.

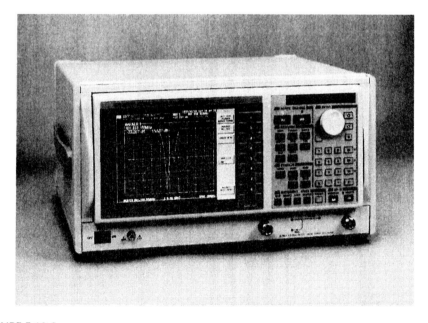

FIGURE E.19-3
High-frequency network analyzer (copyright 1999 Tektronix, Inc. All rights reserved— reproduced by permission.).

E.20 Understand principles and operations of sinusoidal and nonsinusoidal oscillator circuits

INTRODUCTION

An oscillator is a circuit that produces a repetitive waveform on its output with only the DC supply voltage as an input. A repetitive input signal is not required. The output voltage can be either sinusoidal or nonsinusoidal, depending on the type of oscillator.

The basic oscillator concept is illustrated in Figure E.20-1. Essentially, an **oscillator** converts electrical energy in the form of DC to electrical energy in the form of AC. A basic oscillator consists of an amplifier for gain (either discrete transistor or op-amp) and a positive feedback circuit that produces phase shift and provides attenuation, as shown in Figure E.20-2.

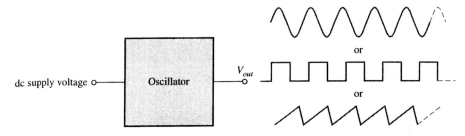

FIGURE E.20-1
The basic oscillator concept showing three possible types of output waveforms: sine wave, square wave, and sawtooth.

FIGURE E.20-2
Basic elements of an oscillator.

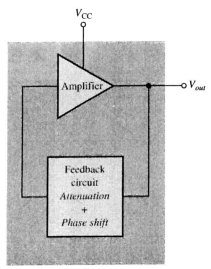

Oscillator

436

OSCILLATOR PRINCIPLES

With the exception of the relaxation oscillator, oscillator operation is based on the principle of positive feedback. In this section, we will examine this concept and look at the general conditions required for oscillation to occur.

Positive Feedback

Positive feedback is characterized by the condition wherein a portion of the output voltage of an amplifier is fed back to the input with no net phase shift, resulting in a reinforcement of the output signal. This basic idea is illustrated in Figure E.20-3. As you can see, the in-phase feedback voltage is amplified to produce the output voltage, which in turn produces the feedback voltage. That is, a loop is created in which the signal sustains itself and a continuous sine wave output is produced. This phenomenon is called *oscillation*.

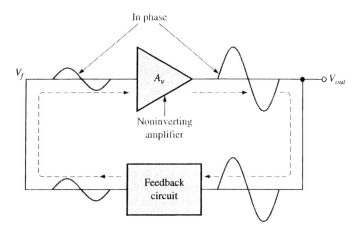

FIGURE E.20-3
Positive feedback produces oscillation.

Conditions for Oscillation

Two conditions are required for a sustained state of oscillation:

1. The phase shift around the feedback loop must be $0°$.

2. The voltage gain around the closed feedback loop must equal 1 (unity).

The voltage gain around the closed feedback loop (A_{cl}) is the product of the amplifier gain (A_v) and the attenuation (B) of the feedback circuit.

$$A_{cl} = A_v B$$

For example, if the amplifier has a gain of 100, the feedback circuit must have an attenuation of 0.01 to make the loop gain equal to 1 ($A_v B = 100 \times 0.01 = 1$). These conditions for oscillation are illustrated in Figure E.20-4.

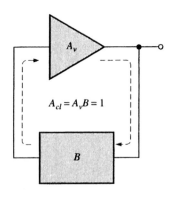

(a) The phase shift around the loop is 0°.   (b) The closed loop gain is 1.

FIGURE E.20-4
Conditions for oscillation.

Start-Up Conditions

So far, you have seen what it takes for an oscillator to produce a continuous sine wave output. Now let's examine the requirements for the oscillation to start when the DC supply voltage is turned on. As you know, the unity-gain condition must be met for oscillation to be sustained. For oscillation to begin, the voltage gain around the positive feedback loop must be greater than 1 so that the amplitude of the output can build up to a desired level. The gain must then decrease to 1 so that the output stays at the desired level.

A question that normally arises is this: If the oscillator is off (no DC voltage) and there is no output voltage, how does a feedback signal originate to start the positive feedback build-up process? Initially, a small positive feedback voltage develops from thermally produced broad-band noise in the resistors or other components or from turn-on transients. The feedback circuit permits only a voltage with a frequency equal to the selected oscillation frequency to appear in-phase on the amplifier's input. This initial feedback voltage is amplified and continually reinforced, resulting in a buildup of the output voltage as previously discussed.

EXAMPLE E.20-1

Determine the frequency of oscillation for the Wien-bridge oscillator in Figure E.20-5. Also, verify that oscillations will start and then continue when the output signal reaches 5.4 V.

FIGURE E.20-5

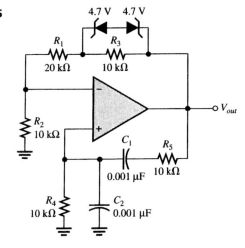

Solution For the lead-lag circuit, $R_4 = R_5 = R = 10 \text{ k}\Omega$ and $C_1 = C_2 = C = 0.001 \text{ }\mu\text{F}$. The resonant frequency is

$$f_r = \frac{1}{2\pi RC} = \frac{1}{2\pi(10 \text{ k}\Omega)(0.001 \text{ }\mu\text{F})} = 15.9 \text{ kHz}$$

Initially, the closed-loop gain is

$$A_{cl} = \frac{R_1 + R_2 + R_3}{R_2} = \frac{40 \text{ k}\Omega}{10 \text{ k}\Omega} = 4$$

Since $A_{cl} > 3$, the start-up condition is met.

When the output reaches 5.4 V (4.7 V + 0.7 V), the zeners conduct (their forward resistance is assumed small, compared to 10 kΩ), and the closed-loop gain is reached. Thus, oscillation is sustained.

$$A_{cl} = \frac{R_1 + R_2}{R_2} = \frac{30 \text{ k}\Omega}{10 \text{ k}\Omega} = 3$$

EXAMPLE E.20-2

(a) Determine the value of R_f necessary for the circuit in Figure E.20-6 to operate as an oscillator.

(b) Determine the frequency of oscillation.

FIGURE E.20-6

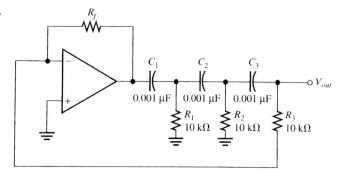

Solution

(a) $A_{cl} = 29$, and $A_{cl} = \dfrac{R_f}{R_3}$. Therefore, $\dfrac{R_f}{R_3} = 29$.

$$R_f = 29R_3 = 29(10 \text{ k}\Omega) = 290 \text{ k}\Omega$$

(b) $R_1 = R_2 = R_3 = R$ and $C_1 = C_2 = C_3 = C$. Therefore,

$$f_r = \frac{1}{2\pi\sqrt{6}RC} = \frac{1}{2\pi\sqrt{6}(10 \text{ k}\Omega)(0.001 \text{ }\mu\text{F})} \cong 6.5 \text{ kHz}$$

EXAMPLE E.20-3

(a) Determine the frequency of oscillation for the Colpitts oscillator in Figure E.20-7. Assume there is negligible loading on the feedback circuit and that its Q is greater than 10.

(b) Find the frequency of oscillation if the oscillator is loaded to a point where the Q drops to 8.

FIGURE E.20-7

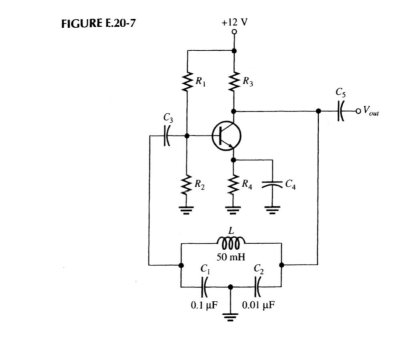

Solution

(a) $C_T = \dfrac{C_1 C_2}{C_1 + C_2} = \dfrac{(0.1\ \mu F)(0.01\ \mu F)}{0.11\ \mu F} = 0.0091\ \mu F$

$f_r \cong \dfrac{1}{2\pi\sqrt{LC_T}} = \dfrac{1}{2\pi\sqrt{(50\ \text{mH})(0.0091\ \mu F)}} = 7.46\ \text{kHz}$

(b) $f_r = \dfrac{1}{2\pi\sqrt{LC_T}}\sqrt{\dfrac{Q^2}{Q^2 + 1}} = (7.46\ \text{kHz})(0.9923) = 7.40\ \text{kHz}$

The Armstrong Oscillator

This type of *LC* oscillator uses transformer coupling to feed back a portion of the signal voltage, as shown in Figure E.20-8. It is sometimes called a "tickler" oscillator in reference to the transformer secondary or "tickler coil" that provides the feedback to keep the oscillation going. The Armstrong is less common than the Colpitts, Clapp, and Hartley, mainly

FIGURE E.20-8
A basic Armstrong oscillator.

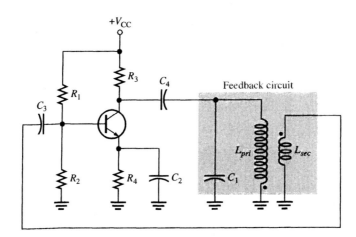

because of the disadvantage of transformer size and cost. The frequency of oscillation is set by the inductance of the primary winding (L_{pri}) in parallel with C_1.

$$f_r = \frac{1}{2\pi\sqrt{L_{pri}C_1}}$$

Crystal-Controlled Oscillators

The most stable and accurate type of oscillator uses a piezoelectric **crystal** in the feedback loop to control the frequency.

The Piezoelectric Effect   Quartz is one type of crystalline substance found in nature that exhibits a property called the **piezoelectric effect.** When a changing mechanical stress is applied across the crystal to cause it to vibrate, a voltage develops at the frequency of mechanical vibration. Conversely, when an AC voltage is applied across the crystal, it vibrates at the frequency of the applied voltage. The greatest vibration occurs at the crystal's natural resonant frequency, which is determined by the physical dimensions and by the way the crystal is cut.

Crystals used in electronic applications typically consist of a quartz wafer mounted between two electrodes and enclosed in a protective "can" as shown in Figure E.20-9(a) and (b). A schematic symbol for a crystal is shown in Figure E.20-9(c) and an equivalent *RLC* circuit for the crystal appears in Figure E.20-9(d). As you can see, the crystal's equivalent circuit is a series-parallel *RLC* circuit and can operate in either series resonance or parallel resonance. At the series resonant frequency, the inductive reactance is cancelled by the reactance of C_s. The remaining series resistor, R_s, determines the impedance of the crystal. Parallel resonance occurs when the inductive reactance and the reactance of the parallel capacitance, C_m, are equal. The parallel resonant frequency is usually at least 1 kHz higher than the series resonant frequency. A great advantage of the crystal is that it exhibits a very high Q (Qs of several thousand are typical).

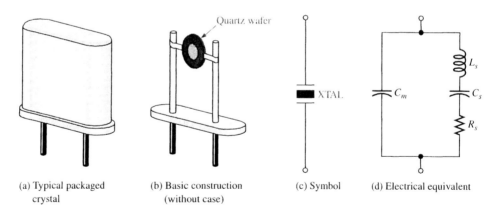

(a) Typical packaged   (b) Basic construction   (c) Symbol   (d) Electrical equivalent
    crystal   (without case)

FIGURE E.20-9
A quartz crystal.

An oscillator that uses a crystal as a series resonant tank circuit is shown in Figure E.20-10(a). The impedance of the crystal is minimum at the series resonant frequency, thus providing maximum feedback. The crystal tuning capacitor, C_C, is used to "fine tune" the oscillator frequency by "pulling" the resonant frequency of the crystal slightly up or down.

A modified Colpitts configuration is shown in Figure E.20-10(b) with a crystal acting as a parallel resonant tank circuit. The impedance of the crystal is maximum at parallel resonance, thus developing the maximum voltage across the capacitors. The voltage across C_1 is fed back to the input.

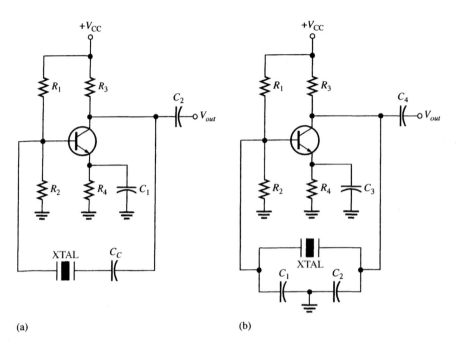

FIGURE E.20-10
Basic crystal oscillators.

Modes of Oscillation in the Crystal Piezoelectric crystals can oscillate in either of two modes—fundamental or overtone. The fundamental frequency of a crystal is the lowest frequency at which it is naturally resonant. The fundamental frequency depends on the crystal's mechanical dimensions, type of cut, and other factors, and is inversely proportional to the thickness of the crystal slab. Because a slab of crystal cannot be cut too thin without fracturing, there is an upper limit on the fundamental frequency. For most crystals, this upper limit is less than 20 MHz. For higher frequencies, the crystal must be operated in the overtone mode. Overtones are approximate integer multiples of the fundamental frequency. The overtone frequencies are usually, but not always, odd multiples (3, 5, 7, . . .) of the fundamental.

A Triangular-Wave Oscillator

The op-amp integrator can be used as the basis for a triangular wave oscillator. The basic idea is illustrated in Figure E.20-11(a) where a dual-polarity, switched input is used. We use the switch only to introduce the concept; it is not a practical way to implement this circuit. When the switch is in position 1, the negative voltage is applied, and the output is a positive-going ramp. When the switch is thrown into position 2, a negative-going ramp is produced. If the switch is thrown back and forth at fixed intervals, the output is a triangular wave consisting of alternating positive-going and negative-going ramps, as shown in Figure E.20-11(b).

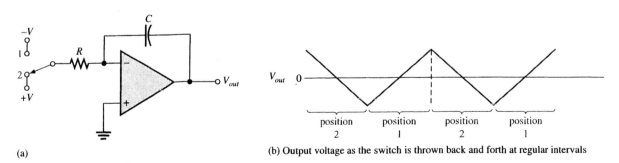

(b) Output voltage as the switch is thrown back and forth at regular intervals

FIGURE E.20-11
Basic triangular-wave oscillator.

A Practical Triangular-Wave Circuit One practical implementation of a triangular-wave oscillator utilizes an op-amp comparator to perform the switching function, as shown in Figure E.20-12. The operation is as follows. To begin, assume that the output voltage of the comparator is at its maximum negative level. This output is connected to the inverting input of the integrator through R_1, producing a positive-going ramp on the output of the integrator. When the ramp voltage reaches the upper trigger point (UTP), the comparator switches to its maximum positive level. This positive level causes the integrator ramp to change to a negative-going direction. The ramp continues in this direction until the lower trigger point (LTP) of the comparator is reached. At this point, the comparator output switches back to the maximum negative level and the cycle repeats. This action is illustrated in Figure E.20-13.

FIGURE E.20-12
A triangular-wave oscillator using two op-amps.

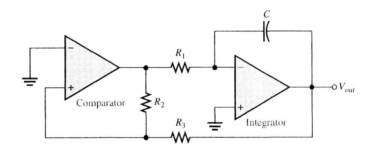

FIGURE E.20-13
Waveforms for the circuit in Figure E.20-12.

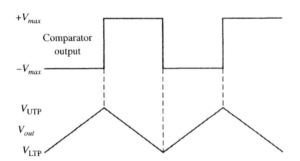

Since the comparator produces a square-wave output, the circuit in Figure E.20-12 can be used as both a triangular-wave oscillator and a square-wave oscillator. Devices of this type are commonly known as *function generators* because they produce more than one output function. The output amplitude of the square wave is set by the output swing of the comparator, and the resistors R_2 and R_3 set the amplitude of the triangular output by establishing the UTP and LTP voltages according to the following formulas:

$$V_{\text{UTP}} = +V_{max}\left(\frac{R_3}{R_2}\right)$$

$$V_{\text{LTP}} = -V_{max}\left(\frac{R_3}{R_2}\right)$$

where the comparator output levels, $+V_{max}$ and $-V_{max}$, are equal. The frequency of both waveforms depends on the R_1C time constant as well as the amplitude-setting resistors, R_2 and R_3. By varying R_1, the frequency of oscillation can be adjusted without changing the output amplitude.

$$f_r = \frac{1}{4R_1C}\left(\frac{R_2}{R_3}\right)$$

Determine the frequency of oscillation of the circuit in Figure E.20-14. To what value must R_1 be changed to make the frequency 20 kHz?

FIGURE E.20-14

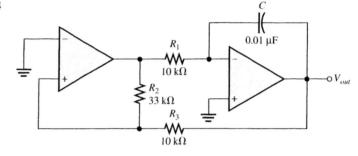

Solution $\quad f_r = \dfrac{1}{4R_1 C}\left(\dfrac{R_2}{R_3}\right) = \left(\dfrac{1}{4(10\ \text{k}\Omega)(0.01\ \mu\text{F})}\right)\left(\dfrac{33\ \text{k}\Omega}{10\ \text{k}\Omega}\right) = 8.25\ \text{kHz}$

To make $f = 20$ kHz,

$$R_1 = \dfrac{1}{4fC}\left(\dfrac{R_2}{R_3}\right) = \left(\dfrac{1}{4(20\ \text{kHz})(0.01\ \mu\text{F})}\right)\left(\dfrac{33\ \text{k}\Omega}{10\ \text{k}\Omega}\right) = 4.13\ \text{k}\Omega$$

A 555 timer configured to run in the astable mode (oscillator) is shown in Figure E.20-15. Determine the frequency of the output and the duty cycle.

FIGURE E.20-15

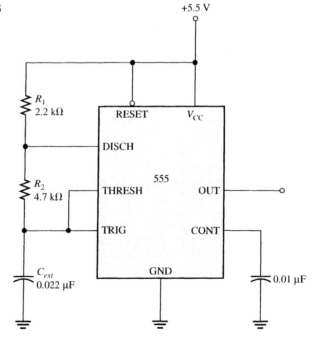

Solution $\quad f_r = \dfrac{1.44}{(R_1 + 2R_2)C_{ext}} = \dfrac{1.44}{(2.2\ \text{k}\Omega + 9.4\ \text{k}\Omega)0.022\ \mu\text{F}} = 5.64\ \text{kHz}$

$$\text{Duty cycle} = \dfrac{R_1 + R_2}{R_1 + 2R_2} \times 100\% = \dfrac{2.2\ \text{k}\Omega + 4.7\ \text{k}\Omega}{2.2\ \text{k}\Omega + 9.4\ \text{k}\Omega} \times 100\% = 59.5\%$$

REVIEW QUESTIONS

True/False

1. An oscillator relies on positive feedback.
2. The closed loop gain of an oscillator is 1.0.

Multiple Choice

3. Which oscillator uses a transformer in the feedback path?

    a. Armstrong.

    b. Colpitts.

    c. Hartley.

4. The op-amp triangle-wave oscillator uses a(n)

    a. Differentiator.

    b. Integrator.

    c. Both a and b.

5. What is the frequency of oscillation in a 555 timer circuit with R_1 and R_2 both equal to 1 k ohms and C equal to 1 μF?

    a. 48 Hz.

    b. 480 Hz.

    c. 4.8 kHz.

E.21 Troubleshoot and repair sinusoidal and nonsinusoidal oscillator circuits

INTRODUCTION

In this exercise we look at several test procedures for common oscillator failures.

Testing a Crystal

An oscillator with a bad crystal may not oscillate at all, may be erratic, or may not oscillate at the correct frequency. One common crystal failure mode is a broken or corroded internal connection. Or if the crystal has been dropped, it may be cracked.

Figure E.21-1 shows how to make a simple test to quickly determine the condition of the crystal. Normally, a crystal oscillator will oscillate at a slightly higher frequency than the crystal's series resonant point. If you can find the series resonant point of the crystal, you know the crystal is good.

FIGURE E.21-1
Crystal test.

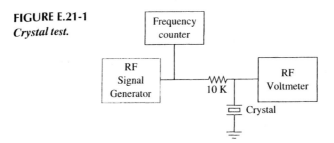

Recall that at the series resonant point, the crystal should have a very low resistance, in the order of 100 ohms. At other frequencies, the crystal impedance should be quite high.

The generator should be very carefully turned across the specified frequency of the crystal. If the crystal is operating properly, the voltmeter will show a dramatic dip at the series resonant point. Remember that the crystal is a very high-Q device, and tuning the signal generator will have to be done very carefully.

Because the impedance of the crystal is very high at the parallel or antiresonant point, perhaps 50,000 ohms, there should be a peak on the voltmeter at a frequency just slightly above the series resonant point. You should look for the series resonant point first because it is easier to find.

The voltage across a broken crystal will not change much as the generator frequency is varied. Internal connection problems could cause erratic operation. Corrosion problems will cause the resonant frequency to shift from the specified value.

Testing Oscillator Capacitors

The capacitors associated with the crystal or inductor, together with the inductor, determine the exact frequency of oscillation. This type of capacitor will seldom show a short, but it can become sensitive to temperature and shock or change value with age.

In the Clapp circuit shown in Figure E.21-2, C3 is primarily responsible for setting the frequency. While observing the frequency with a counter, cool the capacitor with an

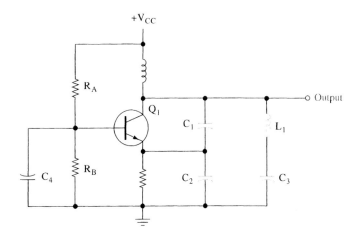

FIGURE E.21-2
Clapp oscillator.

aerosol spray sold for cooling electronic equipment. Defective capacitors will generally change value suddenly and shift the frequency a good bit when cooled. If C3 is open the circuit probably will not oscillate at all.

In the Clapp circuit, C1 and C2 are primarily responsible for providing the proper amount of feedback to allow oscillation. If either of these capacitors fails the oscillator will not work. An oscilloscope connected to the collector of Q1 should show a high-quality sine wave. C1 and C2 do have some effect on the frequency and should not be excluded from suspicion if the frequency is not correct.

Testing Oscillator Inductors

A shorted or open inductor will completely kill an oscillator. Inductors can be easily checked for an open circuit with an ohmmeter, though the ohmmeter will not detect a shorted turn. A short in the inductor is best detected with a Q-meter or impedance bridge.

E.22 Understand principles and operations of fiber-optic circuits using photodiodes or lasers

INTRODUCTION

Recent advances in the development and manufacture of fiber-optic systems have made them the latest frontier in the field of communications. They are being used for both military and commercial data links and have replaced a lot of copper wire. Their use in telecommunications is extensive. They are also expected to take over much of the long-distance communication traffic now handled by satellite links.

A fiber-optic communications system is surprisingly simple, as shown in Figure E.22-1. It is comprised of the following elements:

1. A fiber-optic transmission cable to carry the signal (in the form of a modulated light beam) a few feet or several miles. The cable may be a single, hairlike fiber or a small bundle of hundreds of such fibers.

2. A source of visible or invisible infrared radiation—usually a light-emitting diode (LED) or a solid-state laser—that can be modulated to impress data or an analog signal on the light beam.

3. A photosensitive detector to convert the optical signal back into an electrical signal at the receiver. The most often used detectors are *p-i-n* or avalanche photodiodes.

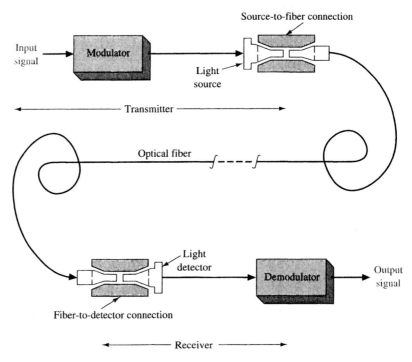

FIGURE E.22-1
Fiber-optic communication system.

4. Efficient optical connectors at the light source-to-cable interface and at the cable-to-photodetector interface. These connectors are also critical when splicing of the optical cable is required due to excessive loss that can occur at connections.

5. Standard communications electronics prior to the light source and following the photodetector.

The optical waveguide propagates the light signal in a fashion similar to the standard metallic waveguide. The light wave travels down the glass fiber by constant reflection off its side walls. Thus, the trapping of light in a fiber results from the phenomenon of **total internal reflection** (TIR).

The advantages of optical communications links compared to waveguides or copper conductors are enormous. Included are the following:

1. *Extremely wide system bandwidths:* The intelligence is impressed by varying the light's amplitude. Since the best LEDs have a 5-ns response time, they provide a maximum bandwidth signal of about 100 MHz. Using laser light sources, however, bandwidths of up to 10 GHz are possible with a single glass fiber. The amount of information multiplexed on such a system is indeed staggering. The higher the carrier frequency in a communication system, the greater its potential signal bandwidth. Since fiber-optics systems have carriers at 10^{13} to 10^{14} Hz compared to radio frequencies of 10^6 to 10^9 Hz, signal bandwidths are potentially many times greater.

2. *Immunity to electromagnetic interference (EMI):* External electrical noise does not affect energy at the frequency of light.

3. *Virtual elimination of crosstalk:* The light on one glass fiber does not interfere with light on an adjacent fiber. This is analogous to one standard waveguide in close proximity to another. Crosstalk can result from two adjacent copper wires, however.

4. *Lower signal attenuation than other propagation systems:* Typical attenuation figures of a 1-GHz bandwidth signal for optical fibers are 0.03 dB per 100 ft compared to 4.0 dB for both RG-58/U coaxial cable and an X-band waveguide. Fewer repeater stations are needed as a result with glass fiber.

5. *Substantially lighter weight and smaller size:* The U.S. Navy replaced conventional wiring on the A-7 airplane that transmitted data between a central computer and all its remote sensors and peripheral avionics with an optical system. In this case, 224 ft of fiber optics weighing 1.52 lb replaced 1900 ft of copper wire weighing 30 lb.

6. *Lower cost:* Optical-fiber costs are continuing to decline while the cost of copper is increasing. Many systems are now cheaper with fiber, and that trend is accelerating.

7. *Conservation of the earth's resources:* The world's supply of copper (and other good electrical conductors) is limited. The principal ingredient in glass is sand, and it is cheap and in virtually unlimited supply.

8. *Safety:* In many wired systems, the potential hazard of short circuits requires precautionary designs. Additionally, the dielectric nature of optic fibers eliminates the spark hazard.

9. *Corrosion:* Since glass is basically inert, the corrosive effects of certain environments are not a problem.

Typical construction of an optical fiber is shown in Figure E.22-2. The central core is the portion that carries the transmitted light. In most cases it is glass, but occasionally it is plastic. The cladding is usually glass, but plastic cladding of a glass fiber is not uncommon. In any event, the refraction index for the core and cladding are different. If they are both glass, their manufacturing processes are varied to provide the desired difference. To provide protection, some sort of rubber or plastic jacket may surround the cladding, as shown in Figure E.22-2. Kevlar is now being used as the protective jacket. A truck can be driven over these cables without causing damage. This type of fiber is termed the **step-index** variety. "Step index" refers to the abrupt change in refractive index from core to clad. Fibers that include a protective covering are called **cables.** Glass fibers are not rigid as you

FIGURE E.22-2
Single-fiber construction.

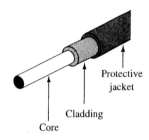

Protective
jacket

Cladding

Core

might expect. A single fiber and its cladding can typically be wound around a pencil without damage. They also have a large tensile strength, greater than equivalently sized steel wire.

Applying the concept of total reflection, propagation of light down the multimode fiber is shown in Figure E.22-3(a). Propagation results from the continuous reflection at the core/clad interface so that the ray "bounces" down the fiber length by the process of total internal reflection (TIR). If we consider point P in Figure E.22-3(a), the critical angle value for θ_3 is, from Snell's law,

$$\theta_c = \theta_3(\text{min}) = \sin^{-1}\frac{n_2}{n_1}$$

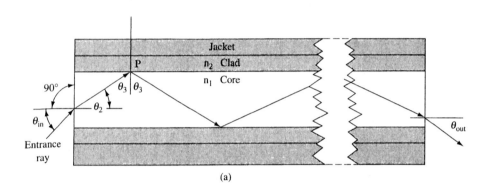

(a)

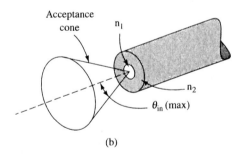

(b)

FIGURE E.22-3
(a) Development of numerical aperture; (b) acceptance cone.

Since θ_2 is the complement of θ_3,

$$\theta_2(\text{max}) = \sin^{-1}\frac{(n_1^2 - n_2^2)^{1/2}}{n_1}$$

Now applying Snell's law at the entrance surface and since $n_{\text{air}} \approx 1$, we obtain

$$\sin \theta_{\text{in}}(\text{max}) = n_1 \sin \theta_2(\text{max})$$

Combining the two preceding equations yields

$$\sin \theta_{\text{in}}(\text{max}) = \sqrt{n_1^2 - n_2^2}$$

Therefore, θ_{in}(max) is the largest angle with the core axis that allows propagation via total internal reflection. Light entering the cable at larger angles will be refracted through the core/clad interface and lost. The value $\sin \theta_{in}$(max) is called the **numerical aperture** (NA) and defines the half-angle of the cone of acceptance for propagated light in the fiber. This is shown in Figure E.22-3(b). The preceding analysis might lead you to think that crossing over θ_{in}(max) causes an abrupt end of light propagation. In practice, however, this is not true; thus, fiber manufacturers usually specify NA as the acceptance angle where the output light is no greater than 10 dB down from the peak value. The NA is a basic specification of a fiber provided by the manufacturer that indicates its ability to accept light and shows how much light can be off-axis and still be propagated.

EXAMPLE E.22-1

An optical fiber and its cladding have refractive indexes of 1.535 and 1.490, respectively. Calculate NA and θ_{in}(max).

Solution
$$\text{NA} = \sin \theta_{in}(\text{max}) = \sqrt{n_1^2 - n_2^2}$$
$$= \sqrt{(1.535)^2 - (1.49)^2} = 0.369$$
$$\theta_{in}(\text{max}) = \sin^{-1} 0.369$$
$$= 21.7°$$

OPTICAL FIBERS

Several types of optical fibers are available, with significant differences in their characteristics. The first communication-grade fibers (early 1970s) had light-carrying core diameters about equal to the wavelength of light. They could carry light in just a single waveguide mode. The difficulty of coupling significant light into such a small fiber led to development of fibers with cores of about 20 to 100 μm. These fibers support many waveguide modes and are called **multimode fibers.** The first commercial fiber-optic systems used multimode fibers with light at 0.8- to 0.9-μm wavelengths. A variation of the multimode fiber was subsequently developed, termed graded-index fiber. This afforded greater bandwidth capability.

As the technology became more mature, the single-mode fibers were found to provide lower losses and even higher bandwidth. This has led to their use at 1.3 and 1.55 μm in many telecommunication applications. The new developments have not made old types of fiber obsolete. The application now determines the type used. The following major criteria affect the choice of fiber type:

1. Signal losses
2. Ease of light coupling and interconnection
3. Bandwidth

A fiber showing three different modes (i.e., multimode) of propagation is presented in Figure E.22-4. The lowest-order mode is seen traveling along the axis of the fiber, and the middle-order mode is reflected twice at the interface. The highest-order mode is reflected many times and makes many trips across the fiber. As a result of these variable path lengths, the light entering the fiber takes a variable length of time to reach the detector. This results in a pulse-stretching characteristic, as shown in Figure E.22-4. The effect is termed **pulse dispersion** and limits the maximum rate at which data (pulses of light) can be practically transmitted. You will also note that the output pulse has reduced amplitude as well as increased width. The greater the fiber length, the worse this effect will be. As a result, manufacturers rate their fiber in bandwidth per length, such as 400 MHz/km. That cable can successfully transmit pulses at the rate of 400 MHz for 1 km, 200 MHz for 2 km, and so on. Of course, longer transmission paths are attained by locating repeaters at appropriate locations.

FIGURE E.22-4
Modes of propagation.

Step-index multimode fibers in common use have core diameters from about 50 to 100 μm. An often used configuration has a 50-μm core and 125-μm cladding. The large core diameter and high NA of these fibers simplifies input coupling and allows the use of relatively inexpensive connectors. Fibers are often specified by the diameters of their core and cladding. For example, the fiber just described would be called 50/125 fiber.

An alternative to glass multimode fibers are all-plastic designs. It should be noted that some glass fibers do use plastic cladding. The all-plastic fibers are inexpensive and easy to handle but have very high losses—about 100 dB/km. This limits their use to very short distance systems, such as transmitting signals throughout the interior of an automobile.

A technique used to minimize pulse dispersion effects is to make the core extremely small—on the order of a few micrometers. This type accepts only the lowest-order modes, thereby allowing operation in high-data-rate, long-distance systems. This fiber is quite expensive and requires high-power, highly directional modulated light sources such as a laser. Fiber cables of this variety are called **single-mode** or monomode fibers. Core diameters of only 5 μm are typical.

Graded-Index Fiber

In an effort to overcome the pulse dispersion problem, the **graded-index fiber** was developed. In the manufacturing process for this fiber, the index of refraction is tailored to follow the parabolic profile shown in Figure E.22-5. This results in low-order modes traveling through the constant-density material in the center. High-order modes see lower index of refraction material farther from the core axis, and thus the velocity of propagation increases away from the center. Therefore, all modes, even though they take various paths and travel different distances, tend to traverse the fiber length in about the same amount of time. These cables can therefore handle higher bandwidths and/or provide longer lengths of transmission before pulse dispersion effects destroy intelligibility.

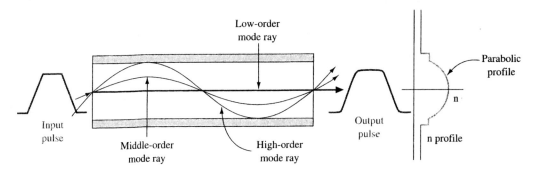

FIGURE E.22-5
Graded-index fiber.

Graded-index multimode fibers with 50-μm diameter cores and 125-μm cladding are used in many telecommunication systems at up to 300 megabits per second over 50-km ranges without repeaters. Graded-index fiber with up to 100-μm core is used in short-distance applications that require easy coupling from the source and high data rates, such as video and high-speed local area networks. The larger core affords better light coupling than the 50-μm core and does not significantly degrade the bandwidth capabilities.

Single-Mode Fibers

The single-mode fiber, by definition, carries light using a single waveguide mode. A single-mode fiber will transmit a single mode for all wavelengths longer than the cut-off wavelength λ_c.

$$\lambda_c = \frac{2\pi a n_1 \sqrt{2\Delta}}{2.405}$$

where $\Delta = \dfrac{n_1 - n_2}{n_1}$

a = core radius

At wavelengths shorter than predicted by the equation, the fiber supports two or more modes. The core of single-mode fiber must be small, about a few times the cut-off wavelength or several micrometers for operation at the standard 1.3 or 1.55 μm.

Single-mode fibers are widely used in long-haul telecommunications. They permit transmission of about 1 Gb/s and repeater spacing of up to 500 km. These bandwidth and repeater spacing capabilities are constantly being upgraded by new developments.

Figure E.22-6 provides a summary of the three types of fiber discussed, including typical dimensions, light paths, refractive index profiles, and pulse dispersion effects.

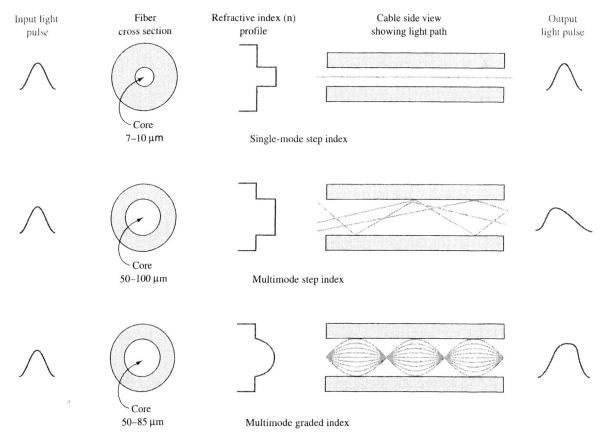

FIGURE E.22-6
Types of optical fiber.

FIBER ATTENUATION AND DISPERSION

Attenuation

The single most important consideration in fiber-optic systems is the loss introduced by the fiber. An optical pulse propagating along a fiber is attenuated exponentially. The optical power (P) at a distance (l) from the power transmitted (P_T) is

$$P = P_T \times 10^{-Al/10}$$

where A is the fiber attenuation in dB/km. The lowest attenuation currently available is about 0.1 dB/km for single-mode fiber at 1.55 μm. In general, multimode fibers tend to have higher loss than single mode because of increased scattering from dopants in the fiber core.

EXAMPLE E.22-2

Calculate the optical power at 50 km from a 0.1-mW source on a single-mode fiber that has 0.25-dB/km loss.

Solution

$$\begin{aligned}
P &= P_T \times 10^{-Al/10} \\
&= 0.1 \times 10^{-3} \times 10^{-(0.25 \times 50)/10} \\
&= 0.1 \times 10^{-3} \times 10^{-1.25} \\
&= 5.62 \ \mu W
\end{aligned}$$

All transparent materials scatter light because of microscopic density fluctuations (nonuniformities). The most familiar example of this effect is the scattering of sunlight by dust particles in the atmosphere, which yields our "blue" sky. The scattering in a fiber decreases rapidly with increasing wavelength, as shown in Figure E.22-7. It is termed **Rayleigh scattering** and is inversely proportional to the wavelength's fourth power. At longer wavelengths the attenuation increases rapidly due to absorption from the tails of infrared resonances in silica and other fiber constituents.

FIGURE E.22-7
Attenuation versus wavelength for typical high-quality fiber.

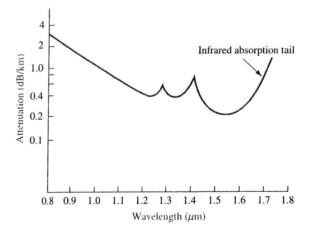

Notice the attenuation peaks at 1.25 and 1.39 μm in Figure E.22-7. They are caused by minute quantities of water trapped in the glass. These hydroxyl (OH^-) ions can be minimized in the manufacturing cycle, but it is simpler and less costly to avoid using the wavelengths most affected.

A comparison of the attenuation versus frequency characteristics for various "guiding" media is shown in Figure E.22-8. The very low loss capabilities of glass are ably shown, especially considering that both scales are logarithmic. Glass has the lowest loss and highest bandwidth capability—and also the smallest physical size!

FIGURE E.22-8
Attenuation comparison.
(From **IEEE**
Communications
Magazine, *May 1985,*
© 1985 IEEE,
"Introduction to
lightwave transmission"
by Paul S. Henry.)

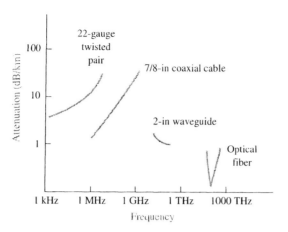

Dispersion

As mentioned previously, dispersion is a particular problem for multimode fiber. The different paths taken by the various propagation modes is the basis of this dispersion. This is called **modal dispersion.** Dispersion is also a factor to consider with single-mode fibers.

Modal dispersion in step-index multimode fibers limits the bandwidth (and bit rate) to about 20 MHz per kilometer of length. The use of graded-index fiber greatly increases this capability to about 1 GHz/km (see Figure E.22-5). Single-mode fiber does not exhibit modal dispersion since only a single mode is transmitted. Two other types of dispersion do limit single-mode fiber bit-rate-distance capability: material and waveguide dispersion. The resultant of these two provides the total dispersion of single-mode fiber.

Material dispersion (also called chromatic dispersion) is caused by the slight variation of refractive index with wavelength for glass. This is the same effect that causes a prism to create the color spectrum. This results in pulse spreading since the light source is not just a single frequency of light. The various Fourier components of a signal thus exhibit slightly different transit times through the fiber.

Waveguide dispersion is caused by a portion of the light energy traveling in the cladding. Typically, 20% of the energy is contained within the cladding. Since the cladding has a lower refractive index than the core, velocity variations result that cause pulse dispersion. Waveguide dispersion is wavelength dependent and can have different polarity than material dispersion. They can completely cancel each other out. This occurs near 1.3 μm and is called the **zero-dispersion wavelength.** The combined effects of these two dispersions become significant at the other commonly used wavelength for single-mode fiber, 1.55 μm. Recall, however, that attenuation is significantly lower at 1.55 μm (Figure E.22-7). Current fiber design is taking place to shift the zero-dispersion wavelength up to 1.55 μm. This appears possible by using a complex graded-index fiber.

LIGHT SOURCES

Two kinds of light sources are used in fiber-optic communication systems: the diode laser (DL) and the high-radiance light-emitting diode (LED). In designing the optimum system, the special qualities of each light source should be considered. Diode lasers and LEDs bring to systems different characteristics:

1. Power levels
2. Temperature sensitivities
3. Response times
4. Lifetimes
5. Characteristics of failure

The diode laser is a preferred source for moderate-band to wideband systems. It offers a fast response time (typically less than 1 ns) and can couple high levels of useful optical power (usually several mW) into an optical fiber with a small core and a small numerical aperture. Recent advances in DL fabrication have resulted in predicted lifetimes of 10^5 to 10^6 hours at room temperature. Earlier DLs were of such limited life as to minimize their use. The DL is usually used as the source for single-mode fiber since LEDs have a low input coupling efficiency.

Some systems operate at a slower bit rate and require more modest levels of fiber-coupled optical power (50 to 250 μW). These applications allow the use of high-radiance LEDs. The LED is cheaper, requires less complex driving circuitry than a DL, and needs no thermal or optical stabilizations. In addition, LEDs have longer operating lives (10^6 to 10^7 h) and fail in a more gradual and predictable fashion than DLs.

Both LEDs and DLs are multilayer devices most frequently fabricated of AlGaAs or GaAs. They both behave electrically as diodes, but their light-emission properties differ substantially. A DL is an optical oscillator; hence it has many typical oscillator characteristics: a threshold of oscillation, a narrow emission bandwidth, a temperature coefficient of threshold and frequency, modulation nonlinearities, and regions of instability.

The light output wavelength spread, or spectrum, of the DL is much narrower than that of LEDs: about 1 nm compared with about 40 nm for an LED. Narrow spectra are advantageous in systems with high bit rates since the dispersion effects of the fiber on pulse width are reduced, and thus pulse degradation over long distances is minimized.

Light is emitted from an LED as a result of the recombining of electrons and holes. Electrically, an LED is a *pn* junction. Under forward bias, minority carriers are injected across the junction. Once across they recombine with majority carriers and give up their energy. The energy given up is about equal to the material's energy gap. This process is radiative for some materials (such as GaAs) but not so for others, such as silicon. LEDs have a distribution of nonradiative sites—usually crystal lattice defects, impurities, and so on. These sites develop over time and explain the finite life/gradual deterioration of light output.

REVIEW QUESTIONS

True/False

1. Fiber optic cable is susceptible to electromagnetic interference.

2. Light is propagated through a fiber via total internal reflection.

Multiple Choice

3. Fiber is composed of a core and a
   a. Cladding.
   b. Core jacket.
   c. Prism.

4. Which is not a type of fiber?
   a. Single mode.
   b. Dual mode.
   c. Multi mode.

5. Two types of dispersion are modal and
   a. Chromatic.
   b. Transwave.
   c. Bi-modal.

E.23 Troubleshoot and repair fiber-optic circuits using photodiodes or lasers

INTRODUCTION

Today optical fiber is the infrastructure of many communications hubs. Fiber carries billions of telephone calls a day. Optical fiber makes up the backbone structure of many local area networks currently in use. In this section we will look at planning an optical-fiber installation and maintaining it once it is in place.

Remember, the diode lasers (DLs) can emit radiation with a far higher energy-density than sunlight, and even though you can't see the radiation, it can easily cause blindness by retinal heating. You should always wear eye protection when working on laser systems. A 1-W or more CW output laser, such as used in medical imaging products, can produce a stunningly high-power density when focused. Even when poorly focused, 1 W across 0.125-in. diameter (like the pupil of your eye) means more than 100 times the power density of sunlight! You need to respect the device and follow the rules for working with it. After completing this section you should be able to

☐ Draw a fiber link showing all components

☐ Explain the use of the radiometer

☐ Describe rise time measurement

☐ Troubleshoot fiber-optic data links

Losses in an Optical-Fiber System

The optical-fiber system in Figure E.23-1 has an emitter, two connectors, the fiber, and the detector. The proper performance of this fiber link depends on the total power losses of the light signal through the link being less than the specified maximum allowable loss. Power is lost in all of the components that make up the system. A connector may have a power loss of 1.5 dB and a splice with 0.5 dB, and the fiber cable itself will also attenuate the light signal. As an example, if a fiber system's maximum allowable losses were 20 dB, and total power losses added up to 17 dB, the system would still have a 3-dB working margin. Of course, this is a small working margin and does not take into account weakening emitters and detectors over a period of time.

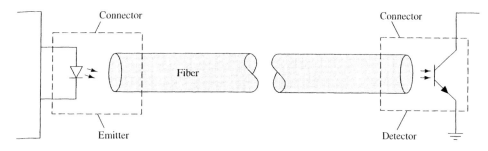

FIGURE E.23-1
A fiber link showing emitter, detector, connectors, and fiber cable.

Calculating Power Requirements

A power budget should be prepared when installing a fiber-based system. The power budget will specify the maximum losses that can be tolerated in the fiber system. This will help ensure that losses stay within the budgeted power allocation. Once the optical fiber system is in place, the **radiometer** would be used to determine the actual power being lost in the system. A calibrated light source injects a known amount of light into the fiber, and the radiometer connected to the other end of the fiber measures the light power reaching it. Periodic checks should be scheduled as preventive maintenance to keep the fiber system in peak performance. Weakening emitters should be replaced before they degrade the system's performance.

Rise Time Measurement

An optical fiber's link performance can be determined by measuring the rise time of injected test pulses. Figure E.23-2 shows the test configuration used to inject pulses into the fiber link. The test pulses must have a very fast rise time, typically in pico- or nanoseconds. At the detector end, connect a fast response oscilloscope to the output of the detector and measure the rise time of the injected pulses. Rise time is measured from the 10% point to the 90% point on the positive going edge of the pulse, as illustrated in Figure E.23-2.

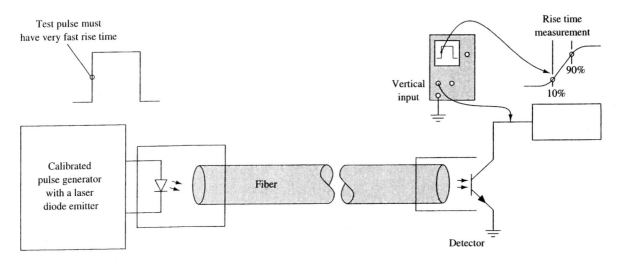

FIGURE E.23-2
Rise time test of an optical-fiber link.

The rise time (R_t) measurement must not be greater than the value provided by the following equation:

$$R_t = \frac{0.35}{BW}$$

A 70-ns rise time output indicates an optical-fiber system bandwidth of 5 MHz, based on this equation.

For this system to function properly the measured rise time would have to be less than 70 ns if a 5-MHz bandwidth is needed. Suppose the 5-MHz bandwidth system has the following signal delays: The emitter has a rise time of 1 ns, the detector a rise time of 20 ns, and a delay of 35 ns exists for 1 km of fiber cable. This adds up to a total system delay of 56 ns. This is a functioning system because the 56-ns delay is less than 70 ns. Scheduling periodic checks to keep an eye on the system rise time ensures that the specified system bandwidth is maintained. Weakening emitters and detectors and tight bends in the fiber cable will cause an increase in delay time, and the rise time measurement will be greater. This is an excellent means for monitoring a fiber system's performance.

Connector and Cable Problems

Some of the problems associated with fiber-optic links are caused by contact of a foreign substance with the fiber (even the oil from your skin can cause serious trouble). Connectors and splices are potential trouble spots. A back-biased photodiode and an op-amp can be used as a relative signal strength indicator. Looking for the signal while gently flexing cables and connectors can help pinpoint problem areas.

Characteristics of LEDs and DLs

These special-purpose diodes are nonetheless diodes and should exhibit the familiar exponential I versus V curve. These diodes do not draw current when forward biased until the voltage reaches about 1.4 V. Some ohmmeters do not put out sufficient voltage to turn a LED on; you may have to use a power supply, a current limiting resistor, and a voltmeter to test the diode.

Reverse voltage ratings are very low compared to ordinary silicon rectifier diodes— as little as 6 V. More voltage may destroy the diode.

A Simple Test Tool

Some systems use visible wavelengths; most use invisible infrared. Another diode of the same type or of similar emission wavelength can be used as a detector to check for output. Use a meter in current mode, not voltage, and compare a good system to the troublesome one.

To increase sensitivity, a simple current-to-voltage converter circuit, made with an op-amp, a feedback resistor, and the detector diode pumping current to the op-amp input, will convert the current from the detector diode to a voltage out of the op-amp. The circuit for this is shown in Figure E.23-3. If signal levels are high, just a resistor across the detector diode is appropriate. Remember to keep the bias voltage small enough so that the voltage developed is well below the maximum reverse voltage allowed for the diode.

FIGURE E.23-3
Light probe.

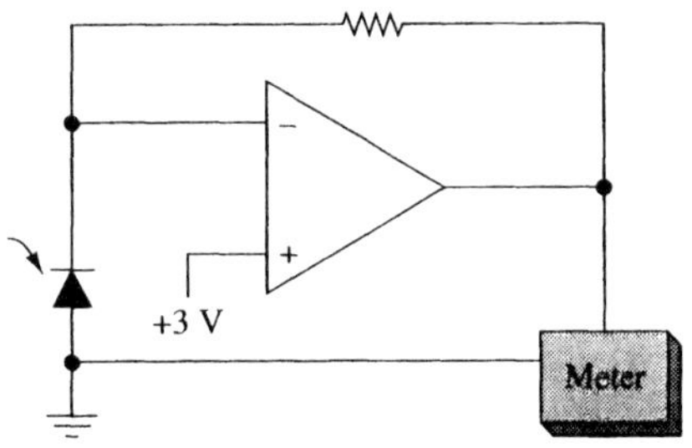

If you wish to see the signal modulation, try using an oscilloscope in place of a simple multimeter. A less quantitative check for emitted output can be made using a test card of the type used in TV-repair shops to check for output from infrared remote controls. These cards are coated with a special chemical that in the simultaneous presence of visible and infrared illumination will emit an orange glow.

E.24 Understand principles and operations of RF circuits

INTRODUCTION

In this section we explore the operation of RF amplifiers in the context of their use in an FM receiver.

FM RECEIVER BLOCK DIAGRAM

The basic FM receiver uses the superheterodyne principle. In block diagram form, it has many similarities to the receivers covered in previous skill topics. In Figure E.24-1, the only apparent differences are the use of the word *discriminator* in place of *detector,* the addition of a deemphasis network, and the fact that AGC may or may not be used as indicated by the dashed lines.

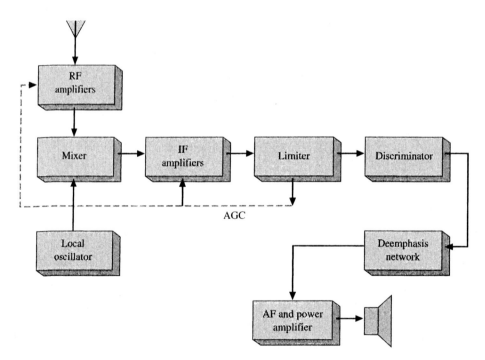

FIGURE E.24-1
FM receiver block diagram.

The **discriminator** extracts the intelligence from the high-frequency carrier and can also be called the detector, as in AM receivers. By definition, however, a discriminator is a device in which amplitude variations are derived in response to frequency or phase variations, and it is the preferred term for describing an FM demodulator.

The deemphasis network following demodulation is required to bring the high-frequency intelligence back to the proper amplitude relationship with the lower frequencies.

The fact that AGC is optional in an FM receiver may be surprising to you. From your understanding of AM receivers, you know that AGC is essential to their satisfactory operation. However, the use of limiters in FM receivers essentially provides an AGC function. Many older FM receivers also included an **automatic frequency control** (AFC) function. This is a circuit that provides a slight automatic control over the local oscillator circuit. It compensates for drift in LO frequency that would otherwise cause a station to become detuned. It was necessary because it had not yet been figured out how to make an economical *LC* oscillator at 100 MHz with sufficient frequency stability. The AFC system is not needed in new designs.

The mixer, local oscillator, and IF amplifiers are basically similar to those discussed for AM receivers and do not require further elaboration. It should be noted that higher frequencies are usually involved, however, because of the fact that FM systems generally function at higher frequencies. The universally standard IF frequency for FM is 10.7 MHz, as opposed to 455 kHz for AM. Because of significant differences in all the other portions of the block diagram shown in Figure E.24-1, they are discussed in the following sections.

RF AMPLIFIERS

Broadcast AM receivers normally operate quite satisfactorily without any RF amplifier. This is rarely the case with FM receivers, however, except for frequencies in excess of 1000 MHz (1 GHz), when it becomes preferable to omit it. The essence of the problem is that FM receivers can function with weaker received signals than AM or SSB receivers because of their inherent noise reduction capability. This means that FM receivers can function with a lower sensitivity, and are called upon to deal with input signals of 1 μV or less as compared with perhaps a 30-μV minimum input for AM. If a 1-μV signal is fed directly into a mixer, the inherently high noise factor of an active mixer stage destroys the intelligibility of the 1-μV signal. It is, therefore, necessary to amplify the 1-μV level in an RF stage to get the signal up to at least 10 to 20 μV before mixing occurs. The FM system can tolerate 1 μV of noise from a mixer on a 20-μV signal but obviously cannot cope with 1 μV of noise with a 1-μV signal.

This reasoning also explains the abandonment of RF stages for the ever-increasing FM systems at the 1-GHz-and-above region. At these frequencies, transistor noise is increasing while gain is decreasing. The frequency is reached where it is advantageous to feed the incoming FM signal directly into a diode mixer so as to immediately step it down to a lower frequency for subsequent amplification. Diode (passive) mixers are less noisy than active mixers.

Of course, the use of an RF amplifier reduces the image frequency problem. Another benefit is the reduction in **local oscillator reradiation** effects. Without an RF amp, the local oscillator signal can more easily get coupled back into the receiving antenna and transmit interference.

FET RF Amplifiers

Virtually all RF amps used in quality FM receivers utilize FETs as the active element. You may think that this is done because of their high input impedance, but this is *not* the reason. In fact, their input impedance at the high frequency of FM signals is greatly reduced because of their input capacitance. The fact that FETs do not offer any significant impedance advantage over other devices at high frequencies is not a deterrent, however, since the impedance that an RF stage works from (the antenna) is only several hundred ohms or less anyway.

The major advantage is that FETs have an input/output square-law relationship while vacuum tubes have a ³⁄₂-power relationship and BJTs have a diode-type exponential characteristic. A square-law device has an output signal at the input frequency and a smaller distortion component at two times the input frequency, whereas the other devices mentioned have many more distortion components, with some of them occurring at frequencies close to the desired signal. The use of an FET at the critical small signal level in a receiver means that the device distortion components are easily filtered out by its tuned circuits, since the closest distortion component is two times the frequency of the desired signal. This becomes an extreme factor when you tune to a weak station that has a very strong adjacent signal. If the high-level adjacent signal gets through the input tuned circuit, even though greatly attenuated, it would probably generate distortion components at the desired signal frequency by a non-square-law device, and the result is audible noise in the speaker output. This form of receiver noise is called **cross-modulation.** This is similar to **intermodulation distortion,** which is characterized by the mixing of *two* undesired signals, resulting in an output component that is equal to the desired signal's frequency. The possibility of intermodulation distortion is also greatly minimized by use of FET RF amplifiers.

MOSFET RF Amplifiers

A dual-gate, common-source MOSFET RF amplifier is shown in Figure E.24-2. The use of a dual-gate device allows a convenient isolated input for an AGC level to control device gain. The MOSFETs also offer the advantage of increased **dynamic range** over JFETs. That is, a wider range of input signal can be tolerated by the MOSFET while still offering the desired square-law input/output relationship. A similar arrangement is often utilized in mixers, since the extra gate allows for a convenient injection point for the local oscillator signal. The accompanying chart in Figure E.24-2 provides component values for operation at 100-MHz and 400-MHz center frequencies. The antenna input signal is coupled into gate 1 via the coupling/tuning network comprised of C_1, L_1, and C_2. The output signal is taken at the drain, which is coupled to the next stage by the L_2–C_3–C_4 combination. The bypass capacitor C_B next to L_2 and the radio-frequency choke (RFC) ensure that the signal frequency is not applied to the DC power supply. The RFC acts as an open to the signal while appearing as a short to DC, and the bypass capacitor acts in the inverse fashion. These precautions are necessary to RF frequencies because while power supply impedance is very low at low frequencies and DC, it looks like a high impedance to RF and can cause appreciable signal power loss. The bypass capacitor from gate 2 to ground provides a short

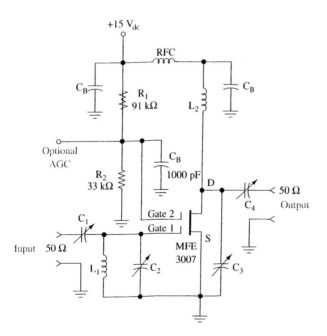

VHF Amplifier
The following component values are used
for the different frequencies:

Component Values	100 MHz	400 MHz
C_1	8.4 pF	4.5 pF
C_2	2.5 pF	1.5 pF
C_3	1.9 pF	2.8 pF
C_4	4.2 pF	1.2 pF
L_1	150 nH	16 nH
L_2	280 nH	22 nH
C_B	1000 pF	250 pF

FIGURE E.24-2
MOSFET RF amplifier. (Courtesy of Motorola Semiconductor Products, Inc.)

to any high-frequency signal that may get to that point. It is necessary to maintain the bias stability set up by R_1 and R_2. The MFE 3007 MOSFET used in this circuit provides a minimum power gain of 18 dB at 200 MHz.

REVIEW QUESTIONS

True/False

1. FETs are typically used for RF amplifiers.
2. RF amplifiers are more sensitive than audio amplifiers.

Multiple Choice

3. MOSFETs in RF amplifiers offer
   a. Wider input voltage range.
   b. Square-law input/output relationship.
   c. Both a and b.
4. RF amplifiers reduce the effects of
   a. Antenna harmonics.
   b. Local oscillator reradiation.
   c. Modulation runaway.
5. Intermodulation noise is
   a. Decreased by an RF amplifier.
   b. Increased by an RF amplifier.
   c. Unchanged by an RF amplifier.

E.25 Fabricate and demonstrate RF circuits

INTRODUCTION

This material is similar to that covered in E.05. The AM or FM sound circuitry contained in the electronic kits will typically contain both types of circuits (IF and RF).

E.26 Troubleshoot and repair RF circuits

INTRODUCTION

In this exercise we examine a number of points to check when troubleshooting an RF amplifier.

Bias Supply

Many RF amplifiers utilize power from the previous stage to provide DC bias. Figure E.26-1 shows how bias for the transistor Q1 is developed. RF from the previous stage is rectified by the base-emitter junction of Q1. The current flows through R1 and the transformer to ground. The reactance of C1 is low at RF, so the RF bypasses the resistor. C1 also serves to filter the RF pulses and develop a DC voltage across R1. At the base of Q1, this DC voltage is negative with respect to ground. Therefore, Q1 will be a class C amplifier conducting only on positive RF peaks. Figure E.26-2 shows the instantaneous voltage at the base of Q1 that you can observe with an oscilloscope.

FIGURE E.26-1
Self-bias circuit.

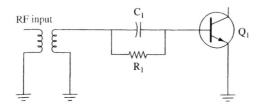

FIGURE E.26-2
Voltage at Q1 base.

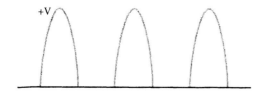

Shorted C1

If C1 were to short, excessive drive would reach Q1. No negative bias for Q1 could be developed. This would cause Q1 to draw excessive current and destroy itself. If Q1 is bad, always check all components ahead of Q1 before replacing it.

Open C1

If C1 were open, the drive reaching Q1 would be greatly reduced. Bias voltage would be low and Q1 would not develop full power output.

465

Open R1

Resistors in these circuits may overheat and fail open. C1 will charge to the negative peak of the RF drive voltage because of the rectifier action of the base-emitter junction. This will cut Q1 off and there will be no power output.

Output Network

Now consider possible faults in components on the output side of Q1. Common faults are shorted blocking capacitors, overheated tuning capacitors, and open chokes.

E.27 Understand principles and operations of signal modulation systems (AM, FM, stereo)

INTRODUCTION

The reasons that modulation is used in electronic communications have previously been explained as:

1. Direct transmission of intelligible signals would result in catastrophic interference problems, since the resulting radio waves would be at approximately the same frequency.

2. Most intelligible signals occur at relatively low frequencies. Efficient transmission and reception of radio waves at low frequencies is not practical due to the large antennas required.

The process of impressing a low-frequency intelligence signal onto a higher-frequency "carrier" signal may be defined as **modulation.** The higher-frequency "carrier" signal will hereafter be referred to as simply the carrier. It is also termed the radio-frequency (RF) signal, since it is at a high-enough frequency to be transmitted through free space as a radio wave. The low-frequency intelligence signal will subsequently be termed the "intelligence." It may also be identified by terms such as modulating signal, information signal, audio signal, or modulating wave.

Three different characteristics of a carrier can be modified so as to allow it to "carry" intelligence. Either the amplitude, frequency, or phase of a carrier are altered by the intelligence signal.

AMPLITUDE MODULATION FUNDAMENTALS

Combining two widely different sine-wave frequencies such as a carrier and intelligence in a linear fashion results in their simple algebraic addition, as shown in Figure E.27-1. A circuit that would perform this function is shown in Figure E.27-1(a)—the two signals combined in a linear device such as a resistor. Unfortunately, the resultant [Figure E.27-1(d)] is *not* suitable for transmission as an AM waveform. If it were transmitted, the receiver antenna would be detecting just the carrier signal [Figure E.27-1(c)], because the low-frequency intelligence component cannot be efficiently propagated as a radio wave.

The method utilized to produce a usable AM signal is to combine the carrier and intelligence through a **nonlinear device.** It can be mathematically proven that the combination of any two sine waves through a nonlinear device produces the following frequency components:

1. A DC level

2. Components at each of the two original frequencies

3. Components at the sum and difference frequencies of the two original frequencies

4. Harmonics of the two original frequencies

Figure E.27-2 shows this process pictorially with the two sine waves, labeled f_c and f_i, to represent the carrier and intelligence. If all but the $f_c - f_i$, f_c, and $f_c + f_i$ components are

467

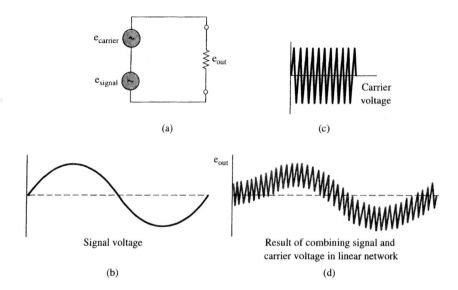

FIGURE E.27-1
Linear addition of two sine waves.

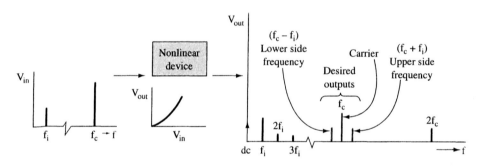

FIGURE E.27-2
Nonlinear mixing.

removed (perhaps with a bandpass filter), the three components left form an AM wave-form. They are referred to as:

1. The *lower side frequency* $(f_c - f_i)$
2. The *carrier frequency* (f_c)
3. The *upper side frequency* $(f_c + f_i)$

AM Waveforms

Figure E.27-3 shows the actual AM waveform under varying conditions of the intelligence signal. Note in Figure E.27-3(a) that the resultant AM waveform is basically a signal at the carrier frequency whose amplitude is changing at the same rate as the intelligence fre-quency. As the intelligence amplitude reaches a maximum positive value, the AM waveform has a maximum amplitude. The AM waveform reaches a minimum value when the intelli-gence amplitude is at a maximum negative value. In Figure E.27-3(b), the intelligence fre-quency remains the same, but its amplitude has been increased. The resulting AM waveform reacts by reaching a larger maximum value and smaller minimum value. In Figure E.27-3(c), the intelligence amplitude is reduced and its frequency has gone up. The resulting AM wave-form, therefore, has reduced maximums and minimums, and the rate at which it swings be-tween these extremes has increased to the same frequency as the intelligence signal.

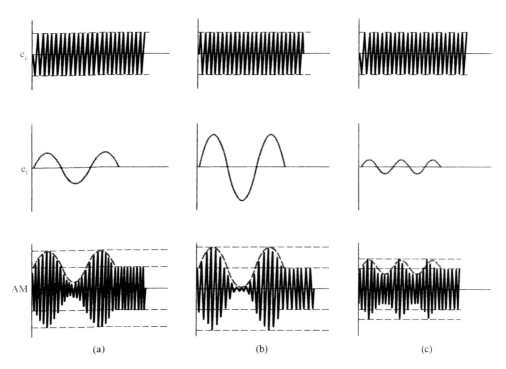

FIGURE E.27-3
AM waveform under varying intelligence signal (e_i) conditions.

It may now be correctly concluded that both the top and bottom envelopes of an AM waveform are replicas of the frequency and amplitude of the intelligence (notice the 180° phase shift). However, the AM waveform does *not* include any components at the intelligence frequency. If a 1-MHz carrier were modulated by a 5-kHz intelligence signal, the AM waveform would include the following components:

$$1 \text{ MHz} + 5 \text{ kHz} = 1,005,000 \text{ Hz (upper side frequency)}$$
$$1 \text{ MHz} = 1,000,000 \text{ Hz (carrier frequency)}$$
$$1 \text{ MHz} - 5 \text{ kHz} = 995,000 \text{ Hz (lower side frequency)}$$

This process is shown in Figure E.27-4. Thus, even though the AM waveform has envelopes that are replicas of the intelligence signal, it *does not* contain a frequency component at the intelligence frequency.

The intelligence envelope is shown in the resultant waveform and results from connecting a line from each RF peak value to the next one for both the top and bottom halves of the AM waveform. The drawn-in envelope is not really a component of the waveform and would not be seen on an oscilloscope display. In addition, the top and bottom envelopes are *not* the upper and lower side frequencies, respectively. The envelopes result from nonlinear combination of a carrier with two lower-amplitude signals spaced in frequency equal amounts above and below the carrier frequency. The increase and decrease in the AM waveform's amplitude is caused by the frequency difference in the side frequencies, which allows them to alternately add to and subtract from the carrier amplitude, depending on their instantaneous phase relationships.

The AM waveform in Figure E.27-4(d) does not show the relative frequencies to scale. The ratio of f_c to the envelope frequency (which is also f_i) is 1 MHz to 5 kHz, or 200:1. Thus, the fluctuating RF should show 200 cycles for every cycle of envelope variation. To do that in a sketch is not possible, and an oscilloscope display of this example, and most practical AM waveforms, results in a well-defined envelope but with so many RF variations that they appear as a blur, as shown in Figure E.27-4(e).

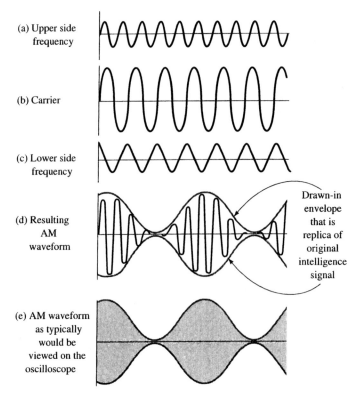

FIGURE E.27-4
Carrier and side-frequency components result in AM waveform.

Modulation of a carrier with a pure sine-wave intelligence signal has thus far been shown. However, in most systems the intelligence is a rather complex waveform that contains many frequency components. For example, the human voice contains components from roughly 200 Hz to 3 kHz and has a very erratic shape. If it were used to modulate the carrier, a whole *band* of side frequencies would be generated. The band of frequencies thus generated above the carrier is termed the **upper sideband,** while those below the carrier are called the **lower sideband.** This situation is illustrated in Figure E.27-5 for a 1-MHz carrier modulated by a whole band of frequencies, which range from 200 Hz up to 3 kHz. The upper sideband is from 1,000,200 to 1,003,000 Hz, and the lower sideband ranges from 997,000 to 999,800 Hz.

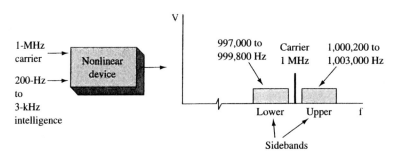

FIGURE E.27-5
Modulation by a band of intelligence frequencies.

EXAMPLE E.27-1

A 1.4-MHz carrier is modulated by a music signal that has frequency components from 20 Hz to 10 kHz. Determine the range of frequencies generated for the upper and lower sidebands.

Solution The upper sideband is equal to the sum of carrier and intelligence frequencies. Therefore, the upper sideband (usb) will include the frequencies from

$$1,400,000 \text{ Hz} + 20 \text{ Hz} = 1,400,020 \text{ Hz}$$

to

$$1,400,000 \text{ Hz} + 10,000 \text{ Hz} = 1,410,000 \text{ Hz}$$

The lower sideband (lsb) will include the frequencies from

$$1,400,000 \text{ Hz} - 10,000 \text{ Hz} = 1,390,000 \text{ Hz}$$

to

$$1,400,000 \text{ Hz} - 20 \text{ Hz} = 1,399,980 \text{ Hz}$$

This result is shown in Figure E.27-6 with a frequency spectrum of the AM modulator's output.

FIGURE E.27-6
Solution for Example E.27-1.

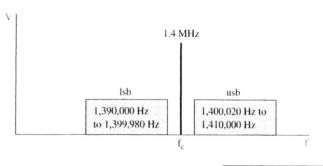

ANGLE MODULATION

As has been stated previously, there are three parameters of a sine-wave carrier that can be varied to allow it to "carry" a low-frequency intelligence signal. They are its amplitude, frequency, and phase. The latter two, frequency and phase, are actually interrelated, as one cannot be changed without changing the other. They both fall under the general category of *angle modulation*. **Angle modulation** is defined as modulation where the angle of a sine wave carrier is varied from its reference value. Angle modulation has two subcategories, phase modulation and frequency modulation, with the following definitions:

> *Phase modulation (PM):* angle modulation where the phase angle of a carrier is caused to depart from its reference value by an amount proportional to the modulating signal amplitude.

> *Frequency modulation (FM):* angle modulation where the instantaneous frequency of a carrier is caused to vary by an amount proportional to the modulating signal amplitude.

The key difference between these two similar forms of modulation is that in PM the amount of phase change is proportional to intelligence amplitude, while in FM it is the frequency change that is proportional to intelligence amplitude. As it turns out, PM is *not* directly used as the transmitted signal in communication systems but does have importance since it is often used to help generate FM, *and* a knowledge of PM helps us to understand the superior noise characteristics of FM as compared to AM systems. In recent years, it has become fairly common practice to denote angle modulation simply as FM instead of specifically referring to FM and PM.

The concept of FM was first practically postulated as an alternative to AM in 1931. At that point, commercial AM broadcasting had been in existence for over 10 years, and the superheterodyne receivers were just beginning to supplant the TRF designs. The goal of research into an alternative to AM at that time was to develop a system less susceptible to external noise pickup. Major E. H. Armstrong developed the first working FM system in 1936, and in July 1939, he began the first regularly scheduled FM broadcast in Alpine, New Jersey.

Radio Emission Classifications

Table E.27-1 gives the codes used to indicate the various types of radio signals. The first letter is A, F, or P to indicate AM, FM, or PM. The next code symbol is one of the numbers 0 through 9 used to indicate the type of transmission. The last code symbol is a subscript. If there is no subscript it means double-sideband, full carrier. Here are some examples of emission codes:

$A3_a$ SSB, reduced carrier

$A3_i$ SSB, no carrier

F3   FM, double-sideband, full carrier

$A7_i$ SSB, no carrier, multiple sidebands with different messages

10A3   AM, double-sideband, full carrier, 10-kHz bandwidth

Notice the last example. If an emission code is preceded by a number, that is the bandwidth of the signal in kHz.

TABLE E.27-1
Radio emission classifications

Modulation		Type	Subscripts	
A	Amplitude	0 Carrier on only	None	Double-sideband, full carrier
F	Frequency	1 Carrier on–off (Morse code, radar)	a	Single-sideband, reduced carrier
P	Phase	2 Carrier on, keyed tone—on–off	b	Two independent sidebands
		3 Telephony, voice, or music	c	Vestigial sideband
		4 Facsimile, nonmoving or slow-scan TV	d	Pulse amplitude modulation (PAM)
		5 Vestigial sideband, commercial TV	e	Pulse width modulation (PWM)
		6 Four-frequency diplex telegraphy	f	Pulse position modulation (PPM)
		7 Multiple sidebands, each with different message	g	Digital video
		8 Unassigned	h	Single-sideband, full carrier
		9 General, all other	i	Single-sideband, no carrier

A Simple FM Generator

To gain an intuitive understanding of FM, consider the system illustrated in Figure E.27-7. This is actually a very simple, yet highly instructive, FM transmitting system. It consists of an *LC* tank circuit, which, in conjunction with an oscillator circuit, generates a sine-wave output. The capacitance section of the *LC* tank is not a standard capacitor but is a capacitor microphone. This popular type of microphone is often referred to as a condenser mike and is, in fact, a variable capacitor. When no sound waves reach its plates, it presents a constant value of capacitance at its two output terminals. However, when sound waves reach the mike, they alternately cause its plates to move in and out. This causes its capacitance to go up and down around its center value. The *rate* of this capacitance change is equal to the

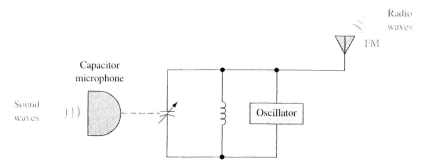

FIGURE E.27-7
Capacitor microphone FM generator.

frequency of the sound waves striking the mike, *and* the *amount* of capacitance change is proportional to the amplitude of the sound waves.

Since this capacitance value has a direct effect on the oscillator's frequency, the following two *important* conclusions can be made concerning the system's output frequency:

1. The frequency of impinging sound waves determines the *rate* of frequency change.

2. The amplitude of impinging sound waves determines the *amount* of frequency change.

Consider the case of the sinusoidal sound wave (the intelligence signal) shown in Figure E.27-8(a). Up until time T_1 the oscillator's waveform at (b) is a constant frequency with constant amplitude. This corresponds to the carrier frequency (f_c) or *rest* frequency in FM systems. At T_1 the sound wave at (a) starts increasing sinusoidally and reaches a maximum positive value at T_2. During this period, the oscillator frequency is gradually increasing and reaches its highest frequency when the sound wave has maximum amplitude at time T_2. From time T_2 to T_4 the sound wave goes from maximum positive to maximum negative and the resulting oscillator frequency goes from a maximum frequency *above* the rest value to a maximum value *below* the rest frequency. At time T_3 the sound wave is passing through zero, and therefore the oscillator output is instantaneously equal to the carrier frequency.

The Two Major Concepts

The amount of oscillator frequency increase and decrease around f_c is called the **frequency deviation, δ**. This deviation is shown in Figure E.27-8(c) as a function of time. Notice that this is a graph of frequency vs. time—not the usual voltage vs. time. It is ideally shown as a sine-wave replica of the original intelligence signal. It shows that the oscillator output is indeed an FM waveform. Recall that FM is defined as a sine-wave carrier that changes in frequency by an *amount* proportional to the instantaneous value of the intelligence wave and at a *rate* equal to the intelligence frequency.

Figure E.27-8(d) shows the AM wave resulting from the intelligence signal shown at (a). This should help you to see the difference between an AM and FM signal. In the case of AM, the carrier's amplitude is varied (by its sidebands) in step with the intelligence, while in FM, the carrier's frequency is varied in step with the intelligence.

The capacitor microphone FM generation system is seldom used in practical applications; its importance is derived from its relative ease of providing an understanding of FM basics. If the sound-wave intelligence striking the microphone were doubled in frequency from 1 kHz to 2 kHz with constant amplitude, the rate at which the FM output swings above and below the center frequency (f_c) would change from 1 kHz to 2 kHz. However, since the intelligence amplitude was not changed, the *amount* of frequency deviation (δ) above and below f_c will remain the same. On the other hand, if the 1-kHz intelligence frequency were kept the same but its amplitude were doubled, the *rate* of deviation

FIGURE E.27-8
FM representation.

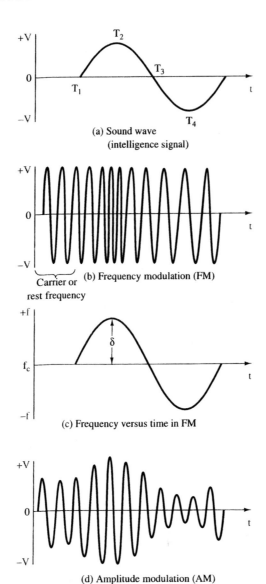

(a) Sound wave
(intelligence signal)

Carrier or
rest frequency

(b) Frequency modulation (FM)

(c) Frequency versus time in FM

(d) Amplitude modulation (AM)

above and below f_c would remain at 1 kHz, but the *amount* of frequency deviation would double.

As you continue through your study of FM, whenever you start getting bogged down on basic theory, it will often be helpful to review the capacitor mike FM generator. *Remember:*

1. The intelligence amplitude determines the *amount* of carrier frequency deviation.

2. The intelligence frequency (f_i) determines the *rate* of carrier frequency deviation.

EXAMPLE E.27-2

An FM signal has a center frequency of 100 MHz but is swinging between 100.001 MHz and 99.999 MHz at a rate of 100 times per second. Determine

(a) The intelligence frequency f_i.
(b) The intelligence amplitude.
(c) What happened to the intelligence amplitude if the frequency deviation changed to between 100.002 and 99.998 MHz.

Solution
(a) Since the FM signal is changing frequency at a 100-Hz rate, f_i = 100 Hz.
(b) There is no way of determining the actual amplitude of the intelligence signal. Every FM system has a different proportionality constant between the intelligence amplitude and the amount of deviation it causes.
(c) The frequency deviation has now been doubled, which means that the intelligence amplitude is now double whatever it originally was.

FM ANALYSIS

The complete mathematical analysis of angle modulation requires the use of high-level mathematics. For our purposes, it will suffice to simply give the solutions and discuss them. For phase modulation (PM), the equation for the instantaneous voltage is

$$e = A \sin(\omega_c t + m_p \sin \omega_i t)$$

where e = instantaneous voltage
  A = peak value of original carrier wave
  ω_c = carrier angular velocity ($2\pi f_c$)
  m_p = maximum phase shift caused by the intelligence signal (radians)
  ω_i = modulating (intelligence) signal angular velocity ($2\pi f_i$)

The maximum phase shift caused by the intelligence signal, m_p, is defined as the **modulation index** for PM.

The following equation provides the equivalent formula for FM:

$$e = A \sin(\omega_c t + m_f \sin \omega_i t)$$

All the terms are defined as they were in the previous equation, with the exception of the new term, m_f. In fact, the two equations are identical except for that term. It is defined as the modulation index for FM, m_f. It is equal to

$$m_f = \text{FM modulation index} = \frac{\delta}{f_i}$$

where δ = maximum frequency shift caused by the intelligence signal (deviation)
  f_i = frequency of the intelligence (modulating) signal

Broadcast FM

Standard broadcast FM uses a 200-kHz bandwidth for each station. This is a very large allocation when one considers that one FM station has a bandwidth that could contain many standard AM stations. Broadcast FM, however, allows for a true high-fidelity modulating signal up to 15 kHz and offers superior noise performance.

Figure E.27-9 shows the FCC allocation for commercial FM stations. The maximum allowed deviation around the carrier is ±75 kHz, and 25-kHz **guard bands** at the upper and lower ends are also provided. The carrier is required to maintain a ±2-kHz stability. Recall that an infinite number of side frequencies are generated during frequency modulation, but their amplitude gradually decreases as you move away from the carrier. In other words, the significant side frequencies exist up to ±75 kHz around the carrier, and the guard bands ensure that adjacent channel interference will not be a problem.

FIGURE E.27-9
Commercial FM bandwidth allocations for two adjacent stations.

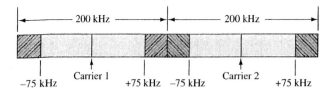

Since full deviation (δ) is 75 kHz, that is 100% modulation. By definition, 100% modulation in FM is when the deviation is the full permissible amount. Recall that the modulation index, m_f, is

$$m_f = \frac{\delta}{f_i}$$

so that the actual modulation index at 100% modulation varies inversely with the intelligence frequency, f_i. This is in contrast with AM, where full or 100% modulation means a modulation index of 1 regardless of intelligence frequency.

STEREO FM

The advent of stereo records and tapes and the associated high-fidelity playback equipment in the 1950s led to the development of stereo FM transmissions as authorized by the FCC in 1961. Stereo systems involve generating two separate signals, as from the left and right sides of a concert hall performance. When played back on left and right speakers, the listener gains greater spatial dimension or directivity.

A stereo radio broadcast requires that two separate 30-Hz to 15-kHz signals be used to modulate the carrier in such a way that the receiver can extract the "left" and "right" channel information and separately amplify them into their respective speakers. In essence, then, the amount of information to be transmitted is doubled in a stereo broadcast. Hartley's law tells us that either the bandwidth or time of transmission must therefore be doubled, but this is not practical. The problem was solved by making more efficient use of the available bandwidth (200 kHz) by **frequency multiplexing** the two required modulating signals. **Multiplex operation** is the simultaneous transmission of two or more signals on one carrier.

Modulating Signal

The system approved by the FCC is *compatible* in that a stereo broadcast received by a normal FM receiver will provide an output equal to the sum of the left plus right channels (L + R), while a stereo receiver can provide separate left and right channel signals. The stereo transmitter has a modulating signal, as shown in Figure E.27-10. Notice that the sum of the L + R modulating signal extends from 30 Hz to 15 kHz just as does the full audio

FIGURE E.27-10
Composite modulating signals.

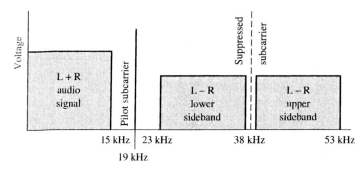

signal used to modulate the carrier in standard FM broadcasts. However, a signal corresponding to the left channel minus right channel (L − R) extends from 23 to 53 kHz. In addition, a 19-kHz pilot subcarrier is included in the composite stereo modulating signal.

Two different signals (L + R and L − R) are used to modulate the carrier. The signal is an example of **frequency-division multiplexing,** in that two different signals are multiplexed together by having them exist in two different frequency ranges.

FM Stereo Generation

The block diagram in Figure E.27-11 shows the method whereby the composite modulating signal is generated and applied to the FM modulator for subsequent transmission. The left and right channels are picked up by their respective microphones and individually preemphasized. They are then applied to a **matrix network** that inverts the right channel, giving a −R signal, and then combines (adds) L and R to provide an (L + R) signal and also combines L and −R to provide the (L − R) signal. The two outputs *are still* 30-Hz to 15-kHz audio signals at this point. The (L − R) signal and a 38-kHz carrier signal are then applied to a balanced modulator that suppresses the carrier but provides a double-sideband (DSB) signal at its output. The upper and lower sidebands extend from 30 Hz to 15 kHz above and below the suppressed 38-kHz carrier and therefore range from 23 kHz (38 kHz − 15 kHz) up to 53 kHz (38 kHz + 15 kHz). Thus, the (L − R) signal has been translated from audio up to a higher frequency so as to keep it separate from the 30-Hz to 15-kHz (L + R) signal. The (L + R) signal is given a slight delay so that both signals are applied to the FM modulator in time phase due to the slight delay encountered by the (L − R) signal in the balanced modulator. The 19-kHz master oscillator in Figure E.27-11 is applied directly to the FM modulator and also doubled in frequency to 38 kHz for the balanced modulator carrier input.

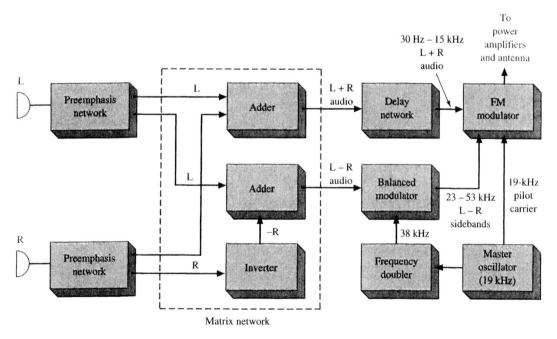

FIGURE E.27-11
Stereo FM transmitter.

Stereo FM is more prone to noise than are monophonic broadcasts. The (L − R) signal is weaker than the (L + R) signal, as shown in Figure E.27-10. The (L − R) signal is also at a higher modulating frequency (23 to 53 kHz), and both of these effects cause poorer noise performance. The net result to the receiver is a *S/N* of about 20 dB less than the monophonic signal. Because of this, some receivers have a mono/stereo switch so that a noisy (weak) stereo signal can be changed to monophonic for improved reception. A stereo signal received by a monophonic receiver is only about 1 dB worse (*S/N*) than an equivalent monophonic broadcast, due to the presence of the 19-kHz pilot carrier.

REVIEW QUESTIONS

True/False

1. AM and FM are both examples of modulation techniques.
2. An AM signal whose carrier is 2 MHz and whose information signal is 2.5 kHz has an upper sideband of 2.25 MHz.

Multiple Choice

3. Which modulation technique employs a variable frequency carrier?
   a. AM.
   b. FM.
   c. PM.
4. The maximum phase shift caused by the intelligence signal is called the
   a. Carrier index.
   b. Modulation index.
   c. Phase index.
5. The guard band on a broadcast FM carrier is
   a. 25 kHz on both ends.
   b. 50 kHz on both ends.
   c. 75 kHz on both ends.

E.28 Troubleshoot and repair signal modulation systems (AM, FM, stereo)

INTRODUCTION

In this exercise we examine a number of techniques and items to check while troubleshooting AM and FM modulation circuitry.

TROUBLESHOOTING AM SYSTEMS

There are two ways to generate SSB signals, but modern manufacturing methods have reduced the cost of filters to the point that nearly all generate the SSB signal with balanced modulators and filters. Most radios even use separate filters to select the upper or lower sideband as desired instead of switching oscillators.

In troubleshooting SSB generators, you will be mainly looking for the presence or absence of various oscillations. A spectrum analyzer is an extremely desirable tool in this regard, but if one is not available, a good general coverage shortwave receiver is the next best tool. It is also desirable to have a frequency counter to measure the exact frequency of the oscillators.

What do we do when faced with a radio receiver that has no reception? Where do we start to look for the trouble? When faced with this kind of problem, how does the technician proceed in formulating a plan of action? This section will show you a popular method used for finding the problem in a receiver with no reception. After completing this section you should be able to:

☐ Troubleshoot SSB generators and demodulators

☐ Test for carrier leakthrough with an oscilloscope or spectrum analyzer

☐ Identify a defective stage in an SSB receiver

☐ Describe the signal injection method of troubleshooting

Balanced Modulators

Things to look for and do:

1. With no audio input, there should be no RF output. An oscilloscope will be helpful here.

2. The voltage from the oscillator must be 6 to 8 times the peak audio voltage. There should be several volts of RF and a few tenths of a volt of audio.

3. The diodes should be well matched. An ohmmeter can be used to select matched pairs or quads.

4. You should be able to null the carrier at the output by adjusting R1 and C1. It may be necessary to adjust each control several times alternately to secure optimum carrier suppression. Further detail on testing for "carrier leakthrough" is provided in the next few paragraphs.

Testing for Carrier Leakthrough

The purpose of the balanced modulator is to suppress or cancel out the carrier. An exactly balanced modulator would totally suppress or remove the carrier. This is an impossibility because there are always imbalances—one diode conducts a little more current than another, perhaps. To achieve a circuit's maximum suppression, balanced modulators usually include one or more balance controls, as shown in Figure E.28-1.

The first thing to do when troubleshooting a balanced modulator is to check the condition of its balance. This can be done by looking for "carrier leakthrough" with an oscilloscope. The circuit for this test is seen in Figure E.28-2. **Carrier leakthrough** simply means the amount of carrier not suppressed by the balanced modulator.

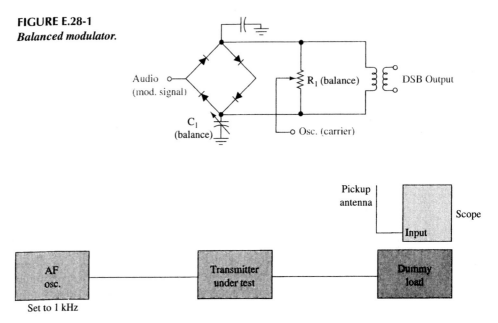

FIGURE E.28-1
Balanced modulator.

FIGURE E.28-2
Checking for carrier leakthrough with an oscilloscope.

The audio signal generator is set to some frequency within the normal audio range of the transmitter, perhaps 1 or 1.5 kHz. Check the manual for the correct RF and audio signal levels into the modulator. (Test points may be included for measuring these.) Typically, the RF input (oscillator output) will be about 4 to 6 times the audio level for proper diode switching. Any DMM can be used to measure the audio, but the meter will require an RF probe for the oscillator output.

Figure E.28-3(a) shows the transmitter signal when there is carrier leakthrough, that is, the carrier is not fully suppressed. Note the similarity to a partially modulated AM signal. Figure E.28-3(b) shows the signal as it should be, a single tone signal; the carrier is fully suppressed.

One of two conditions could cause carrier leakthrough: either the circuit has become unbalanced or there are defective components. To check for unbalance, adjust the balance control(s) for minimum carrier amplitude. Should there be more than one control, it may be necessary to go back and forth more than once between controls because the setting of one often affects the setting of another.

If there is no balance problem and the input signal levels are correct, there is a defective component, most likely one of the diodes in the bridge assembly. Such bridges are usually a sealed unit; you cannot get at individual diodes. If this is the case, replace the suspect unit with a known good one and recheck for proper operation. Be sure to check balance again with the new unit in place.

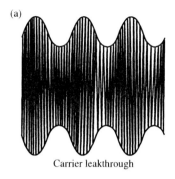
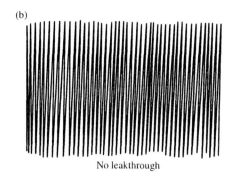

(a) Carrier leakthrough

(b) No leakthrough

FIGURE E.28-3
Single-sideband signal with and without carrier leakthrough.

Figure E.28-4 shows the circuit for checking carrier leakthrough and suppression with a spectrum analyzer.

Having determined that the RF and audio frequencies and levels are correct, observe the screen of the spectrum analyzer. If the modulator is operating correctly, you will see the display in part a of Figure E.28-5. Note the location of the suppressed carrier. Part b shows the same signal with some carrier leakthrough. As before, adjust the balance controls for minimum carrier amplitude (maximum suppression).

An advantage of the spectrum analyzer is that carrier suppression can be measured in dB directly on the analyzer's log scale. Check the transmitter's manual for the suppression figure; it will be in the neighborhood of -60 to -70 dB. The instruction manual for the

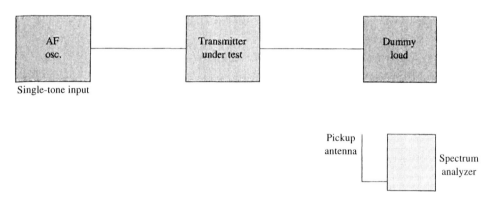

FIGURE E.28-4
Checking carrier suppression with a spectrum analyzer.

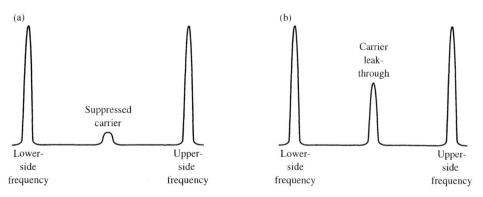

FIGURE E.28-5
Carrier suppression as seen on a spectrum analyzer.

spectrum analyzer will tell you how to set the unit's controls for logarithmic measurements.

Testing Filters

Filters designed for SSB service can be ceramic, crystal, or mechanical, but test methods are the same. The filter can have 6 to 10 dB of loss in the passband. Figure E.28-6 shows how you would set up the equipment to test a filter.

The technician should slowly sweep the generator frequency across the passband of the filter. If the sweep speed is too fast, the filter's time delay will cause misleading results. Response of the filter should fall off rapidly at the band edges. Two or three dB of ripple in the passband is normal. If the ripple is as much as 6 to 10 dB, the signal passing through the filter will be badly distorted.

FIGURE E.28-6
Filter testing.

Testing Linear Amplifiers

The two-tone test is generally used to check amplifier linearity. For example, a 400-Hz tone and a 2500-Hz tone are applied to the input of an SSB transmitter. The output is observed with a spectrum analyzer tuned to the transmitter's carrier frequency. Nonlinearity in the amplifier will cause the amplifier to generate mixer-like products. The proper response is shown in Figure E.28-7. The carrier will not be present if the balanced modulator is properly adjusted. In a good linear amplifier, the distortion products will be at least 30 dB below the two desired tones. If the amplifier is not linear, several spurious outputs will appear and may be only a few dB below the desired signals. Amplifier nonlinearity is usually caused by improper bias points in the amplifier.

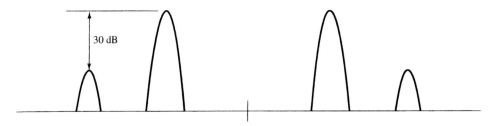

FIGURE E.28-7
Two-tone test.

TROUBLESHOOTING FM SYSTEMS

The most likely types of FM radio a technician will be called to service are an automotive mobile, a fixed base station, or a hand-held portable. Either the mobile or the fixed base station can have power outputs as high as 150 W. Some of these transmitters can be damaged if not operated into the proper load impedance, usually 50 Ω. Some contain automatic circuitry to shut the transmitter off if it is not connected to a proper load. Dummy loads are made for this purpose, so make sure you have one rated for sufficient power.

In an FM transmitter, an oscillator is typically controlled so that its frequency changes when an intelligence signal changes. This control is provided by a modulating circuit. There are several ways that this modulator can work. In this section we will learn some troubleshooting techniques for a reactance modulator circuit.

Frequency-modulated transmitters can be roughly divided into two categories. The first is wideband FM. More popular than AM broadcasting, it is the FM we listen to on our car radios and home stereos (it also produces the audio portion of a TV signal). The second type of FM transmitters is called narrowband FM. It is used in police and fire department radios as well as taxicabs and VHF boat radios and the popular hand-held transceivers called handy talkies, or walkie-talkies. We'll also look at testing a wideband FM generator in this section.

After completing this section you should be able to

☐ Troubleshoot FM transmitter systems

☐ Describe the operation of the reactance modulator

☐ Locate the master oscillator section

☐ Locate the reactance circuit section

☐ Recognize the difference between no modulator output, low output, or oscillator output without FM

☐ Troubleshoot a stereo/SCA FM generator

☐ Measure an FM transmitter's carrier frequency and deviation

FM Transmitter Systems

1. No RF Output   In this case, one should first verify that the oscillator is running. Most of these transmitters multiply the oscillator by 12 or 18 to obtain the output frequency. The service manual will tell you what the multiplier is. It is best to have a spectrum analyzer, but a shortwave receiver will do.

Let's assume that the oscillator is running and move on to the first multiplier stage. Figure E.28-8 shows a simplified schematic of a multiplier stage. If the base–emitter junction is good and there is sufficient input drive, you will find a negative voltage at the base of Q1. This is because the RF input is rectified by the junction and the current flows through R1. When the RF is rectified by the base–emitter junction, current pulses rich in harmonics are amplified and filtered by the tuned circuit in the collector of the transistor.

FIGURE E.28-8
Multiplier stage.

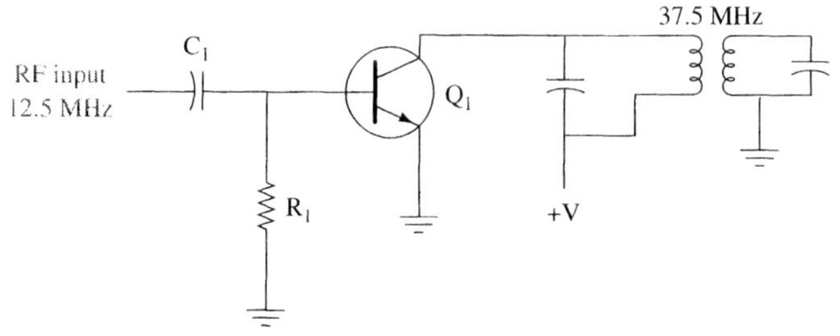

If a spectrum analyzer is available, the technician can verify that the stage is producing the proper multiple of the input frequency by loosely coupling the analyzer to the output coils. A two- or three-turn coil one-half inch in diameter connected to the analyzer will do.

Either the coils or the capacitors in the multiplier stage will be adjustable. You should be able to peak the output on the proper frequency with these adjustments. If not, check the capacitors and the inductors.

If you do not have a spectrum analyzer, simply measuring the bias voltage on the next stage may be sufficient. You can be reasonably sure that the multiplier is working if you can peak up the drive to the next stage with the adjustments. However, there is some possibility of tuning to the wrong harmonic. If the adjustments are all the way to one end, you may have done this or some component has failed. Another indication of improper tuning is that you will probably not be able to tune the next stage.

A typical transmitter will have three multiplier stages. At some point in the chain you will be able to find enough signal to run a frequency counter. Be careful; too much input to the counter will damage it. At the output of the transmitter, you can use a high-power attenuator.

2. Incorrect Frequency   Most of these transmitters will have trimmer capacitors to adjust the frequency, while some will have inductors. If the oscillator is off frequency, check the voltage first, then the capacitors. Intermittent capacitors are the hardest to find. Try cooling the capacitor with an aerosol spray sold for this purpose.

3. Incorrect Deviation   There are usually two adjustments here, one for microphone gain and one on a limiter. The limiter prevents the user from overmodulating the transmitter. When adjusting deviation, make sure the limiter is adjusted so as not to affect the main adjustment. Refer to Figure E.28-9.

FIGURE E.28-9
Audio chain.

Mobile radios are usually set for a peak deviation of 5 kHz. A good service shop will have a deviation meter. If you don't have one, a fair job can be done by simply comparing a known good transmitter with the unit under test by listening to both with any receiver.

If a spectrum analyzer is available, recall Carson's rule and speak into the microphone while adjusting the bandwidth of the transmitted signal to about 16 kHz.

One can also use the zero carrier amplitude method. While viewing the transmitter output on the spectrum analyzer, apply a 2-kHz tone to the microphone input. Adjust the gain from zero up until the carrier is null and you have 5-kHz peak deviation.

Reactance Modulator Circuit Operation

The reactance modulator is efficient and provides a large deviation. It is popular and used often in FM transmitters. Figure E.28-10 illustrates a typical reactance modulator circuit. Refer to this figure throughout the following discussion. The circuit consists of the reactance circuit and the master oscillator. The reactance circuit operates on the master oscillator to cause its resonant frequency to shift up or shift down depending on the modulating signal being applied. The reactance circuit appears capacitive to the master oscillator. In this case, the reactance looks like a variable capacitor in the oscillator's tank circuit.

Transistor Q1 makes up the reactance modulator circuit. Resistors R2 and R3 establish a voltage divider network that biases Q1. Resistor R4 furnishes emitter feedback to

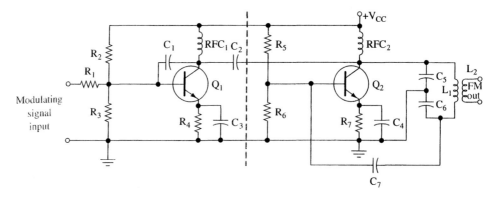

FIGURE E.28-10
Reactance modulator.

thermally stabilize Q1. Capacitor C3 is a bypass component that prevents AC input signal degeneration. Capacitor C1 interacts with transistor Q1's inter-electrode capacitance to cause a varying capacitive reactance directly influenced by the input modulating signal.

The master oscillator is a Colpitts oscillator built around transistor Q2. Coil L1, capacitor C5, and capacitor C6 make up the resonant tank circuit. Capacitor C7 provides the required regenerative feedback to cause the circuit to oscillate. Q1 and Q2 are impedance coupled, and capacitor C2 effectively couples the changes at Q1's collector to the tank circuit of transistor Q2 while blocking DC voltages.

When a modulating signal is applied to the base of transistor Q1 via resistor R1, the reactance of the transistor changes in relation to that signal. If the modulating voltage goes up the reactance of Q1 goes down, and if the modulating voltage goes down the reactance of Q1 goes up. This change in reactance is felt on Q1's collector and also at the tank circuit of the Colpitts oscillator transistor Q2. As capacitive reactance at Q1 goes up the resonant frequency of the master oscillator, Q2, decreases. Conversely, if Q1's capacitive reactance goes down the master oscillator resonant frequency increases.

Troubleshooting the Reactance Modulator

The FM output signal from coil L2 can be lost due to open bias resistors, open RF chokes RFC1 and RFC2, or an open winding at coil L1. In addition, a leaky coupling capacitor C2 may shift the collector voltage of Q2 of the master oscillator, causing a low FM output signal to be present at coil L2. Low collector voltages and weak transistors may produce a low FM signal output condition, as well as changes in bias resistors R2 and R3 in transistor Q1's circuit and resistors R5 and R6 in Q2's circuit. Changes in the emitter resistors of both transistor circuits will lessen the FM output and possibly shut the modulator down.

The master oscillator may operate without being influenced by the reactance circuit. The oscillator output signal at L2 would not be FM. This situation could occur if the modulating signal were missing from the base of transistor Q1. An open R1 would block the modulating signal from getting to the base of Q1. Without the modulating signal present, Q1's reactance would not change. A leaky or shorted C1 could also kill the reactance response of Q1. Transistor Q1 may still be operating perfectly but the reactance changes might not be passed to Q2's tank circuit due to an open C2.

Table E.28-1 is a symptom guide to help you troubleshoot the reactance modulator circuit.

TABLE E.28-1

Symptom	Problem	Probable Cause
No signal out from L2	No FM modulator output	Open bias resistors in Q1 and Q2 circuits; open RFC1 or RFC2; C2 open or leaky; feedback capacitor C7 open
No FM output at L2, master oscillator output only	No FM modulation taking place	Q1 not functioning, check C1, RFC1, and R4; resistor R1 may be open
Amplitude of FM output low	Low modulator output	Changes in bias resistor values, check R2, R3, R5, and R6; change in emitter resistors, check R4 and R7; Q2's gain has decreased

Check the resistors with your DMM for proper values. Resistors in the reactance modulator circuit will be precision types with close tolerances. For low FM signal outputs check capacitors C1, C2, C5, C6, and C7 with a capacitor checker.

E.29 Demonstrate an understanding of motor phase shift control circuits

INTRODUCTION

The speed of a motor can be controlled by varying the phase angle of its applied voltage. Figure E.29-1 shows three different AC waveforms. Each waveform represents one cycle of the voltage applied to a load (such as a motor or a light bulb). Typically, a TRIAC is used to provide the two-way switching required by an AC load. When the TRIAC is off, no voltage gets to the load. When the TRIAC is on, the available AC voltage is applied to the load. By varying the angle θ (called the *delay angle*), we can control the amount of the AC waveform that gets to the load. This in turn controls the speed of the motor (higher energy means faster speed) or brightness of the light bulb.

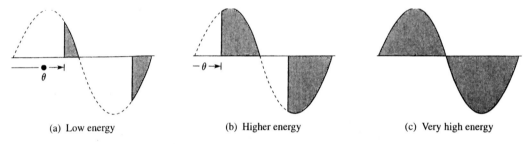

(a) Low energy   (b) Higher energy   (c) Very high energy

FIGURE E.29-1
Phase angle versus energy.

The delay angle represents the phase shift between the beginning of the cycle and the time when the TRIAC turns on. A delay angle of 10 degrees indicates that the TRIAC turns on very early in the cycle. Alternately, you may wish to think of the *conduction angle,* which would be 170 degrees. The conduction angle is proportional to the energy supplied to the load. An RC network is used to provide the associated delay angle. Varying the resistance of the RC circuit changes the delay angle, which in turn controls conduction and the power applied to the load. Figure E.29-2 shows a simple phase shift control circuit. Recall that the TRIAC will not conduct until its gate has sufficient voltage. The RC circuit,

FIGURE E.29-2
Motor speed control circuit.

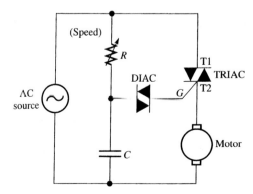

486

together with the DIAC, controls when the AC source is high enough to trigger the TRIAC and energize the load. When the AC source crosses zero, the TRIAC shuts off and the delay begins again for the next half cycle.

REVIEW QUESTIONS

True/False

1. The delay angle is proportional to the energy supplied to the load.

2. The delay angle controls the TRIAC's turn-on time.

Multiple Choice

3. The lower the delay angle,

    a. The lower the motor speed.

    b. The higher the motor speed.

    c. Neither a nor b.

4. The TRIAC shuts off when

    a. The delay angle is reached.

    b. The source voltage crosses zero.

    c. The RC network output is −3 dB.

5. The higher the value of R in Figure E.29-2,

    a. The lower the motor speed.

    b. The higher the motor speed.

    c. Neither a nor b.

E.30 Understand the principles and operations of microwave circuits

INTRODUCTION

Microwave circuits are concerned with very high frequencies of operation (1 GHz and above). In this section we examine the special types of antennas and other devices used for microwave operation.

MICROWAVE ANTENNAS

Microwave antennas actually use optical theory more than standard antenna theory. These antennas tend to be highly directive and therefore provide high gain as compared to the reference half-wavelength dipole. The reasons for this include the following:

1. Because of the short wavelengths involved, the physical sizes required are small enough to allow "peculiar" arrangements not practical at lower frequencies.

2. There is little need for omnidirectional patterns since no broadcasting takes place at these frequencies. Microwave communications are generally of a point-to-point nature.

3. Because of increased device noise at microwave frequencies, receivers require the highest possible input signal. Highly directional antennas (and thus high gain) make this possible.

4. Microwave transmitters are very limited in their output power due to the cost and/or availability problems of microwave power devices. This low output power is compensated for by a highly directional antenna system.

Microwaves are divided into bands as shown in Table E.30-1. The frequencies above 40 GHz are called **millimeter** (mm) **waves** because their wavelength is described in millimeters.

TABLE E.30-1
Microwave frequency designations.

Band	Frequency (GHz)
L	1–2
S	2–4
C	4–8
X	8–12
Ku	12–18
K	18–27
Ka	27–40

Horn Antenna

Open-ended sections of waveguides can be used as radiators of electromagnetic energy. The three basic forms of horn antennas are shown in Figure E.30-1. They all provide a gradual flare to the waveguide so as to allow maximum radiation and thus minimum reflection back into the guide.

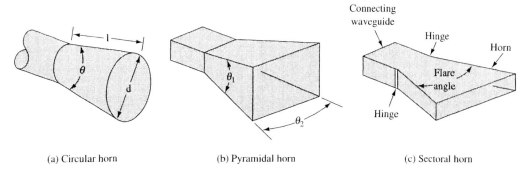

(a) Circular horn   (b) Pyramidal horn   (c) Sectoral horn

FIGURE E.30-1
Horn antennas.

Recall that a plain open-circuited waveguide theoretically reflects 100% of the incident energy. In practice, however, the open-circuited guide "launches" a fair amount of energy, while the short-circuited guide does provide the theoretical 100% reflection. By gradually flaring out the open circuit, the goal of total radiation is nearly attained. The flared end of the horn antenna acts as an impedance transformer between the waveguide and free space. For a proper transformation ratio, the linear dimension of each side must be at least a half-wavelength.

The **circular horn** in Figure E.30-1(a) provides efficient radiation from a circular waveguide. The flare angle θ and length l are important to the amount of gain it can provide. Generally, the greater the l, the greater the gain, while a flare angle of $\theta \approx 50°$ is optimum.

For the **pyramidal horn** in Figure E.30-1(b) there are two flare angles, θ_1 and θ_2, on which the radiation pattern depends. The effect of horn length is similar to that with the circular horn. Wider horizontal patterns are obtained by increasing θ_2, while wider vertical patterns are possible by increasing θ_1. When $\theta_1 = \theta_2$ a symmetrical radiation pattern is realized.

The **sectoral horn** in Figure E.30-1(c) has the top and bottom walls at a 0° flare angle. The side walls are sometimes hinged (as shown) to provide adjustable flare angles. Maximum radiation occurs for angles between 40° and 60°.

The horns just described provide a maximum gain on the order of 20 dB compared to the half-wavelength dipole reference. While they do not provide the amounts of gain of subsequently described microwave antennas, their simplicity and low cost make them popular for noncritical applications.

Parabolic Antenna

The ability of a paraboloid to focus light rays or sound waves at a point is common knowledge. Some common applications include dentists' lights, flashlights, and automobile headlamps. The same ability is applicable to electromagnetic waves of lower frequency than light as long as the paraboloid's mouth diameter is at least 10 wavelengths. This precludes their use at low radio frequencies but allows use at microwave frequencies.

There are various methods of feeding the **microwave dish,** as the paraboloid antenna is commonly called. Figure E.30-2(a) shows the dish being fed with a simple dipole/reflector combination at the paraboloid's focus. A horn-fed version is shown in Figure E.30-2(b). The **Cassegrain feed** in Figure E.30-2(c) is used to shorten the length of feed mechanism in highly critical applications. It uses a hyperboloid secondary reflector whose focus coincides with that of the parabola. Those transmitted rays obstructed by the hyperboloid are generally such a small percentage as to be negligible.

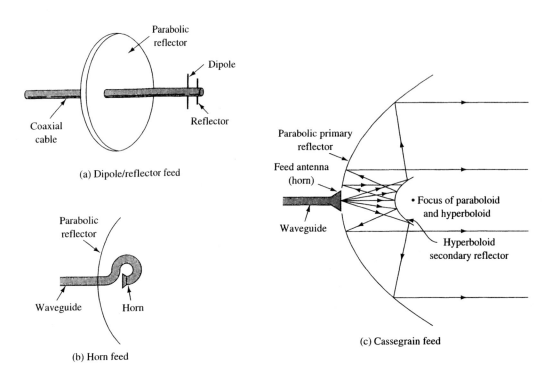

(a) Dipole/reflector feed

(b) Horn feed

(c) Cassegrain feed

FIGURE E.30-2
Microwave dish antennas.

These dish antennas perform equally well transmitting or receiving, as predicted by antenna reciprocity. They provide huge power gains, with a good approximation provided by the following equation:

$$A_p \simeq 6\left(\frac{D}{\lambda}\right)^2$$

where A_p = power gain with respect to a half-wavelength dipole
D = mouth diameter of primary reflector
λ = free-space wavelength of carrier frequency

An accurate approximation of the beamwidth in degrees between half-power points is

$$\text{beamwidth} \simeq \frac{70\lambda}{D}$$

EXAMPLE E.30-1

Calculate the power gain and beamwidth of a microwave dish antenna with a 3-m mouth diameter when used at 10 GHz.

Solution

$$A_p \simeq 6\left(\frac{D}{\lambda}\right)^2$$

$$\lambda = \frac{c}{f} = \frac{3 \times 10^8 \text{ m/s}}{10 \times 10^9} = 0.03 \text{ m}$$

$$A_p = 6 \times \left(\frac{3 \text{ m}}{0.03 \text{ m}}\right)^2 = 60,000 \quad (47.8 \text{ dB})$$

$$\text{beamwidth} \simeq \frac{70\lambda}{D}$$

$$= \frac{70 \times 0.03 \text{ m}}{3 \text{ m}} = 0.7°$$

Example E.30-1 shows the extremely high gain capabilities of these antennas. In this particular case, the dish with a 1-W output is equivalent to a half-wave dipole with 60,000-W output. This power gain is effective, however, only if the receiver is within the 0.7° beamwidth of the dish. Figure E.30-3 shows a **polar pattern** for this antenna. It is typical of parabolic antennas and shows the 47.8-dB gain at the 0° reference. Notice the three side lobes on each side of the main one. As you might expect from the antenna's physical construction, there can be no radiated energy from 90° to 270°.

FIGURE E.30-3
Polar pattern for parabolic antenna in Example E.30-1.

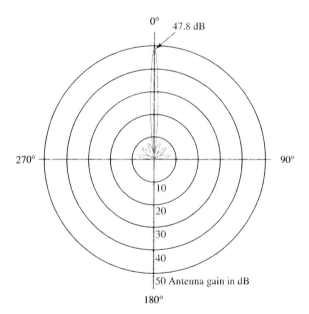

Microwave dish antennas are widely used in satellite communications because of their high gain; they are also used for satellite tracking and radio astronomy. They are also used at 30- to 50-mile intervals (point-to-point line-of-sight conditions) to carry telephone and broadcast TV and other signals throughout the world. Often these antennas have a "cover" over the dish. This is a low-loss dielectric material known as a **radome.** Its purpose may be maintenance of internal pressure or, more simply, environmental protection. The construction of a bird's nest within the dish is undesirable for the bird as well as the antenna user.

Microwave Tubes

Two types of tubes are used for microwave operation: the magnetron and the traveling wave tube.

☐ The magnetron has specially designed resonant cavities in its internal structure that control its frequency of oscillation. A permanent magnet is used to provide the stimulus for oscillation.

☐ The traveling wave tube operates as a high gain, low noise amplifier for microwave frequencies. Microwaves travel along a long single-wire helix enclosed within the tube, in the same direction and speed as the tube's electron beam. RF signals coupled to each end of the tube are amplified by 40 dB or more.

Solid-State Microwave Devices

Several devices are designed for use in microwave circuitry. They are the Gunn oscillator, IMPATT diode, P-I-N diode, and microwave transistor.

☐ The Gunn oscillator uses a DC-biased gallium-arsenide crystal to produce microwave frequencies.

☐ The IMPATT diode is similar to the varactor diode. IMPATT stands for impact ionization avalance transit time. IMPATT diodes operate heavily reverse-biased, in their breakdown region.

☐ P-I-N diodes contain a thin layer of intrinsic (lightly doped) silicon within their *pn* junction. This allows the P-I-N diode to act as a rectifier up to frequencies of 100 MHz. At higher frequencies, the amount of forward current controls the resistance of the P-I-N diode. This allows it to be used as a high-frequency switch.

☐ Bipolar microwave transistors are used for amplification of frequencies up to 5 GHz. The base, collector, and emitter leads are at right angles to each other, to prevent coupling. From 5 GHz to 20 GHz, microwave FETs are used, exhibiting lower noise than the bipolar microwave transistor.

REVIEW QUESTIONS

True/False

1. Microwave antennas operate the same as ordinary antennas.
2. A radome is a required component of a microwave antenna.

Multiple Choice

3. Which is not a microwave tube?
   a. Gunn tube.
   b. Magnetron.
   c. Traveling wave tube.
4. Which device is preferred for amplifying a 10 GHz signal?
   a. Bipolar microwave transistor.
   b. Microwave FET.
   c. Either may be used.
5. Which device uses a crystal to generate microwave frequencies?
   a. Gunn oscillator.
   b. IMPATT diode.
   c. P-I-N diode.

Digital Circuits

F.01 Demonstrate an understanding of the characteristics of integrated circuit (IC) logic families.

INTRODUCTION

In this section, we introduce the two most widely used types of digital integrated circuit families, TTL and CMOS, and look at several specific devices in these IC families. A third type of integrated circuit family, ECL, is also covered. This coverage is important because it gives you basic information on the specific devices you will be using in the lab and in industry.

TTL

The term **TTL** stands for *t*ransistor-*t*ransistor *l*ogic, which refers to the use of **bipolar** junction transistors in the circuit technology used to construct the gates at the chip level.

TTL consists of a series of logic circuits: standard TTL, low-power TTL, **Schottky** TTL, low-power Schottky TTL, advanced low-power Schottky TTL, and advanced Schottky TTL. The differences in these various types of TTL are in their performance characteristics, such as propagation delay times, power dissipation, and fan-out, which are explained later. The TTL family has a number prefix of 54 or 74, followed by a letter or letters that specify the series, as shown in Table F.01-1. The prefix 54 indicates an operating temperature range of $-55°C$ to $125°C$ (generally for military use). The prefix 74 indicates a temperature range of $0°C$ to $70°C$ (for commercial use). We will use the prefix 74 throughout the book. The term *quad* used in the table means "four individual gates per package."

TABLE F.01-1
TTL series designations.

TTL Series	Prefix Designation	Example of Device
Standard TTL	54 or 74 (no letter)	7400 (quad NAND gates)
Low-power TTL	54L or 74L	74L00 (quad NAND gates)
Schottky TTL	54S or 74S	74S00 (quad NAND gates)
Low-power Schottky TTL	54LS or 74LS	74LS00 (quad NAND gates)
Advanced low-power Schottky TTL	54ALS or 74ALS	74ALS00 (quad NAND gates)
Advanced Schottky TTL	54AS or 74AS	74AS00 (quad NAND gates)

CMOS

The term **CMOS** stands for *c*omplementary *m*etal-*o*xide *s*emiconductor. Whereas TTL uses bipolar transistors in its circuit technology, CMOS uses field-effect transistors. Logic functions are the same, however, whether the device is implemented with TTL or CMOS technologies. The circuit technologies make a difference, not in logic function, but only in performance characteristics.

Several series of CMOS logic circuits are available, but they fall basically into two process technology categories: metal-gate CMOS and silicon-gate CMOS. The older, metal-gate technology is the 4000 series. The newer, silicon-gate technology consists of the 74C, the 74HC, and the 74HCT. All of the 74 series CMOS devices are both pin compatible and function compatible with the TTL series. That is, a TTL IC and a CMOS IC of the same number have the inputs, outputs, supply voltage, and ground on the same pins, as well as the same logic gates. In addition, the 74HCT series is voltage-level compatible with TTL and requires no special interfacing as do the 74C and 74HC series. Other differences in the various types of 74 series CMOS are in their performance characteristics.

ECL

The term **ECL** stands for *e*mitter *c*oupled *l*ogic which is a bipolar circuit technology. ECL has the fastest switching speed of any logic family but its power consumption is much higher. The variety of devices that are available in ECL is very limited compared to TTL and CMOS; however, many complex functions and special-purpose circuits are available.

Performance Characteristics

Four important characteristics for the performance of logic circuits are propagation delay time, power dissipation, fan-out, and speed-power product.

Propagation Delay Time   The **propagation delay time** limits the switching speed (frequency) at which logic circuits can operate. The terms *low speed* and *high speed,* when applied to logic circuits, refer to the propagation delays; the shorter the propagation delay, the higher the speed of the circuit.

The propagation delay time of a gate is basically the time interval between the application of an input pulse and the occurrence of the resulting output pulse. There are two different propagation delay times associated with a logic gate:

1. t_{PHL}: The time between a specified reference point on the input pulse and a corresponding reference point on the output pulse, with the output changing from the HIGH level to the LOW level.

2. t_{PLH}: The time between a specified reference point on the input pulse and a corresponding reference point on the output pulse, with the output changing from the LOW level to the HIGH level.

EXAMPLE F.01-1

Show the propagation delay times of the inverter in Figure F.01-1(a).

FIGURE F.01-1

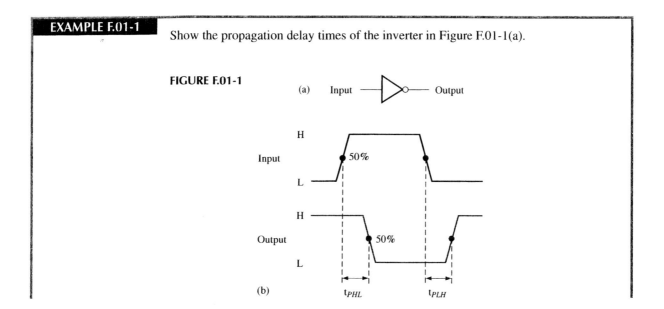

Solution The propagation delay times, t_{PHL} and t_{PLH}, are indicated in part (b) of the figure. In this case, the delays are measured between the 50% points of the corresponding edges of the input and output pulses. The values of t_{PHL} and t_{PLH} are not necessarily equal but in many cases they are the same.

Power Dissipation The **power dissipation** of a logic gate equals the DC supply voltage V_{CC} times the average supply current I_{CC}. Normally, the value of I_{CC} for a LOW gate output is higher than for a HIGH output. The manufacturer's data sheet usually specifies these values as I_{CCL} and I_{CCH}. The average I_{CC} is then determined, based on a 50% duty cycle operation of the gate (LOW half the time and HIGH half the time).

Fan-out The **fan-out** of a gate is the maximum number of inputs of the same series IC family that the gate can drive while maintaining its output levels within specified limits. That is, the fan-out specifies the maximum load that a given gate is capable of handling. For example, a standard TTL gate has a fan-out of 10 **unit loads.** This means that it can drive no more than 10 inputs of other standard TTL gates and still operate reliably. If the fan-out is exceeded, specified operation is not guaranteed. Figure F.01-2 shows a gate driving 10 other gates.

FIGURE F.01-2
The standard TTL NAND gate output fans out to a maximum of ten standard TTL gate inputs.

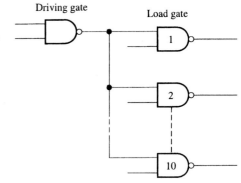

Speed-Power Product The **speed-power product** is sometimes specified by the manufacturer as a measure of the performance of a logic circuit. It is the product of the propagation delay time and the power dissipation at a specified frequency. The smaller the speed-power product is, the better the performance. The speed-power product is expressed as energy in joules, symbolized by J. For example, the speed-power product (SPP) of a 74HC CMOS gate at a frequency of 100 kHz is

$$\text{SPP} = (8 \text{ ns})(0.17 \text{ mW}) = 1.36 \text{ pJ}$$

Table F.01-2 provides a comparison of some performance characteristics of various CMOS, TTL, and ECL logic families

TABLE F.01-2
Comparison of typical performance characteristics of CMOS, TTL, and ECL logic gates.

Technology	CMOS* (silicon-gate)	CMOS* (metal-gate)	TTL Std.	TTL LS	TTL S	TTL ALS	TTL AS	ECL
Device series	74HC	4000B	74	74LS	74S	74ALS	74AS	10KH
Power dissipation: Static @ 100 kHz	2.5 nW 0.17 mW	1 μW 0.1 mW	10 mW 10 mW	2 mW 2 mW	19 mW 19 mW	1 mW 1 mW	8.5 mW 8.5 mW	25 mW 25 mW
Propagation delay time	8 ns	50 ns	10 ns	10 ns	3 ns	4 ns	1.5 ns	1 ns
Fan-out (same series)			10	20	20	20	40	

*Propagation delay is dependent on the DC supply voltage, V_{CC}. Power dissipation and fan-out are functions of frequency.

REVIEW QUESTIONS

True/False

1. An integrated circuit whose part number begins with 54xx is a commercial chip.

2. The fan-out of a TTL gate is based on voltage requirements.

Multiple Choice

3. Which logic family typically uses less power?
   a. CMOS.
   b. ECL.
   c. TTL.

4. Which logic family has the fastest switching speed?
   a. CMOS.
   b. ECL.
   c. TTL.

5. Two logic gates have propagation times of 5 ns and 10 ns. Which gate may operate at a higher frequency?
   a. The 5 ns gate.
   b. The 10 ns gate.
   c. Both gates operate at the same frequency.

F.02 Demonstrate an understanding of minimizing logic circuits using Boolean operations

INTRODUCTION

Many times in the application of Boolean algebra, you have to reduce a particular expression to its simplest form or change its form to a more convenient one to implement the expression most efficiently. The approach taken in this section is to use the basic laws, rules, and theorems of Boolean algebra to manipulate and simplify an expression. This method depends on a thorough knowledge of Boolean algebra and considerable practice in its application, not to mention a little ingenuity and cleverness.

A simplified Boolean expression uses the fewest gates possible to implement a given expression. Four examples follow to illustrate Boolean simplification step by step. The Boolean identities from Table F.02-1 will be used during each example.

TABLE F.02-1
Boolean identities.

1. $A + 0 = A$	**7.** $A \cdot A = A$
2. $A + 1 = 1$	**8.** $A \cdot \overline{A} = 0$
3. $A \cdot 0 = 0$	**9.** $\overline{\overline{A}} = A$
4. $A \cdot 1 = A$	**10.** $A + AB = A$
5. $A + A = A$	**11.** $A + \overline{A}B = A + B$
6. $A + \overline{A} = 1$	**12.** $(A + B)(A + C) = A + BC$

A, B, or *C* can represent a single variable or a combination of variables.

EXAMPLE F.02-1

Using Boolean algebra techniques, simplify this expression:

$$AB + A(B + C) + B(B + C)$$

Solution The following is not necessarily the only approach.

Step 1. Apply the distributive law to the second and third terms in the expression, as follows:

$$AB + AB + AC + BB + BC$$

Step 2. Apply rule 7 ($BB = B$) to the fourth term:

$$AB + AB + AC + B + BC$$

Step 3. Apply rule 5 ($AB + AB = AB$) to the first two terms:

$$AB + AC + B + BC$$

Step 4. Apply rule 10 ($B + BC = B$) to the last two terms:

$$AB + AC + B$$

Step 5. Apply rule 10 ($AB + B = B$) to the first and third terms:

$$B + AC$$

At this point the expression is simplified as much as possible. Once you gain experience in applying Boolean algebra, you can combine many individual steps.

Figure F.02-1 shows that the simplification process in Example F.02-1 has significantly reduced the number of logic gates required to implement the expression. Part (a) shows that five gates are required to implement the expression in its original form; only two gates are needed for the simplified expression, shown in part (b). It is important to realize that these two gate networks are equivalent. That is, for any combination of levels on the A, B, and C inputs, you get the same output from either circuit.

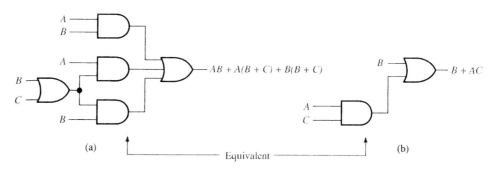

FIGURE F.02-1
Gate networks for Example F.02-1.

EXAMPLE F.02-2

Simplify the Boolean expression:

$$[A\overline{B}(C + BD) + \overline{A}\ \overline{B}]C$$

Note that brackets and parentheses mean the same thing: the term inside is multiplied (ANDed) with the term outside.

Solution

Step 1. Apply the distributive law to the terms within the brackets:

$$(A\overline{B}C + A\overline{B}BD + \overline{A}\ \overline{B})C$$

Step 2. Apply rule 8 ($\overline{B}B = 0$) to the second term within the parentheses:

$$(A\overline{B}C + A \cdot 0 \cdot D + \overline{A}\ \overline{B})C$$

Step 3. Apply rule 3 ($A \cdot 0 \cdot D = 0$) to the second term within the parentheses:

$$(A\overline{B}C + 0 + \overline{A}\ \overline{B})C$$

Step 4. Apply rule 1 (drop the 0) within the parentheses:

$$(A\overline{B}C + \overline{A}\ \overline{B})C$$

Step 5. Apply the distributive law:

$$A\overline{B}CC + \overline{A}\ \overline{B}C$$

Step 6. Apply rule 7 ($CC = C$) to the first term:

$$A\overline{B}C + \overline{A}\ \overline{B}C$$

Step 7.   Factor out $\overline{B}C$:

$$\overline{B}C(A + \overline{A})$$

Step 8.   Apply rule 6 $(A + \overline{A} = 1)$:

$$\overline{B}C \cdot 1$$

Step 9.   Apply rule 4 (drop the 1):

$$\overline{B}C$$

EXAMPLE F.02-3

Simplify the Boolean expression:

$$\overline{A}BC + A\overline{B}\,\overline{C} + \overline{A}\,\overline{B}\,\overline{C} + A\overline{B}C + ABC$$

Solution

Step 1.   Factor BC out of the first and last terms:

$$BC(\overline{A} + A) + A\overline{B}\,\overline{C} + \overline{A}\,\overline{B}\,\overline{C} + A\overline{B}C$$

Step 2.   Apply rule 6 $(\overline{A} + A = 1)$ to the term in parentheses, and factor $A\overline{B}$ from the second and last terms:

$$BC \cdot 1 + A\overline{B}(\overline{C} + C) + \overline{A}\,\overline{B}\,\overline{C}$$

Step 3.   Apply rule 4 (drop the 1) to the first term and rule 6 $(\overline{C} + C = 1)$ to the term in parentheses:

$$BC + A\overline{B} \cdot 1 + \overline{A}\,\overline{B}\,\overline{C}$$

Step 4.   Apply rule 4 (drop the 1) to the second term:

$$BC + A\overline{B} + \overline{A}\,\overline{B}\,\overline{C}$$

Step 5.   Factor $\overline{B}$ from the second and third terms:

$$BC + \overline{B}(A + \overline{A}\,\overline{C})$$

Step 6.   Apply rule 11 $(A + \overline{A}\,\overline{C} = A + \overline{C})$ to the term in parentheses:

$$BC + \overline{B}(A + \overline{C})$$

Step 7.   Use the distributive and commutative laws to get the following expression:

$$BC + A\overline{B} + \overline{B}\,\overline{C}$$

EXAMPLE F.02-4

Simplify the following Boolean expression:

$$\overline{AB + AC} + \overline{A}\,\overline{B}C$$

Solution

Step 1.   Apply DeMorgan's theorem to the first term:

$$(\overline{AB})(\overline{AC}) + \overline{A}\,\overline{B}C$$

Step 2.   Apply DeMorgan's theorem to each term in parentheses:

$$(\overline{A} + \overline{B})(\overline{A} + \overline{C}) + \overline{A}\,\overline{B}C$$

Step 3. Apply the distributive law to the two terms in parentheses:

$$\overline{A}\,\overline{A} + \overline{A}\,\overline{C} + \overline{A}\,\overline{B} + \overline{B}\,\overline{C} + \overline{A}\,BC$$

Step 4. Apply rule 7 ($\overline{A}\,\overline{A} = \overline{A}$) to the first term, and apply rule 10 [$\overline{A}\,\overline{B} + \overline{A}\,BC = \overline{A}\,\overline{B}\,(1 + C) = \overline{A}\,\overline{B}$] to the third and last terms:

$$\overline{A} + \overline{A}\,\overline{C} + \overline{A}\,\overline{B} + \overline{B}\,\overline{C}$$

Step 5. Apply rule 10 [$\overline{A} + \overline{A}\,\overline{C} = \overline{A}\,(1 + \overline{C}) = \overline{A}$] to the first and second terms:

$$\overline{A} + \overline{A}\,\overline{B} + \overline{B}\,\overline{C}$$

Step 6. Apply rule 10 [$\overline{A} + \overline{A}\,\overline{B} = \overline{A}\,(1 + \overline{B}) = \overline{A}$] to the first and second terms:

$$\overline{A} + \overline{B}\,\overline{C}$$

REVIEW QUESTIONS

True/False

1. Factoring a Boolean expression reduces the logic required.

2. The terms $AB + ABC$ cannot be simplified.

Multiple Choice

3. What is the simplified expression for $F = AB + A\overline{B} + C$?

   a. $F = AB + C$.

   b. $F = A + C$.

   c. The expression cannot be simplified.

4. How many times may a Boolean identity be used while reducing an equation?

   a. Once.

   b. Once for every input variable.

   c. As many times as necessary.

5. In addition to Boolean identities, what else is used to simplify an expression?

   a. Factoring.

   b. DeMorgan's theorem.

   c. Both a and b.

F.03 Understand principles and operations of linear integrated circuits

INTRODUCTION

In this section we examine the operation of integrated circuit voltage regulators.

INTEGRATED CIRCUIT VOLTAGE REGULATORS

Fixed Positive Linear Voltage Regulators

Although many types of IC regulators are available, the 7800 series of IC regulators is representative of three-terminal devices that provide a fixed positive output voltage. The three terminals are input, output, and ground as indicated in the standard fixed voltage configuration in Figure F.03-1(a). The last two digits in the part number designate the output

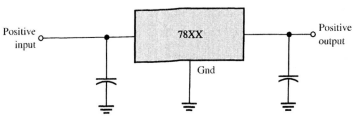

(a) Standard configuration

Type number	Output voltage
7805	+5.0 V
7806	+6.0 V
7808	+8.0 V
7809	+9.0 V
7812	+12.0 V
7815	+15.0 V
7818	+18.0 V
7824	+24.0 V

(b) The 7800 series

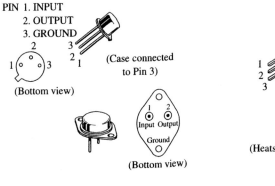

PIN 1. INPUT
2. OUTPUT
3. GROUND

(Case connected to Pin 3)

(Bottom view)

(Bottom view)

Pins 1 and 2 electrically isolated from case. Case is third electrical connection.

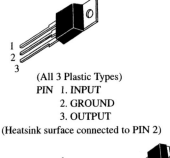

(All 3 Plastic Types)
PIN 1. INPUT
2. GROUND
3. OUTPUT
(Heatsink surface connected to PIN 2)

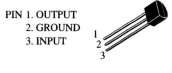

PIN 1. OUTPUT
2. GROUND
3. INPUT

PIN 1. V_{OUT} 5. NC
2. GND   6. GND
3. GND   7. GND
4. NC   8. V_{IN}

(c) Typical metal and plastic packages

FIGURE F.03-1
The 7800 series three-terminal fixed positive voltage regulators.

voltage. For example, the 7805 is a +5.0 V regulator. Other available output voltages are given in part (b) and common packages are shown in part (c).

Capacitors, although not always necessary, are sometimes used on the input and output as indicated in Figure F.03-1(a). The output capacitor acts basically as a line filter to improve transient response. The input capacitor is used to prevent unwanted oscillations when the regulator is some distance from the power supply filter such that the line has a significant inductance.

The 7800 series can produce output current in excess of 1 A when used with an adequate heat sink. The 78L00 series can provide up to 100 mA, the 78M00 series can provide up to 500 mA, and the 78T00 series can provide in excess of 3 A.

The input voltage must be at least 2 V above the output voltage in order to maintain regulation. The circuits have internal thermal overload protection and short-circuit current-limiting features. **Thermal overload** occurs when the internal power dissipation becomes excessive and the temperature of the device exceeds a certain value.

Fixed Negative Linear Voltage Regulators

The 7900 series is typical of three-terminal IC regulators that provide a fixed negative output voltage. This series is the negative-voltage counterpart of the 7800 series and shares most of the same features and characteristics. Figure F.03-2 indicates the standard configuration and part numbers with corresponding output voltages that are available.

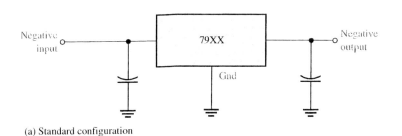

Type number	Output voltage
7905	–5.0 V
7905.2	–5.2 V
7906	–6.0 V
7908	–8.0 V
7912	–12.0 V
7915	–15.0 V
7918	–18.0 V
7924	–24.0 V

(a) Standard configuration

(b) The 7900 series

FIGURE F.03-2
The 7900 series three-terminal fixed negative voltage regulators.

Adjustable Positive Linear Voltage Regulators

The LM317 is an excellent example of a three-terminal positive regulator with an adjustable output voltage. The standard configuration is shown in Figure F.03-3. Input and output capacitors, although not shown, are often used for the reasons discussed previously. Notice that there is an input, an output, and an adjustment terminal. The external fixed resistor R_1 and the external variable resistor R_2 provide the output voltage adjustment. V_{OUT} can be varied from 1.2 V to 37 V depending on the resistor values. The LM317 can provide over 1.5 A of output current to a load.

The LM317 is operated as a "floating" regulator because the adjustment terminal is not connected to ground, but floats to whatever voltage is across R_2. This allows the output voltage to be much higher than that of a fixed-voltage regulator.

Basic Operation As indicated in Figure F.03-4, a constant 1.2 V reference voltage (V_{REF}) is maintained by the regulator between the output terminal and the adjustment terminal. This constant reference voltage produces a constant current (I_{REF}) through R_1, regardless of the value of R_2. I_{REF} also flows through R_2.

$$I_{REF} = \frac{V_{REF}}{R_1} = \frac{1.25 \text{ V}}{R_1}$$

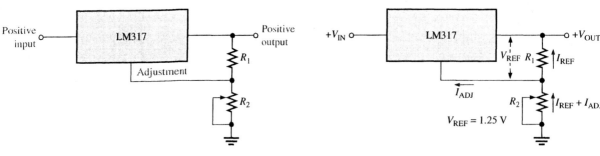

FIGURE F.03-3
The LM317 three-terminal adjustable positive voltage regulator.

FIGURE F.03-4
Operation of the LM317 adjustable voltage regulator.

Also, there is a very small constant current into the adjustment terminal of approximately 50 μA called I_{ADJ}, which flows through R_2. An expression for the output voltage is developed as follows.

$$V_{OUT} = V_{R1} + V_{R2} = I_{REF}R_1 + I_{REF}R_2 + I_{ADJ}R_2$$

$$= I_{REF}(R_1 + R_2) + I_{ADJ}R_2 = \frac{V_{REF}}{R_1}(R_1 + R_2) + I_{ADJ}R_2$$

$$V_{OUT} = V_{REF}\left(1 + \frac{R_2}{R_1}\right) + I_{ADJ}R_2$$

As you can see, the output voltage is a function of both R_1 and R_2. Once the value of R_1 is set, the output voltage is adjusted by varying R_2.

EXAMPLE F.03-1

Determine the minimum and maximum output voltages for the voltage regulator in Figure F.03-5. Assume $I_{ADJ} = 50\ \mu$A.

FIGURE F.03-5

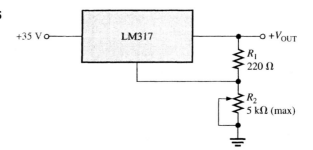

Solution $V_{R1} = V_{REF} = 1.25$ V

When R_2 is set at its maximum of 5 kΩ,

$$V_{OUT} = V_{REF}\left(1 + \frac{R_2}{R_1}\right) + I_{ADJ}R_2 = 1.25\ V\left(1 + \frac{5\ k\Omega}{220\ \Omega}\right) + (50\ \mu A)5\ k\Omega$$

$$= 29.66\ V + 0.25\ V = 29.9\ V$$

When R_2 is set at its minimum of 0 Ω,

$$V_{OUT} = V_{REF}\left(1 + \frac{R_2}{R_1}\right) + I_{ADJ}R_2 = 1.25\ V(1) = 1.25\ V$$

Adjustable Negative Linear Voltage Regulators

The LM337 is the negative output counterpart of the LM317 and is a good example of this type of IC regulator. Like the LM317, the LM337 requires two external resistors for output voltage adjustment as shown in Figure F.03-6. The output voltage can be adjusted from -1.2 V to -37 V, depending on the external resistor values.

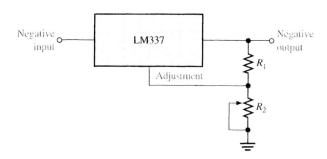

FIGURE F.03-6
The LM337 three-terminal adjustable negative voltage regulator.

REVIEW QUESTIONS

True/False

1. The 78xx series regulators are positive regulators.
2. Voltage regulators provide a constant output voltage under varying load conditions.

Multiple Choice

3. What is the output voltage of the 7805 regulator?
    a. 5 volts.
    b. 12 volts.
    c. 24 volts.
4. The heat sink of a three-terminal regulator is connected internally to
    a. Pin 1 (input).
    b. Pin 2 (ground).
    c. Pin 3 (output).
5. The 78xx and 79xx series regulators are
    a. AC regulators.
    b. DC regulators.
    c. Switching regulators.

F.04 Troubleshoot and repair linear integrated circuits

INTRODUCTION

Some of the typical faults found in op-amp circuits are discussed in this section.

Faults in the Noninverting Amplifier

The first thing to do when you suspect a faulty circuit is to check for the proper supply voltage and ground. Having done that, several other possible faults are as follows.

Open Feedback Resistor If the feedback resistor, R_f, in Figure F.04-1 opens, the op-amp is operating with its very high open-loop gain, which causes the input signal to drive the device into nonlinear operation and results in a severely clipped output signal as shown in part (a).

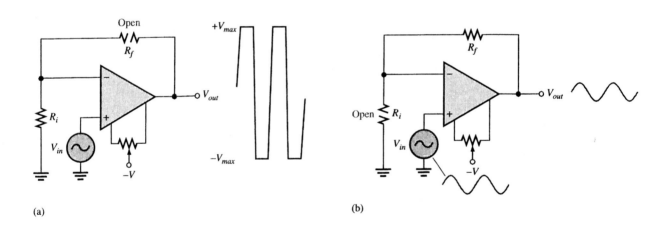

(a)   (b)

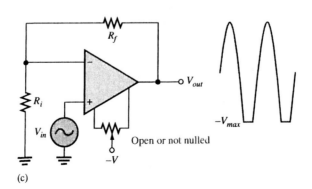

(c)

FIGURE F.04-1
Faults in the noninverting amplifier.

Open Input Resistor In this case, you still have a closed-loop configuration. But, since R_i is open and effectively equal to infinity, the closed-loop gain is

$$A_{cl(\text{NI})} = 1 + \frac{R_f}{R_i} = 1 + \frac{R_f}{\infty} = 1 + 0 = 1$$

This shows that the amplifier acts like a voltage-follower. You would observe an output signal that is the same as the input, as indicated in Figure F.04-1(b).

Open or Incorrectly Adjusted Offset Null Potentiometer In this situation, the output offset voltage will cause the output signal to begin clipping on only one peak as the input signal is increased to a sufficient amplitude. This is indicated in Figure F.04-1(c).

Faulty Op-Amp As mentioned, many things can happen to an op-amp. In general, an internal failure will result in a loss or distortion of the output signal. The best approach is to first make sure that there are no external failures or faulty conditions. If everything else is good, then the op-amp must be bad.

Faults in the Voltage-Follower

The voltage-follower is a special case of the noninverting amplifier. Except for a bad op-amp, a bad external connection, or a problem with the offset null potentiometer, about the only thing that can happen in a voltage-follower circuit is an open feedback loop. This would have the same effect as an open feedback resistor as previously discussed.

Faults in the Inverting Amplifier

Open Feedback Resistor If R_f opens as indicated in Figure F.04-2(a), the input signal still feeds through the input resistor and is amplified by the high open-loop gain of the op-amp. This forces the device to be driven into nonlinear operation, and you will see an output something like that shown. This is the same result as in the noninverting configuration.

Open Input Resistor This prevents the input signal from getting to the op-amp input, so there will be no output signal, as indicated in Figure F.04-2(b).

    Failures in the op-amp itself or the offset null potentiometer have the same effects as previously discussed for the noninverting amplifier.

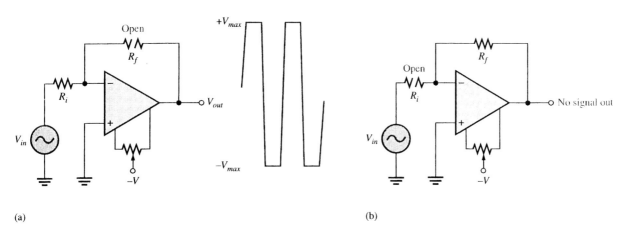

(a)   (b)

FIGURE F.04-2
Faults in the inverting amplifier.

F.05 Understand principles and operations of types of logic gates

INTRODUCTION

Logic gates are the building blocks of digital circuits. In this section we will examine the most basic logic gates, their operation, and their application. Table F.05-1 shows the basic logic gates and their associated information.

TABLE F.05-1
Basic logic functions.

Logical Function	Symbol	Boolean Equation	Truth Table
Inverter (NOT)	A —▷o— F	$F = \overline{A}$	$A \mid F$ $0 \mid 1$ $1 \mid 0$
AND	A, B —⊐— F	$F = AB$	$A\ B \mid F$ $0\ 0 \mid 0$ $0\ 1 \mid 0$ $1\ 0 \mid 0$ $1\ 1 \mid 1$
OR	A, B —⊐— F	$F = A + B$	$A\ B \mid F$ $0\ 0 \mid 0$ $0\ 1 \mid 1$ $1\ 0 \mid 1$ $1\ 1 \mid 1$
NAND	A, B —⊐o— F	$F = \overline{AB}$	$A\ B \mid F$ $0\ 0 \mid 1$ $0\ 1 \mid 1$ $1\ 0 \mid 1$ $1\ 1 \mid 0$
NOR	A, B —⊐o— F	$F = \overline{A + B}$	$A\ B \mid F$ $0\ 0 \mid 1$ $0\ 1 \mid 0$ $1\ 0 \mid 0$ $1\ 1 \mid 0$
Exclusive OR (XOR)	A, B —⊐— F	$F = A \oplus B$	$A\ B \mid F$ $0\ 0 \mid 0$ $0\ 1 \mid 1$ $1\ 0 \mid 1$ $1\ 1 \mid 0$
Exclusive NOR (XNOR)	A, B —⊐o— F	$F = \overline{A \oplus B}$	$A\ B \mid F$ $0\ 0 \mid 1$ $0\ 1 \mid 0$ $1\ 0 \mid 0$ $1\ 1 \mid 1$

EXAMPLE F.05-1

A waveform is applied to an inverter in Figure F.05-1. Determine the output waveform corresponding to the input and sketch the timing diagram. According to the placement of the bubble, what is the active output state?

FIGURE F.05-1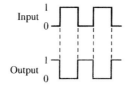

Solution   The output waveform is exactly opposite to the input (inverted), as shown in Figure F.05-2, which is the basic timing diagram. The active output state is 0.

FIGURE F.05-2

EXAMPLE F.05-2

For the 3-input AND gate in Figure F.05-3, determine the output waveform in relation to the inputs.

FIGURE F.05-3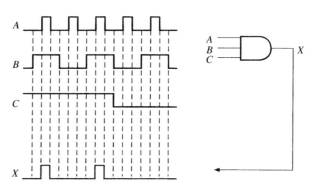

Solution   The output waveform *X* of the 3-input AND gate is HIGH only when all three inputs *A, B,* and *C* are HIGH.

EXAMPLE F.05-3

For the two input waveforms, *A* and *B*, in Figure F.05-4, sketch the output waveform, showing its proper relation to the inputs.

FIGURE F.05-4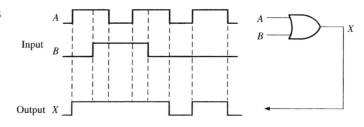

Solution When either or both inputs are HIGH, the output is HIGH as shown by the output waveform X in the timing diagram.

EXAMPLE F.05-4

For the 3-input OR gate in Figure F.05-5, determine the output waveform in proper time relation to the inputs.

FIGURE F.05-5

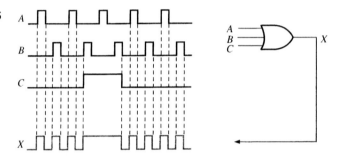

Solution The output is HIGH when one or more of the inputs are HIGH as indicated by the output waveform X in the timing diagram.

REVIEW QUESTIONS

True/False

1. An inverter's input and output may both be low at the same time.

2. A truth table shows all input and output combinations for a logic gate.

Multiple Choice

3. A gate that only outputs a one when all its inputs are high is the
   a. Inverter.
   b. AND gate.
   c. OR gate.

4. Which gate has the following Boolean expression: $X = \overline{A + B}$?
   a. NAND.
   b. OR.
   c. NOR.

5. Which gate outputs a one only when its inputs are different?
   a. AND.
   b. OR.
   c. XOR.

F.06 Fabricate and demonstrate types of logic gates

INTRODUCTION

Using actual components or a software simulation package (such as *Electronics Workbench*), set up each circuit in Figure F.06-1 and determine the associated truth table.

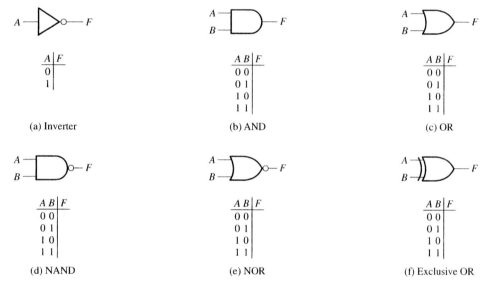

(a) Inverter	(b) AND	(c) OR
(d) NAND	(e) NOR	(f) Exclusive OR

FIGURE F.06-1
Test circuits.

F.07 Troubleshoot and repair types of logic gates

INTRODUCTION

Opens and shorts are the most common types of internal gate failures. These can occur on the inputs or on the output of a gate inside the IC package.

Effects of an Internally Open Input   An internal open is the result of an open component on the chip or a break in the tiny wire connecting the IC chip to the package pin. An open input prevents a pulser signal on that input from getting to the output of the gate, as illustrated in Figure F.07-1(a) for the case of a 2-input NAND gate. An open TTL input acts effectively as a HIGH level so pulses applied to the good input get through to the NAND gate output as shown in Figure F.07-1(b).

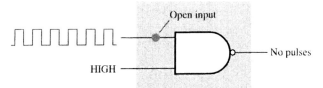

(a) Pulsing the open input will produce no pulses on the output.

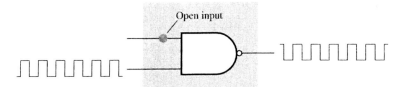

(b) Pulsing the good input will produce output pulses for TTL NAND and AND gates because an open input acts as a HIGH. It is uncertain for CMOS.

FIGURE F.07-1
The effect of an open input on a NAND gate.

Conditions for Testing Gates   When testing a NAND gate or an AND gate, always make sure that the inputs that are not being pulsed are HIGH to enable the gate. When checking a NOR gate or an OR gate, always make sure that the inputs that are not being pulsed are LOW. When checking an XOR or XNOR gate, the level of the nonpulsed input does not matter because the pulses on the other input will force the inputs to alternate between the same level and opposite levels.

Troubleshooting an Open Input   Troubleshooting this type of failure is most easily accomplished with a logic pulser and probe, as demonstrated in Figure F.07-2 for the case of a 2-input NAND gate package.

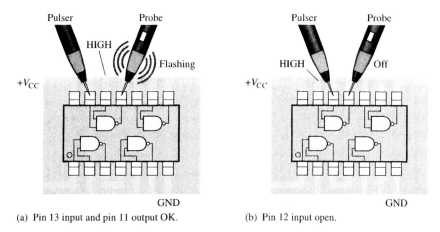

(a) Pin 13 input and pin 11 output OK.   (b) Pin 12 input open.

FIGURE F.07-2
Troubleshooting a NAND gate for an open input.

The first step in troubleshooting an IC suspected of being faulty is to make sure that the DC supply voltage (V_{CC}) and ground are at the appropriate pins of the IC. Next, using a logic pulser to apply continuous pulses to one of the inputs to the gate, make sure that the other inputs are HIGH (in the case of a NAND gate). In Figure F.07-2(a), start by pulsing pin 13, which has been determined to be one of the inputs to the suspected gate. If pulse activity is indicated on the output (pin 11 in this case) by a flashing probe, then the pin 13 input is not open. By the way, this also proves that the output is not open. Next, pulse the other gate input (pin 12). The probe lamp is off, indicating that there are no pulses on the output at pin 11 and that the output is LOW, as shown in Figure F.07-2(b). Notice that the input not being pulsed must be HIGH for the case of a NAND gate or AND gate. If this were a NOR gate, the input not being pulsed would have to be LOW.

Effects of an Internally Open Output   An internally open gate output prevents a signal on any of the inputs from getting to the output. Therefore, no matter what the input conditions are, the output is unaffected. The level at the output pin of the IC will depend upon what it is externally connected to. It could be either HIGH, LOW, or floating (not fixed to any reference). In any case, a logic probe at the output pin will not be flashing.

Troubleshooting an Open Output   Figure F.07-3 illustrates troubleshooting an open NOR gate output with a pulser and probe. In part (a), one of the inputs of the suspected gate (pin 11 in this case) is pulsed, and the probe on the output (pin 13) indicates no pulse

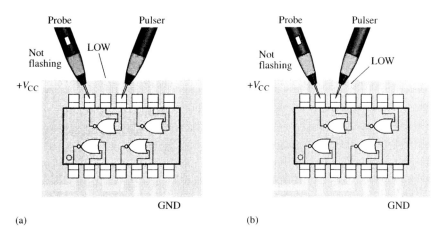

(a)   (b)

FIGURE F.07-3
Troubleshooting a NOR gate for an open output.

activity. In part (b), the other input (pin 12) is pulsed and again there is no indication of pulses on the output. Under the condition that the input that is not being pulsed is at a LOW level, this test shows that the output is internally open.

Shorted Input or Output   Although not as common as an open, an internal short to the DC supply voltage, the ground, another input, or an output can occur. When an input or output is shorted to the supply voltage, it will be stuck in the HIGH state. If an input or output is shorted to the ground, it will be stuck in the LOW state (0 V). If two inputs or an input and an output are shorted together, they will always be at the same level.

External Opens and Shorts

Many failures involving digital ICs are due to faults that are external to the IC package. These include bad solder connections, solder splashes, wire clippings, improperly etched printed circuit (PC) boards, and cracks or breaks in wires or printed circuit interconnections. These open or shorted conditions have the same effect on the logic gate as the internal faults, and troubleshooting is done in basically the same ways. A visual inspection of any circuit that is suspected of being faulty is the first thing a technician should do.

EXAMPLE F.07-1

You are checking a 7410 triple 3-input NAND gate IC that is one of many ICs located on a printed circuit board. You have checked pins 1 and 2 with your logic probe, and they are both HIGH. Now your logic pulser is placed on pin 13, and your logic probe is placed first on pin 12 and then on the connecting PC board trace as indicated in Figure F.07-4. Based on the responses of the probe, what is the most likely problem?

FIGURE F.07-4

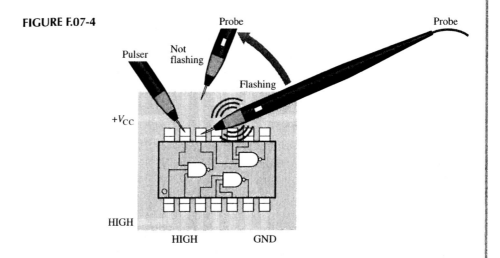

Solution   The flashing indicator on the probe shows that there is pulse activity on the gate output at pin 12 but no activity on the PC board trace. The gate is working properly, but the signal is not getting from pin 12 of the IC to the PC board trace.

Most likely there is a bad solder connection between pin 12 of the IC and the PC board, which is creating an open. You should resolder that point and check it again.

In most cases, you will be troubleshooting ICs that are mounted on printed circuit boards or prototype assemblies and interconnected with other ICs. As you progress through this section, you will learn how different types of digital ICs are used together to perform system functions. At this point, however, we are concentrating on individual IC gates.

This limitation does not prevent us from looking at the system concept at a very basic and simplified level, as we have already done several times.

EXAMPLE F.07-2

After trying to operate the frequency counter shown in Figure F.07-5, you find that it constantly reads out all 0s on its display, regardless of the input frequency. Determine the cause of this malfunction.

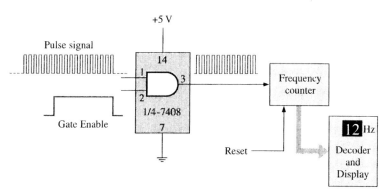

(a) This is how the counter should be working with a 12 Hz input signal.

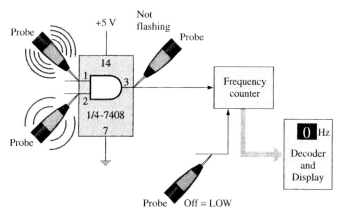

(b) The display is showing a frequency of 0 Hz and the logic probe indications are shown.

FIGURE F.07-5

Solution Here are three possible causes:

1. A constant active or asserted level on the counter reset input, which keeps the counter at zero.

2. No pulse signal on the input to the counter because of an internal open or short in the counter. This problem would keep the counter from advancing after being reset to zero.

3. No pulse signal on the input to the counter because of an open AND gate output or the absence of input signals, again keeping the counter from advancing from zero.

Figure F.07-5(a) gives an example of how the frequency counter should be working with a 12 Hz pulse waveform on the input to the AND gate. Part (b) shows that the display is improperly indicating 0 Hz.

The first step is to make sure that V_{CC} and ground are connected to all the right places; assume that they are found to be OK. Next, check for pulses on both inputs to the AND gate. The probe indicates that there is pulse activity on both of these inputs. A probe check of the counter reset shows a LOW level which is known to be the unasserted level (assume this information was found on the data sheet for the counter) and, therefore, this is not the problem. The next probe check on pin 3 of the 7408 shows that there is no pulse activity on the output of the AND gate, indicating that the gate output is internally open. Replace the 7408 IC and check the operation again.

EXAMPLE F.07-3

The frequency counter shown in Figure F.07-6 appears to measure the frequency of input signals incorrectly. It is found that when a signal with a precisely known frequency is applied to pin 1 of the AND gate, the display indicates a higher frequency. Determine what is wrong.

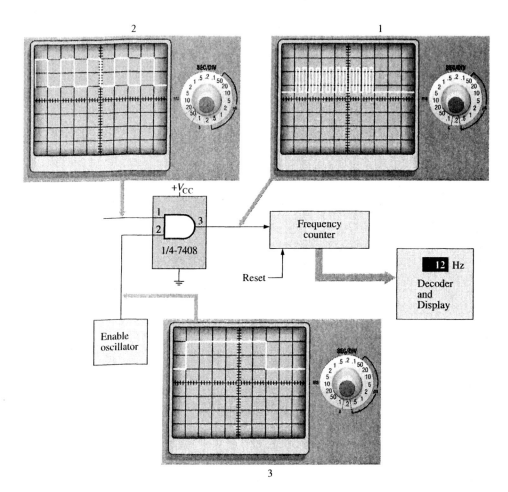

FIGURE F.07-6

Solution   By design, the input pulses are allowed to pass through the AND gate for exactly 1 s. The number of pulses counted in 1 s is equal to the frequency in hertz (cycles per second). Therefore, the 1 s interval, which is produced by the Enable pulse on pin 2 of the AND gate, is very critical to an accurate frequency measurement. The Enable pulses are produced internally by a precision oscillator circuit. The pulse must be exactly 1 s in width and in this case it occurs every 3 s to update the count. Just prior to each Enable pulse, the counter is reset to zero so that it starts a new count each time.

Since the counter appears to be counting more pulses than it should to produce a frequency readout that is too high, the Enable pulse is the primary suspect. Exact time-interval measurements must be made, so an oscilloscope is used instead of a logic probe in this situation. The logic probe indicates only the presence of pulses; it does not provide for frequency or time measurement.

An input pulse waveform of exactly 10 Hz is applied to pin 1 of the AND gate and the display incorrectly shows 12 Hz. The first scope measurement, on the output of the AND gate, shows that there are 12 pulses for each Enable pulse. In the second scope measurement, the input frequency is verified to be precisely 10 Hz (period = 100 ms). In the third scope measurement, the width of the Enable pulse is found to be 1.2 s rather than 1 s.

The conclusion is that the oscillator circuit that produces the Enable pulse is out of calibration for some reason and must be repaired or replaced.

F.08 Understand principles and operations of combinational logic circuits

INTRODUCTION

Combinational logic circuits are the backbone of the digital design process. In this section we examine the operation and reduction of several combinational logic circuits.

EXAMPLE F.08-1

Design a logic circuit to implement the operation specified in the truth table of Table F.08-1.

TABLE F.08-1

Inputs			Output
A	*B*	*C*	*X*
0	0	0	0
0	0	1	0
0	1	0	0
0	1	1	1
1	0	0	0
1	0	1	1
1	1	0	1
1	1	1	0

Solution   Notice that $X = 1$ for only three of the input conditions. Therefore, the logic expression is

$$X = \overline{A}BC + A\overline{B}C + AB\overline{C}$$

The logic gates required are three inverters, three 3-input AND gates and one 3-input OR gate. The logic circuit is shown in Figure F.08-1.

FIGURE F.08-1

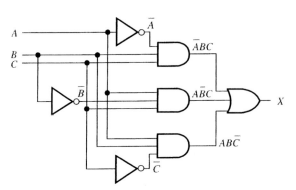

EXAMPLE F.08-2

Develop a logic circuit with four input variables that will only produce a 1 output when three input variables are 1s.

Solution   Out of sixteen possible combinations of four variables, the combinations in which there are three 1s are listed in Table F.08-2, along with the corresponding product term for each.

TABLE F.08-2

A	B	C	D	
0	1	1	1	$\longrightarrow \overline{A}BCD$
1	0	1	1	$\longrightarrow A\overline{B}CD$
1	1	0	1	$\longrightarrow AB\overline{C}D$
1	1	1	0	$\longrightarrow ABC\overline{D}$

The product terms are ORed to get the following expression:

$$X = \overline{A}BCD + A\overline{B}CD + AB\overline{C}D + ABC\overline{D}$$

This expression is implemented in Figure F.08-2 with AND-OR logic.

FIGURE F.08-2

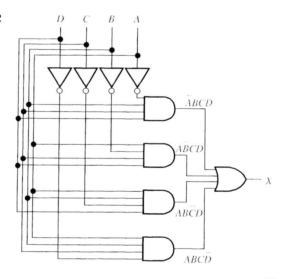

EXAMPLE F.08-3

Reduce the combinational logic circuit in Figure F.08-3 to a minimum form.

FIGURE F.08-3

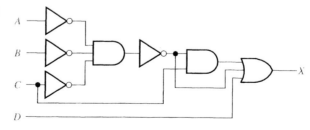

Solution   The expression for the output of the circuit is

$$X = (\overline{\overline{A}\ \overline{B}\ \overline{C}})C + \overline{\overline{A}\ \overline{B}\ \overline{C}} + D$$

Applying DeMorgan's theorem and Boolean algebra,

$$X = (\overline{\overline{A}} + \overline{\overline{B}} + \overline{\overline{C}})C + \overline{\overline{A}} + \overline{\overline{B}} + \overline{\overline{C}} + D$$
$$= AC + BC + CC + A + B + C + D$$
$$= AC + BC + C + A + B + \cancel{C} + D$$
$$= C(A + B + 1) + A + B + D$$
$$X = A + B + C + D$$

The simplified circuit is a 4-input OR gate as shown in Figure F.08-4.

FIGURE F.08-4

Minimize the combinational logic circuit in Figure F.08-5. Inverters are not shown.

FIGURE F.08-5

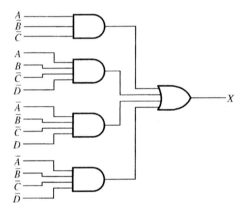

Solution   The output expression is

$$X = A\overline{B}\,\overline{C} + AB\overline{C}\,\overline{D} + \overline{A}\,\overline{B}\,CD + \overline{A}\,B\,\overline{C}\,D$$

Expanding the first term to include the missing variables D and $\overline{D}$,

$$X = A\overline{B}\,\overline{C}(D + \overline{D}) + AB\overline{C}\,\overline{D} + \overline{A}\,\overline{B}\,CD + \overline{A}\,B\,\overline{C}\,D$$
$$= A\overline{B}\,\overline{C}D + A\overline{B}\,\overline{C}\,\overline{D} + AB\overline{C}\,\overline{D} + \overline{A}\,\overline{B}\,CD + \overline{A}\,B\,\overline{C}\,D$$

This expanded SOP expression is mapped and simplified on the Karnaugh map in Figure F.08-6(a). The simplified implementation is shown in part (b). Inverters are not shown.

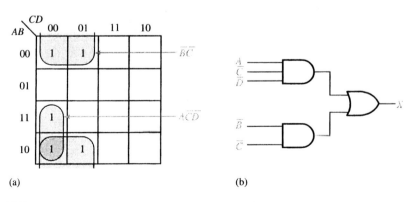

(a)   (b)

FIGURE F.08-6

REVIEW QUESTIONS

True/False

1. Combinational logic circuits are implemented from Boolean equations.

2. Propagation time may be ignored in a combinational logic circuit.

Multiple Choice

3. Assuming each gate has 5 nS of delay, what is the total delay time from input to output for a circuit implementing the equation $F = A + B(C + D)$?

   a. 5 nS.

   b. 10 nS.

   c. 15 nS.

4. Repeat Question 3 for this expression: $F = A + BC + BD$.

   a. 5 nS.

   b. 10 nS.

   c. 15 nS.

5. How many inverters are required while implementing the equation $F = \overline{A}BCD + \overline{AB}CD + \overline{ABC}D + AB\overline{CD}$?

   a. 2.

   b. 4.

   c. 7.

F.09 Fabricate and demonstrate combinational logic circuits

INTRODUCTION

Use actual components or software simulation (such as *Electronics Workbench*) to set up and examine the operation of each combinational logic circuit shown in Figure F.09-1. Determine the truth table outputs for each circuit, and also determine the worst case delay between input and output.

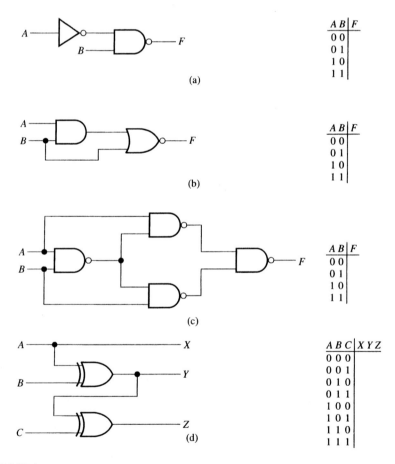

A B	F
0 0	
0 1	
1 0	
1 1	

(a)

A B	F
0 0	
0 1	
1 0	
1 1	

(b)

A B	F
0 0	
0 1	
1 0	
1 1	

(c)

A B C	X Y Z
0 0 0	
0 0 1	
0 1 0	
0 1 1	
1 0 0	
1 0 1	
1 1 0	
1 1 1	

(d)

FIGURE F.09-1
Test circuits.

F.10 Troubleshoot and repair combinational logic circuits

INTRODUCTION

In this section, the logic pulser, the probe, and the current tracer are used to troubleshoot a logic circuit when a gate output is connected to several gate inputs. Also, an example of signal tracing and waveform analysis methods is presented using an oscilloscope for locating a fault in a combinational logic circuit.

In a combinational logic circuit, the output of one gate may be connected to two or more gate inputs as shown in Figure F.10-1. The interconnecting paths share a common electrical point known as a **node.**

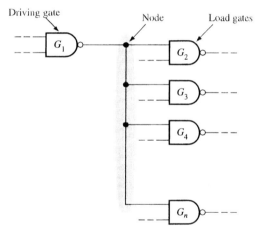

FIGURE F.10-1
Illustration of a node in a logic circuit.

Gate G_1 in Figure F.10-1 is driving the node, and the other gates represent loads connected to the node. A driving gate can drive a number of load gate inputs up to its specified fan-out. Several types of failures are possible in this situation. Some of these failure modes are difficult to isolate to a single bad gate because all the gates connected to the node are affected. Common types of failures are the following:

1. *Open output in driving gate.* This failure will cause a loss of signal to all load gates.

2. *Open input in a load gate.* This failure will not affect the operation of any of the other gates connected to the node, but it will result in loss of signal output from the faulty gate.

3. *Shorted output in driving gate.* This failure can cause the node to be stuck in the LOW state.

4. *Shorted input in a load gate.* This failure can also cause the node to be stuck in the LOW state.

Some approaches to troubleshooting each of the faults just listed are presented next.

Open Output in Driving Gate

In this situation there is no pulse activity on the node. With circuit power on, an open node will normally result in a "floating" level and may be indicated by a dim lamp on the logic probe, as illustrated in Figure F.10-2.

Open Input in a Load Gate

If the check for an open driver output is negative, then a check for an open input in a load gate should be performed. Apply the logic pulser tip to the node with all nonpulsed inputs HIGH. Then check the output of each gate for pulse activity with the logic probe, as illustrated in Figure F.10-3. If one of the inputs that is normally connected to the node is open, no pulses will be detected on that gate's output.

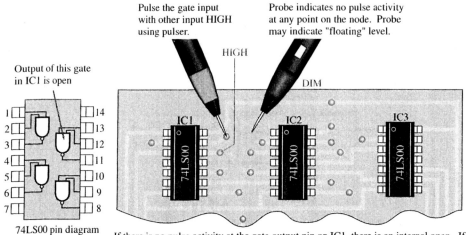

FIGURE F.10-2
Open output in driving gate.

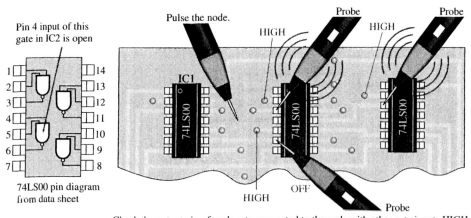

FIGURE F.10-3
Open input in a load gate.

Shorted Output in Driving Gate

This fault can cause the node to be stuck LOW, as previously mentioned. A quick check with a pulser and a logic probe will indicate this, as shown in Figure F.10-4(a). A short to ground in the driving gate's output or in any load gate input will cause this symptom, and further checks must therefore be made to isolate the short to a particular gate.

If the driving gate's output is internally shorted to ground, then, essentially, no current activity will be present on any of the connections to the node. Thus, a current tracer will indicate no activity with circuit power on, as illustrated in Figure F.10-4(b).

To further verify a shorted output, a pulser and a current tracer can be used with the circuit power off, as shown in Figure F.10-4(c). When current pulses are applied to the node with the pulser, all of the current will be into the shorted output, and none through the circuit paths into the load gate inputs.

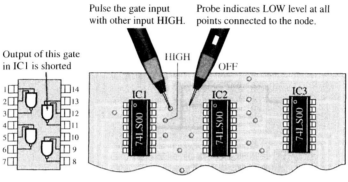

(a) Node is "stuck" LOW.

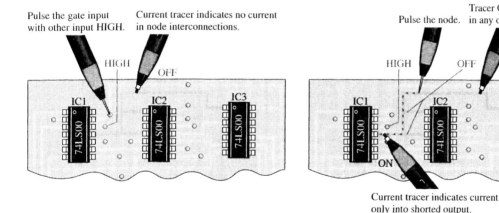

(b) No current in any node interconnection

(c) There is current from point of pulser application directly to short. There is no current in other paths connected to node.

FIGURE F.10-4
Shorted output in driving gate.

Shorted Input in a Load Gate

If one of the load gate inputs is internally shorted to ground, the node will be stuck in the LOW state. Again, as in the case of a shorted output, the logic pulser and current tracer can be used to isolate the faulty gate.

When the node is pulsed with circuit power off, essentially all the current will be into the shorted input, and tracing its path with the current tracer will lead to the shorted input, as illustrated in Figure F.10-5.

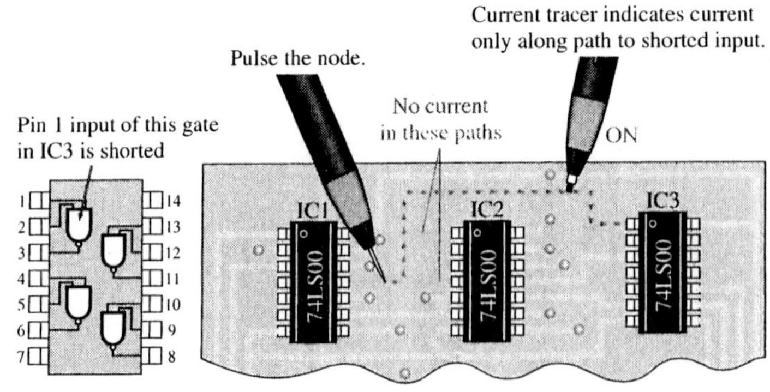

FIGURE F.10-5
Shorted input in a load gate.

Signal Tracing and Waveform Analysis

Although the methods of isolating an open or a short at a node point are very useful from time to time, a more general troubleshooting technique called **signal tracing** is of great value to the technician or technologist in just about every troubleshooting situation. Waveform measurement is accomplished with an oscilloscope or a logic analyzer.

Basically, the signal tracing method requires that you observe the waveforms and their time relationships at all accessible points in the logic circuit. You can begin at the inputs and, from an analysis of the waveform timing diagram for each point, determine where an incorrect waveform first occurs. With this procedure you can usually isolate the fault to a specific gate. A procedure beginning at the output and working back toward the inputs can also be used.

The general procedure for signal tracing starting at the inputs is outlined as follows:

☐ Within a system, define the section of logic that is suspected of being faulty.

☐ Start at the inputs to the section of logic under examination. We assume, for this discussion, that the input waveforms coming from other sections of the system have been found to be correct.

☐ For each gate, beginning at the input and working toward the output of the logic network, observe the output waveform of the gate and compare it with the input waveforms by using the oscilloscope or the logic analyzer.

☐ Determine if the output waveform is correct, using your knowledge of the logical operation of the gate.

☐ If the output is incorrect, the gate under test may be faulty. Pull the IC containing the gate that is suspected of being faulty, and test it out-of-circuit. If the gate is found to be faulty, replace the IC. If it works correctly, the fault is in the external circuitry or in another IC to which the tested one is connected.

☐ If the output is correct, go to the next gate. Continue checking each gate until an incorrect waveform is observed.

Figure F.10-6 is an example that illustrates the general procedure for a specific logic circuit in the following steps:

Step 1. Observe the output of gate G_1 relative to the inputs. If it is correct, check the inverter next. If the output is not correct, the gate or its connections are bad; or, if the output is LOW, the input to gate G_2 may be shorted.

Step 2. Observe the output of the inverter relative to the input. If it is correct, check gate G_2 next. If the output is not correct, the inverter or its connections are bad; or, if the output is LOW, the input to gate G_3 may be shorted.

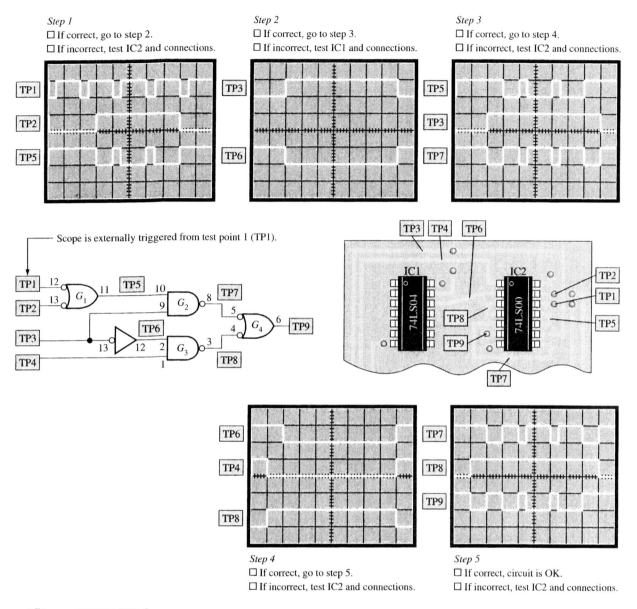

FIGURE F.10-6

Example of signal tracing and waveform analysis in a portion of a digital system.

Step 3. Observe the output of gate G_2 relative to the inputs. If it is correct, check gate G_3 next. If the output is not correct, the gate or its connections are bad; or, if the output is LOW, the input to gate G_4 may be shorted.

Step 4. Observe the output of gate G_3 relative to the inputs. If it is correct, check gate G_4 next. If the output is not correct, the gate or its connections are bad; or, if the output is LOW, the input to gate G_4 may be shorted.

Step 5. Observe the output of gate G_4 relative to the inputs. If it is correct, the circuit is OK. If the output is not correct, the gate or its connections are bad.

EXAMPLE F.10-1

Determine the fault in the TTL logic network of Figure F.10-7(a) by using waveform analysis. You have observed the waveforms shown in Figure F.10-7(b). Note that the waveforms for G_3 and G_4 contain both expected and actual levels.

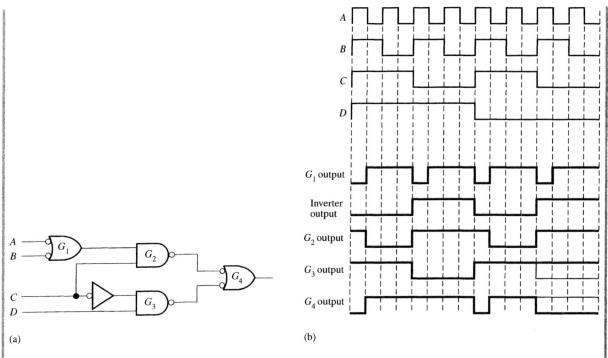

(a)   (b)

FIGURE F.10-7

Solution

1. Determine what the correct waveform should be for each gate.

2. Compare waveforms gate by gate until you find a measured waveform that does not match the correct waveform.

   In this example, everything tested is correct until gate G_3. The output of this gate is not correct as the differences in the waveforms indicate. An analysis of the waveforms indicates that if the D input to gate G_3 is open and acting as a HIGH, you will get the output waveform indicated in Figure F.10-7. Notice that the output of G_4 is correct for the inputs measured, although the input from G_3 is incorrect.

   Replace the IC containing G_3, and check the circuit's operation again.

F.11 Understand principles and operations of types of flip-flop circuits

INTRODUCTION

A **flip-flop** is a synchronous bistable device. An **edge-triggered flip-flop** changes state either at the positive edge (rising edge) or at the negative edge (falling edge) of the clock pulse and is sensitive to its inputs only at this transition of the clock. Three types of edge-triggered flip-flops are covered in this section: S-R, D, and J-K. The logic symbols for all of these are shown in Figure F.11-1. Notice that each type can be either positive edge-triggered (no bubble at C input) or negative edge-triggered (bubble at C input). The key to identifying an edge-triggered flip-flop by its logic symbol is the small triangle inside the block at the clock (C) input. This triangle is called the *dynamic input indicator.*

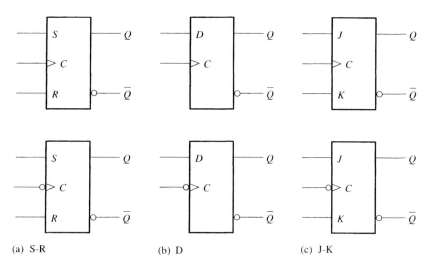

(a) S-R   (b) D   (c) J-K

FIGURE F.11-1
Edge-triggered flip-flop logic symbols (top—positive edge-triggered; bottom—negative edge-triggered).

The Edge-Triggered S-R Flip-Flop

The S and R inputs of the **S-R flip-flop** are called **synchronous** inputs because data on these inputs are transferred to the flip-flop's output only on the triggering edge of the clock pulse. When S is HIGH and R is LOW, the Q output goes HIGH on the triggering edge of the clock pulse, and the flip-flop is SET. When S is LOW and R is HIGH, the Q output goes LOW on the triggering edge of the clock pulse, and the flip-flop is RESET. When both S and R are LOW, the output does not change from its prior state. An invalid condition exists when both S and R are HIGH.

This basic operation of a positive edge-triggered flip-flop is illustrated in Figure F.11-2, and Table F.11-1 is the truth table for this type of flip-flop. Remember, *the flip-flop cannot change state except on the triggering edge of a clock pulse.* The S and R inputs can be changed at any time when the clock input is LOW or HIGH (except for a very short interval around the triggering transition of the clock) without affecting the output.

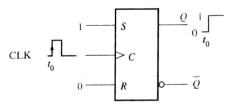
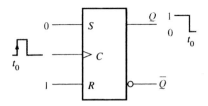

(a) $S = 1$, $R = 0$ flip-flop SETS on positive clock edge. (If already SET, it remains SET.)

(b) $S = 0$, $R = 1$ flip-flop RESETS on positive clock edge. (If already RESET, it remains RESET.)

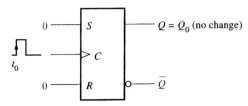

(c) $S = 0$, $R = 0$ flip-flop does not change. (If SET, it remains SET; if RESET, it remains RESET.)

FIGURE F.11-2
Operation of a positive edge-triggered S-R flip-flop.

TABLE F.11-1
Truth table for a positive edge-triggered S-R flip-flop.

Inputs			Outputs		
S	R	CLK	Q	$\overline{Q}$	Comments
0	0	X	Q_0	$\overline{Q}_0$	No change
0	1	↑	0	1	RESET
1	0	↑	1	0	SET
1	1	↑	?	?	Invalid

↑ = clock transition LOW to HIGH

X = irrelevant ("don't care")

Q_0 = output level prior to clock transition

The operation and truth table for a negative edge-triggered S-R flip-flop are the same as those for a positive edge-triggered device except that the falling edge of the clock pulse is the triggering edge.

EXAMPLE F.11-1

Determine the Q and $\overline{Q}$ output waveforms of the flip-flop in Figure F.11-3 for the S, R, and CLK inputs in Figure F.11-4(a). Assume that the positive edge-triggered flip-flop is initially RESET.

FIGURE F.11-3

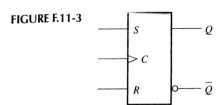

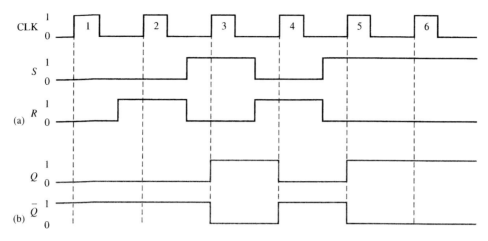

FIGURE F.11-4

Solution

1. At clock pulse 1, S is LOW and R is LOW, so Q does not change.

2. At clock pulse 2, S is LOW and R is HIGH, so Q remains LOW (RESET).

3. At clock pulse 3, S is HIGH and R is LOW, so Q goes HIGH (SET).

4. At clock pulse 4, S is LOW and R is HIGH, so Q goes LOW (RESET).

5. At clock pulse 5, S is HIGH and R is LOW, so Q goes HIGH (SET).

6. At clock pulse 6, S is HIGH and R is LOW, so Q stays HIGH.

Once Q is determined, $\overline{Q}$ is easily found since it is simply the complement of Q. The resulting waveforms for Q and $\overline{Q}$ are shown in Figure F.11-4(b) for the input waveforms in part (a).

A Method of Edge-Triggering

A simplified implementation of an edge-triggered S-R flip-flop is illustrated in Figure F.11-5(a) and is used to demonstrate the concept of edge-triggering. This coverage of the S-R flip-flop does not imply that it is the most important type. Actually, the D flip-flop and

FIGURE F.11-5
Edge triggering.

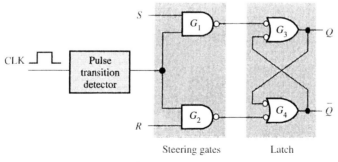

(a) A simplified logic diagram for a positive edge-triggered S-R flip-flop

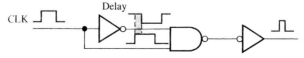

(b) A type of pulse transition detector

the J-K flip-flop are more widely used and more available in IC form than is the S-R type. However, understanding the S-R is important because both the D and the J-K flip-flops are derived from the S-R flip-flop. Notice that the S-R flip-flop differs from the gated S-R latch only in that it has a pulse transition detector. This circuit produces a very short-duration spike on the positive-going transition of the clock pulse.

One basic type of pulse transition detector is shown in Figure F.11-5(b). As you can see, there is a small delay on one input to the NAND gate so that the inverted clock pulse arrives at the gate input a few nanoseconds after the true clock pulse. This produces an output spike with a duration of only a few nanoseconds. In a negative edge-triggered flip-flop the clock pulse is inverted first, thus producing a narrow spike on the negative-going edge.

Notice that the circuit in Figure F.11-5 is partitioned into two sections, one labeled Steering gates, and the other Latch. The steering gates direct, or steer, the clock spike either to the input to gate G_3 or to the input to gate G_4, depending on the state of the S and R inputs. To understand the operation of this flip-flop, begin with the assumptions that it is in the RESET state ($Q = 0$) and that the S, R, and CLK inputs are all LOW. For this condition, the outputs of gate G_1 and gate G_2 are both HIGH. The LOW on the Q output is coupled back into one input of gate G_4, making the $\overline{Q}$ output HIGH. Because $\overline{Q}$ is HIGH, both inputs to gate G_3 are HIGH (remember, the output of gate G_1 is HIGH), holding the Q output LOW. If a pulse is applied to the CLK input, the outputs of gates G_1 and G_2 remain HIGH because they are disabled by the LOWs on the S input and the R input; therefore, there is no change in the state of the flip-flop—it remains RESET.

Let's now make S HIGH, leave R LOW, and apply a clock pulse. Because the S input to gate G_1 is now HIGH, the output of gate G_1 goes LOW for a very short time (spike) when CLK goes HIGH, causing the Q output to go HIGH. Both inputs to gate G_4 are now HIGH (remember, gate G_2 output is HIGH because R is LOW), forcing the $\overline{Q}$ output LOW. This LOW on $\overline{Q}$ is coupled back into one input of gate G_3, ensuring that the Q output will remain HIGH. The flip-flop is now in the SET state. Figure F.11-6 illustrates the logic level transitions that take place within the flip-flop for this condition.

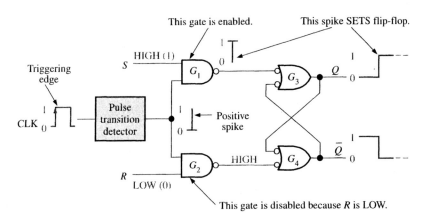

FIGURE F.11-6
Flip-flop making a transition from the RESET state to the SET state on the positive-going edge of the clock pulse.

Next, let's make S LOW and R HIGH and apply a clock pulse. Because the R input is now HIGH, the positive-going edge of the clock produces a negative-going spike on the output of gate G_2, causing the $\overline{Q}$ output to go HIGH. Because of this HIGH on $\overline{Q}$, both inputs to gate G_3 are now HIGH (remember, the output of gate G_1 is HIGH because of the LOW on S), forcing the Q output to go LOW. This LOW on Q is coupled back into one input of gate G_4, ensuring that $\overline{Q}$ will remain HIGH. The flip-flop is now in the RESET state. Figure F.11-7 illustrates the logic level transitions that occur within the flip-flop for this

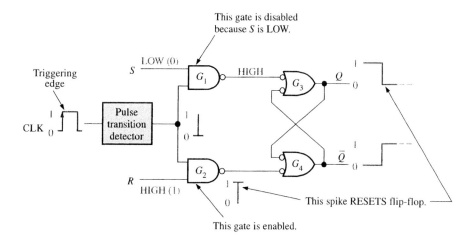

FIGURE F.11-7

Flip-flop making a transition from the SET state to the RESET state on the positive-going edge of the clock pulse.

condition. As with the gated latch, an invalid condition exists when both S and R are HIGH at the same time. This is the major drawback of the S-R flip-flop.

The Edge-Triggered D Flip-Flop

The **D flip-flop** is useful when a single data bit (1 or 0) is to be stored. The addition of an inverter to an S-R flip-flop creates a basic D flip-flop, as in Figure F.11-8, where a positive edge-triggered type is shown.

Notice that the flip-flop in Figure F.11-8 has only one input, the D input, in addition to the clock. If there is a HIGH on the D input when a clock pulse is applied, the flip-flop will SET, and the HIGH on the D input is stored by the flip-flop on the positive-going edge of the clock pulse. If there is a LOW on the D input when the clock pulse is applied, the flip-flop will RESET, and the LOW on the D input is stored by the flip-flop on the leading edge of the clock pulse. In the SET state the flip-flop is storing a 1, and in the RESET state it is storing a 0.

The operation of the positive edge-triggered D flip-flop is summarized in Table F.11-2. The operation of a negative edge-triggered device is, of course, the same, except that triggering occurs on the falling edge of the clock pulse. *Remember, Q follows D at the clock edge.*

FIGURE F.11-8

A positive edge-triggered D flip-flop formed with an S-R flip-flop and an inverter.

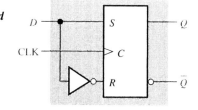

TABLE F.11-2

Truth table for a positive edge-triggered D flip-flop.

Inputs		Outputs		
D	CLK	Q	$\overline{Q}$	Comments
1	↑	1	0	SET (stores a 1)
0	↑	0	1	RESET (stores a 0)

↑ = clock transition LOW to HIGH

EXAMPLE F.11-2

Given the waveforms in Figure F.11-9(a) for the *D* input and the clock, determine the *Q* output waveform if the flip-flop starts out RESET.

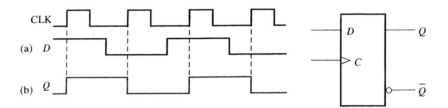

FIGURE F.11-9

Solution The *Q* output goes to the state of the *D* input at the time of the positive-going clock edge. The resultant output is shown in Figure F.11-9(b).

The Edge-Triggered J-K Flip-Flop

The **J-K flip-flop** is versatile and is perhaps the most widely used type of flip-flop. The *J* and *K* designations for the inputs have no known significance except that they are adjacent letters in the alphabet.

The functioning of the J-K flip-flop is identical to that of the S-R flip-flop in the SET, RESET, and no-change conditions of operation. *The difference is that the J-K flip-flop has no invalid state as does the S-R flip-flop.*

Figure F.11-10 shows the basic internal logic for a positive edge-triggered J-K flip-flop. Notice that it differs from the S-R edge-triggered flip-flop in that the *Q* output is connected back to the input of gate G_2, and the $\overline{Q}$ output is connected back to the input of gate G_1. The two inputs are labeled *J* and *K*. A J-K flip-flop can also be of the negative edge-triggered type, in which case the clock input is inverted.

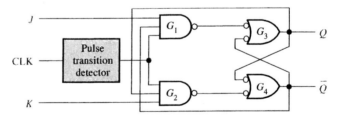

FIGURE F.11-10
A simplified logic diagram for a positive edge-triggered J-K flip-flop.

Let's assume that the flip-flop in Figure F.11-11 is RESET and that the *J* input is HIGH and the *K* input is LOW rather than as shown. When a clock pulse occurs, a leading-edge spike indicated by ① is passed through gate G_1 because $\overline{Q}$ is HIGH and *J* is HIGH. This will cause the latch portion of the flip-flop to change to the SET state.

The flip-flop is now SET. If we now make *J* LOW and *K* HIGH, the next clock spike indicated by ② will pass through gate G_2 because *Q* is HIGH and *K* is HIGH. This will cause the latch portion of the flip-flop to change to the RESET state.

Now if a LOW is applied to both the *J* and *K* inputs, the flip-flop will stay in its present state when a clock pulse occurs. So, a LOW on both *J* and *K* results in a *no-change* condition.

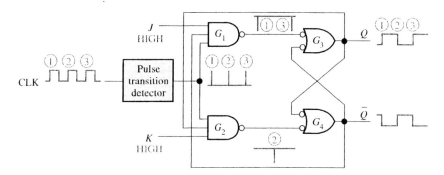

FIGURE F.11-11
Transitions illustrating the toggle operation when J = 1 and K = 1.

So far, the logical operation of the J-K flip-flop is the same as that of the S-R type in the SET, RESET, and no-change modes. The difference in operation occurs when both the *J* and *K* inputs are HIGH. To see this, assume that the flip-flop is RESET. The HIGH on the $\overline{Q}$ enables gate G_1, so the clock spike indicated by ③ passes through to SET the flip-flop. Now there is a HIGH on *Q*, which allows the next clock spike to pass through gate G_2 and RESET the flip-flop.

As you can see, on each successive clock spike, the flip-flop changes to the opposite state. This mode is called **toggle** operation. Figure F.11-11 illustrates the transitions when the flip-flop is in the toggle mode.

Table F.11-3 summarizes the operation of the edge-triggered J-K flip-flop in truth table form. *Notice that there is no invalid state as there is with an S-R flip-flop.* The truth table for a negative edge-triggered device is identical except that it is triggered on the falling edge of the clock pulse.

TABLE F.11-3
Truth table for a positive edge-triggered J-K flip-flop.

	Inputs		Outputs		
J	*K*	**CLK**	*Q*	*Q*	**Comments**
0	0	↑	Q_0	$\overline{Q}_0$	No change
0	1	↑	0	1	RESET
1	0	↑	1	0	SET
1	1	↑	$\overline{Q}_0$	Q_0	Toggle

↑ = clock transition LOW to HIGH

Q_0 = output level prior to clock transition

EXAMPLE F.11-3

The waveforms in Figure F.11-12(a) are applied to the *J*, *K*, and clock inputs as indicated. Determine the *Q* output, assuming that the flip-flop is initially RESET.

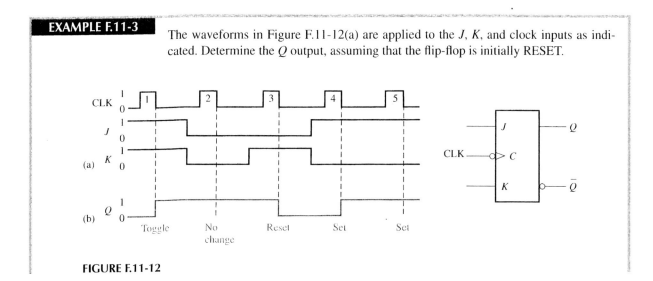

FIGURE F.11-12

Solution

1. First, since this is a negative edge-triggered flip-flop, as indicated by the "bubble" at the clock input, the Q output will change only on the negative-going edge of the clock pulse.

2. At the first clock pulse, both J and K are HIGH; and because this is a toggle condition, Q goes HIGH.

3. At clock pulse 2, a no-change condition exists on the inputs, keeping Q at a HIGH level.

4. When clock pulse 3 occurs, J is LOW and K is HIGH, resulting in a RESET condition; Q goes LOW.

5. At clock pulse 4, J is HIGH and K is LOW, resulting in a SET condition; Q goes HIGH.

6. A SET condition still exists on J and K when clock pulse 5 occurs, so Q will remain HIGH.

The resulting Q waveform is indicated in Figure F.11-12(b).

Asynchronous Inputs

For the flip-flops just discussed, the S-R, D, and J-K inputs are called *synchronous inputs* because data on these inputs are transferred to the flip-flop's output only on the triggering edge of the clock pulse; that is, the data are transferred synchronously with the clock.

Most integrated circuit flip-flops also have **asynchronous** inputs. These are inputs that affect the state of the flip-flop independent of the clock. They are normally labeled **preset** (*PRE*) and **clear** (*CLR*), or *direct set* (S_D) and *direct reset* (R_D) by some manufacturers. An active level on the preset input will SET the flip-flop, and an active level on the clear input will RESET it. A logic symbol for a J-K flip-flop with preset and clear inputs is shown in Figure F.11-13. These inputs are active-LOW, as indicated by the bubbles. These preset and clear inputs must both be kept HIGH for synchronous operation.

Figure F.11-14 shows the logic diagram for an edge-triggered J-K flip-flop with active-LOW preset ($\overline{PRE}$) and clear ($\overline{CLR}$) inputs. This figure illustrates basically how

FIGURE F.11-13

Logic symbol for a J-K flip-flop with active-LOW preset and clear inputs.

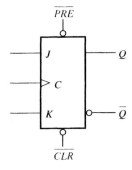

FIGURE F.11-14

Logic diagram for a basic J-K flip-flop with active-LOW preset and clear.

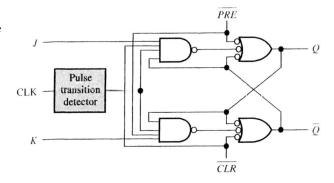

these inputs work. As you can see, they are connected so that they override the effect of the synchronous inputs, *J*, *K*, and the clock.

EXAMPLE F.11-4

For the positive edge-triggered J-K flip-flop with preset and clear inputs in Figure F.11-15(a), determine the *Q* output for the inputs shown in the timing diagram if *Q* is initially LOW.

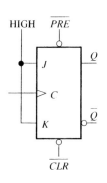

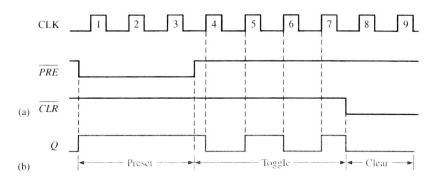

FIGURE F.11-15

Solution

1. During clock pulses 1, 2, and 3, the preset ($\overline{PRE}$) is LOW, keeping the flip-flop SET regardless of the synchronous *J* and *K* inputs.

2. For clock pulses 4, 5, 6, and 7, toggle operation occurs because *J* is HIGH, *K* is HIGH, and both $\overline{PRE}$ and $\overline{CLR}$ are HIGH.

3. For clock pulses 8 and 9, the clear ($\overline{CLR}$) input is LOW, keeping the flip-flop RESET regardless of the synchronous inputs. The resulting *Q* output is shown in Figure F.11-15(b).

FLIP-FLOP OPERATING CHARACTERISTICS

Propagation Delay Times

A **propagation delay time** is the interval of time required after an input signal has been applied for the resulting output change to occur. Several categories of propagation delay are important in the operation of a flip-flop:

1. Propagation delay t_{PLH} as measured from the triggering edge of the clock pulse to the LOW-to-HIGH transition of the output. This delay is illustrated in Figure F.11-16(a).

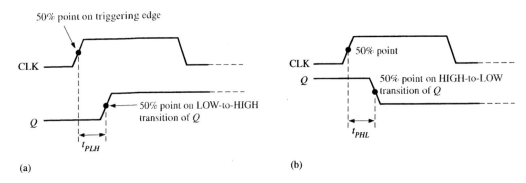

FIGURE F.11-16
Propagation delays, clock to output.

2. Propagation delay t_{PHL} as measured from the triggering edge of the clock pulse to the HIGH-to-LOW transition of the output. This delay is illustrated in Figure F.11-16(b).

3. Propagation delay t_{PLH} as measured from the preset input to the LOW-to-HIGH transition of the output. This delay is illustrated in Figure F.11-17(a) for an active-LOW preset input.

4. Propagation delay t_{PHL} as measured from the clear input to the HIGH-to-LOW transition of the output. This delay is illustrated in Figure F.11-17(b) for an active-LOW clear input.

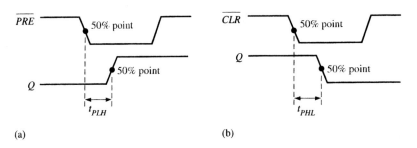

FIGURE F.11-17
Propagation delays, preset input to output and clear input to output.

Set-up Time

The **set-up time** (t_s) is the minimum interval required for the logic levels to be maintained constantly on the inputs (*J* and *K*, or *S* and *R*, or *D*) prior to the triggering edge of the clock pulse in order for the levels to be reliably clocked into the flip-flop. This interval is illustrated in Figure F.11-18 for a D flip-flop.

Hold Time

The **hold time** (t_h) is the minimum interval required for the logic levels to remain on the inputs after the triggering edge of the clock pulse in order for the levels to be reliably clocked into the flip-flop. This is illustrated in Figure F.11-19 for a D flip-flop.

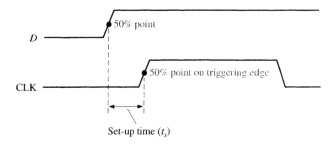

FIGURE F.11-18
Set-up time (t_s).

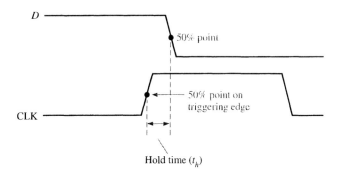

FIGURE F.11-19 .
Hold time (t_h).

Maximum Clock Frequency

The maximum clock frequency (f_{max}) is the highest rate at which a flip-flop can be reliably triggered. At clock frequencies above the maximum, the flip-flop would be unable to respond quickly enough, and its operation would be impaired.

Pulse Widths

Minimum pulse widths (t_W) for reliable operation are usually specified by the manufacturer for the clock, preset, and clear inputs. Typically, the clock is specified by its minimum HIGH time and its minimum LOW time.

Power Dissipation

The **power dissipation** of any digital circuit is the total power consumption of the device. For example, if the flip-flop operates on a +5 V DC source and draws 50 mA of current, the power dissipation is

$$P = V_{CC} \times I_{CC} = 5 \text{ V} \times 50 \text{ mA} = 250 \text{ mW}$$

The power dissipation is very important in most applications in which the capacity of the DC supply is a concern. As an example, let's assume that we have a digital system that requires a total of ten flip-flops, and each flip-flop dissipates 250 mW of power. The total power requirement is

$$P_T = 10 \times 250 \text{ mW} = 2500 \text{ mW} = 2.5 \text{ W}$$

This tells us the output capacity required of the DC supply. If the flip-flops operate on +5 V DC, then the amount of current that the supply must provide is as follows:

$$I = \frac{2.5 \text{ W}}{5 \text{ V}} = 0.5 \text{ A}$$

We must use a +5 V DC supply that is capable of providing at least 0.5 A of current.

Comparison of Specific Flip-Flops

Table F.11-4 provides a comparison, in terms of the operating parameters discussed in this section, of several TTL devices, as well as a CMOS device.

TABLE F.11-4
Comparison of operating parameters for several types of flip-flops.

Parameter (Times in ns)	TTL					CMOS
	7474	74LS74A	74S74	74LS76A	74LS112A	74HC112
t_{PHL} (CLK *to* Q)	40	30	9	20	20	21
t_{PLH} (CLK to Q)	25	25	9	20	20	21
t_{PHL} ($\overline{CLR}$ to Q)	40	30	13.5	20	20	26
t_{PLH} ($\overline{PRE}$ to Q)	25	25	6	20	20	28
t_s (set-up time)	20	20	3	20	20	20
t_h (hold time)	5	0	2	0	0	0
t_W (CLK HIGH)	30	18	6	20	20	16
t_W (CLK LOW)	37	—	7.3	—	25	—
t_W ($\overline{CLR/PRE}$)	30	15	7	25	30	—
f_{max} (MHz)	15	25	75	45	30	30
Power (mW/F-F)	60	14	105	10	14	0.14

REVIEW QUESTIONS

True/False

1. Flip-flops only trigger on the rising edge of the clock pulse.

2. A synchronous input is not recognized until a clock edge is applied.

Multiple Choice

3. When Q equals one, we say the flip-flop is
   a. Clear.
   b. Set.
   c. Invalid.

4. The J-K flip-flop has four modes of operation: no change, set,
   a. Unset, and invalid.
   b. Clear, and invalid.
   c. Clear, and toggle.

5. What types of inputs are preset and clear?
   a. Asynchronous.
   b. Synchronous.
   c. Complementary.

F.12 Fabricate and demonstrate types of flip-flop circuits

INTRODUCTION

Use actual components or software simulation (such as *Electronics Workbench*) to examine the operation of the flip-flop circuits shown in Figure F.12-1.

The CLK input should be connected to the output of a bounceless pushbutton or a low-speed (1 Hz or less) digital oscillator. This will allow the results of each clock pulse to be recorded. The DATA input is simply a 1 or 0 from a logic switch.

For the 3-bit counter, examine and record the patterns seen on the ABC outputs (A is the LSB). For the shift-register, adjust the DATA input as you clock the register so that a 4-bit pattern is loaded into the shift register. Then connect the D output (MSB) to the DATA input to *recirculate* the 4-bit pattern every four clock pulses.

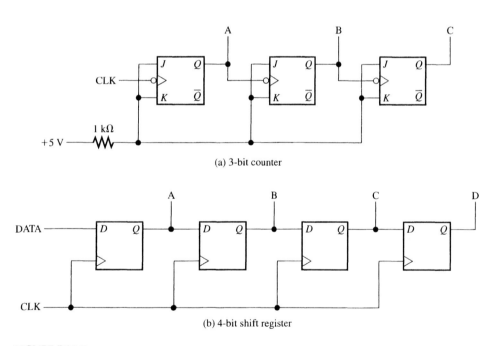

(a) 3-bit counter

(b) 4-bit shift register

FIGURE F.12-1
Test circuits.

F.13 Troubleshoot and repair flip-flop circuits

INTRODUCTION

In this section we examine the typical timing problems associated with flip-flops.

Troubleshooting Two-Phase CLocks

The circuit shown in Figure F.13-1(a) generates two clock waveforms (CLK A and CLK B) having an alternating occurrence of pulses. Each waveform is to be one-half the frequency of the original clock (CLK), as shown in the ideal timing diagram in part (b).

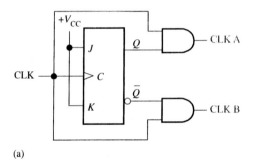

(a)

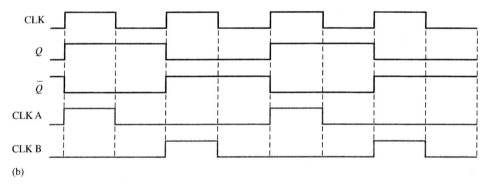

(b)

FIGURE F.13-1
Two-phase clock generator with ideal waveforms.

When the circuit is tested, the CLK A and CLK B waveforms appear on the oscilloscope as shown in Figure F.13-2(a). Since glitches are observed on both waveforms, something is wrong with the circuit either in its basic design or in the way it is connected. Further investigation reveals that the glitches are caused by a **race** condition between the CLK signal and the Q and $\overline{Q}$ signals at the inputs of the AND gates. As displayed in Figure F.13-2(b), the propagation delays between CLK and Q and $\overline{Q}$ create a short-duration

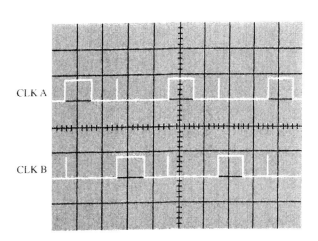

(a) Oscilloscope display of CLK A and CLK B waveforms

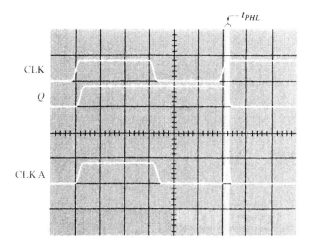

(b) Oscilloscope display showing propagation delay that creates glitch on CLK A waveform

FIGURE F.13-2
Oscilloscope displays for the circuit in Figure F.13-1.

coincidence of HIGH levels at the leading edges of alternate clock pulses. Thus, there is a basic design flaw.

The problem can be corrected by using a negative edge-triggered flip-flop in place of the positive edge-triggered device, as shown in Figure F.13-3(a). Although the propagation delays between CLK and Q and $\overline{Q}$ still exist, they are initiated on the trailing edges of the clock (CLK), thus eliminating the glitches, as shown in the timing diagram of Figure F.13-3(b).

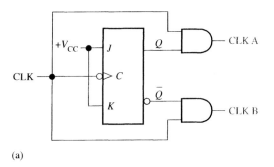

(a)

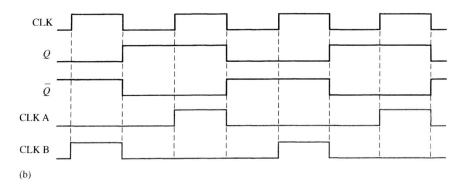

(b)

FIGURE F.13-3
Two-phase clock generator using negative edge-triggered flip-flop to eliminate glitches.

F.14 Understand principles and operations of types of registers and counters

INTRODUCTION

In this section we examine how groups of flip-flops are used to make registers and counters.

EXAMPLE F.14-1

A 4-bit asynchronous binary counter is shown in Figure F.14-1(a). Each flip-flop is negative edge-triggered and has a propagation delay of 10 nanoseconds (ns). Draw a timing diagram showing the Q output of each flip-flop, and determine the total propagation delay time from the triggering edge of a clock pulse until a corresponding change can occur in the state of Q_3. Also determine the maximum clock frequency at which the counter can be operated.

FIGURE F.14-1
Four-bit asynchronous binary counter and its timing diagram.

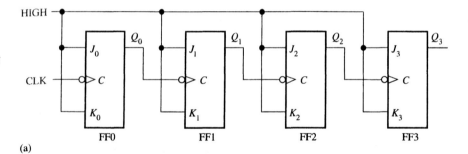

(a)

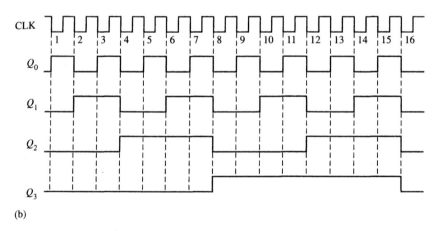

(b)

Solution The timing diagram with delays omitted is as shown in Figure F.14-1(b). For the total delay time, the effect of CLK8 or CLK16 must propagate through four flip-flops before Q_3 changes, so

$$t_{p(tot)} = 4 \times 10 \text{ ns} = 40 \text{ ns}$$

The maximum clock frequency is

$$f_{max} = \frac{1}{t_{p(tot)}} = \frac{1}{40 \text{ ns}} = 25 \text{ MHz}$$

EXAMPLE F.14-2

Show how an asynchronous counter can be implemented having a modulus of twelve with a straight binary sequence from 0000 through 1011.

Solution Since three flip-flops can produce a maximum of eight states, four flip-flops are required to produce any modulus greater than eight but less than or equal to sixteen.

When the counter gets to its last state, 1011, it must recycle back to 0000 rather than going to its normal next state of 1100, as illustrated in the following sequence chart:

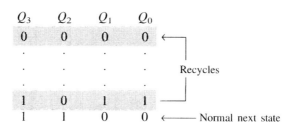

Observe that Q_0 and Q_1 both go to 0 anyway, but Q_2 and Q_3 must be forced to 0 on the twelfth clock pulse. Figure F.14-2(a) shows the modulus-12 counter. The NAND gate partially decodes count twelve (1100) and resets flip-flop 2 and flip-flop 3. Thus, on

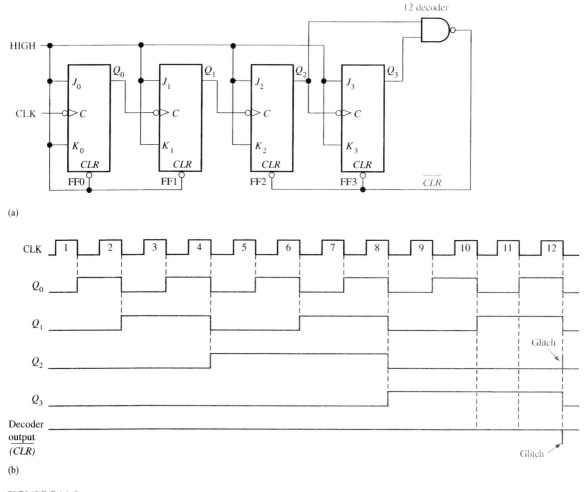

(a)

(b)

FIGURE F.14-2
Asynchronously clocked modulus-12 counter with asynchronous recycling.

the twelfth clock pulse, the counter is forced to recycle from count eleven to count zero, as shown in the timing diagram of Figure F.14-2(b). (It is in count twelve for only a few nanoseconds before it is reset by the glitch on $\overline{CLR}$.)

EXAMPLE F.14-3

Show how the 74LS93A can be used as a modulus-12 counter.

Solution   Use the gated reset inputs, $RO(1)$ and $RO(2)$, to partially decode count 12 (remember, there is an internal NAND gate associated with these inputs). The count-12 decoding is accomplished by connecting Q_3 to $RO(1)$ and Q_2 to $RO(2)$, as shown in Figure F.14-3. Output Q_0 is connected to CLK B to create a 4-bit counter.

FIGURE F.14-3
74LS93A connected as a modulus-12 counter.

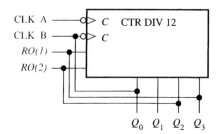

Immediately after the counter goes to count 12 (1100), it is reset to 0000. The recycling, however, results in a glitch on Q_2 because the counter must go into the 1100 state for several nanoseconds before recycling.

EXAMPLE F.14-4

Sketch the timing diagram and determine the sequence of a synchronous 4-bit binary up/down counter if the clock and $UP/\overline{DOWN}$ control inputs have waveforms as shown in Figure F.14-4(a). The counter starts in the all 0s state and is positive edge-triggered.

FIGURE F.14-4

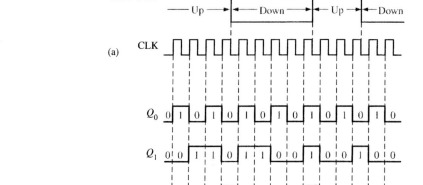

(a)

(b)

Solution   The timing diagram showing the Q outputs is shown in Figure F.14-4(b). From these waveforms, the counter sequence is as shown in Table F.14-1.

TABLE F.14-1

Q_3	Q_2	Q_1	Q_0	
0	0	0	0	⎫
0	0	0	1	
0	0	1	0	⎬ UP
0	0	1	1	
0	1	0	0	⎭
0	0	1	1	⎫
0	0	1	0	
0	0	0	1	⎬ DOWN
0	0	0	0	
1	1	1	1	⎭
0	0	0	0	⎫
0	0	0	1	⎬ UP
0	0	1	0	⎭
0	0	0	1	⎫ DOWN
0	0	0	0	⎭

EXAMPLE F.14-5

Show the states of the 5-bit register in Figure F.14-5(a) for the specified data input and clock waveforms. Assume that the register is initially cleared (all 0s).

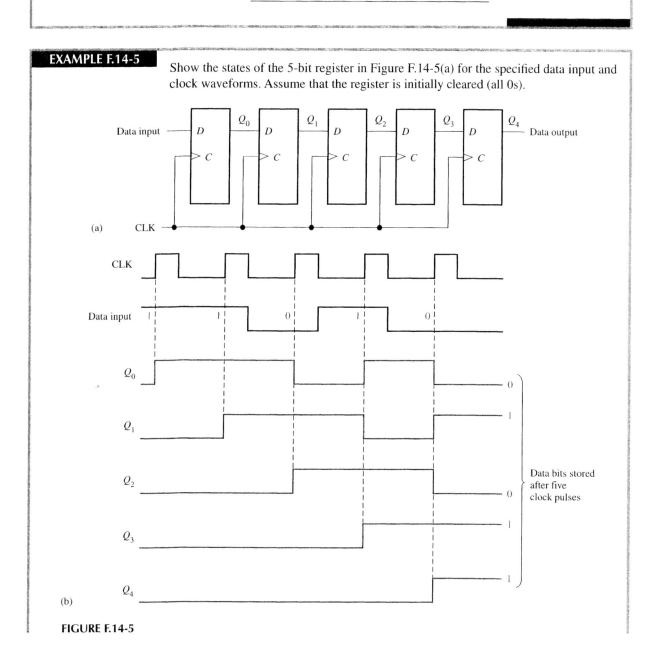

(a)

(b)

Data bits stored after five clock pulses

FIGURE F.14-5

Solution The first data bit (1) is entered into the register on the first clock pulse and then shifted from left to right as the remaining bits are entered and shifted. The register contains 11010 after five clock pulses. See Figure F.14-5(b).

EXAMPLE F.14-6

Show the states of the 4-bit register for the data input and clock waveforms in Figure F.14-6(a). The register initially contains all 1s.

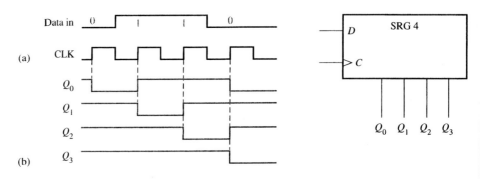

FIGURE F.14-6

Solution The register contains 0110 after four clock pulses. See Figure F.14-6(b).

REVIEW QUESTIONS

True/False

1. All outputs of a ripple counter become valid at the same time.

2. A 4-bit shift register requires two clock cycles to fully load.

Multiple Choice

3. How fast may a 4-bit ripple counter be clocked if each flip-flop has 8 nS of delay?

    a. 31.25 MHz.

    b. 62.5 MHz.

    c. 125 MHz.

4. In a synchronous counter, all flip-flops
   a. Are clocked at the same time.
   b. Operate in toggle mode.
   c. Both a and b.

5. How fast may a 4-bit shift register be clocked if each flip-flop has 8 nS of delay?
   a. 31.25 MHz.
   b. 62.5 MHz.
   c. 125 MHz.

F.15 Fabricate and demonstrate types of registers and counters

INTRODUCTION

Use actual components or simulation software (such as *Electronics Workbench*) to examine the operation of each circuit in Figure F.15-1.

For each circuit, apply a 1 KHz clock and record the frequency of each output (A through D and Q_0 through Q_3). Use a logic analyzer to capture a complete set of output waveforms for each circuit. For the decade and binary counters, try connecting one or two outputs back to the *RO* inputs (remove ground first) and note the changes in the counting sequence.

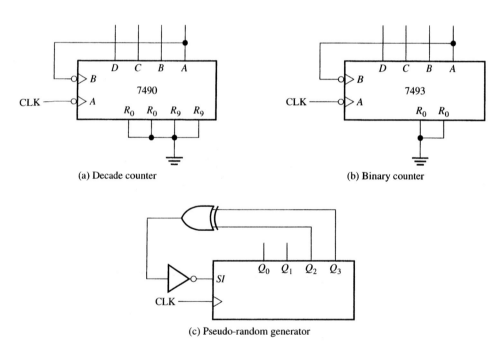

(a) Decade counter   (b) Binary counter

(c) Pseudo-random generator

FIGURE F.15-1
Test circuits.

F.16 Troubleshoot and repair types of registers and counters

INTRODUCTION

For an IC counter with a straightforward sequence that is not controlled by external logic, about the only thing to check (other than V_{CC} and ground) is the possibility of open or shorted inputs or outputs. An IC counter almost never alters its sequence of states because of an internal fault, so you need only check for pulse activity on the Q outputs to detect the existence of an open or a short. The absence of pulse activity on one of the Q outputs indicates an internal short or open. Absence of pulse activity on all the Q outputs indicates that the clock input is faulty or the clear input is stuck in its active state.

To check the clear input, apply a constant active level while the counter is clocked. You will observe a LOW on each of the Q outputs if it is functioning properly.

The parallel load feature on a counter can be checked by activating the $\overline{LOAD}$ input and exercising each state as follows: Apply LOWs to the parallel data inputs, pulse the clock input, and check for LOWs on all the Q outputs. Next, apply HIGHs to all the parallel data inputs, pulse the clock input, and check for HIGHs on all the Q outputs.

Cascaded IC Counters with Maximum Modulus

A failure in one of the counters in a chain of cascaded counters can affect all the counters that follow it. For example, if a count enable input opens, it effectively acts as a HIGH (TTL), and the counter is always enabled. This type of failure in one of the counters will cause that counter to run at the full clock rate and will also cause all the succeeding counters to run at higher than normal rates. This is illustrated in Figure F.16-1 for a divide-by-

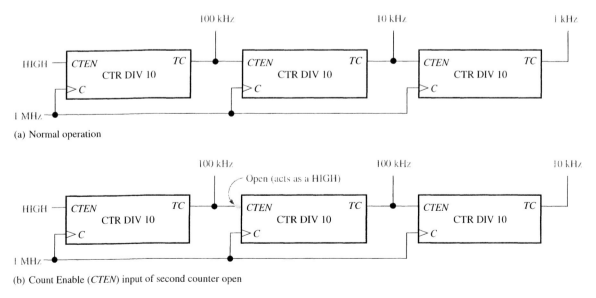

(a) Normal operation

(b) Count Enable (*CTEN*) input of second counter open

FIGURE F.16-1
Example of a failure that affects following counters in a cascaded arrangement.

1000 cascaded counter arrangement where an open enable (*CTEN*) input acts as a TTL HIGH and continuously enables the second counter. Other faults that can affect "downstream" counter stages are open or shorted clock inputs or terminal count outputs. In some of these situations, pulse activity can be observed, but it may be at the wrong frequency. Exact frequency measurements must be made.

Cascaded Counters with Truncated Sequences

The count sequence of a cascaded counter with a truncated sequence, such as that in Figure F.16-2, can be affected by other types of faults in addition to those mentioned for maximum-modulus cascaded counters. For example, a failure in one of the parallel data inputs, the $\overline{LOAD}$ input, or the inverter can alter the preset count and thus change the modulus of the counter.

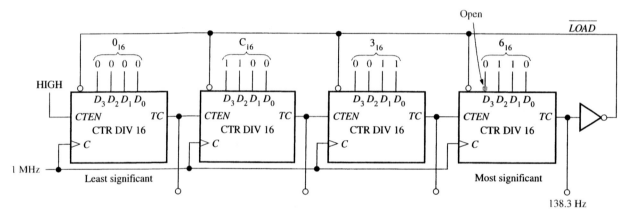

FIGURE F.16-2
Example of a failure in a cascaded counter with a truncated sequence.

For example, suppose the D_3 input of the most significant counter in Figure F.16-2 is open and acts as a HIGH. Instead of 6_{16} (0110) being preset into the counter, E_{16} (1110) is preset in. So, instead of beginning with $63C0_{16}$ ($25,536_{10}$) each time the counter recycles, the sequence will begin with $E3C0_{16}$ ($58,304_{10}$). This changes the modulus of the counter from 40,000 to $65,536 - 58,304 = 7232$.

To check this counter, apply a known clock frequency, say 1 MHz, and measure the output frequency at the final terminal count output. If the counter is operating properly, the output frequency is

$$f_{out} = \frac{f_{in}}{\text{modulus}} = \frac{1 \text{ MHz}}{40,000} = 25 \text{ Hz}$$

In this case, the specific failure described in the preceding paragraph will cause the output frequency to be

$$f_{out} = \frac{f_{in}}{\text{modulus}} = \frac{1 \text{ MHz}}{7232} = 138.3 \text{ Hz}$$

EXAMPLE F.16-1

Frequency measurements are made on the truncated counter in Figure F.16-3 as indicated. Determine if the counter is working properly, and if not, isolate the fault.

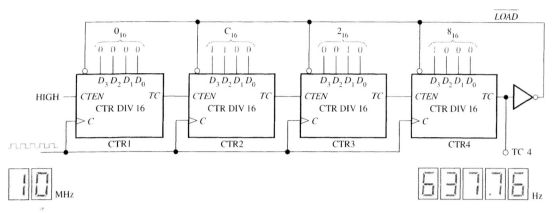

FIGURE F.16-3

Solution   Check to see if the frequency measured at TC 4 is correct. If it is, the counter is working properly.

$$\text{truncated modulus} = \text{full modulus} - \text{preset count}$$
$$= 16^4 - 82C0_{16}$$
$$= 65,536 - 33,472 = 32,064$$

The correct frequency at TC 4 is

$$f_4 = \frac{10 \text{ MHz}}{32,064} = 311.88 \text{ Hz}$$

Uh-oh! There is a problem. The measured frequency of 637.76 Hz does not agree with the correct calculated frequency of 311.88 Hz.

To find the faulty counter, determine the actual truncated modulus as follows:

$$\text{modulus} = \frac{f_{out}}{f_{in}} = \frac{10 \text{ MHz}}{637.76 \text{ Hz}} = 15,680$$

Because the truncated modulus should be 32,064, most likely the counter is being preset to the wrong count when it recycles. The actual preset count is determined as follows:

$$\text{truncated modulus} = \text{full modulus} - \text{preset count}$$
$$\text{preset count} = \text{full modulus} - \text{truncated modulus}$$
$$= 65,536 - 15,680$$
$$= 49,856$$
$$= C2C0_{16}$$

This shows that the counter is being preset to $C2C0_{16}$ instead of $82C0_{16}$ each time it recycles.

Counters 1, 2, and 3 are being preset properly but counter 4 is not. Since $C_{16} = 1100_2$, the D_2 input to counter 4 is HIGH when it should be LOW. This is most likely caused by an open input. Check for an external open caused by a bad solder connection or a broken conductor; if none can be found, replace the IC and the counter should work properly.

Counters Implemented with Individual Flip-Flops

Counters implemented with individual flip-flop and gate ICs are sometimes more difficult to troubleshoot because there are many more inputs and outputs with external connections than there are in an IC counter. The sequence of a counter can be altered by a single open or short on an input or output, as Example F.16-2 illustrates.

EXAMPLE F.16-2

Suppose that you observe the output waveforms that are indicated for the counter in Figure F.16-4. Determine if there is a problem with the counter.

FIGURE F.16-4

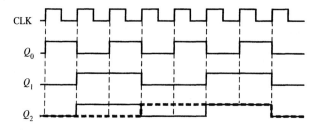

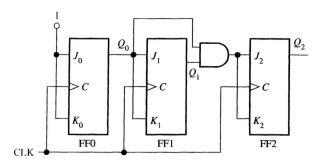

Solution The Q_2 waveform is incorrect. The correct waveform is shown as a dashed line. You can see that the Q_2 waveform looks exactly like the Q_1 waveform. So whatever is causing FF1 to toggle appears to also be controlling FF2.

Checking the J and K inputs to FF2, you find a waveform that looks like Q_0. This result indicates that Q_0 is somehow getting through the AND gate. The only way this can happen is if the Q_1 input to the AND gate is always HIGH. But you have seen that Q_1 has a correct waveform. This observation leads to the conclusion that the lower input to the AND gate must be internally open and acting as a HIGH. Replace the AND gate and retest the circuit.

F.17 Understand principles and operations of clock and timing circuits

INTRODUCTION

Binary information that is handled by digital systems appears as waveforms that represent sequences of bits. When the waveform is HIGH, a binary 1 is present; when the waveform is LOW, a binary 0 is present. Each bit in a sequence occupies a defined time interval called a **bit time.**

The Clock In many digital systems, all waveforms are synchronized with a basic timing waveform called the **clock.** The clock is a periodic waveform in which each interval between pulses (the period) equals one bit time.

An example of a clock waveform is shown in Figure F.17-1. Notice that, in this case, each change in level of waveform *A* occurs at the leading edge of the clock waveform. In other cases, level changes occur at the trailing edge of the clock. During each bit time of the clock, waveform *A* is either HIGH or LOW. These HIGHs and LOWs represent a sequence of bits as indicated. A group of several bits can be used as a piece of binary information, such as a number or a letter.

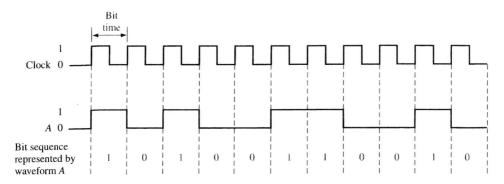

FIGURE F.17-1
Example of a clock waveform synchronized with a waveform representation of a sequence of bits.

Timing Diagrams

A **timing diagram** is a graph of digital waveforms showing the proper time relationship of all the waveforms and how each waveform changes in relation to the others. Figure F.17-1 is an example of a simple timing diagram that shows how the clock waveform and waveform *A* are related.

A timing diagram can consist of any number of related waveforms. By looking at a timing diagram, you can determine the states (HIGH or LOW) of all the waveforms at any specified point in time and the exact time that a waveform changes state relative to the other waveforms. Figure F.17-2 is an example of a timing diagram made up of four waveforms. From this timing diagram you can see, for example, that all three waveforms (*A, B,* and *C*) are HIGH only during bit time 7 and they all change back LOW at the end of bit time 7.

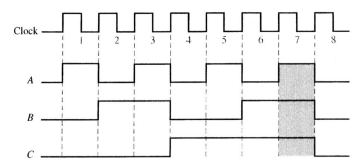

FIGURE F.17-2
Example of a timing diagram.

Data Transfer

Data refers to groups of bits that convey some type of information. Binary data, which are represented by digital waveforms, must be transferred from one circuit to another within a digital system or from one system to another in order to accomplish a given purpose. For example, numbers stored in binary form in the memory of a computer must be transferred to the central processing unit in order to be added. The sum of the addition must then be transferred to a monitor for display and/or transferred back to the memory. In digital systems as in Figure F.17-3, binary data are transferred in two ways: serial and parallel.

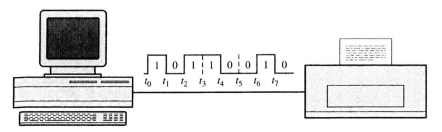

(a) Serial transfer of binary data from computer to printer. Interval t_0 to t_1 is first.

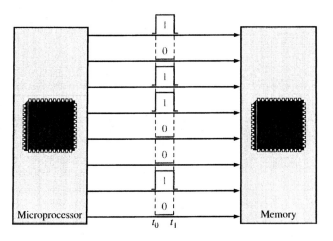

(b) Parallel transfer of binary data from microprocessor to memory in a computer system. t_0 is the beginning time.

FIGURE F.17-3
Illustration of serial and parallel transfer of binary data.

When bits are transferred in **serial** form from one point to another, they are sent one bit at a time along a single conductor, as illustrated in Figure F.17-3(a) for the case of a computer-to-printer transfer. During the time interval from t_0 to t_1, the first bit is transferred. During the time interval from t_1 to t_2, the second bit is transferred, and so on. To transfer eight bits in series, it takes eight time intervals.

When bits are transferred in **parallel** form, all the bits in a group are sent out on separate lines at the same time. There is one line for each bit, as shown in Figure F.17-3(b) for the case of eight bits being transferred from the microprocessor to the memory in a computer. To transfer eight bits in parallel, it takes one time interval compared to eight time intervals for the serial transfer.

To summarize, the advantage of serial transfer of binary data is that only one line is required. In parallel transfer, a number of lines equal to the number of bits to be transferred at one time is required. The disadvantage of serial transfer is that it takes longer to transfer a given number of bits than with parallel transfer. For example, if one bit can be transferred in 1 μs, then it takes 8 μs to serially transfer eight bits but only 1 μs to parallel transfer eight bits. The disadvantage of parallel transfer is that it takes more lines.

EXAMPLE F.17-1

(a) Determine the total time required to serially transfer the eight bits contained in waveform A of Figure F.17-4, and indicate the sequence of bits. The left-most bit is the first to be transferred. The 100 kHz clock is used as reference.

(b) What is the total time to transfer the same eight bits in parallel?

FIGURE F.17-4

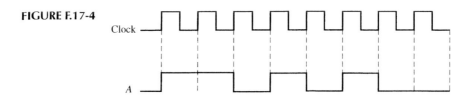

Solution

(a) Since the frequency of the clock is 100 kHz, the period is

$$T = \frac{1}{f} = \frac{1}{100 \text{ kHz}} = 10 \ \mu\text{s}$$

It takes 10 μs to transfer each bit in the waveform. The total transfer time is

$$8 \times 10 \ \mu\text{s} = 80 \ \mu\text{s}$$

To determine the sequence of bits, examine the waveform in Figure F.17-4 during each bit time. If waveform A is HIGH during the bit time, a 1 is transferred. If waveform A is LOW during the bit time, a 0 is transferred. The bit sequence is illustrated in Figure F.17-5. The left-most bit is the first to be transferred.

FIGURE F.17-5

(b) A parallel transfer would take 10 μs for all eight bits.

True/False

1. One bit time is typically the same as the period of the clock.

2. Data transfers may be performed in parallel or serial.

Multiple Choice

3. Multiple waveforms displayed relative to each other constitute a
   a. Waveform diagram.
   b. Timing sequence.
   c. Timing diagram.

4. The bit time of a waveform is 250 nS. What is the frequency of the waveform?
   a. 400 KHz.
   b. 4 MHz.
   c. 40 MHz.

5. Data transfers are aligned with the
   a. High level of the clock signal.
   b. Low level of the clock signal.
   c. Rising or falling edge of the clock signal.

F.18 Fabricate and demonstrate clock and timing circuits

INTRODUCTION

Use actual components or simulation software (such as *Electronics Workbench*) to examine the 555 timer shown in Figure F.18-1. Determine the frequency of the output and the duty cycle. Sketch and label the output waveform.

FIGURE F.18-1

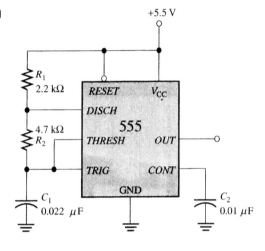

F.19 Troubleshoot and repair clock and timing circuits

INTRODUCTION

One of the most useful pieces of equipment one can use when troubleshooting timing-related problems is the logic analyzer. In this section we examine the features of a typical logic analyzer and another helpful instrument, the logic probe.

The Logic Analyzer

A typical logic analyzer is shown in Figure F.19-1. This instrument can detect and display digital data in several formats.

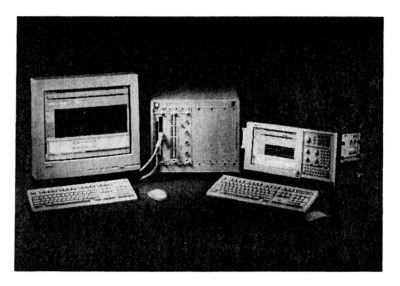

FIGURE F.19-1
A typical logic analyzer (copyright 1999 Tektronix, Inc. All rights reserved—reproduced by permission.)

Oscilloscope Format   The logic analyzer can be used to display single or dual waveforms on the screen, as indicated in Figure F.19-2(a), so that characteristics of individual pulses or waveform parameters can be measured.

Timing Diagram Format   The logic analyzer can display typically up to sixteen waveforms in proper time relationship, as indicated in Figure F.19-2(b), so that you can analyze sets of waveforms and determine how they change in time with respect to each other.

Oscilloscope/Timing Diagram Combination   In this format, as indicated in Figure F.19-2(c), both individual waveforms and a complete timing diagram can be displayed simultaneously. This allows you to examine the details of a certain waveform while having a timing diagram available.

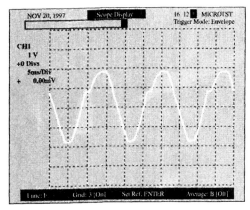

(a) Oscilloscope format

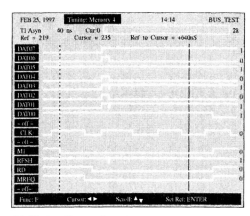

(b) Timing diagram format

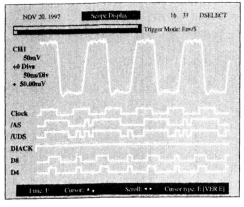

(c) Oscilloscope/Timing diagram

(d) State table format

FIGURE F.19-2
Illustration of typical logic analyzer display formats.

State Table Format The logic analyzer can display binary data in tabular form, as illustrated in Figure F.19-2(d). For example, various memory locations in a microprocessor-based system can be examined to determine the contents. The data can be displayed in a variety of number systems and codes such as binary, hexadecimal, octal, binary coded decimal (BCD), and ASCII.

The Logic Probe, Pulser, and Current Probe

The logic probe is a convenient, hand-held tool that provides a means of troubleshooting a digital circuit by sensing various conditions at a point in a circuit, as illustrated in Figure F.19-3. The probe can detect high-level voltage, low-level voltage, single pulses, repetitive pulses, and opens on a circuit board. The probe lamp indicates the condition that exists at a certain point, as indicated in the figure.

The logic pulser is a pulse source that produces a repetitive pulse waveform that can be used to force a condition in a circuit. You can apply pulses at one point in a circuit with the pulser and check another point for resulting pulses with a logic probe. Also the pulser can be used in conjunction with the current probe as indicated in Figure F.19-4. The current probe senses when there is pulsating current in a line and is particularly useful for locating shorts on a circuit board.

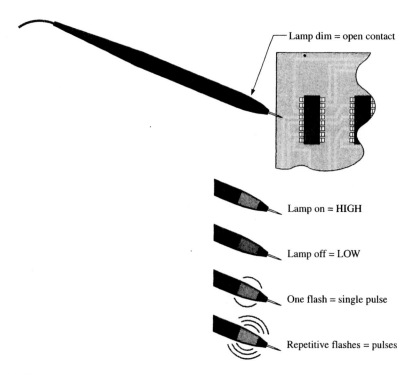

FIGURE F.19-3
Illustration of how a logic probe is used to detect various voltage conditions at a given point in a circuit.

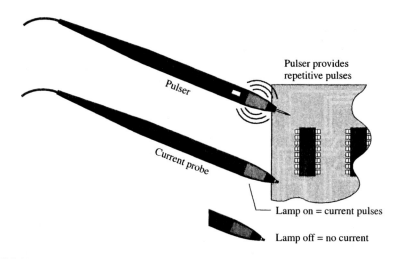

FIGURE F.19-4
Illustration of how a logic pulser and a current probe can be used to pulse a given point and check for resulting current in another part of the circuit.

F.20 Understand principles and operations of types of arithmetic-logic circuits

INTRODUCTION

Adders are important not only in computers, but in many types of digital systems in which numerical data are processed. An understanding of the basic adder operation is fundamental to the study of digital systems. In this section, the half-adder and the full-adder are introduced.

The Half-Adder

Recall the basic rules for binary addition:

$$0 + 0 = 0$$
$$0 + 1 = 1$$
$$1 + 0 = 1$$
$$1 + 1 = 10$$

These operations are performed by a logic circuit called a **half-adder.**

> **The half-adder accepts two binary digits on its inputs and produces two binary digits on its outputs, a sum bit and a carry bit.**

A half-adder is represented by the logic symbol in Figure F.20-1.

FIGURE F.20-1
Logic symbol for a half-adder.

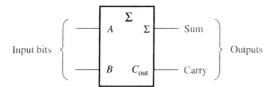

From the logical operation of the half-adder as stated in Table F.20-1, expressions can be derived for the sum and the output carry as functions of the inputs. Notice that the output carry (C_{out}) is a 1 only when both A and B are 1s; therefore, C_{out} can be expressed as the AND of the input variables.

$$C_{out} = AB$$

Now observe that the sum output (Σ) is a 1 only if the input variables, A and B, are not equal. The sum can therefore be expressed as the exclusive-OR of the input variables.

$$\Sigma = A \oplus B$$

From these equations, the logic implementation required for the half-adder function can be developed. The output carry is produced with an AND gate with A and B on the inputs, and the sum output is generated with an exclusive-OR gate, as shown in Figure F.20-2.

TABLE F.20-1
Half-adder truth table.

A	B	C_{out}	Σ
0	0	0	0
0	1	0	1
1	0	0	1
1	1	1	0

Σ = sum

C_{out} = output carry

A and B = input variables (operands)

FIGURE F.20-2
Half-adder logic diagram.

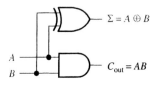

$$\Sigma = A \oplus B$$

$$C_{out} = AB$$

The Full-Adder

The second basic category of adder is the **full-adder.**

> **The full-adder accepts three inputs including an input carry and generates a sum output and an output carry.**

The basic difference between a full-adder and a half-adder is that the full-adder accepts an input carry. A logic symbol for a full-adder is shown in Figure F.20-3, and the truth table in Table F.20-2 shows the operation of a full-adder.

FIGURE F.20-3
Logic symbol for a full-adder.

Input bits {
A
B
Input carry — C_{in}

Σ

Σ — Sum

C_{out} — Output carry

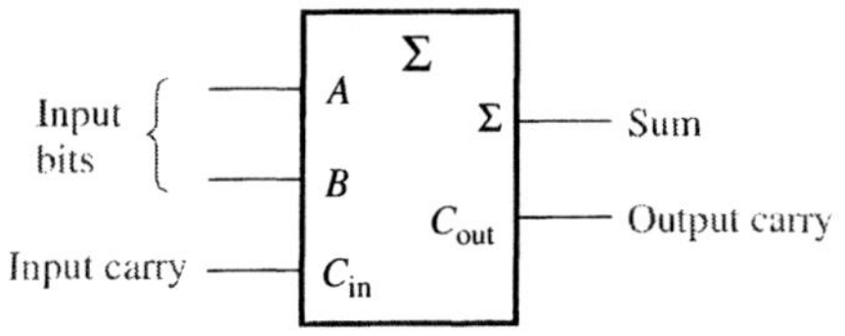

TABLE F.20-2
Full-adder truth table.

A	B	C_{in}	C_{out}	Σ
0	0	0	0	0
0	0	1	0	1
0	1	0	0	1
0	1	1	1	0
1	0	0	0	1
1	0	1	1	0
1	1	0	1	0
1	1	1	1	1

C_{in} = input carry, sometimes designated as CI

C_{out} = output carry, sometimes designated as CO

Σ = sum

A and B = input variables (operands)

The full-adder must add the two input bits and the input carry. From the half-adder we know that the sum of the input bits A and B is the exclusive-OR of those two variables, $A \oplus B$. For the input carry (C_{in}) to be added to the input bits, it must be exclusive-ORed with $A \oplus B$, yielding the equation for the sum output of the full-adder.

$$\Sigma = (A \oplus B) \oplus C_{in}$$

This means that to implement the full-adder sum function, two exclusive-OR gates can be used. The first must generate the term $A \oplus B$, and the second has as its inputs the output of the first XOR gate and the input carry, as illustrated in Figure F.20-4(a).

The output carry is a 1 when both inputs to the first XOR gate are 1s or when both inputs to the second XOR gate are 1s. You can verify this fact by studying Table F.20-2. The output carry of the full-adder is therefore produced by the inputs A ANDed with B and $A \oplus B$ ANDed with C_{in}. These two terms are then ORed. This function is implemented and combined with the sum logic to form a complete full-adder circuit, as shown in Figure F.20-4(b).

$$C_{out} = AB + (A \oplus B)C_{in}$$

Notice in Figure F.20-4(b) there are two half-adders, connected as shown in the block diagram of Figure F.20-5(a), with their output carries ORed. The logic symbol shown in Figure F.20-5(b) will normally be used to represent the full-adder.

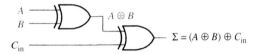

(a) Logic required to form the sum of three bits

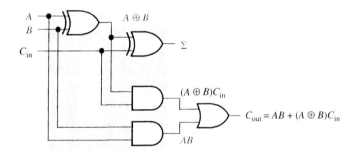

(b) Complete logic circuit for a full-adder (each half-adder is enclosed by a shaded area)

FIGURE F.20-4
Full-adder logic.

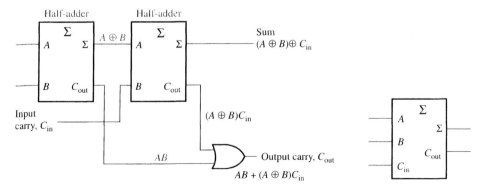

(a) Arrangement of two half-adders to form a full-adder

(b) Full-adder logic symbol

FIGURE F.20-5
Full-adder implemented with half-adders.

EXAMPLE F.20-1

Determine an alternative method for implementing the full-adder.

Solution Referring to Table F.20-2, you can write sum-of-products expressions for both Σ and C_{out} by observing the input conditions that make them 1s. The expressions are as follows:

$$\Sigma = \overline{A}\,\overline{B}C_{in} + \overline{A}B\overline{C}_{in} + A\overline{B}\,\overline{C}_{in} + ABC_{in}$$

$$C_{out} = \overline{A}BC_{in} + A\overline{B}C_{in} + AB\overline{C}_{in} + ABC_{in}$$

Mapping these two expressions on the Karnaugh maps in Figure F.20-6, you find that the sum (Σ) expression cannot be simplified. The output carry expression (C_{out}) is reduced as indicated.

FIGURE F.20-6

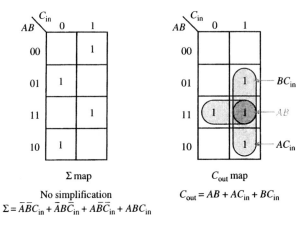

Σ map

No simplification

$\Sigma = \overline{A}\overline{B}C_{in} + \overline{A}B\overline{C}_{in} + A\overline{B}\overline{C}_{in} + ABC_{in}$

C_{out} map

$C_{out} = AB + AC_{in} + BC_{in}$

These two expressions are implemented with AND-OR logic as shown in Figure F.20-7 to form a complete full-adder.

FIGURE F.20-7

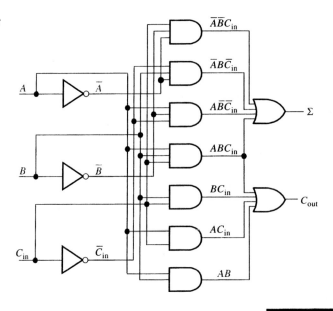

PARALLEL BINARY ADDERS

As you have seen, a single full-adder is capable of adding two 1-bit numbers and an input carry. To add binary numbers with more than one bit, additional full-adders must be used. When one binary number is added to another, each column generates a sum bit and a 1 or 0 carry bit to the next column to the left, as illustrated here with 2-bit numbers.

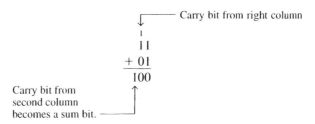

To implement the addition of binary numbers, a full-adder is required for each bit in the numbers. So for 2-bit numbers, two adders are needed; for 4-bit numbers, four adders are used; and so on. The carry output of each adder is connected to the carry input of the next higher-order adder, as shown in Figure F.20-8 for a 2-bit adder. Notice that either a half-adder can be used for the least significant position or the carry input of a full-adder can be made 0 (grounded) because there is no carry input to the least significant bit position.

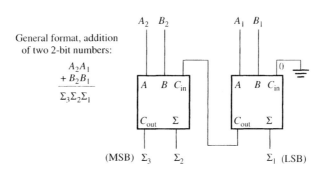

FIGURE F.20-8
Block diagram of a basic 2-bit parallel adder.

In Figure F.20-8 the least significant bits (LSB) of the two numbers are represented by A_1 and B_1. The next higher-order bits are represented by A_2 and B_2. The three sum bits are Σ_1, Σ_2, and Σ_3. Notice that the output carry from the left-most full-adder becomes the most significant bit (MSB) in the sum, Σ_3.

EXAMPLE F.20-2

Verify that the 2-bit parallel adder in Figure F.20-9 properly performs the following addition:

$$
\begin{array}{r}
11 \\
+\ 10 \\
\hline
101
\end{array}
$$

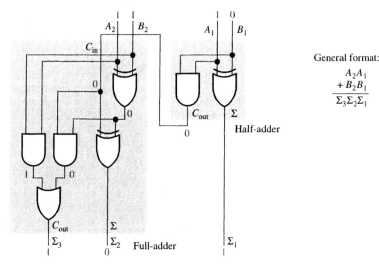

General format:
$$A_2A_1$$
$$+ B_2B_1$$
$$\overline{\Sigma_3\Sigma_2\Sigma_1}$$

FIGURE F.20-9

Solution The logic level at each point in the circuit for the given input numbers is determined from the truth table of the relevant gate. By following these levels through the circuit as indicated on the logic diagram, you find that the proper levels appear on the sum outputs.

Four-Bit Parallel Adders

A basic 4-bit parallel adder is implemented with four full-adders as shown in Figure F.20-10. Again, the LSBs (A_1 and B_1) in each number being added go into the right-most full-adder; the higher-order bits are applied as shown to the successively higher-order adders, with the MSBs (A_4 and B_4) in each number being applied to the left-most full-adder. The carry output of each adder is connected to the carry input of the next higher-order adder as indicated. These are called *internal carries.*

In keeping with most manufacturers' data sheets, the input labeled C_0 is the input carry to the least significant bit adder; C_4, in the case of four bits, is the output carry of the most significant bit adder; and Σ_1 (LSB) through Σ_4 (MSB) are the sum outputs. The logic symbol is shown in Figure F.20-10(b).

Truth Table for a 4-Bit Parallel Adder

Table F.20-3 is the truth table for a 4-bit adder. On some data sheets, truth tables may be called *function tables.* At first glance, this table may be confusing but, with a little practice, you will learn to interpret it easily. The internal carry, which is designated C_2 in the table, is the output carry from the addition, $A_2A_1 + B_2B_1$. This internal carry bit is then used to find the sum of $A_4A_3 + B_4B_3$, as well as the output carry, C_4.

The 74LS83A and 74LS283 MSI Adders

Examples of 4-bit parallel adders that are available as medium-scale integrated (MSI) circuits are the 74LS83A and the 74LS283 low-power Schottky TTL devices. These devices are also available in other logic families such as standard TTL (7483A and 74283) and CMOS (74HC283). The 74LS83A and the 74LS283 are functionally identical to each other but not pin compatible; that is, the pin numbers for the inputs and outputs are different due to different power and ground pin connections. For the 74LS83A, V_{CC} is pin 5 and ground is pin 12 on the 16-pin package. For the 74LS283, V_{CC} is pin 16 and ground is pin

FIGURE F.20-10
A 4-bit parallel adder.

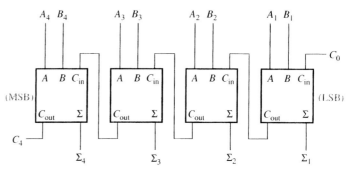

(a) Block diagram

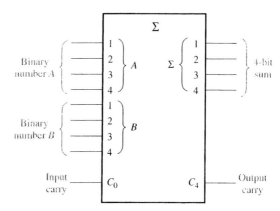

(b) Logic symbol

TABLE F.20-3
Truth table for a 4-bit parallel adder.

				Outputs					
				When $C_0 = 0$			When $C_0 = 1$		
Inputs						When $C_2 = 0$			When $C_2 = 1$
A_1 / A_3	B_1 / B_3	A_2 / A_4	B_2 / B_4	Σ_1 / Σ_3	Σ_2 / Σ_4	C_2 / C_4	Σ_1 / Σ_3	Σ_2 / Σ_4	C_2 / C_4
0	0	0	0	0	0	0	1	0	0
1	0	0	0	1	0	0	0	1	0
0	1	0	0	1	0	0	0	1	0
1	1	0	0	0	1	0	1	1	0
0	0	1	0	0	1	0	1	1	0
1	0	1	0	1	1	0	0	0	1
0	1	1	0	1	1	0	0	0	1
1	1	1	0	0	0	1	1	0	1
0	0	0	1	0	1	0	1	1	0
1	0	0	1	1	1	0	0	0	1
0	1	0	1	1	1	0	0	0	1
1	1	0	1	0	0	1	1	0	1
0	0	1	1	0	0	1	1	0	1
1	0	1	1	1	0	1	0	1	1
0	1	1	1	1	0	1	0	1	1
1	1	1	1	0	1	1	1	1	1

NOTE: Input conditions at A_1, B_1, A_2, B_2, and C_0 are used to determine outputs Σ_1 and Σ_2 and the value of the internal carry C_2. The values at C_2, A_3, B_3, A_4, and B_4 are then used to determine outputs Σ_3, Σ_4, and C_4.

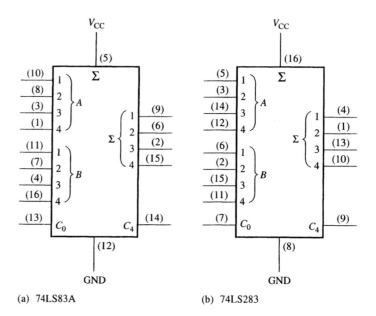

(a) 74LS83A   (b) 74LS283

FIGURE F.20-11
MSI 4-bit parallel adders.

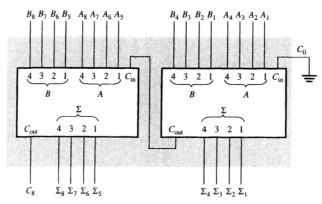

(a) Cascading of 4-bit adders to form an 8-bit adder

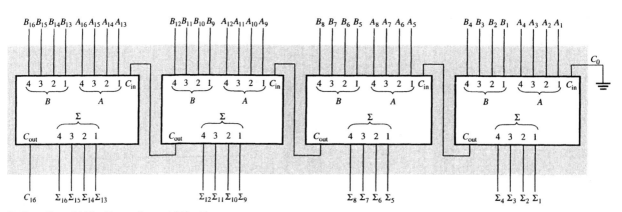

(b) Cascading of 4-bit adders to form a 16-bit adder

FIGURE F.20-12
Examples of adder expansion.

8, which is a more standard configuration. Logic symbols for both of these devices are shown, with pin numbers in parentheses, in Figure F.20-11.

Adder Expansion

The 4-bit parallel adder can be expanded to handle the addition of two 8-bit numbers by using two 4-bit adders and connecting the carry input of the low-order adder (C_0) to ground because there is no carry into the least significant bit position and by connecting the carry output of the low-order adder to the carry input of the high-order adder as shown in Figure F.20-12(a). This process is known as **cascading.** Notice that, in this case, the output carry is designated C_8 because it is generated from the eighth bit position. The low-order adder is the one that adds the lower or less significant four bits in the numbers and the high-order adder is the one that adds the higher or more significant four bits in the 8-bit numbers.

Similarly, four 4-bit adders can be cascaded to handle two 16-bit numbers as shown in Figure F.20-12(b). Notice that the output carry is designated C_{16} because it is generated from the sixteenth bit position.

EXAMPLE F.20-3

Show how two 74LS83A adders can be connected to form an 8-bit parallel adder. Show output bits for the following 8-bit input numbers:

$$A_8A_7A_6A_5A_4A_3A_2A_1 = 10111001$$

and

$$B_8B_7B_6B_5B_4B_3B_2B_1 = 10011110$$

Solution   Two 74LS83A 4-bit parallel adders are used to implement the 8-bit adder. The only connection between the two 74LS83As is the carry output (pin 14) of the low-order adder to the carry input (pin 13) of the high-order adder, as shown in Figure F.20-13. Pin 13 of the low-order adder is grounded (no carry input).

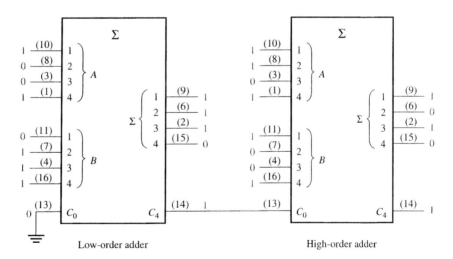

FIGURE F.20-13
Two 74LS83A adders connected as an 8-bit parallel adder (pin numbers are in parentheses).

An Adder Application

An example of full-adder and parallel adder application is a simple voting system that can be used to simultaneously provide the number of "yes" votes and the number of "no" votes. For example, this type of system can be used where a group of people are assembled and there is a need for immediately determining opinions (for or against), making decisions, or voting on certain issues or other matters.

In its simplest form, the system includes a switch for "yes" or "no" selection at each position in the assembly and a digital display for the number of yes votes and one for the number of no votes. The basic system is shown in Figure F.20-14 for a 6-position setup, but it can be expanded to any number of positions with additional 6-position modules and additional parallel adder and display circuits.

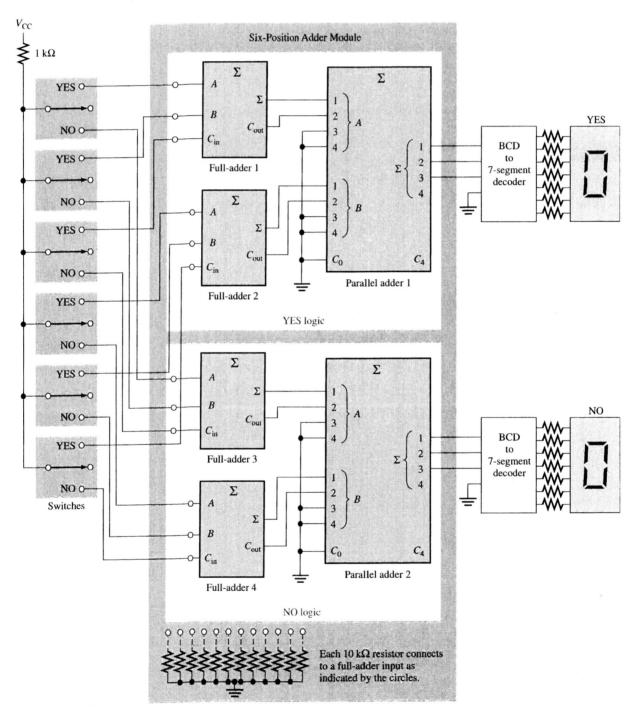

FIGURE F.20-14
A voting system using full-adders and parallel binary adders.

In Figure F.20-14, each full-adder can produce the sum of up to three votes. The sum and output carry of each full-adder then goes to the two lower-order inputs of a parallel binary adder. The two higher-order inputs of the parallel adder are connected to ground (0) because there is never a case where the binary input exceeds 0011 (decimal 3). For this basic 6-position system, the outputs of the parallel adder go to a BCD-to-7-segment decoder which drives the 7-segment display. As mentioned, additional circuits must be included when the system is expanded.

The resistors from the inputs of each full-adder to ground assure that each input is LOW when the switch is in the neutral position (CMOS logic is used). When a switch is moved to the "yes" or to the "no" position, a HIGH level (V_{CC}) is applied to the associated full-adder input.

RIPPLE CARRY VERSUS LOOK-AHEAD CARRY ADDERS

The Ripple Carry Adder

A **ripple carry** adder is one in which the carry output of each full-adder is connected to the carry input of the next higher-order stage (a stage is one full-adder). The sum and the output carry of any stage cannot be produced until the input carry occurs; this causes a time delay in the addition process, as illustrated in Figure F.20-15. The carry propagation delay for each full-adder is the time from the application of the input carry until the output carry occurs, assuming that the A and B inputs are already present.

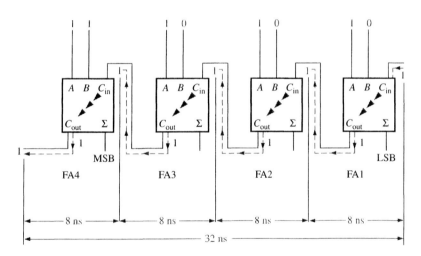

FIGURE F.20-15
A 4-bit parallel ripple carry adder showing "worst-case" carry propagation delays.

Full-adder 1 (FA1) cannot produce a potential output carry until an input carry is applied. Full-adder 2 (FA2) cannot produce a potential output carry until full-adder 1 produces an output carry. Full-adder 3 (FA3) cannot produce a potential output carry until an output carry is produced by FA1 followed by an output carry from FA2, and so on. As you can see in Figure F.20-15, the input carry to the least significant stage has to ripple through all the adders before a final sum is produced. The cumulative delay through all the adder stages is a "worst-case" addition time. The total delay can vary, depending on the carry bit produced by each full-adder. If two numbers are added such that no carries (0) occur between stages, the addition time is simply the propagation time through a single full-adder from the application of the data bits on the inputs to the occurrence of a sum output.

The Look-Ahead Carry Adder

The speed with which an addition can be performed is limited by the time required for the carries to propagate, or ripple, through all the stages of a parallel adder. One method of speeding up the addition process by eliminating this ripple carry delay is called **look-ahead carry** addition. The look-ahead carry adder anticipates the output carry of each stage, and based on the inputs, produces the output carry by either carry generation or carry propagation.

Carry generation occurs when an output carry is produced (generated) internally by the full-adder. A carry is generated only when both input bits are 1s. The generated carry, C_g, is expressed as the AND function of the two input bits, A and B.

$$C_g = AB$$

Carry propagation occurs when the input carry is rippled to become the output carry. An input carry may be propagated by the full-adder when either or both of the input bits are 1s. The propagated carry, C_p, is expressed as the OR function of the input bits.

$$C_p = A + B$$

The conditions for carry generation and carry propagation are illustrated in Figure F.20-16. The three arrowheads symbolize ripple (propagation).

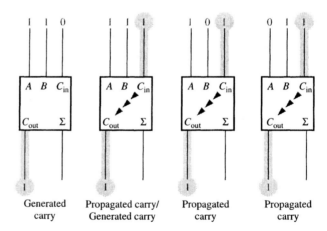

FIGURE F.20-16
Illustration of conditions for carry generation and carry propagation.

The output carry of a full-adder can be expressed in terms of both the generated carry (C_g) and the propagated carry (C_p). The output carry (C_{out}) is a 1 if the generated carry is a 1 OR if the propagated carry is a 1 AND the input carry (C_{in}) is a 1. In other words, we get an output carry of 1 if it is generated by the full-adder ($A = 1$ AND $B = 1$) or if the adder propagates the input carry ($A = 1$ OR $B = 1$) AND $C_{in} = 1$. This relationship is expressed as

$$C_{out} = C_g + C_p C_{in}$$

Now let's see how this concept can be applied to a parallel adder, whose individual stages are shown in Figure F.20-17 for a 4-bit example. For each full-adder, the output carry is dependent on the generated carry (C_g), the propagated carry (C_p), and its input carry (C_{in}). The C_g and C_p functions for each stage are *immediately* available as soon as the input bits A and B and the input carry to the LSB adder are applied, because they are

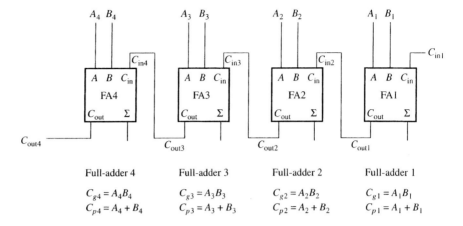

FIGURE F.20-17
Carry generation and carry propagation in terms of the input bits to a 4-bit adder.

dependent only on these bits. The input carry to each stage is the output carry of the previous stage.

Based on this analysis, we can now develop expressions for the output carry, C_{out}, of each full-adder stage for the 4-bit example.

Full-adder 1:

$$C_{out1} = C_{g1} + C_{p1}C_{in1}$$

Full-adder 2:

$$C_{in2} = C_{out1}$$
$$C_{out2} = C_{g2} + C_{p2}C_{in2}$$
$$= C_{g2} + C_{p2}C_{out1}$$
$$= C_{g2} + C_{p2}(C_{g1} + C_{p1}C_{in1})$$

$$C_{out2} = C_{g2} + C_{p2}C_{g1} + C_{p2}C_{p1}C_{in1}$$

Full-adder 3:

$$C_{in3} = C_{out2}$$
$$C_{out3} = C_{g3} + C_{p3}C_{in3}$$
$$= C_{g3} + C_{p3}C_{out2}$$
$$= C_{g3} + C_{p3}(C_{g2} + C_{p2}C_{g1} + C_{p2}C_{p1}C_{in1})$$

$$C_{out3} = C_{g3} + C_{p3}C_{g2} + C_{p3}C_{p2}C_{g1} + C_{p3}C_{p2}C_{p1}C_{in1}$$

Full-adder 4:

$$C_{in4} = C_{out3}$$
$$C_{out4} = C_{g4} + C_{p4}C_{in4}$$
$$= C_{g4} + C_{p4}C_{out3}$$
$$= C_{g4} + C_{p4}(C_{g3} + C_{p3}C_{g2} + C_{p3}C_{p2}C_{g1} + C_{p3}C_{p2}C_{p1}C_{in1})$$

$$C_{out4} = C_{g4} + C_{p4}C_{g3} + C_{p4}C_{p3}C_{g2} + C_{p4}C_{p3}C_{p2}C_{g1} + C_{p4}C_{p3}C_{p2}C_{p1}C_{in1}$$

Notice that in each of these expressions, the output carry for each full-adder stage is dependent only on the initial input carry (C_{in1}), the C_g and C_p functions of that stage, and the C_g and C_p functions of the preceding stages. Since each of the C_g and C_p functions can be expressed in terms of the A and B inputs to the full-adders, all the output carries are immediately available (except for gate delays), and you do not have to wait for a carry to ripple through all the stages before a final result is achieved. Thus, the look-ahead carry technique speeds up the addition process.

The adder equations are implemented with logic gates and connected to the full-adders to create a 4-bit look-ahead carry adder, as shown in Figure F.20-18.

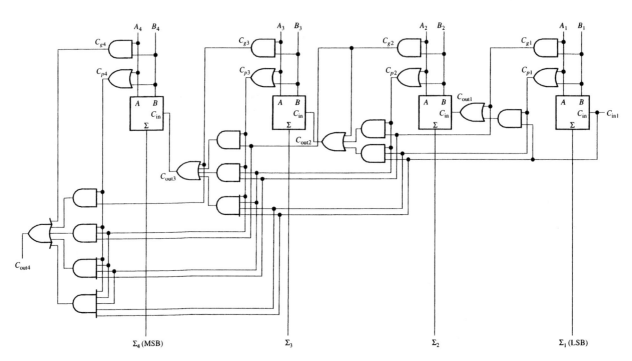

FIGURE F.20-18
Logic diagram for a 4-stage look-ahead carry adder.

Subtraction Subtraction is also performed by a digital circuit. A **subtracter** requires three inputs: the two numbers that are to be subtracted and a borrow input. The two outputs are the difference and the borrow output. When, for instance, 5 is subtracted from 8 with no borrow input, the difference is 3 with no borrow output. Subtraction can actually be performed by an adder because subtraction is simply a special case of addition (using 2's complement representation).

Multiplication Multiplication is performed by a digital circuit called a *multiplier.* Since numbers are always multiplied two at a time, two inputs are required. The output of the multiplier is the product. Since multiplication is simply a series of additions with shifts in the positions of the partial products, it can be performed by using an adder in conjunction with other circuits.

Division Division can be performed with a series of subtractions, comparisons, and shifts, and thus it can also be done using an adder in conjunction with other circuits. Two inputs to the divider are required, and the outputs generated are the quotient and the remainder.

REVIEW QUESTIONS

True/False

1. A half-adder adds three bits, producing a sum and carry.

2. The exclusive-OR gate acts like a simple two-bit adder.

Multiple Choice

3. How many gates are required for a full-adder?

   a. 2.

   b. 3.

   c. 5.

4. Using more than one parallel adder to add large binary numbers is called

   a. Addition grouping.

   b. Cascading.

   c. Parallel propagation.

5. Which is faster, an 8-bit ripple adder or an 8-bit carry look-ahead adder?

   a. Ripple adder.

   b. Carry look-ahead adder.

   c. Both take the same time.

F.21 Troubleshoot and repair types of arithmetic-logic circuits

INTRODUCTION

Troubleshooting arithmetic circuitry may involve looking at the circuit several different ways. For example, in a ripple adder, is the data being supplied too fast? What is the worst case delay between valid inputs and a stable output? This information is typically available on the manufacturer's data sheet for the device. Data arriving too fast at the inputs may never get a chance to affect the outputs in the correct way.

Furthermore, are the overall equations for the arithmetic circuitry correct? Have the equations been properly implemented in hardware? Are some results correct and others invalid? This may be due to improper cascading of the required signals (carry, borrow, etc.), or swapped bits on a parallel input (*ABCD* actually connected as *ACBD,* for instance).

A logic analyzer will go a long way toward determining what is going wrong in the circuit. Take a snapshot of the circuit input and output waveforms. Locate incorrect output patterns and try to determine their cause. Are the inputs correct? Is the device being used properly?

F.22 Understand principles and operations of types of multiplexer and demultiplexer circuits

INTRODUCTION

Multiplexer and demultiplexer circuits are the backbone of many modern communication systems. In this section we examine their basic operation.

MULTIPLEXERS (DATA SELECTORS)

A multiplexer (MUX) is a device that allows digital information from several sources to be routed onto a single line for transmission over that line to a common destination. The basic multiplexer has several data-input lines and a single output line. It also has data-select inputs, which permit digital data on any one of the inputs to be switched to the output line. Multiplexers are also known as data selectors.

A logic symbol for a 4-input **multiplexer (MUX)** is shown in Figure F.22-1. Notice that there are two data-select lines because, with two select bits, any one of the four data-input lines can be selected.

FIGURE F.22-1

Logic symbol for a 1-of-4 data selector/multiplexer.

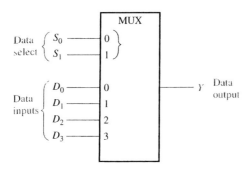

In Figure F.22-1, a 2-bit binary code on the data-select (S) inputs will allow the data on the selected data input to pass through to the data output. If a binary 0 ($S_1 = 0$ and $S_0 = 0$) is applied to the data-select lines, the data on input D_0 appear on the data-output line. If a binary 1 ($S_1 = 0$ and $S_0 = 1$) is applied to the data-select lines, the data on input D_1 appear on the data output. If a binary 2 ($S_1 = 1$ and $S_0 = 0$) is applied, the data on D_2 appear on the output. If a binary 3 ($S_1 = 1$ and $S_0 = 1$) is applied, the data on D_3 are switched to the output line. A summary of this operation is given in Table F.22-1.

TABLE F.22-1

Data selection for a 1-of-4 multiplexer.

Data-select Inputs		Input Selected
S_1	S_0	
0	0	D_0
0	1	D_1
1	0	D_2
1	1	D_3

Now let's look at the logic circuitry required to perform this multiplexing operation. The data output is equal to the state of the *selected* data input. We can therefore derive a logic expression for the output in terms of the data input and the select inputs.

The data output is equal to D_0 if and only if $S_1 = 0$ and $S_0 = 0$: $Y = D_0\overline{S}_1\overline{S}_0$.

The data output is equal to D_1 if and only if $S_1 = 0$ and $S_0 = 1$: $Y = D_1\overline{S}_1 S_0$.

The data output is equal to D_2 if and only if $S_1 = 1$ and $S_0 = 0$: $Y = D_2 S_1\overline{S}_0$.

The data output is equal to D_3 if and only if $S_1 = 1$ and $S_0 = 1$: $Y = D_3 S_1 S_0$.

When these terms are ORed, the total expression for the data output is

$$Y = D_0\overline{S}_1\overline{S}_0 + D_1\overline{S}_1 S_0 + D_2 S_1\overline{S}_0 + D_3 S_1 S_0$$

The implementation of this equation requires four 3-input AND gates, a 4-input OR gate, and two inverters to generate the complements of S_1 and S_0, as shown in Figure F.22-2. Because data can be selected from any one of the input lines, this circuit is also referred to as a **data selector.**

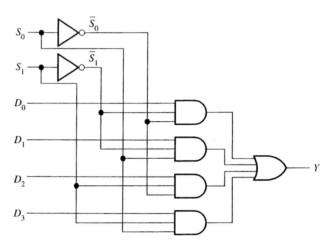

FIGURE F.22-2
Logic diagram for a 4-input multiplexer.

EXAMPLE F.22-1

The data-input and data-select waveforms in Figure F.22-3(a) are applied to the multiplexer in Figure F.22-2. Determine the output waveform in relation to the inputs.

FIGURE F.22-3

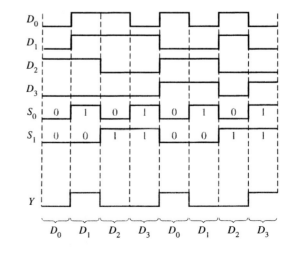

(a)

(b)

Solution The binary state of the data-select inputs during each interval determines which data input is selected. Notice that the data-select inputs go through a repetitive binary sequence 00, 01, 10, 11, 00, 01, 10, 11, and so on. The resulting output waveform is shown in Figure F.22-3(b).

The 74157 Quadruple 2-Input Data Selector/Multiplexer

The 74157, as well as its LS and CMOS versions, consists of four separate 2-input multiplexers. Each of the four multiplexers shares a common data-select line and a common *Enable,* as shown in Figure F.22-4(a). Because there are only two inputs to be selected in each multiplexer, a single data-select input is sufficient.

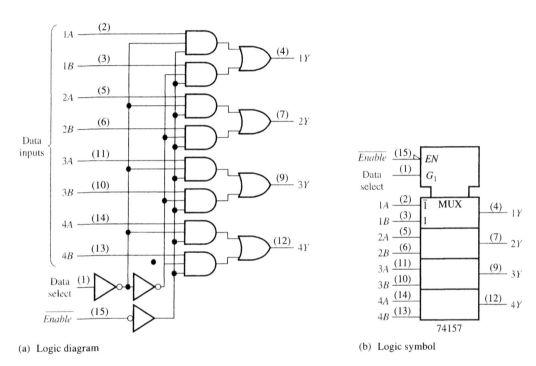

(a) Logic diagram

(b) Logic symbol

FIGURE F.22-4
The 74157 quadruple 2-input data selector/multiplexer.

As you can see in the logic diagram, the data-select input is ANDed with the *B* input of each 2-input multiplexer, and the complement of data-select is ANDed with each *A* input.

A LOW on the $\overline{Enable}$ input allows the selected input data to pass through to the output. A HIGH on the $\overline{Enable}$ input prevents data from going through to the output; that is, it disables the multiplexers.

The ANSI/IEEE Logic Symbol

The ANSI/IEEE logic symbol for the 74157 is shown in Figure F.22-4(b). Notice that the four multiplexers are indicated by the partitioned outline and that the inputs common to all four multiplexers are indicated as inputs to the notched block at the top, which is called the *common control block.* All labels within the upper MUX block apply to the other blocks below it.

Notice the 1 and $\overline{1}$ labels in the MUX blocks and the G1 label in the common control block. These labels are an example of the **dependency notation** system specified in the ANSI/IEEE Standard 91–1984. In this case G1 indicates an AND relationship between the data-select input and the data inputs with 1 or $\overline{1}$ labels. (The $\overline{1}$ means that the AND relationship applies to the complement of the G1 input.) In other words, when the data-select input is HIGH, the *B* inputs of the multiplexers are selected; and when the data-select input is LOW, the *A* inputs are selected. A "G" is always used to denote AND dependency. Other aspects of dependency notation will be introduced when appropriate.

The 74151A 8-Input Data Selector/Multiplexer

The 74151A has eight data inputs and, therefore, three data-select input lines. Three bits are required to select any one of the eight data inputs ($2^3 = 8$). A LOW on the $\overline{Enable}$ input allows the selected input data to pass through to the output. Notice that the data output and its complement are both available. The logic diagram is shown in Figure F.22-5(a), and the ANSI/IEEE logic symbol is shown in part (b). In this case there is no need for a common control block on the logic symbol because there is only one multiplexer to be controlled, not four as in the 74157. The G_7^0 label within the logic symbol indicates the AND relationship between the data-select inputs and each of the data inputs 0 through 7.

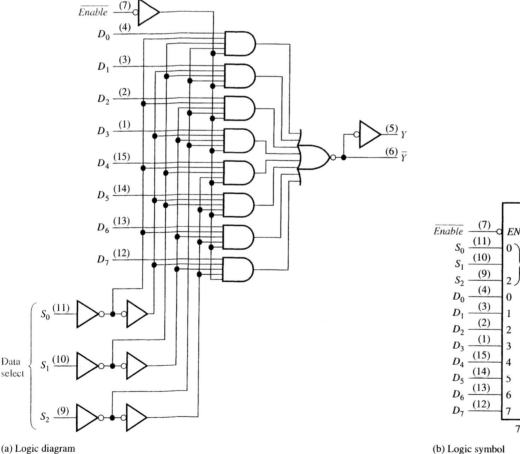

(a) Logic diagram

(b) Logic symbol

FIGURE F.22-5
The 74151A 8-input data selector/multiplexer.

The 74150 16-Input Data Selector/Multiplexer

The 74150 has sixteen data inputs and four data-select lines. In this case four bits are required to select any one of the sixteen data inputs ($2^4 = 16$). There is also an active-LOW $\overline{Enable}$ input. On this particular device, only the complement of the output is available. The logic symbol is shown in Figure F.22-6.

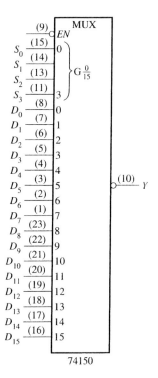

FIGURE F.22-6
The 74150 16-input data selector/multiplexer.

EXAMPLE F.22-2

Use 74150s and any other logic necessary to multiplex 32 data lines onto a single data-output line.

Solution An implementation of this system is shown in Figure F.22-7. Five bits are required to select one of 32 data inputs ($2^5 = 32$). In this application the $\overline{Enable}$ input is used as the most significant data-select bit. When the MSB in the data-select code is LOW, the left 74150 is enabled, and one of the data inputs (D_0 through D_{15}) is selected by the other four data-select bits. When the data-select MSB is HIGH, the right 74150 is enabled, and one of the data inputs (D_{16} through D_{31}) is selected. The selected input data are then passed through to the negative-OR gate and onto the single output line.

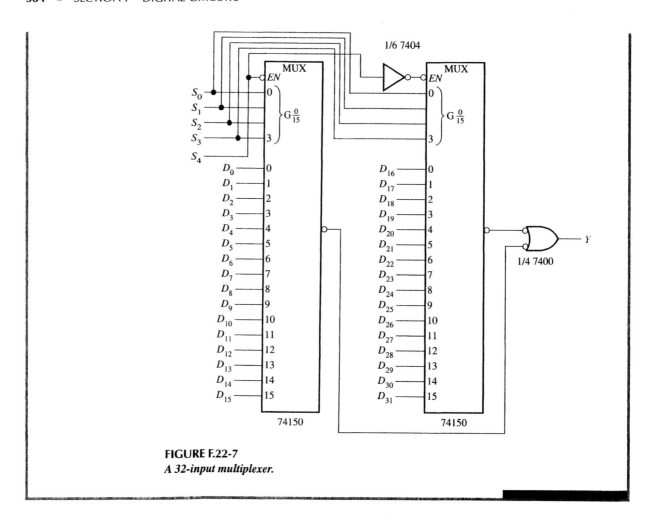

FIGURE F.22-7
A 32-input multiplexer.

Data Selector/Multiplexer Applications

A 7-Segment Display Multiplexer   Figure F.22-8 shows a simplified method of multiplexing BCD numbers to a 7-segment display. In this example, 2-digit numbers are displayed on the 7-segment readout by the use of a single BCD-to-7-segment decoder. This basic method of display multiplexing can be extended to displays with any number of digits. The basic operation is as follows.

Two BCD digits ($A_3A_2A_1A_0$ and $B_3B_2B_1B_0$) are applied to the multiplexer inputs. A square wave is applied to the data-select line, and when it is LOW, the A bits ($A_3A_2A_1A_0$) are passed through to the inputs of the 74LS48 BCD-to-7-segment decoder. The LOW on the data-select also puts a LOW on the 1 input of the 74LS139 2-line-to-4-line decoder, thus activating its 0 output and enabling the A-digit display by effectively connecting its common terminal to ground. The A digit is now on and the B digit is off.

When the data-select line goes HIGH, the B bits ($B_3B_2B_1B_0$) are passed through to the inputs of the BCD-to-7-segment decoder. Also, the 74LS139 decoder's 1 output is activated, thus enabling the B-digit display. The B digit is now on and the A digit is off. The cycle repeats at the frequency of the data-select square wave. This frequency must be high enough (about 30 Hz) to prevent visual flicker as the digit displays are multiplexed.

A Logic Function Generator   A useful application of the data selector/multiplexer is in the generation of combinational logic functions in sum-of-products form. When used in this way, the device can replace discrete gates, can often greatly reduce the number of ICs, and can make design changes much easier.

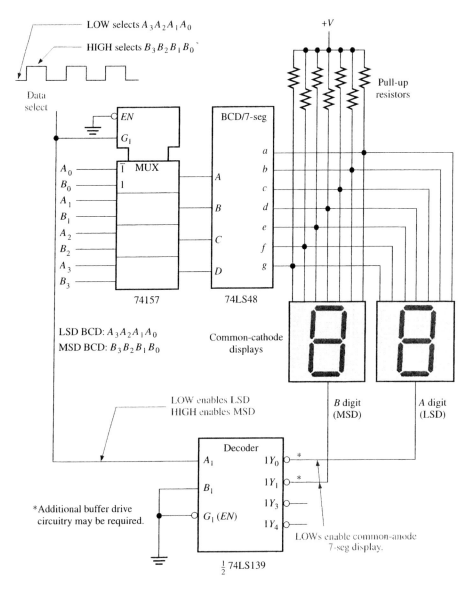

FIGURE F.22-8
Simplified 7-segment display multiplexing logic.

To illustrate, a 74151A 8-input data selector/multiplexer can be used to implement any specified 3-variable logic function if the variables are connected to the data-select inputs, and each data input is set to the logic level required in the truth table for that function. For example, if the function is a 1 when the variable combination is $\overline{A}_2 A_1 \overline{A}_0$, the 2 input (selected by 010) is connected to a HIGH. This HIGH is passed through to the output when this particular combination of variables occurs on the data-select lines. An example will help clarify this application.

EXAMPLE F.22-3

Implement the logic function specified in Table F.22-2 by using a 74151A 8-input data selector/multiplexer. Compare this method with a discrete logic gate implementation.

TABLE F.22-2

	Inputs		Output
A_2	A_1	A_0	Y
0	0	0	0
0	0	1	1
0	1	0	0
0	1	1	1
1	0	0	0
1	0	1	1
1	1	0	1
1	1	1	0

Solution   Notice from the truth table that Y is a 1 for the following input variable combinations: 001, 011, 101, and 110. For all other combinations, Y is 0. For this function to be implemented with the data selector, the data input selected by each of the above-mentioned combinations must be connected to a HIGH (5 V) through a pull-up resistor. All the other data inputs must be connected to a LOW (ground), as shown in Figure F.22-9.

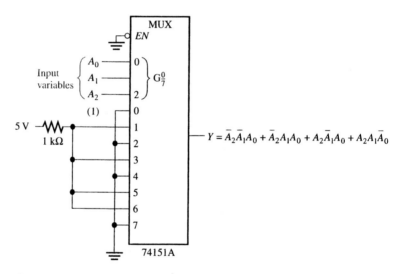

$$Y = \bar{A_2}\bar{A_1}A_0 + \bar{A_2}A_1A_0 + A_2\bar{A_1}A_0 + A_2A_1\bar{A_0}$$

FIGURE F.22-9
Data selector/multiplexer connected as a 3-variable logic function generator.

The implementation of this function with logic gates would require four 3-input AND gates, one 4-input OR gate, and three inverters unless the expression can be simplified.

Example F.22-3 illustrated how the 8-input data selector can be used as a logic function generator for three variables. Actually, this device can be also used as a 4-variable logic function generator by the utilization of one of the bits (A_0) in conjunction with the data inputs.

A 4-variable truth table has sixteen combinations of input variables. When an 8-bit data selector is used, each input is selected twice: the first time when A_0 is 0 and the second time when A_0 is 1. With this in mind, the following rules can be applied (Y is the output, and A_0 is the least significant bit):

1. If $Y = 0$ both times a given data input is selected by a certain combination of the input variables, $A_3A_2A_1$, connect that data input to ground (0).

2. If $Y = 1$ both times a given data input is selected by a certain combination of the input variables, $A_3A_2A_1$, connect that data input to $+V$ (1).

3. If Y is different the two times a given data input is selected by a certain combination of the input variables, $A_3A_2A_1$, and if $Y = A_0$, connect that data input to A_0.

4. If Y is different the two times a given data input is selected by a certain combination of the input variables, $A_3A_2A_1$, and if $Y = \overline{A}_0$, connect that data input to $\overline{A}_0$.

The following example illustrates this method.

EXAMPLE F.22-4

Implement the logic function in Table F.22-3 by using a 74151A 8-input data selector/multiplexer. Compare this method with a discrete logic gate implementation.

TABLE F.22-3

Decimal Digit	Inputs				Output Y
	A_3	A_2	A_1	A_0	
0	0	0	0	0	0
1	0	0	0	1	1
2	0	0	1	0	1
3	0	0	1	1	0
4	0	1	0	0	0
5	0	1	0	1	1
6	0	1	1	0	1
7	0	1	1	1	1
8	1	0	0	0	1
9	1	0	0	1	0
10	1	0	1	0	1
11	1	0	1	1	0
12	1	1	0	0	1
13	1	1	0	1	1
14	1	1	1	0	0
15	1	1	1	1	1

Solution The data-select inputs are $A_3A_2A_1$. In the first row of the table, $A_3A_2A_1 = 000$ and $Y = A_0$. In the second row, where $A_3A_2A_1$ again is 000, $Y = A_0$. Thus, A_0 is connected to the 0 input. In the third row of the table, $A_3A_2A_1 = 001$ and $Y = \overline{A}_0$. Also, in the fourth row, when $A_3A_2A_1$ again is 001, $Y = \overline{A}_0$. Thus, A_0 is inverted and connected to the 1 input. This analysis is continued until each input is properly connected according to the specified rules. The implementation is shown in Figure F.22-10.

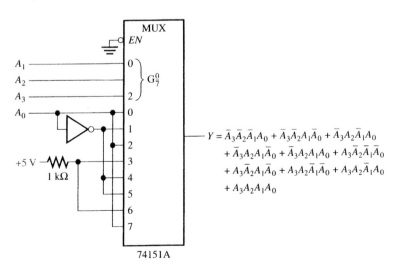

$$Y = \overline{A}_3\overline{A}_2\overline{A}_1A_0 + \overline{A}_3\overline{A}_2A_1\overline{A}_0 + \overline{A}_3A_2\overline{A}_1A_0$$
$$+ \overline{A}_3A_2A_1\overline{A}_0 + \overline{A}_3A_2A_1A_0 + A_3\overline{A}_2\overline{A}_1\overline{A}_0$$
$$+ A_3\overline{A}_2A_1\overline{A}_0 + A_3A_2\overline{A}_1\overline{A}_0 + A_3A_2\overline{A}_1A_0$$
$$+ A_3A_2A_1A_0$$

74151A

FIGURE F.22-10
Data selector/multiplexer connected as a 4-variable logic function generator.

> If implemented with logic gates, the function would require as many as ten 4-input AND gates, one 10-input OR gate, and four inverters, although possible simplification would reduce this requirement.

DEMULTIPLEXERS

A demultiplexer (DEMUX) basically reverses the multiplexing function. It takes data from one line and distributes them to a given number of output lines. For this reason, the demultiplexer is also known as a data distributor. As you will learn, decoders can also be used as demultiplexers.

Figure F.22-11 shows a 1-line-to-4-line **demultiplexer (DEMUX)** circuit. The data-input line goes to all of the AND gates. The two data-select lines enable only one gate at a time, and the data appearing on the data-input line will pass through the selected gate to the associated data-output line.

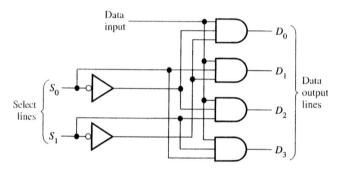

FIGURE F.22-11
A 1-line-to-4-line demultiplexer.

EXAMPLE F.22-5

The serial data-input waveform (Data in) and data-select inputs (S_0 and S_1) are shown in Figure F.22-12. Determine the data-output waveforms on D_0 through D_3 for the demultiplexer in Figure F.22-11.

FIGURE F.22-12

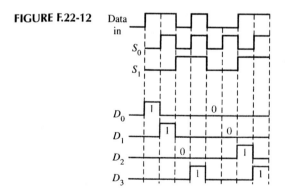

Solution   Notice that the select lines go through a binary sequence so that each successive input bit is routed to D_0, D_1, D_2, and D_3 in sequence, as shown by the output waveforms in Figure F.22-12.

The 74154 as a Demultiplexer

We have already discussed the 74154 in its application as a 4-line-to-16-line decoder. This device and other decoders are also used in demultiplexing applications. The logic symbol for this device when used as a demultiplexer is shown in Figure F.22-13. In demultiplexer applications, the input lines are used as the data-select lines. One of the *Enable* inputs is used as the data-input line, with the other *Enable* input held LOW to enable the internal negative-AND gate at the bottom of the diagram.

FIGURE F.22-13
The 74154 used as a demultiplexer.

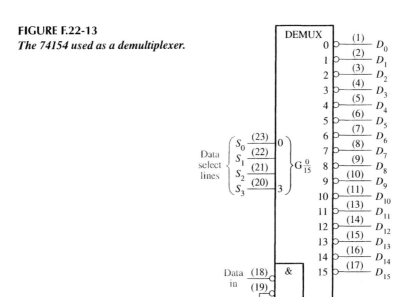

REVIEW QUESTIONS

True/False

1. A multiplexer selects one signal from many inputs.

2. A demultiplexer sends a signal to one of many outputs.

Multiple Choice

3. How many select inputs does an 8-input multiplexer require?

   a. 2.

   b. 3.

   c. 8.

4. How many quad 2-input multiplexers are required to select four bits from a group of 32 bits (eight groups of four bits)?

   a. 5.

   b. 6.

   c. 7.

5. A decoder has four select inputs. How many outputs are selectable?

   a. 4.

   b. 8.

   c. 16.

F.23 Troubleshoot and repair types of multiplexer and demultiplexer circuits

INTRODUCTION

In this section, the problem of decoder glitches is introduced and examined from a troubleshooting standpoint. A glitch is any undesired voltage or current spike (pulse) of very short duration. A glitch can be interpreted as a valid signal by a logic circuit and may cause improper operation.

A 74LS138 3-line-to-8-line decoder (binary-to-octal) is used in Figure F.23-1 to illustrate how **glitches** occur and how to identify their cause. The $A_2A_1A_0$ inputs of the decoder are sequenced through a binary count, and the resulting waveforms of the inputs and outputs can be displayed on the screen of a logic analyzer as shown in Figure F.23-1.

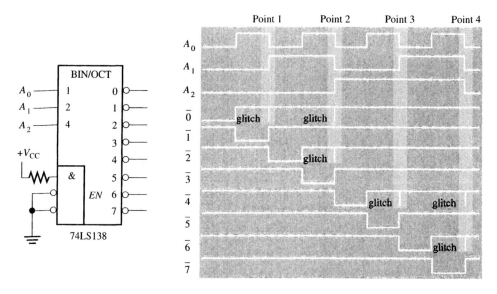

FIGURE F.23-1
Decoder waveforms with output glitches.

The output waveforms are correct except for the glitches that occur on some of the output signals. An oscilloscope or the oscilloscope function of a logic analyzer can be used to examine the critical timing details of the input waveforms in an effort to pinpoint the cause of the glitches. The oscilloscope is useful in this case because the waveforms can be "magnified" so that very small time differences between waveform transitions can be seen. A_2 transitions are delayed from A_1 transitions and A_1 transitions are delayed from A_0 transitions. This commonly occurs when waveforms are generated by a binary counter.

The points of interest indicated by the highlighted areas on the input waveforms in Figure F.23-1 are displayed on an oscilloscope as shown in Figure F.23-2. At point 1 there is a transitional state of 000 due to delay differences in the waveforms. This causes the first

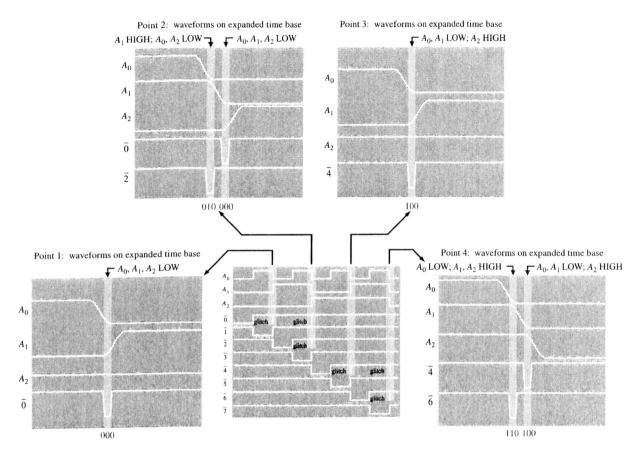

FIGURE F.23-2
Decoder waveform displays showing how transitional input states produce glitches in the output waveforms.

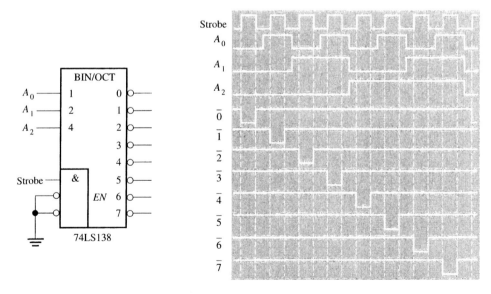

FIGURE F.23-3
Application of a strobe waveform to eliminate glitches on decoder outputs.

glitch on the $\overline{0}$ output of the decoder. At point 2 there are two transitional states, 010 and 000. These cause the glitch on the $\overline{2}$ output of the decoder and the second glitch on the $\overline{0}$ output, respectively. At point 3 the transitional state is 100, which causes the first glitch on the $\overline{4}$ output of the decoder. At point 4 the two transitional states, 110 and 100, result in the glitch on the $\overline{6}$ output and the second glitch on the $\overline{4}$ output, respectively.

One way to eliminate the glitch problem is a method called **strobing,** in which the decoder is enabled by a strobe pulse only during the times when the waveforms are not in transition. This method is illustrated in Figure F.23-3.

F.24 Understand principles and operations of types of digital to analog and analog to digital circuits

INTRODUCTION

Digital to analog and analog to digital converters are the interfaces between the digital world and the analog world. In this section we examine the operation of several common converters.

EXAMPLE F.24-1

Determine the output of the DAC in Figure F.24-1(a) if the waveforms representing a sequence of 4-bit numbers in part (b) are applied to the inputs. Input D_0 is the least significant bit (LSB).

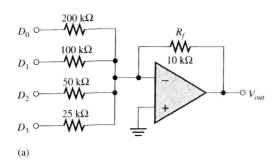

(a)

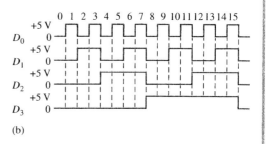

(b)

FIGURE F.24-1

Solution First, determine the current for each of the weighted inputs. Since the inverting input ($-$) of the op-amp is at 0 V (virtual ground) and a binary 1 corresponds to $+5$ V, the current through any of the input resistors is 5 V divided by the resistance value:

$$I_0 = \frac{5 \text{ V}}{200 \text{ k}\Omega} = 0.025 \text{ mA}$$

$$I_1 = \frac{5 \text{ V}}{100 \text{ k}\Omega} = 0.05 \text{ mA}$$

$$I_2 = \frac{5 \text{ V}}{50 \text{ k}\Omega} = 0.1 \text{ mA}$$

$$I_3 = \frac{5 \text{ V}}{25 \text{ k}\Omega} = 0.2 \text{ mA}$$

Almost no current goes into the inverting op-amp input because of its extremely high impedance. Therefore, assume that all of the current goes through the feedback resistor R_f. Since one end of R_f is at 0 V (virtual ground), the drop across R_f equals the output voltage, which is negative with respect to virtual ground.

$$V_{out(D0)} = (10 \text{ k}\Omega)(-0.025 \text{ mA}) = -0.25 \text{ V}$$
$$V_{out(D1)} = (10 \text{ k}\Omega)(-0.05 \text{ mA}) = -0.5 \text{ V}$$
$$V_{out(D2)} = (10 \text{ k}\Omega)(-0.1 \text{ mA}) = -1 \text{ V}$$
$$V_{out(D3)} = (10 \text{ k}\Omega)(-0.2 \text{ mA}) = -2 \text{ V}$$

593

From Figure F.24-1(b), the first binary input code is 0000, which produces an output voltage of 0 V. The next input code is 0001, which produces an output voltage of −0.25 V. The next code is 0010, which produces an output voltage of −0.5 V. The next code is 0011, which produces an output voltage of −0.25 V + −0.5 V = −0.75 V. Each successive binary code increases the output voltage by −0.25 V, so for this particular straight binary sequence on the inputs, the output is a stairstep waveform going from 0 V to −3.75 V in −0.25 V steps. This is shown in Figure F.24-2.

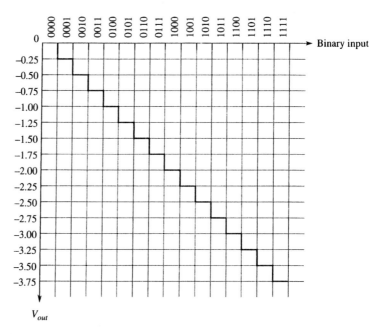

FIGURE F.24-2
Output of the DAC in Figure F.24-1.

The R/2R Ladder Digital-to-Analog Converter

Another method of D/A conversion is the $R/2R$ ladder, as shown in Figure F.24-3 for four bits. It overcomes one of the problems in the binary-weighted-input DAC in that it requires only two resistor values.

Start by assuming that the D_3 input is HIGH (+5 V) and the others are LOW (ground, 0 V). This condition represents the binary number 1000. A circuit analysis will show that this reduces to the equivalent form shown in Figure F.24-4(a). Essentially no current goes through the $2R$ equivalent resistance because the inverting input is at virtual ground. Thus,

FIGURE F.24-3
An R/2R ladder DAC.

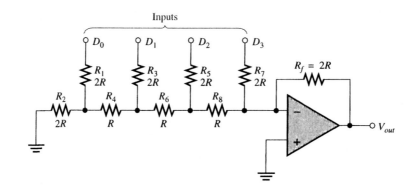

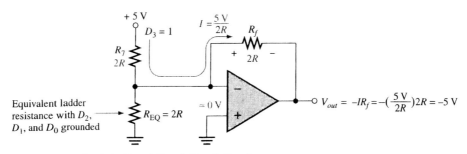

(a) Equivalent circuit for $D_3 = 1$, $D_2 = 0$, $D_1 = 0$, $D_0 = 0$

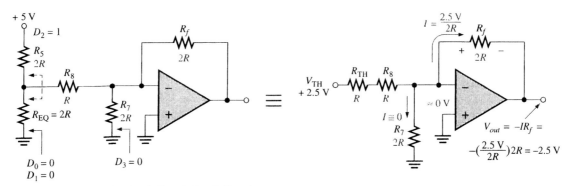

(b) Equivalent circuit for $D_3 = 0$, $D_2 = 1$, $D_1 = 0$, $D_0 = 0$

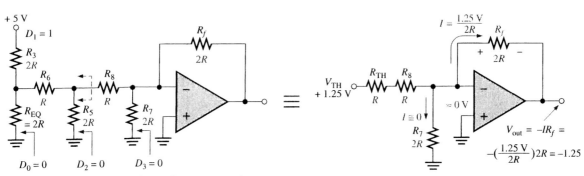

(c) Equivalent circuit for $D_3 = 0$, $D_2 = 0$, $D_1 = 1$, $D_0 = 0$

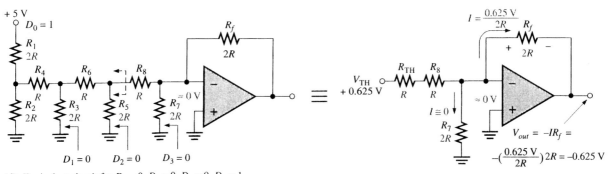

(d) Equivalent circuit for $D_3 = 0$, $D_2 = 0$, $D_1 = 0$, $D_0 = 1$

FIGURE F.24-4
Analysis of the R/2R ladder DAC.

all of the current ($I = 5$ V/$2R$) through R_7 also goes through R_f, and the output voltage is -5 V. The operational amplifier keeps the inverting ($-$) input near zero volts (≈ 0 V) because of negative feedback. Therefore, all current goes through R_f rather than into the inverting input.

Figure F.24-4(b) shows the equivalent circuit when the D_2 input is at $+5$ V and the others are at ground. This condition represents 0100. If we thevenize looking from R_8, we get 2.5 V in series with R, as shown. This results in a current through R_f of $I = 2.5$ V/$2R$, which gives an output voltage of -2.5 V. Keep in mind that there is no current into the op-amp inverting input and that there is no current through the equivalent resistance to ground because it has 0 V across it, due to the virtual ground.

Figure F.24-4(c) shows the equivalent circuit when the D_1 input is at $+5$ V and the others are at ground. This condition represents 0010. Again thevenizing looking from R_8, we get 1.25 V in series with R as shown. This results in a current through R_f of $I = 1.25$ V/$2R$, which gives an output voltage of -1.25 V.

In part (d) of the figure, the equivalent circuit representing the case where D_0 is at $+5$ V and the other inputs are at ground is shown. This condition represents 0001. Thevenizing from R_8 gives an equivalent of 0.625 V in series with R as shown. The resulting current through R_f is $I = 0.625$ V/$2R$, which gives an output voltage of -0.625 V.

Notice that each successively lower weighted input produces an output voltage that is halved, so that the output voltage is proportional to the binary weight of the input bits.

Performance Characteristics of Digital-to-Analog Converters

The performance characteristics of a DAC include resolution, accuracy, linearity, monotonicity, and settling time, each of which is discussed in the following list:

☐ *Resolution.* The resolution of a DAC is the reciprocal of the number of discrete steps in the output. This, of course, is dependent on the number of input bits. For example, a 4-bit DAC has a resolution of one part in $2^4 - 1$ (one part in fifteen). Expressed as a percentage, this is $(1/15)100 = 6.67\%$. The total number of discrete steps equals $2^n - 1$, where n is the number of bits. Resolution can also be expressed as the number of bits that are converted.

☐ *Accuracy.* Accuracy is a comparison of the actual output of a DAC with the expected output. It is expressed as a percentage of a full-scale, or maximum, output voltage. For example, if a converter has a full-scale output of 10 V and the accuracy is $\pm 0.1\%$, then the maximum error for any output voltage is $(10$ V$)(0.001) = 10$ mV. Ideally, the accuracy should be, at most, $\pm \frac{1}{2}$ of an LSB. For an 8-bit converter, 1 LSB is $1/256 = 0.0039$ (0.39% of full scale). The accuracy should be approximately $\pm 0.2\%$.

☐ *Linearity.* A linear error is a deviation from the ideal straight-line output of a DAC. A special case is an offset error, which is the amount of output voltage when the input bits are all zeros.

☐ *Monotonicity.* A DAC is **monotonic** if it does not take any reverse steps when it is sequenced over its entire range of input bits.

☐ *Settling time.* Settling time is normally defined as the time it takes a DAC to settle within $\pm \frac{1}{2}$ LSB of its final value when a change occurs in the input code.

EXAMPLE F.24-2

Determine the resolution, expressed as a percentage, of the following:
(a) an 8-bit DAC   (b) a 12-bit DAC

Solution
(a) For the 8-bit converter,

$$\frac{1}{2^8 - 1} \times 100 = \frac{1}{255} \times 100 = 0.392\%$$

(b) For the 12-bit converter,

$$\frac{1}{2^{12} - 1} \times 100 = \frac{1}{4095} \times 100 = 0.0244\%$$

ANALOG-TO-DIGITAL (A/D) CONVERSION

Flash (Simultaneous) Analog-to-Digital Converter

The flash method utilizes comparators that compare reference voltages with the analog input voltage. When the analog voltage exceeds the reference voltage for a given comparator, a HIGH is generated. Figure F.24-5 shows a 3-bit converter that uses seven comparator circuits; a comparator is not needed for the all-0s condition. A 4-bit converter of this type requires fifteen comparators. In general, $2^n - 1$ comparators are required for conversion to an n-bit binary code. The large number of comparators necessary for a reasonable-sized binary number is one of the disadvantages of the flash ADC. Its chief advantage is that it provides a fast conversion time.

The reference voltage for each comparator is set by the resistive voltage-divider circuit. The output of each comparator is connected to an input of the priority encoder. The encoder is sampled by a pulse on the enable input, and a 3-bit code representing the value of the analog input appears on the encoder's outputs. The binary code is determined by the highest-order input having a HIGH level.

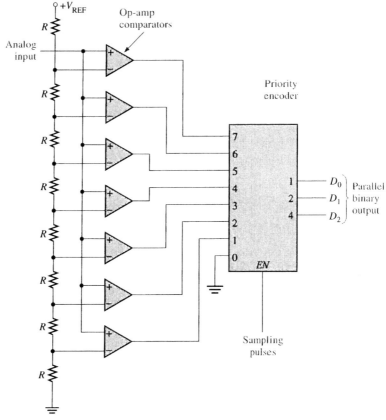

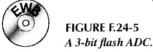

FIGURE F.24-5
A 3-bit flash ADC.

The sampling rate determines the accuracy with which the sequence of digital codes represents the analog input of the ADC. The more samples taken in a given unit of time, the more accurately the analog signal is represented in digital form.

Example F.24-3 illustrates the basic operation of the flash ADC in Figure F.24-5.

EXAMPLE F.24-3

Determine the binary code output of the 3-bit flash ADC for the analog input signal in Figure F.24-6 and the sampling pulses (encoder enabled) shown. For this example, $V_{REF} = +8$ V.

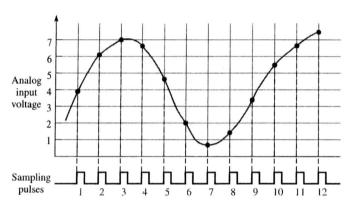

FIGURE F.24-6
Sampling of values on an analog waveform for conversion to digital form.

Solution The resulting digital output sequence is listed as follows and shown in the waveform diagram of Figure F.24-7 in relation to the sampling pulses:

100, 110, 111, 110, 100, 010, 000, 001, 011, 101, 110, 111

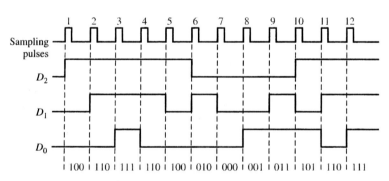

FIGURE F.24-7
Resulting digital outputs for sampled values. Output D_0 is the LSB.

Digital-Ramp Analog-to-Digital Converter

The digital-ramp method of A/D conversion is also known as the *stairstep-ramp* or the *counter* method. It employs a DAC and a binary counter to generate the digital value of an analog input. Figure F.24-8 shows a diagram of this type of converter.

Assume that the counter begins RESET and the output of the DAC is zero. Now assume that an analog voltage is applied to the input. When it exceeds the reference voltage

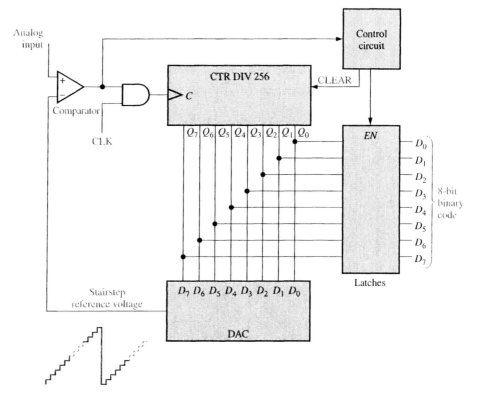

FIGURE F.24-8
Digital-ramp ADC (8 bits).

(output of DAC), the comparator switches to a HIGH output state and enables the AND gate. The clock pulses begin advancing the counter through its binary states, producing a stairstep reference voltage from the DAC. The counter continues to advance from one binary state to the next, producing successively higher steps in the reference voltage. When the stairstep reference voltage reaches the analog input voltage, the comparator output will go LOW and disable the AND gate, thus cutting off the clock pulses to stop the counter. The binary state of the counter at this point equals the number of steps in the reference voltage required to make the reference equal to or greater than the analog input. This binary number, of course, represents the value of the analog input. The control logic loads the binary count into the latches and resets the counter, thus beginning another count sequence to sample the input value.

The digital-ramp method is slower than the flash method because, in the worst case of maximum input, the counter must sequence through its maximum number of states before a conversion occurs. For an 8-bit conversion, this means a maximum of 256 counter states. Figure F.24-9 illustrates a conversion sequence for a 4-bit conversion. Notice that for each sample, the counter must count from zero up to the point at which the stairstep reference voltage reaches the analog input voltage. The conversion time varies, depending on the analog voltage.

Tracking Analog-to-Digital Converter

The tracking method uses an up/down counter and is faster than the digital-ramp method because the counter is not reset after each sample, but rather tends to track the analog input. Figure F.24-10 shows a typical 8-bit tracking ADC.

As long as the output reference voltage is less than the analog input, the comparator output is HIGH, putting the counter in the UP mode, which causes it to produce an up sequence of binary counts. This causes an increasing stairstep reference voltage out of the DAC, which continues until the ramp reaches the value of the input voltage.

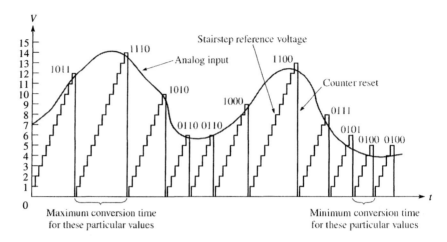

FIGURE F.24-9

Example of a 4-bit conversion, showing an analog input and the stairstep reference voltage for a digital-ramp ADC.

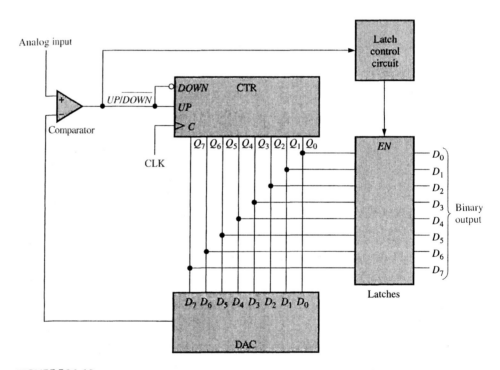

FIGURE F.24-10

An 8-bit tracking ADC.

When the reference voltage equals the analog input, the comparator's output switches LOW and puts the counter in the DOWN mode, causing it to back down one count. If the analog input is decreasing, the counter will continue to back down in its sequence and effectively track the input. If the input is increasing, the counter will back down one count after the comparison occurs and then will begin counting up again. When the input is constant, the counter backs down one count when a comparison occurs. The reference output is now less than the analog input, and the comparator output goes HIGH, causing the counter to count up. As soon as the counter increases one state, the reference voltage becomes greater than the input, switching the comparator to its LOW state. This causes the counter to back down one count. This back-and-forth action continues as long as

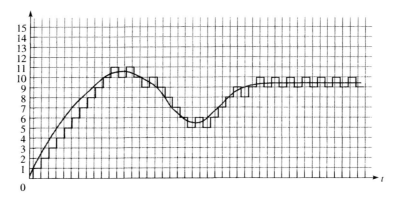

FIGURE F.24-11
Tracking action of a tracking ADC.

the analog input is a constant value, thus causing an oscillation between two binary states in the output. This is a disadvantage of this type of converter.

Figure F.24-11 illustrates the tracking action of this type of ADC for a 4-bit conversion.

Single-Slope Analog-to-Digital Converter

Unlike the digital-ramp and tracking methods, the single-slope converter does not require a DAC. It uses a linear ramp generator to produce a constant-slope reference voltage. A diagram is shown in Figure F.24-12. At the beginning of a conversion cycle, the counter is RESET and the ramp generator output is 0 V. The analog input is greater than the reference

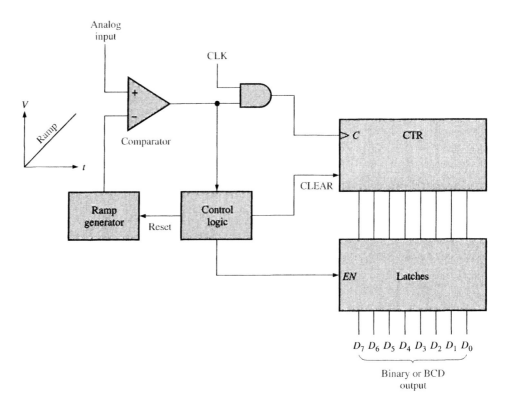

FIGURE F.24-12
Single-slope ADC.

voltage at this point and therefore produces a HIGH output from the comparator. This HIGH enables the clock to the counter and starts the ramp generator.

Assume that the slope of the ramp is 1 V/ms. The ramp will increase until it equals the analog input; at this point the ramp is RESET, and the binary or BCD count is stored in the latches by the control logic. Let's assume that the analog input is 2 V at the point of comparison. This means that the ramp is also 2 V and has been running for 2 ms. Since the comparator output has been HIGH for 2 ms, 200 clock pulses have been allowed to pass through the gate to the counter (assuming a clock frequency of 100 kHz). At the point of comparison, the counter is in the binary state that represents decimal 200. With proper scaling and decoding, this binary number can be displayed as 2.00 V. This basic concept is used in some digital voltmeters.

Dual-Slope Analog-to-Digital Converter

The operation of the dual-slope ADC is similar to that of the single-slope type except that a variable-slope ramp and a fixed-slope ramp are both used. This type of converter is common in digital voltmeters and other types of measurement instruments.

A ramp generator (integrator), A_1, is used to produce the dual-slope characteristic. A block diagram of a dual-slope ADC is shown in Figure F.24-13 for reference.

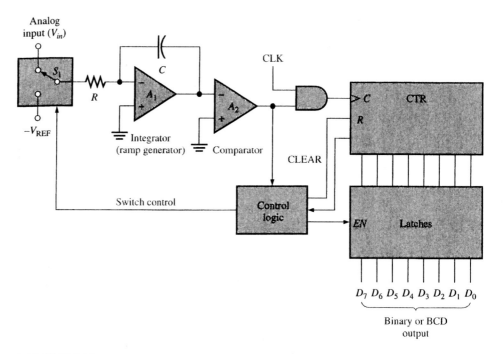

FIGURE F.24-13
Dual-slope ADC.

Figure F.24-14 illustrates dual-slope conversion. Start by assuming that the counter is RESET and the output of the integrator is zero. Now assume that a positive input voltage is applied to the input through the switch (S_1) as selected by the control logic. Since the inverting input of A_1 is at virtual ground, and assuming that V_{in} is constant for a period of time, there will be constant current through the input resistor R and therefore through the capacitor C. Capacitor C will charge linearly because the current is constant, and as a result, there will be a negative-going linear voltage ramp on the output of A_1, as illustrated in Figure F.24-14(a).

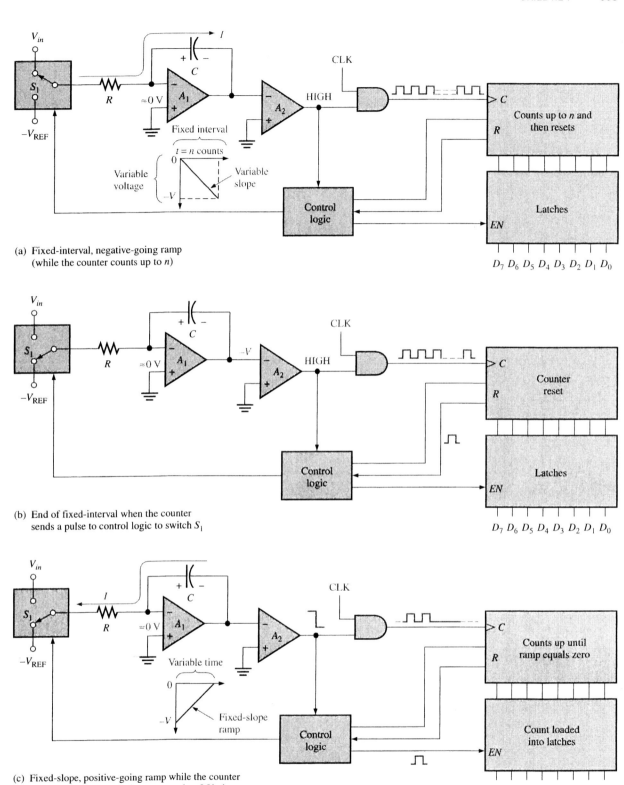

(a) Fixed-interval, negative-going ramp
(while the counter counts up to n)

(b) End of fixed-interval when the counter
sends a pulse to control logic to switch S_1

(c) Fixed-slope, positive-going ramp while the counter
counts up again. When the ramp reaches 0 V, the counter
stops, and the counter output is loaded into latches.

FIGURE F.24-14
Illustration of dual-slope conversion.

When the counter reaches a specified count, it will be RESET, and the control logic will switch the negative reference voltage ($-V_{REF}$) to the input of A_1 as shown in Figure F.24-14(b). At this point the capacitor is charged to a negative voltage ($-V$) proportional to the input analog voltage.

Now the capacitor discharges linearly because of the constant current from the $-V_{REF}$ as shown in Figure F.24-14(c). This linear discharge produces a positive-going ramp on the A_1 output, starting at $-V$ and having a constant slope that is independent of the charge voltage. As the capacitor discharges, the counter advances from its RESET state. The time it takes the capacitor to discharge to zero depends on the initial voltage $-V$ (proportional to V_{in}) because the discharge rate (slope) is constant. When the integrator (A_1) output voltage reaches zero, the comparator (A_2) switches to the LOW state and disables the clock to the counter. The binary count is latched, thus completing one conversion cycle. The binary count is proportional to V_{in} because the time it takes the capacitor to discharge depends only on $-V$, and the counter records this interval of time.

Successive-Approximation Analog-to-Digital Converter

Successive-approximation is perhaps the most widely used method of A/D conversion. It has a much shorter conversion time than the other methods with the exception of the flash method. It also has a fixed conversion time that is the same for any value of the analog input.

Figure F.24-15 shows a basic block diagram of a 4-bit successive-approximation ADC. It consists of a DAC, a successive-approximation register (SAR), and a comparator. The basic operation is as follows: The input bits of the DAC are enabled (made equal to a 1) one at a time, starting with the MSB. As each bit is enabled, the comparator produces an output that indicates whether the analog input voltage is greater or less than the output of the DAC. If the DAC output is greater than the analog input, the comparator's output is LOW, causing the bit in the register to RESET. If the output is less than the analog input, the 1 bit is retained in the register. The system does this with the MSB first, then the next most significant bit, then the next, and so on. After all the bits of the DAC have been tried, the conversion cycle is complete.

FIGURE F.24-15
Successive-approximation ADC.

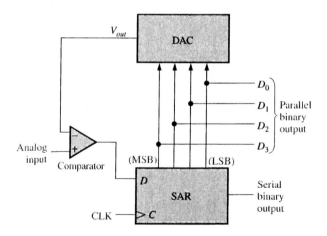

In order to better understand the operation of the successive-approximation ADC, let's take a specific example of a 4-bit conversion. Figure F.24-16 illustrates the step-by-step conversion of a constant analog input voltage (5 V in this case). Let's assume that the DAC has the following output characteristic: $V_{out} = 8$ V for the 2^3 bit (MSB), $V_{out} = 4$ V for the 2^2 bit, $V_{out} = 2$ V for the 2^1 bit, and $V_{out} = 1$ V for the 2^0 bit (LSB).

Figure F.24-16(a) shows the first step in the conversion cycle with the MSB = 1. The output of the DAC is 8 V. Since this is greater than the analog input of 5 V, the output of the comparator is LOW, causing the MSB in the SAR to be RESET to a 0.

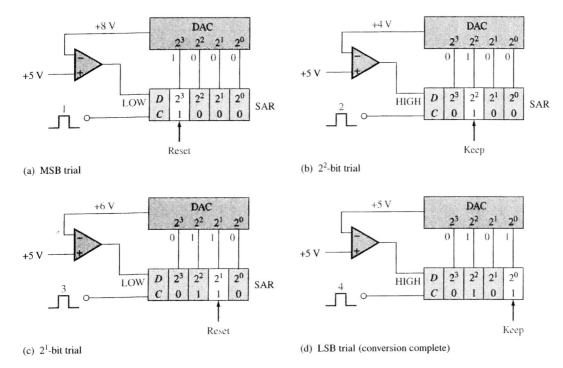

FIGURE F.24-16
Illustration of the successive-approximation conversion process.

(a) MSB trial

(b) 2^2-bit trial

(c) 2^1-bit trial

(d) LSB trial (conversion complete)

Figure F.24-16(b) shows the second step in the conversion cycle with the 2^2 bit equal to a 1. The output of the DAC is 4 V. Since this is less than the analog input of 5 V, the output of the comparator switches to a HIGH, causing this bit to be retained in the SAR.

Figure F.24-16(c) shows the third step in the conversion cycle with the 2^1 bit equal to a 1. The output of the DAC is 6 V because there is a 1 on the 2^2 bit input and on the 2^1 bit input; 4 V + 2 V = 6 V. Since this is greater than the analog input of 5 V, the output of the comparator switches to a LOW, causing this bit to be RESET to a 0.

Figure F.24-16(d) shows the fourth and final step in the conversion cycle with the 2^0 bit equal to a 1. The output of the DAC is 5 V because there is a 1 on the 2^2 bit input and on the 2^0 bit input; 4 V + 1 V = 5 V.

The four bits have all been tried, thus completing the conversion cycle. At this point the binary code in the register is 0101, which is the binary value of the analog input of 5 V. Another conversion cycle now begins, and the basic process is repeated. The SAR is cleared at the beginning of each cycle.

A Specific Analog-to-Digital Converter

The ADC0804 is an example of a successive-approximation ADC. A block diagram is shown in Figure F.24-17. This device operates from a +5 V supply and has a resolution of eight bits with a conversion time of 100 μs. Also, it has guaranteed monotonicity and an on-chip clock generator. The data outputs are tristate so that it can be interfaced with a microprocessor bus system.

The basic operation of the device is as follows: The ADC0804 contains the equivalent of a 256 resistor DAC network. The successive-approximation logic sequences the network to match the analog differential input voltage ($V_{in+} - V_{in-}$) with an output from the resistive network. The MSB is tested first. After eight comparisons (sixty-four clock-periods), an 8-bit binary code is transferred to output latches, and the interrupt ($\overline{INTR}$) output goes LOW. The device can be operated in a free-running mode by connecting the $\overline{INTR}$ output to the write ($\overline{WR}$) input and holding the conversion start ($\overline{CS}$) LOW. To ensure

FIGURE F.24-17
The ADC0804 analog-to-digital converter.

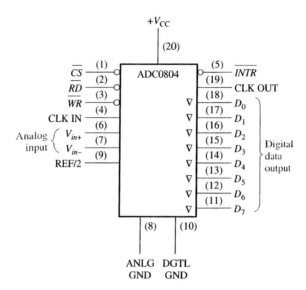

start-up under all conditions, a LOW $\overline{WR}$ input is required during the power-up cycle. Taking $\overline{CS}$ low anytime after that will interrupt the conversion process.

When the $\overline{WR}$ input goes LOW, the internal successive-approximation register (SAR) and the 8-bit shift register are RESET. As long as both $\overline{CS}$ and $\overline{WR}$ remain LOW, the ADC remains in a RESET state. One to eight clock-periods after $\overline{CS}$ or $\overline{WR}$ makes a LOW-to-HIGH transition, conversion starts.

When a LOW is at both the $\overline{CS}$ and $\overline{RD}$ inputs, the tristate output latch is enabled and the output code is applied to the D_0 through D_7 lines. When either the $\overline{CS}$ or the $\overline{RD}$ input returns to a HIGH, the D_0 through D_7 outputs are disabled.

REVIEW QUESTIONS

True/False

1. The SAR type of analog-to-digital converter takes a fixed time for each conversion.
2. The resolution of a DAC is the range of its output voltage.

Multiple Choice

3. Which is not a type of DAC?
    a. R/2R ladder.
    b. Dual slope.
    c. Binary weighted input.
4. Which is not a type of ADC?
    a. Voltage lock.
    b. Flash.
    c. Single slope.
5. How many different output voltages are possible with a 10-bit DAC?
    a. 10.
    b. 256.
    c. 1024.

F.25 Troubleshoot and repair types of digital to analog and analog to digital circuits

INTRODUCTION

Basic testing of DACs and ADCs includes checking their performance characteristics—such as monotonicity, offset, linearity, and gain—and checking for missing or incorrect codes. In this section, the fundamentals of testing these analog interfaces are introduced.

Testing Digital-to-Analog Converters

The concept of DAC testing is illustrated in Figure F.25-1. In this basic method, a sequence of binary codes is applied to the inputs, and the resulting output is observed. The binary code sequence extends over the full range of values from 0 to $2^n - 1$ in ascending order, where n is the number of bits.

FIGURE F.25-1
Basic test setup for a DAC.

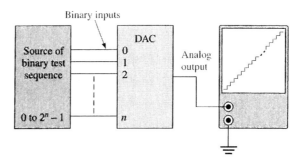

The ideal output is a straight-line stairstep as indicated. As the number of bits in the binary code is increased, the resolution is improved. That is, the number of discrete steps increases, and the output approaches a straight-line linear ramp.

D/A Conversion Errors

Several D/A conversion errors to be checked for are shown in Figure F.25-2, which uses a 4-bit conversion for illustration purposes. A 4-bit conversion produces fifteen discrete steps. Each graph in the figure includes an ideal stairstep ramp for comparison with the faulty outputs.

Nonmonotonicity   The step reversals in Figure F.25-2(a) indicate nonmonotonic performance, which is a form of nonlinearity. In this particular case, the error occurs because the 2^1 bit in the binary code is interpreted as a constant 0. That is, a short is causing the bit input line to be stuck LOW.

Differential Nonlinearity   Figure F.25-2(b) illustrates differential nonlinearity in which the step amplitude is less than it should be for certain input codes. This particular output could be caused by the 2^2 bit having an insufficient weight, perhaps because of a faulty input resistor. We could also see steps with amplitudes greater than normal, if a particular binary weight were greater than it should be.

607

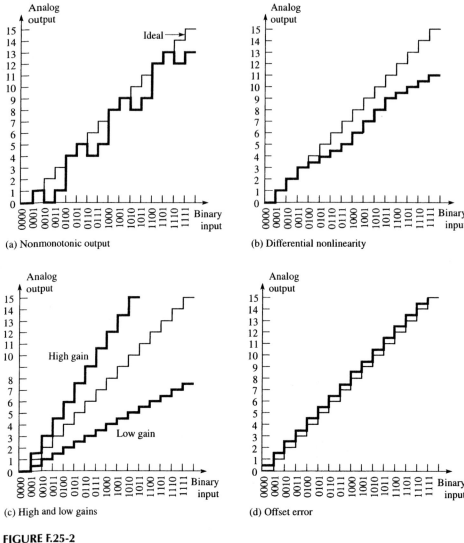

FIGURE F.25-2
Illustrations of several D/A conversion errors.

Low or High Gain   Output errors caused by low or high gain are illustrated in Figure F.25-2(c). In the case of low gain, all of the step amplitudes are less than ideal. In the case of high gain, all of the step amplitudes are greater than ideal. This situation may be caused by a faulty feedback resistor in the op-amp circuit.

Offset Error   An offset error is illustrated in Figure F.25-2(d). Notice that when the binary input is 0000, the output voltage is nonzero and that this amount of offset is the same for all steps in the conversion. A faulty op-amp may be the culprit in this situation.

EXAMPLE F.25-1

The DAC output in Figure F.25-3 is observed when a straight 4-bit binary sequence is applied to the inputs. Identify the type of error, and suggest an approach to isolate the fault.

FIGURE F.25-3

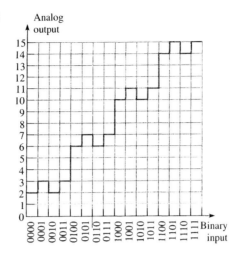

Solution The DAC in this case is nonmonotonic. Analysis of the output reveals that the device is converting the following sequence, rather than the actual binary sequence applied to the inputs.

0010, 0011, 0010, 0011, 0110, 0111, 0110, 0111, 1010, 1011, 1010, 1011, 1110, 1111, 1110, 1111

Apparently, the 2^1 bit is stuck in the HIGH (1) state. To find the problem, first monitor the bit input pin to the device. If it is changing states, the fault is internal, most likely an open. If the external pin is not changing states and is always HIGH, check for an external short to $+V$ that may be caused by a solder bridge somewhere on the circuit board. If no problem is found here, disconnect the source output from the DAC input pin, and see if the output signal is correct. If these checks produce no results, the fault is most likely internal to the DAC, perhaps a short to the supply voltage.

Testing Analog-to-Digital Converters

One method for testing ADCs is shown in Figure F.25-4. A DAC is used as part of the test setup to convert the ADC output back to analog form for comparison with the test input.

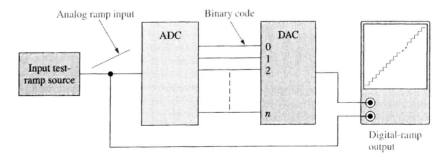

FIGURE F.25-4
A method for testing ADCs.

A test input in the form of a linear ramp is applied to the input of the ADC. The resulting binary output sequence is then applied to the DAC test unit and converted to a stairstep ramp. The input and output ramps are compared for any deviation.

A/D Conversion Errors

Again, a 4-bit conversion is used to illustrate the principles. Let's assume that the test input is an ideal linear ramp.

Missing Code The stairstep output in Figure F.25-5(a) indicates that the binary code 1001 does not appear on the output of the ADC. Notice that the 1000 value stays for two intervals and then the output jumps to the 1010 value.

In a flash ADC, for example, a failure of one of the comparators can cause a missing-code error.

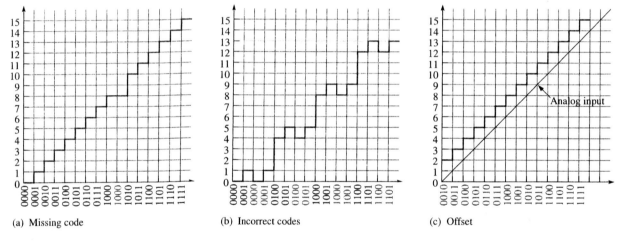

(a) Missing code   (b) Incorrect codes   (c) Offset

FIGURE F.25-5
Illustrations of A/D conversion errors.

Incorrect Codes The stairstep output in Figure F.25-5(b) indicates that several of the binary code words coming out of the ADC are incorrect. Analysis indicates that the 2^1-bit line is stuck in the LOW (0) state in this particular case.

Offset Offset conditions are shown in F.25-5(c). In this situation the ADC interprets the analog input voltage as greater than its actual value. This error is probably due to a faulty comparator circuit.

EXAMPLE F.25-2

A 4-bit flash ADC is shown in Figure F.25-6(a). It is tested with a setup like the one in Figure F.25-4. The resulting reconstructed analog output is shown in Figure F.25-6(b). Identify the problem and the most probable fault.

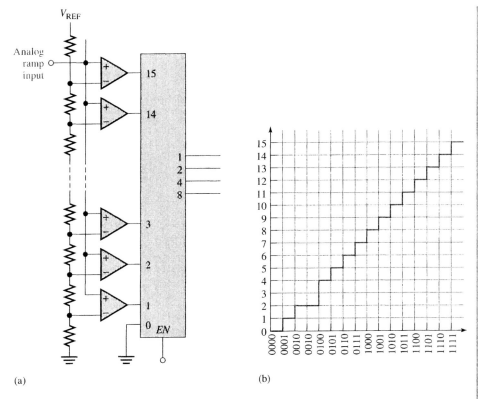

(a)

(b)

FIGURE F.25-6

Solution The binary code 0011 is missing from the ADC output, as indicated by the missing step. Most likely, the output of comparator 3 is stuck in its inactive state (LOW).

F.26 Understand principles and operations of types of digital display circuits

This material is covered in F.20, F.22, F.30, and F.32.

F.27 Troubleshoot and repair types of digital display circuits

INTRODUCTION

Digital display circuits typically utilize LEDs, LCD and seven-segment displays, and dot-matrix and segmented alphanumeric displays. Here are a number of points to keep in mind while troubleshooting a faulty display circuit.

☐ LEDs must be installed correctly and have sufficient forward current to illuminate. The same applies to common-cathode and common-anode seven-segment displays.

☐ The control signals on a seven-segment decoder/driver must be set up properly or the display will not function correctly.

☐ LCD displays require a backplane oscillator. Without it the display will not update. LCD displays can also crack, as they are sensitive to pressure. Cold temperature also interferes with correct operation.

☐ Dot-matrix and segmented alphanumeric displays usually contain their own built-in decoder/drivers. It is important to verify that all control signals are at their proper levels.

☐ Displays that are interfaced to a microprocessor may require a logic analyzer or oscilloscope to verify that they are receiving the correct control signals.

F.28 Understand principles and operations of power distribution noise problems

INTRODUCTION

Noise is unwanted voltage that is induced in electrical circuits and can present a threat to the proper operation of the circuit. Wires and other conductors within a system can pick up stray high-frequency electromagnetic radiation from adjacent conductors in which currents are changing rapidly or from many other sources external to the system. Also, power-line voltage fluctuation is a form of low-frequency noise.

In order not to be adversely affected by noise, a logic circuit must have a certain amount of **noise immunity.** This is the ability to tolerate a certain amount of unwanted voltage fluctuation on its inputs without changing its output state. For example, if noise

FIGURE F.28-1
Input and output logic levels for (a) TTL and (b) CMOS.

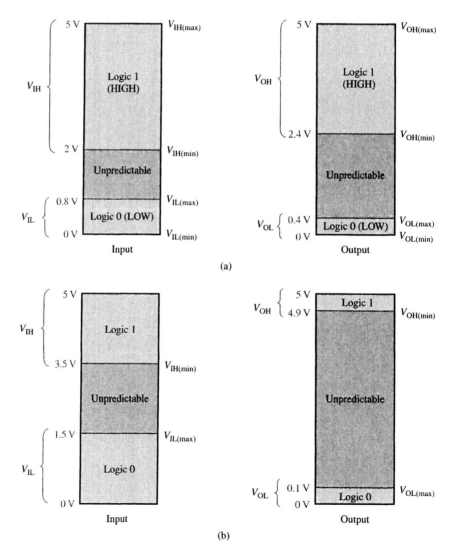

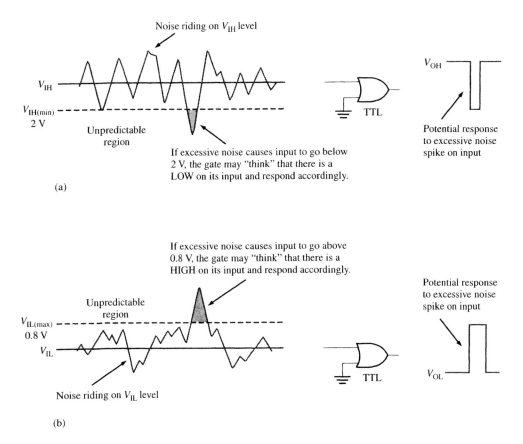

FIGURE F.28-2
Illustration of the effects of input noise on gate operation.

voltage causes the input of a TTL gate to drop below 2 V in the HIGH state, the operation is in the unpredictable region (see Figure F.28-1[a]). Thus, the gate may interpret the fluctuation below 2 V as a LOW level, as illustrated in Figure F.28-2(a). Similarly, if noise causes a gate input to go above 0.8 V in the LOW state, an uncertain condition is created, as illustrated in part (b).

Noise Margin

A measure of a circuit's noise immunity is called the **noise margin,** which is expressed in volts. There are two values of noise margin specified for a given logic circuit: the HIGH-level noise margin (V_{NH}) and the LOW-level noise margin (V_{NL}). These parameters are defined by the following equations:

$$V_{NH} = V_{OH(min)} - V_{IH(min)}$$

$$V_{NL} = V_{IL(max)} - V_{OL(max)}$$

From the equations, V_{NH} is the difference between the lowest possible HIGH output from a driving gate ($V_{OH(min)}$) and the lowest possible HIGH input that the load gate can tolerate ($V_{IH(min)}$). Noise margin V_{NL} is the difference between the maximum possible LOW input that a gate can tolerate ($V_{IL(max)}$) and the maximum possible LOW output of the driving gate ($V_{OL(max)}$). Noise margins are illustrated in Figure F.28-3.

FIGURE F.28-3
Noise margins.

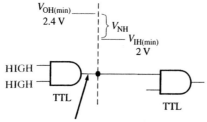

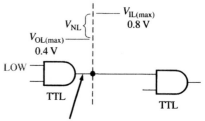

The voltage on this line will never be less than 2.4 V unless noise or improper operation is introduced.

(a) HIGH-level noise margin

The voltage on this line will never exceed 0.4 V unless noise or improper operation is introduced.

(b) LOW-level noise margin

EXAMPLE F.28-1

Determine the HIGH-level and LOW-level noise margins for TTL and for CMOS by using the information in Figure F.28-1.

Solution   For TTL,

$$V_{IH(min)} = 2 \text{ V}$$
$$V_{IL(max)} = 0.8 \text{ V}$$
$$V_{OH(min)} = 2.4 \text{ V}$$
$$V_{OL(max)} = 0.4 \text{ V}$$
$$V_{NH} = V_{OH(min)} - V_{IH(min)} = 2.4 \text{ V} - 2 \text{ V} = 0.4 \text{ V}$$
$$V_{NL} = V_{IL(max)} - V_{OL(max)} = 0.8 \text{ V} - 0.4 \text{ V} = 0.4 \text{ V}$$

A TTL gate is immune to up to 0.4 V of noise for both the HIGH and LOW input states. For CMOS,

$$V_{IH(min)} = 3.5 \text{ V}$$
$$V_{IL(max)} = 1.5 \text{ V}$$
$$V_{OH(min)} = 4.9 \text{ V}$$
$$V_{OL(max)} = 0.1 \text{ V}$$
$$V_{NH} = V_{OH(min)} - V_{IH(min)} = 4.9 \text{ V} - 3.5 \text{ V} = 1.4 \text{ V}$$
$$V_{NL} = V_{IL(max)} - V_{OL(max)} = 1.5 \text{ V} - 0.1 \text{ V} = 1.4 \text{ V}$$

BYPASS CAPACITORS

The amount of noise riding on the 5 V power line of a digital circuit increases as more chips are added or as the frequency of operation is increased. One simple solution is to place 0.1 μF (or some other suitably small value) capacitors, called *bypass capacitors*, between the power pins of each chip. The high frequency noise will cause the bypass capacitors to exhibit a low impedance, thus shorting the noise directly to ground. It is important for the bypass capacitor to be physically close to the chip it is protecting.

REVIEW QUESTIONS

True/False

1. Digital ICs have some noise immunity built in.
2. Noise margin is a measure of a circuit's noise immunity.

Multiple Choice

3. The TTL high-level noise margin is
    a. 0.4 V.
    b. 2 V.
    c. 5 V.
4. The low-level TTL noise margin is
    a. 0 V.
    b. 0.4 V.
    c. 0.8 V.
5. A bypass capacitor
    a. Reduces high frequency noise.
    b. Increases the gain of the IC.
    c. Increases the noise margin.

F.29 Troubleshoot and repair power distribution noise problems

INTRODUCTION

Digital circuits containing more than five ICs should have bypass capacitors on every IC to eliminate the possibility of switching noise on the power supply. In addition, the ground returns from each IC should have very good connections back to the power supply. A wire-wrapped circuit may have the power pins of each IC daisy-chained back to the supply. This is a bad idea, since some resistance is associated with each wrap. Eventually, ICs far away from the supply have their ground reference floating above the ground level at the power supply. A better method would be to only daisy-chain two-to-four ICs at a time, running separate ground wires back to the supply from each group.

F.30 Understand principles and operations of types of digital encoders and decoders

INTRODUCTION

Decoders and encoders provide the means of converting from one type of data format to another. For example, we may convert binary data into BCD. In this section we examine several types of decoders and encoders.

EXAMPLE F.30-1

Determine the logic required to decode the binary number 1011 by producing a HIGH level on the output.

Solution   The decoding function can be formed by complementing only the variables that appear as 0 in the binary number, as follows:

$$X = A_3\bar{A}_2A_1A_0 \quad (1011)$$

This function can be implemented by connecting the true (uncomplemented) variables A_0, A_1, and A_3 directly to the inputs of an AND gate, and inverting the variable A_2 before applying it to the AND gate input. The decoding logic is shown in Figure F.30-1.

FIGURE F.30-1
Decoding logic for producing a HIGH output when 1011 is on the inputs.

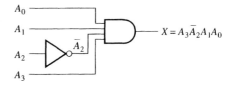

The 4-Bit Decoder

In order to decode all possible combinations of four bits, sixteen decoding gates are required ($2^4 = 16$). This type of decoder is commonly called a *4-line-to-16-line decoder* because there are four inputs and sixteen outputs or a *1-of-16 decoder* because for any given code on the inputs, one of the sixteen outputs is activated. A list of the sixteen binary codes and their corresponding decoding functions is given in Table F.30-1.

If an active-LOW output is desired for each decoded number, the entire decoder can be implemented with NAND gates and inverters as follows. First, since each variable and its complement are required in the decoder, as seen from Table F.30-1, each complement can be generated once and then used for all decoding gates as required, rather than be generated by a separate inverter for each place that complement is used. This arrangement is shown in Figure F.30-2. In order to decode each of the sixteen binary codes, sixteen NAND gates are required (AND gates can be used to produce active-HIGH outputs). The general decoding gate arrangement is illustrated in Figure F.30-2. Rather than reproducing the complex logic diagram for the decoder each time it is required in a schematic, a simpler representation is normally used.

TABLE 30–1
Decoding functions and truth table for a 4-line-to-16-line decoder with active-LOW outputs.

Decimal Digit	Binary Inputs				Decoding Function	Outputs															
	A_3	A_2	A_1	A_0		0	1	2	3	4	5	6	7	8	9	10	11	12	13	14	15
0	0	0	0	0	$\overline{A_3}\,\overline{A_2}\,\overline{A_1}\,\overline{A_0}$	0	1	1	1	1	1	1	1	1	1	1	1	1	1	1	1
1	0	0	0	1	$\overline{A_3}\,\overline{A_2}\,\overline{A_1}A_0$	1	0	1	1	1	1	1	1	1	1	1	1	1	1	1	1
2	0	0	1	0	$\overline{A_3}\,\overline{A_2}A_1\overline{A_0}$	1	1	0	1	1	1	1	1	1	1	1	1	1	1	1	1
3	0	0	1	1	$\overline{A_3}\,\overline{A_2}A_1A_0$	1	1	1	0	1	1	1	1	1	1	1	1	1	1	1	1
4	0	1	0	0	$\overline{A_3}A_2\overline{A_1}\,\overline{A_0}$	1	1	1	1	0	1	1	1	1	1	1	1	1	1	1	1
5	0	1	0	1	$\overline{A_3}A_2\overline{A_1}A_0$	1	1	1	1	1	0	1	1	1	1	1	1	1	1	1	1
6	0	1	1	0	$\overline{A_3}A_2A_1\overline{A_0}$	1	1	1	1	1	1	0	1	1	1	1	1	1	1	1	1
7	0	1	1	1	$\overline{A_3}A_2A_1A_0$	1	1	1	1	1	1	1	0	1	1	1	1	1	1	1	1
8	1	0	0	0	$A_3\overline{A_2}\,\overline{A_1}\,\overline{A_0}$	1	1	1	1	1	1	1	1	0	1	1	1	1	1	1	1
9	1	0	0	1	$A_3\overline{A_2}\,\overline{A_1}A_0$	1	1	1	1	1	1	1	1	1	0	1	1	1	1	1	1
10	1	0	1	0	$A_3\overline{A_2}A_1\overline{A_0}$	1	1	1	1	1	1	1	1	1	1	0	1	1	1	1	1
11	1	0	1	1	$A_3\overline{A_2}A_1A_0$	1	1	1	1	1	1	1	1	1	1	1	0	1	1	1	1
12	1	1	0	0	$A_3A_2\overline{A_1}\,\overline{A_0}$	1	1	1	1	1	1	1	1	1	1	1	1	0	1	1	1
13	1	1	0	1	$A_3A_2\overline{A_1}A_0$	1	1	1	1	1	1	1	1	1	1	1	1	1	0	1	1
14	1	1	1	0	$A_3A_2A_1\overline{A_0}$	1	1	1	1	1	1	1	1	1	1	1	1	1	1	0	1
15	1	1	1	1	$A_3A_2A_1A_0$	1	1	1	1	1	1	1	1	1	1	1	1	1	1	1	0

FIGURE F.30-2
Abbreviated logic for a 4-line-to-16-line decoder.

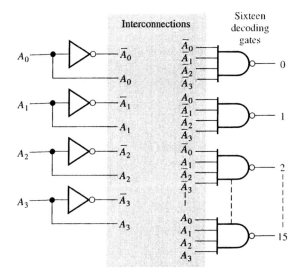

A logic symbol for a 4-line-to-16-line decoder is shown in Figure F.30-3. The BIN/DEC label indicates that a binary input makes the corresponding decimal output active. The input labels 1, 2, 4, and 8 represent the binary weights of the input bits $(2^3 2^2 2^1 2^0)$.

The 74154 4-Line-to-16-Line Decoder

The 74154 is a good example of an MSI decoder that is available in the LS TTL family as well as others. The CMOS equivalent is designated 74HC154. The logic diagram for the 74154 is shown in Figure F.30-4(a), and the logic symbol is shown in part (b). The additional inverters on the inputs are required to prevent excessive loading of the driving source(s). Each input is connected to the input of only one inverter, rather than to the inputs of several NAND gates as in Figure F.30-2. There is also an enable function provided on this particular device, which is implemented with a NOR gate used as a negative-AND. A

FIGURE F.30-3
Logic symbol for a 4-line-to-16-line decoder.

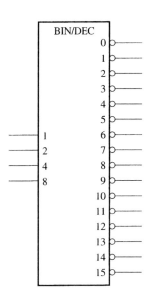

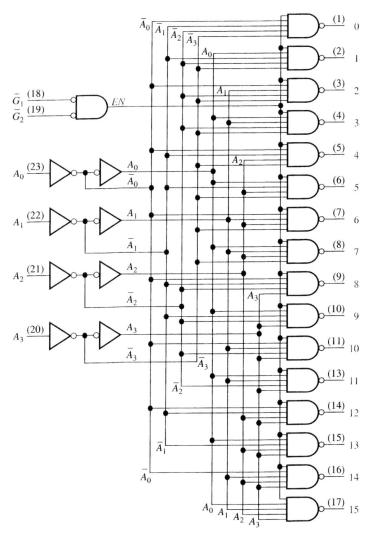

(a) Logic diagram

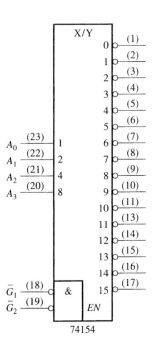

(b) Logic symbol

FIGURE F.30-4
The 74154 4-line-to-16-line decoder.

LOW level on each input, $\overline{G}_1$ and $\overline{G}_2$, is required in order to make the enable gate output (*EN*) HIGH. The enable gate output is connected to an input of *each* NAND gate, so it must be HIGH for the NAND gates to be enabled. If the enable gate is not activated by a LOW on both inputs, then all sixteen decoder outputs will be HIGH regardless of the states of the four input variables, A_0, A_1, A_2, and A_3.

EXAMPLE F.30-2

A certain application requires that a 5-bit number be decoded. Use 74154 decoders to implement the logic. The binary number is represented by the format $A_4A_3A_2A_1A_0$.

Solution Since the 74154 can handle only four bits, two decoders must be used to decode five bits. The fifth bit, A_4, is connected to the enable inputs, $\overline{G}_1$ and $\overline{G}_2$, of one decoder, and $\overline{A}_4$ is connected to the enable inputs of the other decoder, as shown in Figure F.30-5. When the decimal number is 15 or less, $A_4 = 0$, and the low-order decoder is enabled and the high-order decoder is disabled. When the decimal number is greater than 15, $A_4 = 1$ so $\overline{A}_4 = 0$, and the high-order decoder is enabled and the low-order decoder is disabled.

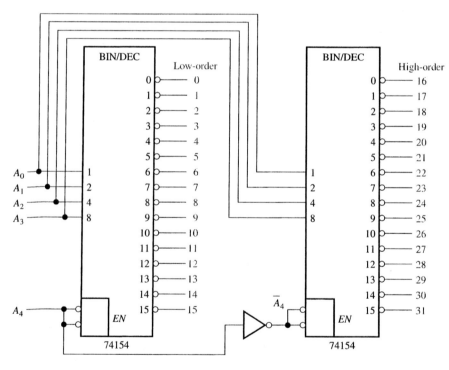

FIGURE F.30-5
A 5-bit decoder using 74154s.

EXAMPLE F.30-3

The 7442A is an integrated circuit BCD-to-decimal decoder. The logic symbol is shown in Figure F.30-6. If the input waveforms in Figure F.30-7(a) are applied to the inputs of the 7442A, sketch the output waveforms.

FIGURE F.30-6
The 7442A BCD-to-decimal decoder.

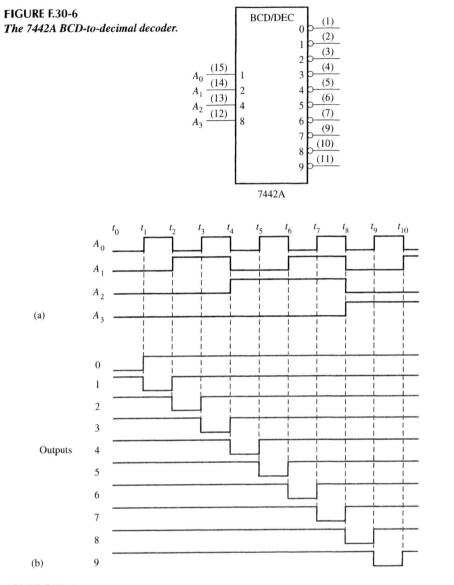

(a)

Outputs

(b)

FIGURE F.30-7

Solution The output waveforms are shown in Figure F.30-7(b). As you can see, the inputs are sequenced through the BCD for digits 0 through 9. The output waveforms in the timing diagram indicate that sequence.

The BCD-to-7-Segment Decoder

This type of decoder accepts the BCD code on its inputs and provides outputs to energize 7-segment display devices to produce a decimal readout. The logic diagram for a basic 7-segment decoder is shown in Figure F.30-8.

The 7447 BCD-to-7-Segment Decoder/Driver

The 7447 (also 74LS47) is an example of an MSI device that decodes a BCD input and drives a 7-segment display. In addition to its decoding and segment drive capability, the 7447 has several additional features as indicated by the *LT*, *RBI*, and *BI/RBO* functions in

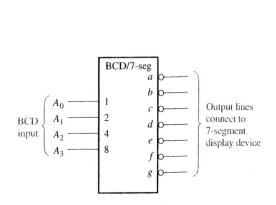

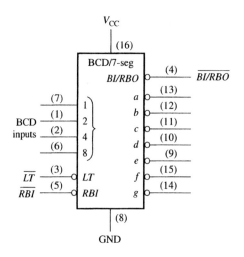

FIGURE F.30-8
Logic symbol for a BCD-to-7-segment decoder/driver with active-LOW outputs.

FIGURE F.30-9
The 7447 BCD-to-7-segment decoder/driver.

the logic symbol of Figure F.30-9. As indicated by the bubbles on the logic symbol, all of the outputs (*a* through *g*) are active-LOW as are the *LT* (lamp test), *RBI* (ripple blanking input), and *BI/RBO* (blanking input/ripple blanking output) functions. The outputs can drive a common-anode 7-segment display directly.

In addition to decoding a BCD input and producing the appropriate 7-segment outputs, the 7447 has lamp test and zero suppression capability.

Lamp Test   When a LOW is applied to the *LT* input and the *BI/RBO* is HIGH, all of the 7 segments in the display are turned on. Lamp test is used to verify that no segments are burned out.

Zero Suppression   **Zero suppression** is a feature used for multidigit displays to blank out unnecessary zeros. For example, in a 6-digit display the number 6.4 may be displayed as 006.400 if the zeros are not blanked out. Blanking the zeros at the front of a number is called *leading zero suppression* and blanking the zeros at the back of the number is called *trailing zero suppression*. Keep in mind that only nonessential zeros are blanked. With zero suppression, the number 030.080 will be displayed as 30.08 (the essential zeros remain).

Zero suppression in the 7447 is accomplished using the *RBI* and *BI/RBO* functions. *RBI* is the ripple blanking input and *RBO* is the ripple blanking output on the 7447; these are used for zero suppression. *BI* is the blanking input which shares the same pin with *RBO*; in other words, the *BI/RBO* pin can be used as an input or an output. When used as a *BI* (blanking input), all segment outputs are HIGH (nonactive) when *BI* is LOW, which overrides all other inputs. The *BI* function is not part of the zero suppression capability of the device.

All of the segment outputs of the decoder are nonactive (HIGH) if a zero code (0000) is on its BCD inputs and if its *RBI* is LOW. This causes the display to be blank and produces a LOW *RBO*.

The logic diagram in Figure F.30-10(a) illustrates leading zero suppression for a whole number. The highest-order digit position (left-most) is always blanked if a zero code is on its BCD inputs because the *RBI* of the most-significant decoder is made LOW by connecting it to ground. The *RBO* of each decoder is connected to the *RBI* of the next lowest-order decoder so that all zeros to the left of the first nonzero digit are blanked. For example, in part (a) of the figure the two highest-order digits are zeros and therefore are blanked. The remaining two digits, 3 and 9 are displayed.

The logic diagram in Figure F.30-10(b) illustrates trailing zero suppression for a fractional number. The lowest-order digit (right-most) is always blanked if a zero code is on its BCD inputs because the *RBI* is connected to ground. The *RBO* of each decoder is

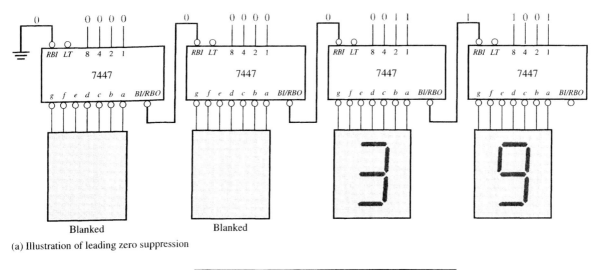

(a) Illustration of leading zero suppression

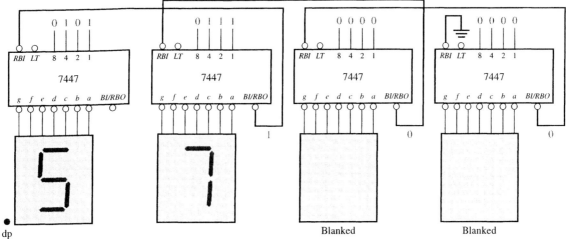

(b) Illustration of trailing zero suppression

FIGURE F.30-10
Examples of zero suppression using the 7447 BCD to 7-segment decoder/driver.

connected to the *RBI* of the next highest-order decoder so that all zeros to the right of the first nonzero digit are blanked. In part (b) of the figure, the two lowest-order digits are zeros and therefore are blanked. The remaining two digits, 5 and 7, are displayed. To combine both leading and trailing zero suppression in one display and to have decimal point capability, additional logic is required.

ENCODERS

An encoder is a combinational logic circuit that essentially performs a "reverse" decoder function. An encoder accepts an active level on one of its inputs representing a digit, such as a decimal or octal digit, and converts it to a coded output, such as BCD or binary. Encoders can also be devised to encode various symbols and alphabetic characters.

The Decimal-to-BCD Encoder

This type of **encoder** has ten inputs—one for each decimal digit—and four outputs corresponding to the BCD code, as shown in Figure F.30-11. This is a basic 10-line-to-4-line encoder.

FIGURE F.30-11
Logic symbol for a decimal-to-BCD encoder.

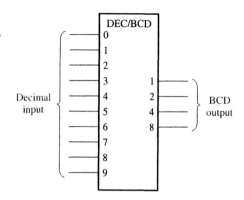

The BCD (8421) code is listed in Table F.30-2. From this table you can determine the relationship between each BCD bit and the decimal digits in order to analyze the logic. For instance, the most significant bit of the BCD code, A_3, is a 1 for decimal digit 8 or 9. The OR expression for bit A_3 in terms of the decimal digits can therefore be written

$$A_3 = 8 + 9$$

Bit A_2 is a 1 for decimal digit 4, 5, 6, or 7 and can be expressed as an OR function as follows:

$$A_2 = 4 + 5 + 6 + 7$$

Bit A_1 is a 1 for decimal digit 2, 3, 6, or 7 and can be expressed as

$$A_1 = 2 + 3 + 6 + 7$$

Finally, A_0 is a 1 for digit 1, 3, 5, 7, or 9. The expression for A_0 is

$$A_0 = 1 + 3 + 5 + 7 + 9$$

TABLE F.30-2

Decimal Digit	BCD Code			
	A_3	A_2	A_1	A_0
0	0	0	0	0
1	0	0	0	1
2	0	0	1	0
3	0	0	1	1
4	0	1	0	0
5	0	1	0	1
6	0	1	1	0
7	0	1	1	1
8	1	0	0	0
9	1	0	0	1

Now let's implement the logic circuitry required for encoding each decimal digit to a BCD code by using the logic expressions just developed. It is simply a matter of ORing the appropriate decimal digit input lines to form each BCD output. The basic encoder logic resulting from these expressions is shown in Figure F.30-12.

The basic operation of the circuit in Figure F.30-12 is as follows: When a HIGH appears on *one* of the decimal digit input lines, the appropriate levels occur on the four BCD output lines. For instance, if input line 9 is HIGH (assuming all other input lines are LOW), this condition will produce a HIGH on outputs A_0 and A_3 and LOWs on outputs A_1 and A_2, which is the BCD code (1001) for decimal 9.

FIGURE F.30-12
Basic logic diagram of a decimal-to-BCD encoder. A 0-digit input is not needed because the BCD outputs are all LOW when there are no HIGH inputs.

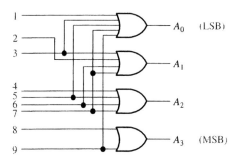

The Decimal-to-BCD Priority Encoder

The decimal-to-BCD **priority encoder** performs the same basic encoding function as previously discussed. It also offers additional flexibility in that it can be used in applications that require priority detection. The priority function means that the encoder will produce a BCD output corresponding to the highest-order decimal digit input that is active and will ignore any other active inputs. For instance, if the 6 and the 3 inputs are both active, the BCD output is 0110 (which represents decimal 6).

Priority Logic   Now let's look at the requirements for the priority detection logic. This logic circuitry prevents a lower-order digit input from disrupting the encoding of a higher-order digit. We start by examining each BCD output (beginning with output A_0). Referring to Figure F.30-12, notice that A_0 is HIGH when 1, 3, 5, 7, or 9 is HIGH. Digit input 1 should activate the A_0 output only if no higher-order digits other than those that also activate A_0 are HIGH. This requirement can be stated as follows:

1. A_0 is HIGH if 1 is HIGH *and* 2, 4, 6, and 8 are LOW.

2. A_0 is HIGH if 3 is HIGH *and* 4, 6, and 8 are LOW.

3. A_0 is HIGH if 5 is HIGH *and* 6 and 8 are LOW.

4. A_0 is HIGH if 7 is HIGH *and* 8 is LOW.

5. A_0 is HIGH if 9 is HIGH.

These five statements describe the priority of encoding for the BCD bit A_0. The A_0 output is HIGH if any of the conditions listed occur; that is, A_0 is true if statement 1, statement 2, statement 3, statement 4, or statement 5 is true. This can be expressed in the form of the following logic equation:

$$A_0 = (1 \cdot \overline{2} \cdot \overline{4} \cdot \overline{6} \cdot \overline{8}) + (3 \cdot \overline{4} \cdot \overline{6} \cdot \overline{8}) + (5 \cdot \overline{6} \cdot \overline{8}) + (7 \cdot \overline{8}) + 9$$

From this expression the logic circuitry required for the A_0 output with priority inhibits can be implemented, as shown in Figure F.30-13.

The same reasoning process can be applied to output A_1, and the following logical statements can be made:

1. A_1 is HIGH if 2 is HIGH *and* 4, 5, 8, and 9 are LOW.

2. A_1 is HIGH if 3 is HIGH *and* 4, 5, 8, and 9 are LOW.

FIGURE F.30-13
Logic for the A_0 output of a decimal-to-BCD priority encoder.

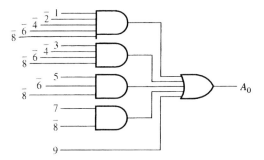

3. A_1 is HIGH if 6 is HIGH *and* 8 and 9 are LOW.

4. A_1 is HIGH if 7 is HIGH *and* 8 and 9 are LOW.

These statements are summarized in the following equation, and the logic implementation is shown in Figure F.30-14.

$$A_1 = (2 \cdot \overline{4} \cdot \overline{5} \cdot \overline{8} \cdot \overline{9}) + (3 \cdot \overline{4} \cdot \overline{5} \cdot \overline{8} \cdot \overline{9}) + (6 \cdot \overline{8} \cdot \overline{9}) + (7 \cdot \overline{8} \cdot \overline{9})$$

Next, output A_2 can be described as follows:

1. A_2 is HIGH if 4 is HIGH *and* 8 and 9 are LOW.

2. A_2 is HIGH if 5 is HIGH *and* 8 and 9 are LOW.

3. A_2 is HIGH if 6 is HIGH *and* 8 and 9 are LOW.

4. A_2 is HIGH if 7 is HIGH *and* 8 and 9 are LOW.

In equation form, output A_2 is

$$A_2 = (4 \cdot \overline{8} \cdot \overline{9}) + (5 \cdot \overline{8} \cdot \overline{9}) + (6 \cdot \overline{8} \cdot \overline{9}) + (7 \cdot \overline{8} \cdot \overline{9})$$

The logic circuitry for the A_2 output appears in Figure F.30-15.

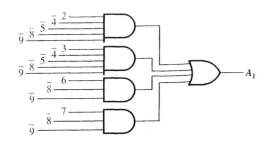

FIGURE F.30-14
Logic for the A_1 output of a decimal-to-BCD priority encoder.

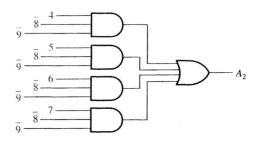

FIGURE F.30-15
Logic for the A_2 output of a decimal-to-BCD priority encoder.

Finally, for the A_3 output,

A_3 is HIGH if 8 is HIGH *or* if 9 is HIGH.

This statement is expressed in equation form as follows:

$$A_3 = 8 + 9$$

The logic for the A_3 output is shown in Figure F.30-16. No inhibits are required for this one.

We now have developed the basic logic for the decimal-to-BCD priority encoder. All the complements of the input digits are realized by inverting the inputs.

FIGURE F.30-16
Logic for the A_3 output of a decimal-to-BCD priority encoder.

8 ⎓⎓⎓⎓
9 ⎓⎓⎓⎓ A_3

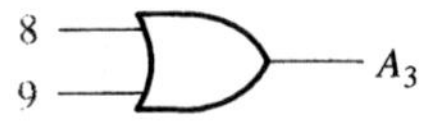

The 74LS147 Decimal-to-BCD Encoder

The 74LS147 is a priority encoder with active-LOW inputs for decimal digits 1 through 9 and active-LOW BCD outputs as indicated in the logic symbol in Figure F.30-17. A BCD zero output is represented when none of the inputs is active. The device pin numbers are in parentheses.

FIGURE F.30-17
Logic symbol for the 74LS147 decimal-to-BCD priority encoder (HPRI means highest value input has priority).

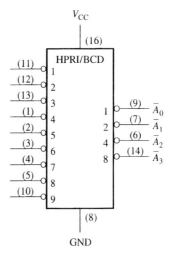

EXAMPLE F.30-4

If LOW levels appear on pins 1, 4, and 13 of the 74LS147 illustrated in Figure F.30-17, indicate the state of the four outputs. All other inputs are HIGH.

Solution Pin 4 is the highest-order decimal digit input having a LOW level and represents decimal 7. Therefore, the output levels indicate the BCD code for decimal 7 where $\overline{A}_0$ is the LSB and $\overline{A}_3$ is the MSB. Output $\overline{A}_0$ is LOW, $\overline{A}_1$ is LOW, $\overline{A}_2$ is LOW, and $\overline{A}_3$ is HIGH.

The 74148 Octal-to-Binary Encoder

The 74148 is a priority encoder that has eight active-LOW inputs and three active-LOW binary outputs, as shown in Figure F.30-18. This device can be used for converting octal inputs (recall that the octal digits are 0 through 7) to a 3-bit binary code. To enable the device, the *EI* (enable input) must be LOW. It also has the *EO* (enable output) and *GS* output for expansion purposes. The *EO* is LOW when the *EI* is LOW and none of the inputs (0 through 7) is active. *GS* is LOW when *EI* is LOW and any of the inputs is active.

The 74148 can be expanded to a 16-line-to-4-line encoder by connecting the *EO* of the higher-order encoder to the *EI* of the lower-order encoder and negative-ORing the

FIGURE F.30-18
Logic symbol for the 74148 octal-to-binary encoder.

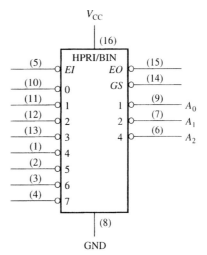

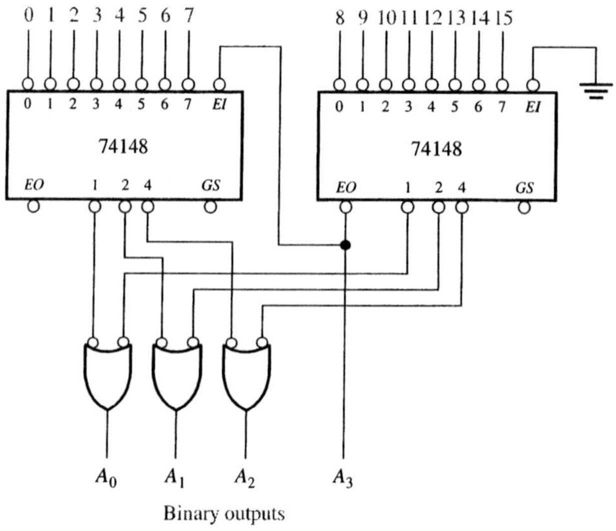

FIGURE F.30-19
A 16-line-to-4-line encoder using 74148s and external logic.

corresponding binary outputs as shown in Figure F.30-19. The EO is used as the fourth and most-significant bit. This particular configuration produces active-HIGH outputs for the 4-bit binary number.

REVIEW QUESTIONS

True/False

1. A decoder activates a single output in a group of outputs.
2. BCD stands for binary coded decimal.

Multiple Choice

3. How many 74154s are required to control 100 outputs?

    a. 2.

    b. 6.

    c. 7.

4. What does the number 0140 look like when zero-blanking is enabled?

    a. 14.

    b. 014.

    c. 140.

5. An encoder with eight inputs requires

    a. 1 output.

    b. 2 outputs.

    c. 3 outputs.

F.31 Troubleshoot and repair types of digital encoders and decoders

This material is covered in F.23.

F.32 Understand principles and operations of types of digital display devices

INTRODUCTION

Seven-segment displays are used in everything from automobile instruments to Z-meters. The tablet counting and control system that has been the focus of the system applications in the previous skill topics has two 7-segment displays. These displays are used with logic circuits that decode a binary coded decimal (BCD) number and activate the appropriate digits on the display. In this section, we will focus on a minimum-gate design using decoder logic to illustrate applications of Boolean expressions and the Karnaugh map.

The 7-Segment Display

Figure F.32-1 shows a common display format composed of seven elements or segments. Energizing certain combinations of these segments can cause each of the ten decimal digits to be displayed. Figure F.32-2 illustrates this method of digital display for each of the ten digits by using a darkened segment to represent one that is energized. To produce a 1, segments b and c are energized; to produce a 2, segments a, b, g, e, and d are used; and so on.

FIGURE F.32-1
Seven-segment display format showing arrangement of segments.

FIGURE F.32-2
Display of decimal digits with a 7-segment device.

LED Displays   One common type of 7-segment display consists of light-emitting diodes **(LED)** arranged as shown in Figure F.32-3. Each segment is an LED that emits light when there is current through it. In Figure F.32-3(a), the common-anode arrangement requires the driving circuit to provide a low-level voltage in order to activate a given segment. When a LOW is applied to a segment input, the LED is turned on, and there is current through it. In Figure F.32-3(b) the common-cathode arrangement requires the driver to provide a high-level voltage to activate a segment. When a HIGH is applied to a segment input, the LED is turned on and there is current through it.

LCD Displays   Another common type of 7-segment display is the liquid crystal display **(LCD)**. LCDs operate by polarizing light so that a nonactivated segment reflects incident light and thus appears invisible against its background. An activated segment does not

FIGURE F.32-3
Arrangements of 7-segment LED displays.

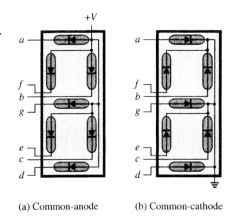

(a) Common-anode   (b) Common-cathode

reflect incident light and thus appears dark. LCDs consume much less power than LEDs, but they cannot be seen in the dark, while LEDs can.

LCDs operate from a low-frequency signal voltage (30 Hz to 60 Hz) applied between the segment and a common element called the *backplane* (bp). The basic operation is as follows. Figure F.32-4 shows a square wave used as the source signal. Each segment in the display is driven by an exclusive-OR gate with one input connected to an output of the 7-segment decoder/driver and the other input connected to the signal source. When the decoder/driver output is HIGH (1), the exclusive-OR output is a square wave that is 180° out-of-phase with the source signal, as shown in Figure F.32-4(a). You can verify this by reviewing the truth table operation of the exclusive-OR. The resulting voltage between the LCD segment and the backplane is also a square wave because when $V_{seg} = 1$, $V_{bp} = 0$, and vice versa. The voltage difference turns the segment on.

When the decoder/driver output is LOW (0), the exclusive-OR output is a square wave that is in-phase with the source signal, as shown in Figure F.32-4(b). The resulting voltage difference between the segment and the backplane is 0 because $V_{seg} = V_{bp}$. This condition turns the segment off.

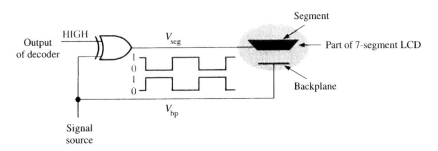

(a) Segment activated (on)

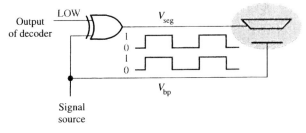

(b) Segment not activated (off)

FIGURE F.32-4
Basic operation of an LCD.

For driving LCDs, TTL is not recommended, because its low-level voltage is typically a few tenths of a volt, thus creating a DC component across the LCD, which degrades its performance. Therefore, CMOS is used in LCD applications.

Segment Decoding Logic

Each segment is used for various decimal digits, but no one segment is used for all ten digits. Therefore, each segment must be activated by its own decoding circuit that detects the occurrence of any of the numbers in which the segment is used. From Figures F.32-1 and F.32-2, the segments that are required to be activated for each digit are determined and listed in Table F.32-1.

TABLE F.32-1
Active segments for each decimal digit.

Digit	Segments Activated
0	*a, b, c, d, e, f*
1	*b, c*
2	*a, b, d, e, g*
3	*a, b, c, d, g*
4	*b, c, f, g*
5	*a, c, d, f, g*
6	*a, c, d, e, f, g*
7	*a, b, c*
8	*a, b, c, d, e, f, g*
9	*a, b, c, d, f, g*

Truth Table for the Segment Logic The segment decoding logic requires four binary coded decimal (BCD) inputs and seven outputs, one for each segment in the display, as indicated in the block diagram of Figure F.32-5. The multiple-output truth table, shown in Table F.32-2, is actually seven truth tables in one and could be separated into a separate table for each segment. A 1 in the segment output columns of the table indicates an activated segment.

Since the BCD code does not include the binary values 1010, 1011, 1100, 1101, 1110, and 1111, these combinations will never appear on the inputs and can therefore be treated as "don't care" (X) conditions, as indicated in the truth table. To conform with the practice of most IC manufacturers, *A* represents the least significant bit and *D* represents the most significant bit in this application.

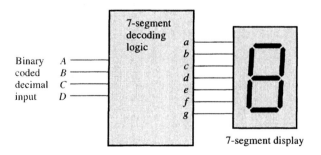

FIGURE F.32-5
Block diagram of 7-segment logic and display.

TABLE F.32-2
Truth table for 7-segment logic.

Decimal Digit	D	C	B	A	a	b	c	d	e	f	g
	Inputs				**Segment Outputs**						
0	0	0	0	0	1	1	1	1	1	1	0
1	0	0	0	1	0	1	1	0	0	0	0
2	0	0	1	0	1	1	0	1	1	0	1
3	0	0	1	1	1	1	1	1	0	0	1
4	0	1	0	0	0	1	1	0	0	1	1
5	0	1	0	1	1	0	1	1	0	1	1
6	0	1	1	0	1	0	1	1	1	1	1
7	0	1	1	1	1	1	1	0	0	0	0
8	1	0	0	0	1	1	1	1	1	1	1
9	1	0	0	1	1	1	1	1	0	1	1
10	1	0	1	0	X	X	X	X	X	X	X
11	1	0	1	1	X	X	X	X	X	X	X
12	1	1	0	0	X	X	X	X	X	X	X
13	1	1	0	1	X	X	X	X	X	X	X
14	1	1	1	0	X	X	X	X	X	X	X
15	1	1	1	1	X	X	X	X	X	X	X

Output = 1 means segment is activated (on)

Output = 0 means segment is not activated (off)

Output = X means "don't care"

Boolean Expressions for the Segment Logic   From the truth table, a standard SOP or POS expression can be written for each segment. For example, the standard SOP expression for segment a is

$$a = \overline{D}\ \overline{C}\ \overline{B}\ \overline{A} + \overline{D}\ \overline{C}B\overline{A} + \overline{D}\ \overline{C}BA + \overline{D}C\overline{B}A + \overline{D}CB\overline{A} + \overline{D}CBA + D\overline{C}\ \overline{B}\ \overline{A} + D\overline{C}\ \overline{B}A$$

and the standard SOP expression for segment e is

$$e = \overline{D}\ \overline{C}\ \overline{B}\ \overline{A} + \overline{D}\ \overline{C}B\overline{A} + \overline{D}CB\overline{A} + D\overline{C}\ \overline{B}\ \overline{A}$$

Expressions for the other segments can be similarly developed. As you can see, the expression for segment a has eight product terms and the expression for segment e has four product terms representing each of the BCD inputs that activate that segment. This means that the standard SOP implementation of segment-a logic requires an AND-OR circuit consisting of eight 4-input AND gates and one 8-input OR gate. The implementation of segment-e logic requires four 4-input AND gates and one 4-input OR gate. In both cases, four inverters are required to produce the complement of each variable.

REVIEW QUESTIONS

True/False

1. LCD displays consume less power than LED displays.
2. LED displays can be seen in the dark.

Multiple Choice

3. LCD displays require

   a. A backplane oscillator.

   b. High current drivers.

   c. Both a and b.

4. What type of display requires a positive voltage to turn a segment on?

   a. Common anode.

   b. Common cathode.

   c. Both a and b.

5. In a common-cathode display, the common pin must be

   a. Grounded.

   b. Taken positive through a limiting resistor.

   c. Clocked at 30 Hz or faster.

F.33 Troubleshoot and repair digital display devices

This material is covered in F.27.

Microprocessors

G.01 Demonstrate an understanding of microprocessor interfaces

INTRODUCTION

Any microprocessor must be interfaced to various support circuits in order to be useful. These circuits include bus controllers, I/O, memory, and interrupt circuits. These, and other interfaces, are examined in this section.

THE INPUT/OUTPUT (I/O) PORT

A block diagram of a basic microcomputer system, including I/O ports and typical peripherals, is shown in Figure G.01-1. Notice that a **port** can be strictly for input, strictly for output, or bidirectional for both input and output. Although only four I/O ports are shown in the figure, the CPU is capable of handling many more than that.

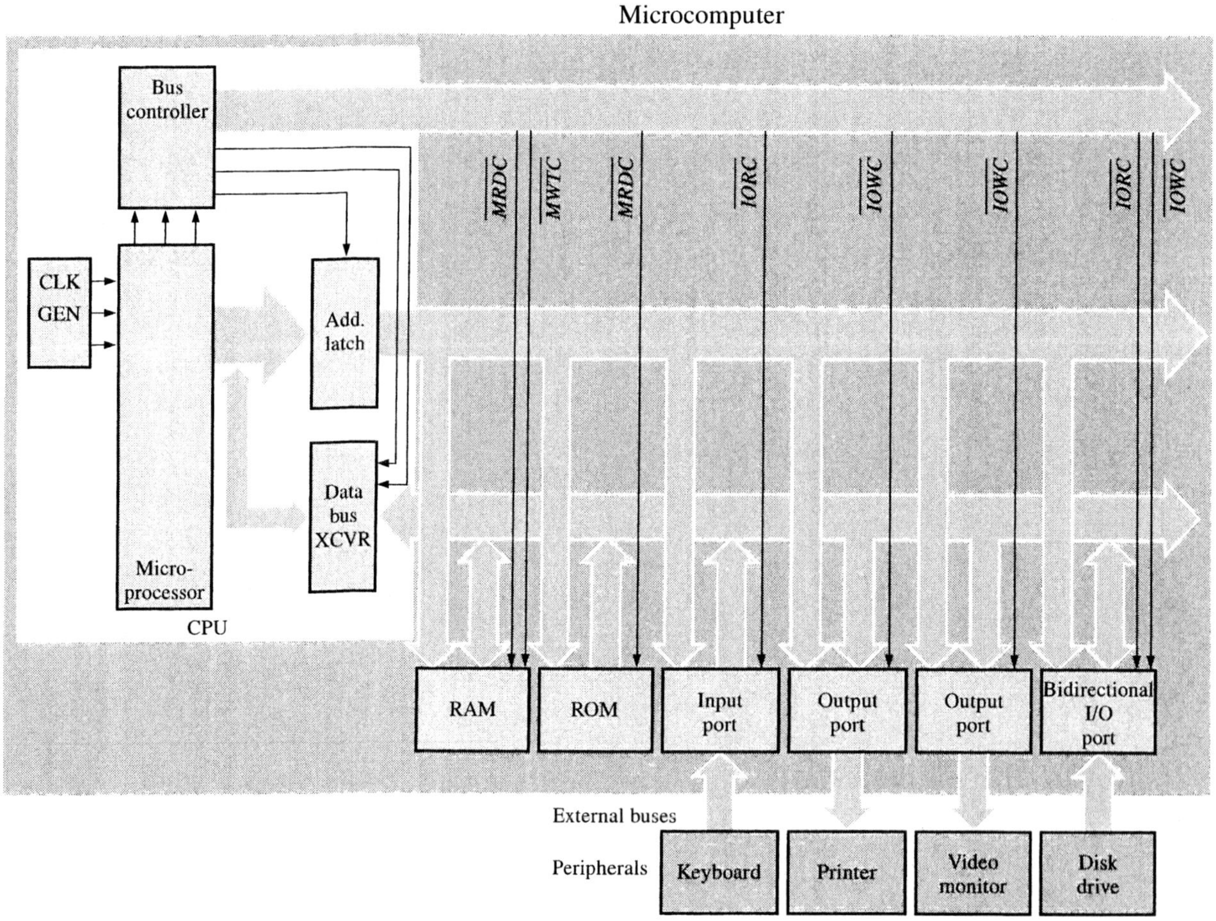

Microcomputer

FIGURE G.01-1
Basic microcomputer system with I/O ports and peripherals.

As mentioned, an I/O port provides an interface between the computer and the "outside world" of peripheral equipment. Depending on the system design, the I/O ports can be either *dedicated* or *memory mapped.* These terms describe the ways in which the ports are accessed by the CPU.

Dedicated I/O Ports

A dedicated I/O port is assigned a unique address within the I/O address space of the computer, and dedicated I/O commands are used for communication. Dedicated I/O ports are accessed by the I/O read and write commands, $\overline{IORC}$ and $\overline{IOWC}$ in an 8086/8088-based system.

Output Port   The basic operation of the output port is as follows: The port occupies one location within the I/O address space of the system; that is, it has a unique address different from the addresses of other ports. Typically, only a portion of the 20-bit address bus is used for port addresses because the actual number of I/O ports required is small compared with the number of memory locations.

During a CPU write cycle, the port address is placed on the address bus, and the CPU issues a low $\overline{IOWC}$ (I/O write command) signal to enable the address decoder. The decoder then produces a signal to latch the data byte onto the port output lines. Figure G.01-2 shows how a dedicated output port might be implemented.

Input Port   A dedicated input port is not quite as simple as an output port because the port output lines are connected to the system data bus. This arrangement requires that the port output lines be disabled when the port is not in use to prevent interference with other activity on the data bus. Figure G.01-3 shows an example of a basic input port implementation.

FIGURE G.01-2
A basic dedicated output port.

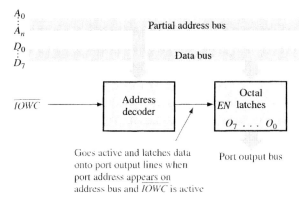

FIGURE G.01-3
A basic dedicated input port.

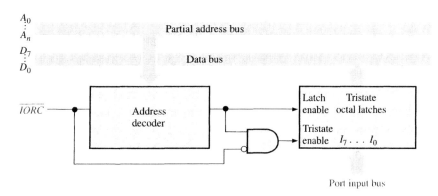

Bidirectional Port The third type of dedicated I/O port is the bidirectional port, which is basically a combination of the input and output ports with additional enabling controls. A bidirectional port allows data to flow to or from a peripheral device such as a disk drive.

Memory-Mapped I/O Ports

A second approach to accessing I/O ports is the memory-mapped port. The port implementation is essentially the same as for the dedicated ports in Figures G.01-2 and G.01-3 except for two things:

1. The memory-mapped ports are assigned addresses within the computer memory address space and are actually viewed as memory locations by the CPU. All twenty address lines are used for memory-mapped I/O.

2. The CPU accesses a memory-mapped I/O port with the $\overline{MRDC}$ and $\overline{MWTC}$ commands rather than the $\overline{IORC}$ and $\overline{IOWC}$ commands. In other words, the memory-mapped I/O ports are treated exactly as memory locations.

Programmable Peripheral Interface

In many systems, special ICs are used to implement the I/O ports. One such device is the programmable peripheral interface **(PPI),** shown in Figure G.01-4. The address decoder is a separate device for enabling the PPI when a port address is decoded. This particular PPI provides three I/O ports, labeled A, B, and C. Ports A and B have 8 lines each, and port C is split into two 4-line groups. The ports can be programmed for various combinations of input ports, output ports, and bidirectional ports and for control and handshaking lines. The PPI can be used for either dedicated or memory-mapped ports.

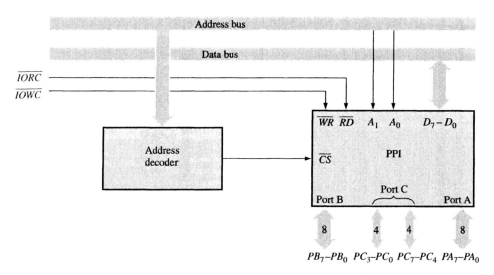

FIGURE G.01-4
A basic programmable peripheral interface (PPI) configuration.

I/O INTERRUPTS

In microprocessor-based systems such as the microcomputer, peripheral devices require periodic service from the CPU. The term *service* generally means sending data to or taking data from the device or performing some updating process.

In general, peripheral devices are very slow compared with the CPU. For example, the 8086/8088 CPU can perform about 1,000,000 read or write operations in one second (based on a 4.77 MHz clock frequency). A printer may average only a few characters per second (one character is represented by eight bits), depending on the type of material being printed. A keyboard input may be about one or two characters per second, depending on the operator. So, in between the times that the CPU is required to service a peripheral, it can do a lot of processing, and in most systems this processing time must be maximized by using an efficient method for servicing the peripherals.

Polled I/O

One method of servicing the peripherals is called **polling.** In this method the CPU must test each peripheral device in sequence at certain intervals to see if it needs or is ready for servicing. Figure G.01-5 illustrates the basic polled I/O method.

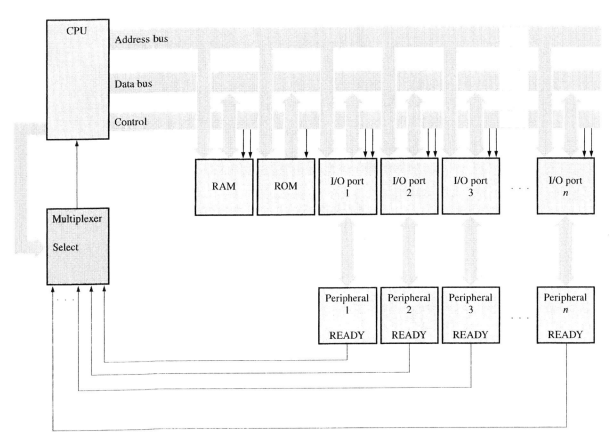

FIGURE G.01-5
The basic polled I/O configuration.

The CPU sequentially selects each peripheral device via the multiplexer to see if it needs service by checking the state of its ready line. Certain peripherals may need service at irregular and unpredictable intervals, that is, more frequently on some occasions than on others. Nevertheless, the CPU must poll the device at the highest rate. For example, let's say that a certain peripheral occasionally needs service every 1000 μs but most of the time requires service only once every 100 ms. As you can see, precious processing time is wasted if the CPU polls the device, as it must, at its maximum rate (every 1000 μs) because most of the time the device will not need service when it is polled.

Each time the CPU polls a device, it must stop the program that it is currently processing, go through the polling sequence, provide service if needed, and then return to the point where it left off in its current program.

Another problem with the sequentially polled I/O approach is that if two or more devices need service at the same time, the first one polled will be serviced first; the other devices will have to wait even though they may need servicing much more urgently than the first device polled.

As you can see, polling is suitable only for devices that can be serviced at regular and predictable intervals and only in situations in which there are no priority considerations.

Interrupt-Driven I/O

This approach overcomes the disadvantages of the polling method. In the interrupt method the CPU responds to a need for service only when service is requested by a peripheral device. Thus, the CPU can concentrate on running the current program without having to break away unnecessarily to see if a device needs service.

A request for service from a peripheral device is called an **interrupt.** When the CPU receives an I/O interrupt, it temporarily stops its current program and fetches a special program (service routine) from memory for the particular device that has issued the interrupt. When the service routine is complete, the CPU returns where it left off.

A device called a *programmable interrupt controller* (**PIC**) handles the interrupts on a priority basis. It accepts service requests from the peripherals. If two or more devices request service at the same time, the one assigned the highest priority is serviced first, then the one with the next highest priority, and so on. After issuing an interrupt (*INTR*) signal to the CPU, the PIC provides the CPU with information that "points" the CPU to the beginning memory address of the appropriate service routine. This process is called *vectoring.* Figure G.01-6 shows a basic interrupt-driven I/O configuration.

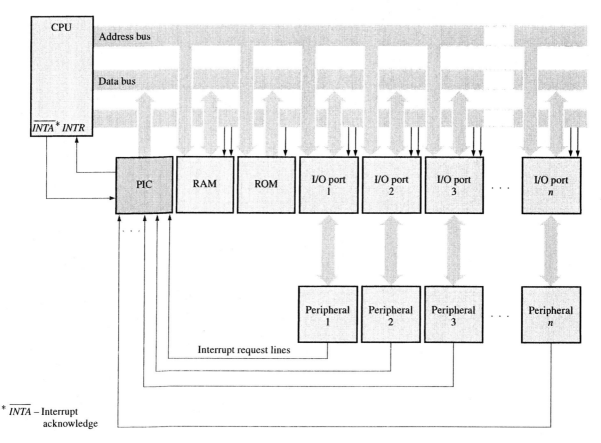

* $\overline{INTA}$ – Interrupt
    acknowledge

FIGURE G.01-6
A basic interrupt-driven I/O configuration.

Standard Buses

Several so-called standard buses are used to provide interfacing for the various internal components of digital systems. These standard buses assure compatibility of printed circuit board size, pin numbers, type of signal on each pin, and input and output characteristics. The standard bus allows for system expansion by specifying the conditions under which expansion or replacement units or modules must operate in order to interface properly with existing modules. To introduce the concept of standard buses, two examples will be discussed briefly.

The Multibus   The Multibus is a general-purpose bus system developed by Intel but used in industry. Some manufacturers offer products that are compatible with this particular bus system. The Multibus provides a flexible interface that can be used to interconnect a wide variety of microcomputer devices or modules. The modules in a Multibus system are designated as *masters* or *slaves*. Masters obtain use of the bus and initiate data transfers. Slaves are devices that cannot transfer data themselves or control the bus. A major feature of the Multibus is that several processors (masters) can be connected to the bus at the same time to implement a multiprocessing operation, in which each processor performs a dedicated task.

The Multibus provides a total of 86 lines. There are 16 bidirectional data lines, 20 address lines, 8 interrupt lines, and various control lines, command lines, and power lines. The Multibus standard defines the physical and electrical parameters for devices using the bus. A generalized Multibus system configuration is shown in Figure G.01-7.

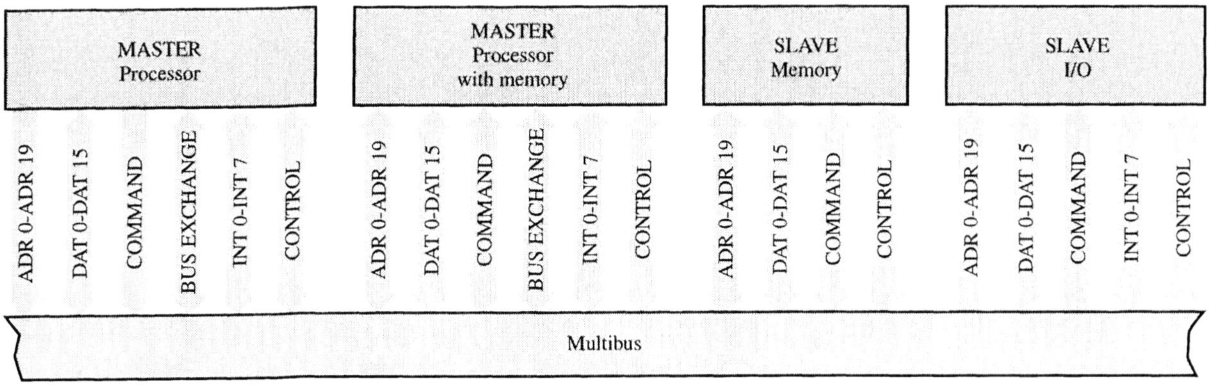

FIGURE G.01-7
A Multibus system configuration.

The PC Bus   The IBM PC bus structure actually includes three buses—the address bus, the data bus, and the control bus—as shown in Figure G.01-8 in a greatly simplified representation. The bus runs along the PC system board and connects the computer to expansion slots for adding various compatible boards for memory expansion, video interface, disk interface, and other peripheral interfaces. The original PC bus standard is called the ISA (industry standard architecture) bus.

A diagram of the basic 62-pin PC bus connector is shown in Figure G.01-9. There are twenty address lines (A0–A19) and eight data lines (D0–D7). The control bus and various power, ground, and timing signals occupy the remaining lines.

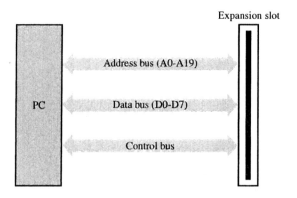

FIGURE G.01-8
Basic PC bus.

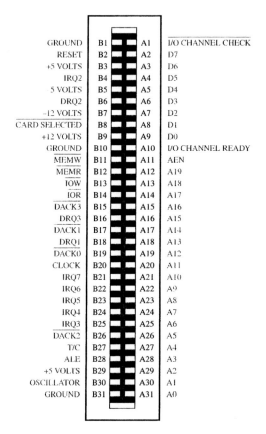

FIGURE G.01-9
I/O expansion connector for basic PC bus.

REVIEW QUESTIONS

True/False

1. The CPU performs all activities associated with controlling the bus.

2. Ports can only be used to input data to the processor.

Multiple Choice

3. The PPI contains
   a. Several programmable I/O ports.
   b. Programmable interrupts.
   c. Polled parallel inputs.

4. Constantly reading an input device is called
   a. Status testing.
   b. Auto-read looping.
   c. Polling.

5. What signals are provided on an ISA connector?
   a. Address/data signals.
   b. Control signals.
   c. Both a and b.

G.02 Troubleshoot and repair microprocessor interfaces

INTRODUCTION

Troubleshooting microprocessor interfaces requires skill and patience. In this section we will cover a series of tips that can be used for many troubleshooting situations.

MEMORY CIRCUITRY

As we have seen before, a good knowledge of binary numbers is both beneficial when working with microprocessors and necessary to the efficient design of address decoders (and other interfacing circuitry). Once the decoders are designed, however, they must be tested to see whether they perform as required. Testing the operation of a memory address decoder can be accomplished any number of ways. The circuit can be set up on a breadboard, simulated via software, or just plain stared at on paper until it seems correct.

In situations like the troubleshooting phase of a single-board computer project, testing the memory address decoder is often necessary. In addition to checking the wiring connections visually (or via a continuity tester or DMM), a logic analyzer is connected so that the waveforms can be examined. Even an old eight-channel logic analyzer can be used to diagnose difficult problems in a microprocessor-based system. Oscilloscopes typically fall short in showing the associated high-speed timing relationships (unless they are storage scopes).

The logic analyzer is connected so that all of the inputs to the address decoder are sampled, and as many of the outputs as necessary. In addition, the logic analyzer is set up so that it triggers (and begins capturing data) at RESET, and the initial activity of the processor's address bus can be observed. If the single-board's EPROM is not enabled, the system will not function. The logic analyzer will show how the address decoder responds at power on.

The logic analyzer can also be used to examine the data coming out of the EPROM or RAM. Sampling the data and a few of the address lines should be enough to verify whether the data is correct. It is sometimes possible to spot switched data or address lines this way.

These techniques also apply to I/O circuitry.

I/O CIRCUITRY

Finding the cause of a faulty I/O device can be tricky. Try some of these when you encounter an I/O problem.

☐ Write a short loop that continually accesses the I/O device. This should allow you to use an oscilloscope to look for a stream of pulses on the output of the address decoder. Use something like this to test an 8-bit output port:

```
        MOV    DX, ⟨I/O address⟩
        SUB    AL, AL
PTEST:  OUT    DX, AL
        INC    AL
        JMP    PTEST
```

In addition to the steady stream of pulses that should appear on the address decoder output, there should be a binary count appearing at the output port. An oscilloscope shows whether the waveform periods double (or halve) as you step from bit to bit. This is a good way to check for stuck or crossed outputs.

When checking an input port, use these instructions (or something similar):

```
        MOV    DX, ⟨I/O address⟩
PTEST:  IN     AL, DX
        JMP    PTEST
```

This loop is good for checking the operation of the address decoder. If possible, combine both loops so that data read from the input port is echoed to the output port.

☐ Check for easily overlooked mistakes, such as using AD_0 through AD_7 to connect to the I/O device, but not using IO/$\overline{M}$ in the address decoder.

☐ Verify that the enable signals on the I/O device all go to their active states when accessed.

☐ For a serial device, examine the serial output for activity. Check for valid transmitter and receiver clocks. If the serial device is connected to a keyboard, press the keys and watch the serial input of the device. Make sure the TTL-to-RS232 driver is working correctly.

Other I/O devices may require you to test the interrupt system, or write a special initialization code to program a peripheral. Keeping track of the new software and hardware designs you develop or troubleshoot will save you time and effort in the future.

PERIPHERALS

When you come across a new device and are faced with the task of getting it to work, keep these suggestions in mind:

☐ Look over the data manual on the new peripheral. If you do not have a data manual, try searching the Web. Intel has a very useful site, offering downloads of manuals (80x86 series plus many others) in PDF format. Many other educational institutions also post important information on the Web.

Skim the figures and captions, look at the register and bit assignments, and read the tables. Read about the hardware signals. Study the timing diagrams. Look at any sample interface designs provided by the manufacturer. Be sure you understand why the signals are used the way they are.

Read about the software architecture of the peripheral. How is it controlled? How do you send data to it, or read data from it? How many different functions does it perform?

☐ Get the hardware interface working properly. This requires your skill in designing I/O hardware. Some software may be required to fully test the interface.

☐ If the peripheral has many modes of operation, begin with the simplest. Program the peripheral to operate in this mode to be sure you have control over it. Expand to other modes of operation as you learn more about the device.

Even if all you are doing is modifying someone else's code, written long ago, for a peripheral that is already operational, it is still good to learn as much about the device as possible. This will help you avoid typical problems, such as forgetting to issue the master reset command, even though power was just applied.

SYSTEM TROUBLESHOOTING

Here are some things we can do if an 80×86 single-board computer does not work when we turn it on:

□ Feel around the board for hot components. A chip that is incorrectly wired or placed backwards in its socket can get very hot. You may even smell the hot component.

□ Make sure all the ICs have power by measuring with a DMM or oscilloscope. Put the probe right on the pin of the IC, not on the socket lead.

□ Use an oscilloscope to examine the CLK output of the 8284. Push the RESET button to verify that the RESET signal is being generated properly.

□ Look at the address, data, and control lines with an oscilloscope or logic analyzer. Activity is a good sign; there may be something as simple as a missing address or data line, or crossed lines. No activity means the processor is not receiving the right information. By examining the logic analyzer traces, you should be able to determine if the memory and I/O address decoders are working properly, as well as the address and data bus drivers.

□ Verify that the EPROM was burned correctly, and it is in the right socket and not switched with the RAM. You should be able to connect a logic analyzer to verify that the processor fetches the first instruction from address FFFF0H. You should also be able to see the first instruction byte come out of the EPROM as well.

□ Examine the TxD output of the 8251. Activity at power on or RESET is a good sign, since the monitor program is designed to output a short greeting to let us know it is alive. If TxD wiggles around, but the serial output of the MAX232 does not, there could be a wiring problem there. Putting the capacitors in backward is bad for the MAX chip.

□ Check every connection again from a fresh schematic. Many times, a missing connection is found, no matter how careful the construction.

□ Change all the chips, one by one. Look for bent or missing pins when you remove them.

□ When all else fails, tell someone else everything you've done and see if they can suggest anything else.

□ You could also set the project aside for a while to get your mind off it. The problem may present itself to you when you least expect it. You may suddenly remember that you commented out an important I/O routine in the monitor because you were having trouble with the assembler.

G.03 Demonstrate an understanding of essential microprocessor components

INTRODUCTION

The microprocessor is a digital integrated circuit device that can be programmed with a series of instructions to perform specified functions on data. When a microprocessor is connected to a memory device and provided with a means of transferring data to and from the "outside world," you have a microcomputer.

Basic Elements of a Microprocessor

In its basic form, a **microprocessor** consists of three elements as shown in Figure G.03-1: an arithmetic logic unit **(ALU)**, a register unit, and a control unit.

FIGURE G.03-1
The basic elements of a microprocessor.

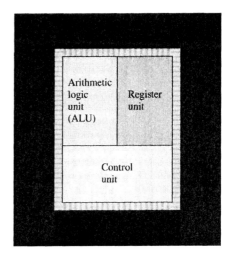

Arithmetic Logic Unit The ALU performs arithmetic operations such as addition and subtraction and logic operations such as NOT, AND, OR, and exclusive-OR.

Register Unit During the execution of a program (series of instructions), data are temporarily stored in any of the many registers that make up this unit.

Control Unit This unit provides the timing and control signals for getting data into and out of the microprocessor, for performing programmed instructions, and for all other operations.

Microprocessor Buses

Typically a microprocessor has three buses for information transfer internally and externally, as shown in Figure G.03-2. These buses are the address bus, the data bus, and the control bus.

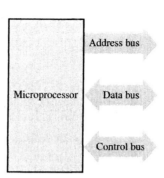

FIGURE G.03-2
Microprocessor buses.

The Address Bus   The address bus is a "one-way street" over which the microprocessor sends an address code to a memory or other external device. The size or "width" of the address bus is specified by the number of bits that it can handle. Early microprocessors had 4-bit address buses. This number has increased to 8, 16, 20, 24, and 32 bits as microprocessor technology has advanced.

The more bits there are in the address bus, the more memory locations a given microprocessor can access. With 8 bits, 256 memory locations can be accessed. With 16 bits, 65,536 memory locations can be accessed. With 32 bits, 4,294,967,296 memory locations can be accessed.

The Data Bus   The data bus is a "two-way street" on which data or instruction codes are transferred into the microprocessor or on which the result of an operation or computation is sent out from the microprocessor. Depending on the particular microprocessor, the data bus can handle 8 bits, 16 bits, 32 bits, or 64 bits.

The Control Bus   The control bus is used by the microprocessor to coordinate its operations and to communicate with external devices.

A Summary of the Pentium

The newest, and fastest, chip in the Intel high-performance microprocessor line is the Pentium. As usual, upward compatibility has been maintained. The Pentium will run all programs written for any machine in the 80x86 line, though it does so at a speed several times faster than the fastest 80486. And the Pentium does so with a radically new architecture!

There are two major computer architectures in use: CISC and RISC. CISC stands for Complex Instruction Set Computer. RISC stands for Reduced Instruction Set Computer. All of the 80x86 machines prior to the Pentium can be considered CISC machines. The Pentium itself is a mixture of both CISC and RISC technologies. The CISC aspect of the Pentium provides for upward compatibility with the other 80x86 architectures. The RISC aspects lead to additional performance improvements. Some of these improvements are separate 8KB data and instruction caches, dual integer pipelines, and branch prediction.

Refer to Figure G.03-3 for a look at the Pentium architecture.

Splitting the 80486's integrated instruction/data cache into two separate caches prevents data and instruction accesses from interfering with each other. This helps keep a steady stream of data flowing into the instruction and integer pipelines.

Adding a second integer pipeline (the 80486 has a single integer pipeline) leads to times when two instructions may execute *at once*. The Pentium has special internal circuitry to recognize when both pipelines may be used. This architectural improvement is borrowed from the RISC world of microprocessing, where multiple pipelines are employed to gain a performance increase.

Another significant addition is that of a branch prediction unit. This circuit keeps track of branch instructions in an executing program and predicts when they will be taken and not taken. By loading special internal data buffers with prefetched instructions, the branch prediction unit can keep the instruction pipeline running smoothly, even when a branch instruction changes the flow of instructions. This is another technique borrowed

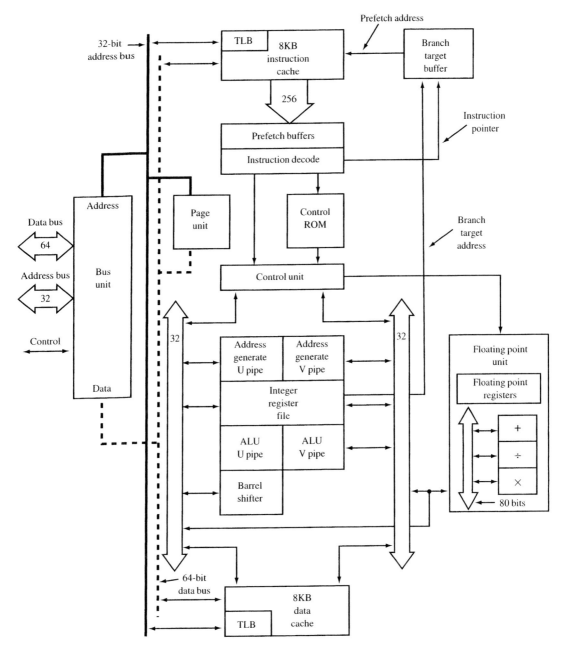

FIGURE G.03-3
Pentium architecture block program.

from RISC technology, and leads to a speeding up of execution in programs that employ many branch instructions.

These differences between the Pentium and the earlier 80x86 machines give it a significant speed improvement. This was possible by blending CISC and RISC technology together. The benefit to us as programmers lies in the fact that all Intel processors from the 8086 up, including the Pentium, run the same basic instruction set; they just do it faster and faster.

Microprocessor Programming

The **assembly language** that is used to program a microprocessor is classified as a *low-level language* because the English-like instructions contained in a given assembly language represent binary codes that directly control the microprocessor. The binary code instructions are called **machine language** and are all that the microprocessor recognizes. A program called an assembler converts the English-like instructions, called **mnemonics,** in the assembly language into binary patterns called machine language for use by the microprocessor as illustrated in Figure G.03-4. Assembly language and the corresponding machine language is specific to the type of microprocessor or microprocessor family.

FIGURE G.03-4
Block diagram of microprocessor programming.

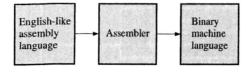

A sample 8086 assembly language program appears as follows:

```
        MOV  CX,05      ;CX holds the loop count
        MOV  BX,0200H   ;BX hold the offset data address
        MOV  AL,00      ;initialize AL
ADD_LP: ADD  AL,[BX]    ;add the next byte to AL
        INC  BX         ;increment the data pointer
        DEC  CX         ;decrement the loop counter
        JNZ  ADD_LP     ;jump to next iteration if counter not zero
```

On the other hand, high-level programming languages such as BASIC, Pascal, C, or FORTRAN are independent of the type of microprocessor in a computer system. A program called a compiler or interpreter translates the high-level program statements into machine language.

REVIEW QUESTIONS

True/False

1. A typical microprocessor consists of three parts: ALU, memory, and buses.

2. Assembly language and machine language are the same thing.

Multiple Choice

3. The advanced Intel processors (80286 and up) operate in real-mode or
   a. Imaginary-mode.
   b. Accelerated-mode.
   c. Protected-mode.

4. Cache is
   a. A low-speed memory used for buffer I/O devices.
   b. A high-speed memory used to buffer frequently used instructions and data.
   c. The only type of memory used by a microprocessor.

5. RISC stands for
   a. Reduced instruction set computer.
   b. Redundant Intel system core.
   c. Register integrated system controller.

G.04 Demonstrate an understanding of microprocessor bus concepts

INTRODUCTION

The internal working of every computer can be broken down into three parts: *CPU* (central processing unit), *memory,* and *I/O* (input/output) devices (see Figure G.04-1). The function of the CPU is to execute (process) information stored in memory. The function of I/O devices such as the keyboard and video monitor is to provide a means of communicating with the CPU. The CPU is connected to memory and I/O through strips of wire called a *bus.* The bus inside a computer carries information from place to place just as a street bus carries people from place to place. In every computer there are three types of buses: *address bus, data bus,* and *control bus.*

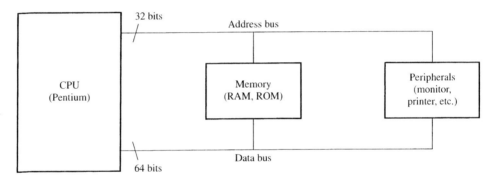

FIGURE G.04-1
Inside the computer.

For a device (memory or I/O) to be recognized by the CPU, it must be assigned an *address.* The address assigned to a given device must be unique; no two devices are allowed to have the same address. The CPU puts the address (in binary) on the address bus, and the decoding circuitry finds the device. Then the CPU uses the data bus either to get data from that device or to send data to it. The control buses are used to provide read or write signals to the device to indicate if the CPU is asking for information or sending it information. Of the three buses, the address bus and data bus determine the capability of a given CPU.

More About the Data Bus

Since data buses are used to carry information in and out of a CPU, the more data buses available, the better the CPU. If one thinks of data buses as highway lanes, it is clear that more lanes provide a better pathway between the CPU and its external devices (such as printers, RAM, ROM, etc.; see Figure G.04-1). By the same token, that increase in the number of lanes increases the cost of construction. More data buses mean a more expensive CPU and computer. The average size of data buses in CPUs varies between 8 and 64. Data buses are bidirectional, since the CPU must use them either to receive or to send data. The processing power of a computer is related to the size of its buses, since an 8-bit bus can send out 1 byte at a time, but a 16-bit bus can send out 2 bytes at a time, which is twice as fast. Multiple devices connected to the data bus are enabled through tristate outputs.

More About the Address Bus

Since the address bus is used to identify the devices and memory connected to the CPU, the more address buses available, the larger the number of devices that can be addressed. In other words, the number of address buses for a CPU determines the number of locations with which it can communicate. The number of locations is always equal to 2^x, where x is the number of address lines, regardless of the size of the data bus. For example, a CPU with 16 address lines can provide a total of 65,536 (2^{16}) or 64K bytes of addressable memory. Each location can have a maximum of 1 byte of data. This is due to the fact that all general-purpose microprocessors are *byte addressable*. As another example, the IBM PC

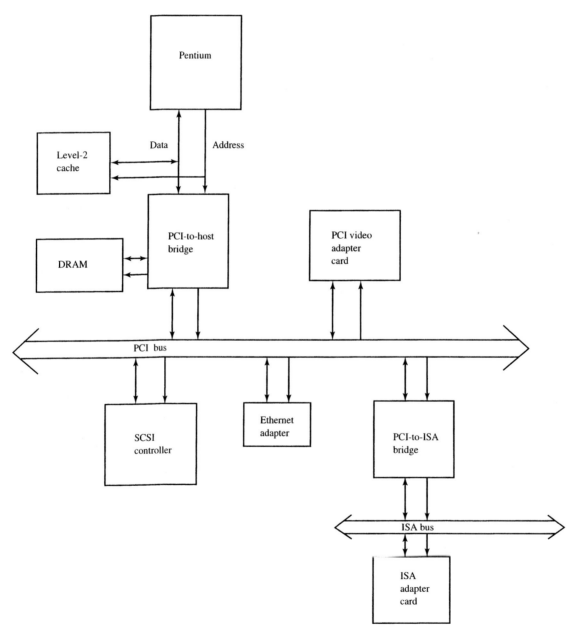

FIGURE G.04-2
PCI bridge in a Pentium system.

AT uses a CPU with 24 address lines and 16 data lines. In this case the total accessible memory is 16 megabytes (2^{24} = 16 megabytes). In this example there would be 2^{24} locations, and since each location is one byte, there would be 16 megabytes of memory. The address bus is a *unidirectional* bus, which means that the CPU uses the address bus only to send out addresses. To summarize: The total number of memory locations addressable by a given CPU is always equal to 2^x where x is the number of address bits, regardless of the size of the data bus.

The PCI Bus

PCI stands for peripheral component interconnect, and it is Intel's offering in the world of standardized buses. The PCI bus uses a *bridge* IC to control data transfers between the processor and the system bus, as indicated in Figure G.04-2.

In essence, the PCI bus is not strictly a local bus, since connections to the PCI bus are not connections to the processor, but rather a special PCI-to-Host controller chip. Other chips, such as PCI-to-ISA bridges, interface the older ISA bus with the PCI bus, allowing both types of buses on one motherboard, with a single chip controlling them all. The PCI bus is designed to be processor independent, plug-and-play compatible, and capable of 64-bit transfers at 33 MHz (with faster versions becoming available). Figure G.04-3 shows the pinout for a 32-bit PCI connector.

FIGURE G.04-3
32-bit PCI connector.

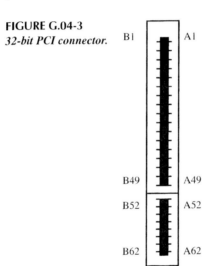

PCMCIA

The PCMCIA (personal computer memory card international association) bus, now referred to as *PC Card Bus,* evolved from the need to expand the memory available on early laptop computers. The standard has since expanded to include almost any kind of peripheral you can imagine, from hard drives, to LAN adapters and modem/fax cards. Figure G.04-4 on page 659 shows a typical PCMCIA Ethernet card.

The PCMCIA bus supports four styles of cards, as shown in Table G.04-1.

TABLE G.04-1
PCMCIA slot styles.

Slot Type	Meaning
Type I	Original standard. Supports 3.3 mm cards. Memory cards only.
Type II	Supports 3.3 mm and 5 mm cards.
Type III	Supports 10.5 mm cards, as well as Type I and II.
Type IV	Greater than 10.5 mm supported.

TABLE G.04-2
PCMCIA pin assignments (available at card insertion).

Pin	Signal	Pin	Signal	Pin	Signal
1	GND	24	A_5	47	A_{18}
2	D_3	25	A_4	48	A_{19}
3	D_4	26	A_3	49	A_{20}
4	D_5	27	A_2	50	A_{21}
5	D_6	28	A_1	51	Vcc
6	D_7	29	A_0	52	Vpp2
7	CE1	30	D_0	53	A_{22}
8	A_{10}	31	D_1	54	A_{23}
9	OE	32	D_2	55	A_{24}
10	A_{11}	33	WP	56	A_{25}
11	A_9	34	GND	57	RFU
12	A_8	35	GND	58	RESET
13	A_{13}	36	CD1	59	WAIT
14	A_{14}	37	D_{11}	60	RFU
15	WE/PGM	38	D_{12}	61	REG
16	RDY/BSY	39	D_{13}	62	BVD2
17	Vcc	40	D_{14}	63	BVD1
18	Vpp1	41	D_{15}	64	D_8
19	A_{16}	42	CS2	65	D_9
20	A_{15}	43	RFSH	66	D_{10}
21	A_{12}	44	RFU	67	CD2
22	A_7	45	RFU	68	GND
23	A_6	46	A_{17}		

TABLE G.04-3
PCMCIA signal differences.

Pin	Memory Card	I/O Card
16	RDY/BSY	IREQ
33	WP	IOIS16
44	RFU	IORD
45	RFU	IOWR
60	RFU	INPACK
62	BVD2	SPKR
63	BVD1	STSCHG

All PCMCIA cards allow *hot swapping,* removing and inserting the card with power on.

A type I connector is shown in Figure G.04-5 on the next page. The signal assignments are illustrated in Tables G.04-2 and G.04-3. The popularity of laptop and notebook computers suggests the continued use of this bus.

AGP

The AGP (accelerated graphics port) is a new technology that improves multimedia performance on Pentium II computers. Figure G.04-6 shows where the AGP technology fits into the other bus hardware.

The heart of the AGP is the 440LX AGPset hardware, a *quad-ported* data switch that controls transfers between the processor, main memory, graphics memory, and the PCI bus. AGP technology uses a connector similar to a PCI connector.

With growing emphasis on multimedia applications, AGP technology sets the stage for improved performance.

FIGURE G.04-4
PCMCIA ethernet card (photograph by John T. Butchko).

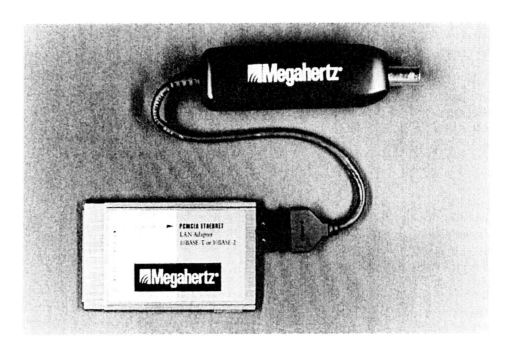

FIGURE G.04-5
PCMCIA connector.

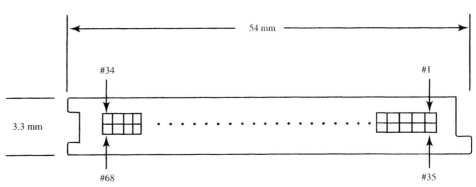

FIGURE G.04-6
AGP interface.

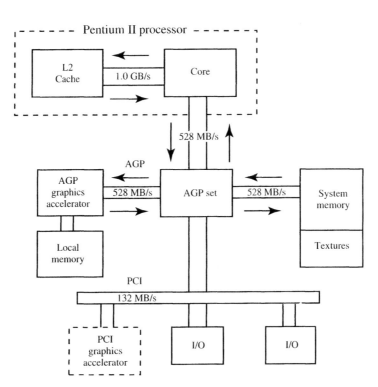

REVIEW QUESTIONS

True/False

1. Buses carry address, data, and control signals.

2. A data bus is bidirectional.

Multiple Choice

3. An address bus of 32 bits is capable of accessing
   a. 4 MB.
   b. 32 MB.
   c. 4096 MB.

4. A tristate device has three states: off, on, and
   a. Low-impedance.
   b. High-impedance.
   c. Buffer-mode.

5. Data transfers on the PCI bus are controlled by
   a. The CPU.
   b. The I/O devices.
   c. A bridge IC.

G.05 Demonstrate an understanding of microprocessor components and terminology

INTRODUCTION

In this section we examine additional details concerning microprocessors and coprocessors (FPUs) and their use in modern motherboards.

Definition of the Motherboard

The main system board of the computer is commonly referred to as the **motherboard.** A typical motherboard is shown in G.05-1. Sometimes the motherboard is referred to as the **system board,** or the **planar.**

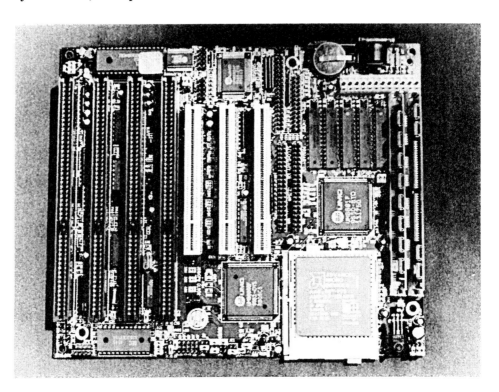

FIGURE G.05-1
Typical PC motherboard (photo by John T. Butchko).

Contents of the Motherboard

The motherboard holds and electrically interconnects all the major components of a personal computer. The motherboard contains the following:

☐ The microprocessor

☐ The math coprocessor (only on older 386 motherboards)

☐ BIOS ROM

□ RAM (Dynamic RAM, or DRAM, as well as Level-2 cache)

□ The expansion slots

□ Connectors for IDE drives, floppies, and COM ports.

Table G.05-1 lists these major parts and gives a brief overview of the purpose of each part.

Figure G.05-2 on the facing page shows a typical motherboard layout and the locations of the major motherboard parts.

In this exercise, you will have the opportunity to learn more details about the microprocessor and the coprocessor. In the following exercises, you will learn about the other areas of the motherboard.

The Microprocessor

You can think of the **microprocessor** in a computer as the central processing unit (CPU), or the "brain," so to speak, of the computer. The microprocessor sets the stage for everything else in the computer system. Several major features distinguish one microprocessor from another. These features are listed in Table G.05-2.

TABLE G.05-1
Purposes of major motherboard parts.

Part	Purpose
Microprocessor	Interprets the instructions for the computer and performs the required process for each of these instructions.
Math coprocessor	Used to take over arithmetic functions from the microprocessor.
BIOS ROM	Read-only memory. Memory programmed at the factory that cannot be changed or altered by the user.
RAM	Read/write memory. Memory used to store computer programs and interact with them.
Expansion slots	Connectors used for the purpose of interconnecting adapter cards to the motherboard.
Connectors	Integrated controller on motherboard provides signals for IDE and floppy drives, the printer, and the COM ports.

TABLE G.05-2
Microprocessor features.

Feature	Description
Bus structure	The number of connectors used for specific tasks.
Word size	The largest number that can be used by the microprocessor in one operation.
Data path size	The largest number that can be copied to or from the microprocessor in one operation.
Maximum memory	The largest amount of memory that can be used by the microprocessor.
Speed	The number of operations that can be performed per unit time.
Code efficiency	The number of steps required for the microprocessor to perform its processes.

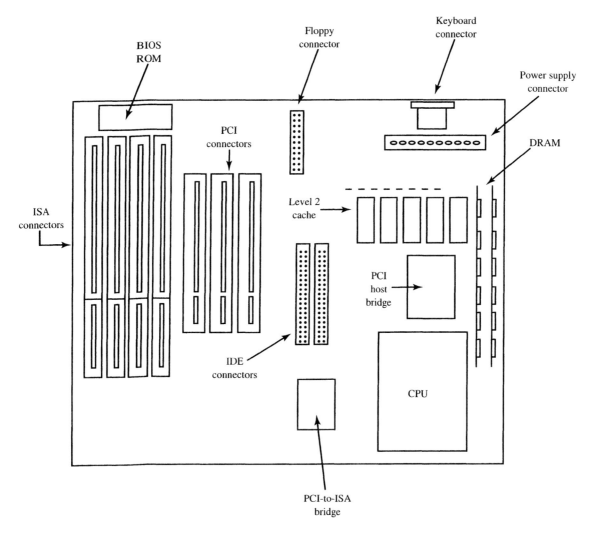

FIGURE G.05-2
Motherboard layout.

You can think of a **bus** as nothing more than a group of wires all dedicated to a specific task. For example, all microprocessors have the following buses:

Data bus	Group of wires for handling data. This determines the data path size.
Address bus	Group of wires for getting and placing data in different locations. This helps determine the maximum memory that can be used by the microprocessor.
Control bus	Group of wires for exercising different controls over the microprocessor.
Power bus	Group of wires for supplying electrical power to the microprocessor.

Figure G.05-3 shows the bus structure of a typical microprocessor.

Since all the data that goes in and out of a microprocessor is in the form of 1s and 0s, the more wires used in the data bus, the more information the microprocessor can handle at one time. For example, some microprocessors have eight lines (wires or pins) in their data buses, others have 16, and some have 32 or 64.

The number of lines used for the address bus determines how many different places the microprocessor can use for getting and placing data. The *places* that the microprocessor

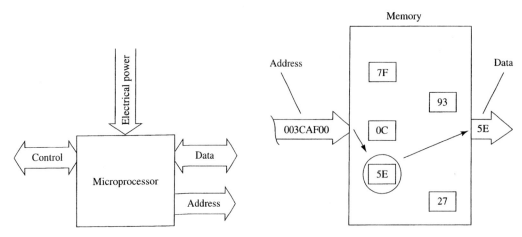

FIGURE G.05-3
Typical microprocessor bus structure.

FIGURE G.05-4
Relationship between data and address.

uses for getting and placing data are referred to as **memory locations.** The relationship between the data and the address is shown in Figure G.05-4.

The greater the number of lines used in the address bus of a microprocessor, the greater the number of memory locations the microprocessor can use.

About the 80x86 Architecture   The advanced nature of the Pentium microprocessor requires us to think differently about the nature of computing. The Pentium architecture contains exotic techniques such as branch prediction, pipelining, and superscalar processing to pave the way for improved performance. Let us take a quick look at some other improvements from Intel:

☐ Intel has added MMX Technology to its line of Pentium processors (Pentium, Pentium Pro, and Pentium II). A total of 57 new instructions enhance the processors' ability to manipulate audio, graphic, and video data. Intel accomplished this major architectural addition by *reusing* the 80-bit floating-point registers in the FPU. Using a method called **SIMD** (single instruction multiple data), one MMX instruction is capable of operating on 64-bits of data stored in an FPU register.

☐ The Pentium Pro processor (and the newer Pentium II) employs a technique called *speculative execution.* In this technique, multiple instructions are fetched and executed, possibly out of order, in order to keep the pipeline busy. The results of each instruction are speculative until the processor determines that they are needed (based on the result of branch instructions and other program variables). Overall, a high level of parallelism is maintained.

☐ First used in the Pentium Pro, a new bus technology called Dual Independent Bus architecture utilizes two data buses to transfer data between the processor and main memory (including the level-2 cache). One bus is for main memory, the second for the level-2 cache. The buses may be used independently or in parallel, significantly improving the bus performance over that of a single-bus machine.

☐ The 5-stage Pentium pipeline was redesigned for the Pentium Pro into a *superpipelined* 14-stage pipeline. By adding more stages, less logic can be used in each stage, which allows the pipeline to be clocked at a higher speed. This is easily verified by the over 400-MHz processors currently available. Although there are drawbacks to superpipelining, such as bigger branch penalties during an incorrect prediction, its benefits are well worth the price.

Spend some time on the Web reading material about these changes, and others. It will be time well invested.

The Coprocessor

Each Intel microprocessor released before the 80486 has a companion to help it do arithmetic calculations. This companion is called a **coprocessor.** For most software, the coprocessor is optional, and most computer systems are sold without it. However, some programs (such as CAD—computer-aided design—programs) have so many math calculations to perform that they need the assistance of the math coprocessor; the main microprocessor simply cannot keep up with the math demand.

These **math chips,** as they are sometimes called, are capable of performing mathematical calculations 10 to 100 times faster than their companion microprocessors and with a higher degree of accuracy. This doesn't mean that if your system is without a coprocessor it can't do math; it simply means that your microprocessor will be handling all the math along with everything else, such as displaying graphics, reading the keyboard, etc.

Table G.05-3 lists the math coprocessors that go with various microprocessors. Note that the 80486 and above processors have built-in coprocessors.

TABLE G.05-3
Matching math coprocessors.

Microprocessor	Math Coprocessor
8086	8087
8088	8087
80286	80287
80386	80387
80386SX	80387SX
80486DX	Built-in coprocessor enabled
80486SX	Coprocessor disabled
Pentium, Pentium Pro, and Pentium II	Built-in coprocessor enabled always

For a math coprocessor chip to be used by software, that software must be specifically designed to look for the chip and use it if it is there. Some spreadsheet programs look for the presence of this chip and use the microprocessor for math if the coprocessor is not present. If the coprocessor is present, the software uses it instead. Some programs, such as word processing programs, have no use for the math functions of the coprocessor and do not use the coprocessor at all. Therefore, the fact that a system has a coprocessor doesn't necessarily mean that the coprocessor will improve the overall system performance. Improvement will take place only if the software is specifically designed to use the coprocessor and there are many complex math functions involved in the program.

TROUBLESHOOTING TECHNIQUES

Windows 95 identifies the processor it is running on. Use System Properties in Control Panel to check the processor type, as shown in Figure G.05-5. Notice that Windows 95 has identified a Pentium as the CPU. It is interesting to note that other Pentium-compatible CPUs, such as a Cyrix 586, are not recognized by Windows 95 as Pentiums, and are reported in System Properties as 80486 CPUs. This kind of information is not easily found in existing documentation. It is good to talk directly to the manufacturer of the motherboard to determine how your performance might be affected.

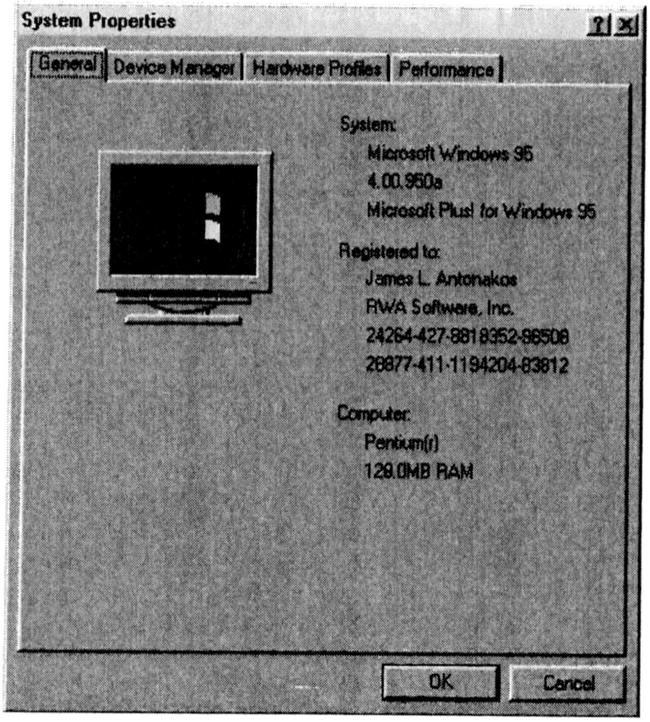

FIGURE G.05-5
Processor identification.

REVIEW QUESTIONS

True/False

1. The motherboard is sometimes referred to as the system board.
2. All the major components in the computer are interconnected through the motherboard.

Multiple Choice

3. You can think of the microprocessor in the computer as the
   a. CPU.
   b. "Brains of the computer."
   c. Both of the above.
4. The largest number that can be used by the computer in one operation is determined by the
   a. Amount of available memory.
   b. Speed of the computer.
   c. Word size.
5. In a computer, a group of wires dedicated to a specific task is called the
   a. Bus.
   b. Track.
   c. Data path.

G.06 Understand principles and operations of types of microprocessor memory circuits

INTRODUCTION

In the design of all computers, semiconductor memories are used as primary storage for code and data. Semiconductor memories are connected directly to the CPU and they are the memory that the CPU first asks for information (code and data). For this reason, semiconductor memories are sometimes referred to as *primary memory*. The main requirement of primary memory is that it must be fast in responding to the CPU; only semiconductor memories can do that. Among the most widely used semiconductor memories are ROM and RAM. Before we discuss different types of RAM and ROM, we discuss some important terminology common to all semiconductor memories, such as capacity, organization, and speed.

Memory Capacity

The number of bits that a semiconductor memory chip can store is called its *chip capacity*. It can be in units of Kbits (kilobits), Mbits (megabits), and so on. This must be distinguished from the storage capacity of computers. While the memory capacity of a memory IC chip is always given in bits, the memory capacity of a computer is given in bytes. For example, an article in a technical journal may state that the 4M chip has become popular. In that case, although it is not mentioned that 4M means 4 megabits, it is understood since the article is referring to an IC memory chip. However, if an advertisement states that a computer comes with 4M memory, since it is referring to a computer it is understood that 4M means 4 megabytes.

Memory Organization

Memory chips are organized into a number of locations within the IC. Each location can hold 1 bit, 4 bits, 8 bits, or even 16 bits, depending on how it is designed internally. The number of bits that each location within the memory chip can hold is always equal to the number of data pins on the chip. How many locations exist inside a memory chip? That depends on the number of address pins. The number of locations within a memory IC always equals 2 to the power of the number of address pins (there is a slight modification of this rule for dynamic memory, as will be explained later). Therefore, the total number of bits that a memory chip can store is equal to the number of locations times the number of data bits per location. To summarize:

1. Each memory chip contains 2^x locations, where x is the number of address pins on the chip.

2. Each location contains y bits, where y is the number of pins on the chip.

3. The entire chip will contain 2^x x y bits, where x is the number of address pins and y is the number of data pins on the chip.

Speed

One of the most important characteristics of a memory chip is the speed at which data can be accessed from it. To access the data, the address is presented to the address pins, and after a certain amount of time has elapsed, the data show up at the data pins. The shorter this elapsed time, the better and, consequently, the more expensive, the memory chip. The speed of the memory chip is commonly referred to as its *access time*. The access time of memory chips varies from a few nanoseconds to hundreds of nanoseconds, depending on the IC technology used in the design and fabrication.

The three important memory characteristics of capacity, organization, and access time will be used extensively in this skill topic. Many of these characteristics will be explored in more detail in the context of applications in this and future skill topics. Table G.06-1 serves as a reference for the calculation of memory organization.

TABLE G.06-1
Powers of 2.

x	2^x
10	1K
11	2K
12	4K
13	8K
14	16K
15	32K
16	64K
17	128K
18	256K
19	512K
20	1M
21	2M
22	4M
23	8M
24	16M

ROM (Read-Only Memory)

ROM is the type of memory that does not lose its contents when the power is turned off. For this reason, ROM is also called *nonvolatile memory*. There are different types of read-only memory, such as PROM, EPROM, EEPROM, flash EPROM, and mask ROM.

PROM (Programmable ROM)

PROM refers to the kind of ROM that the user can burn information into. In other words, PROM is a user-programmable memory. For every bit of the PROM, there exists a fuse. The PROM is programmed by blowing the fuses. If the information burned into the PROM is wrong, that PROM must be discarded since internal fuses are blown permanently. For this reason, PROM is also referred to as OTP (one-time programmable). The process of programming ROM is also called burning ROM and requires special equipment called a ROM burner or ROM programmer.

Table G.06-2 shows examples of some popular ROM chips and their characteristics. Notice the patterns of the IC numbers. For example, 27128-20 refers to UV-EPROM that has the capacity of 128K bits and access time of 200 nanoseconds. The capacity of the memory chip is indicated in the part number, and the access time is given with a zero dropped. In part numbers, C refers to CMOS technology.

EPROM (Erasable Programmable ROM)

EPROM was invented to allow making changes in the contents of PROM after it is burned. In EPROM, one can program the memory chip and erase it thousands of times. This is especially useful during development of the prototype of a microprocessor-based

TABLE G.06-2
Examples of ROM memory chips.

Type	Part Number	Speed (ns)	Capacity	Organization	Number of Pins	VPP (V)
PROM	74S188	35	256	32x8	16	5
	74S472	60	4K	512x8	20	5
	74S573	60	4K	1Kx4	18	5
UV-EPROM	2716	450	16K	2Kx8	24	25
	2716-1	350	16K	2Kx8	24	25
	2716B	450	16K	2Kx8	24	12.5
	2732A-45	450	32K	4Kx8	24	21
	2732A-20	200	32K	4Kx8	24	21
	27C32	450	32K	4Kx8	24	25
	2764A-25	250	64K	8Kx8	28	12.5
	27C64-15	150	64K	8Kx8	28	12.5
	27128-20	200	128K	16Kx8	28	12.5
	27C128-25	250	128K	16Kx8	28	12.5
	27256-20	200	256K	32Kx8	28	12.5
	27C256-20	200	256K	32Kx8	28	12.5
	27512-25	250	512K	64Kx8	28	12.5
	27C512-25	250	512K	64Kx8	28	12.5
	27C010-12	120	1M	128Kx8	32	12.5
	27C201-12	120	2M	256Kx8	32	12.5
	27C401-12	120	4M	512Kx8	32	12.5
EEPROM	28C16A-25	250	16K	2Kx8	24	none
	2864A	250	64K	8Kx8	28	none
Flash ROM	28F256-20	200	256K	32Kx8	32	12
	28F256-15	150	256K	32Kx8	32	12
	28F010-20	200	1M	128Kx8	32	12
	28F020-15	150	2M	256Kx8	32	12

project. The only problem with EPROM is that erasing its contents can take up to 20 minutes. All EPROM chips have a window that is used to shine ultraviolet (UV) radiation to erase its contents. For this reason, EPROM is also referred to as UV-erasable EPROM or simply UV-EPROM. Figure G.06-1 shows the pins for a UV-EPROM chip.

To program a UV-EPROM chip, the following steps must be taken:

1. Its contents must be erased. To erase a chip, it is removed from its socket on the system board and placed in EPROM erasure equipment to expose it to UV radiation for 15–20 minutes.

2. Program the chip. To program a UV-EPROM chip, place it in the ROM burner (programmer). To burn code or data into EPROM, the ROM burner uses 12.5 volts or higher, depending on the EPROM type. This voltage is referred to as VPP in the UV-EPROM data sheet.

3. Place the chip back into its socket on the system board.

As can be seen from the above steps, in the same way that there is an EPROM programmer (burner), there is also separate EPROM erasure equipment. The main problem, and indeed the major disadvantage of UV-EPROM, is that it cannot be programmed while in the system board (motherboard). To find a solution to this problem, EEPROM was invented.

FIGURE G.06-1
UV ROM chip. (Copyright 1987 Intel Corporation.)

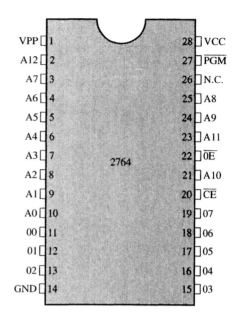

EEPROM (Electrically Erasable Programmable ROM)

EEPROM has several advantages over EPROM, such as the fact that its method of erasure is electrical and therefore instant, as opposed to the 20-minute erasure time required for UV-EPROM. In addition, in EEPROM, one can select which byte to be erased, in contrast to UV-EPROM, in which the entire contents of ROM are erased. However, the main advantage of EEPROM is the fact that one can program and erase its contents while it is still in the system board. It does not require physical removal of the memory chip from its socket. In other words, unlike UV-EPROM, EEPROM does not require an external erasure and programming device. To utilize EEPROM fully, the designer must incorporate into the system board the circuitry to program the EEPROM, using 12.5 V for VPP. EEPROM with VPP of 5–7 V is available, but it is expensive. In general, the cost per bit for EEPROM is much higher than for UV-EPROM.

Flash Memory EPROM

Since the early 1990s, flash EPROM has become a popular user-programmable memory chip, and for good reasons. First, the process of erasure of the entire contents takes less than a second, or one might say in a flash, hence its name: flash memory. In addition, the erasure method is electrical and for this reason it is sometimes referred to as flash EEPROM. To avoid confusion it is commonly called flash memory. The major difference between EEPROM and flash memory is the fact that when flash memory's contents are erased the entire device is erased, in contrast to EEPROM, where one can erase a desired section or byte. Although there are some flash memories recently made available in which the contents are divided into blocks and the erasure can be done block by block, unlike EEPROM, no byte erasure option is available. Due to the fact that flash memory can be programmed while it is in its socket on the system board, it is becoming widely used as a way to upgrade the BIOS ROM of the PC. Some designers believe that flash memory will replace the hard disk as a mass storage medium. This would increase the performance of the computer tremendously, since flash memory is semiconductor memory with access time in the range of 100 ns compared with disk access time in the range of tens of milliseconds. For this to happen, flash memory's program/erase cycles must become infinite, just like hard disks. *Program/erase cycle* refers to the number of times that a chip can be erased and programmed before it becomes unusable. At this time, the program/erase cycle is 10,000 for flash and EEPROM, 1000 for UV-EPROM, and for RAM and disks it is infinite.

Mask ROM

Mask ROM refers to a kind of ROM whose contents are programmed by the IC manufacturer. In other words, it is not a user-programmable ROM. The terminology *mask* is used in IC fabrication. Since the process is costly, mask ROM is used when the needed volume is high and it is absolutely certain that the contents will not change. It is common practice to use UV-EPROM for the development phase of a project, and only after the code/data have been finalized is mask ROM ordered. The main advantage of mask ROM is its cost, since it is significantly cheaper than other kinds of ROM, but if an error in the data/code is found, the entire batch must be thrown away.

RAM (Random Access Memory)

RAM memory is called *volatile memory* since cutting off the power to the IC will mean the loss of data. Sometimes RAM is also referred to as RAWM (read and write memory), in contrast to ROM, which cannot be written to. There are three types of RAM: static RAM (SRAM), dynamic RAM (DRAM), and NV-RAM (nonvolatile RAM). Each is explained separately.

SRAM (Static RAM)

Storage cells in static RAM memory are made of flip-flops and therefore do not require refreshing in order to keep their data. This is in contrast to DRAM, discussed shortly. The problem with the use of flip-flops for storage cells is that each cell requires at least 6 transistors to build, and the cell holds only 1 bit of data. In recent years, the cells have been made of 4 transistors, which still is too many. The use of 4-transistor cells plus the use of CMOS technology has given birth to a high-capacity SRAM, but the capacity of SRAM is far below DRAM. Table G.06-3 shows some examples of SRAM. SRAMs are widely used for cache memory.

TABLE G.06-3
Examples of RAM chips.

RAM Type	Part Number	Speed (ns)	Capacity	Organization	Number of Pins
SRAM	6116-1	100	16K	2Kx8	24
	6116LP-70*	70	16K	2Kx8	24
	6264-10	100	64K	8Kx8	28
	62256LP-10*	100	256K	32Kx8	28
DRAM	4116-20	200	16K	16Kx1	16
	4116-15	150	16K	16Kx1	16
	4116-12	120	16K	16Kx1	16
	4416-12	120	64K	16Kx4	18
	4416-15	150	64K	16Kx4	18
	4164-15	150	64K	64Kx1	16
	41464-8	80	256K	64Kx4	18
	41256-15	150	256K	256Kx1	16
	41256-6	60	256K	256Kx1	16
	414256-10	100	1M	256Kx4	20
	511000P-8	80	1M	1Mx1	18
	514100-7	70	4M	4Mx1	20
NV-SRAM	DS1220	100	16K	2Kx8	24
	DS1225	150	64K	8Kx8	28
	DS1230	70	256K	32Kx8	28

*LP indicates low power.

DRAM (Dynamic RAM)

Since the early days of the computer, the need for huge, inexpensive read/write memory was a major preoccupation of computer designers. In 1970, Intel Corporation introduced the first dynamic RAM (random access memory). Its density (capacity) was 1024 bits, and it used a capacitor to store each bit. The use of a capacitor as a means to store data cuts down the number of transistors needed to build the cell; however, it requires constant refreshing due to leakage. This is in contrast to SRAM (static RAM), whose individual cells are made of flip-flops. Since each bit in SRAM uses a single flip-flop and each flip-flop requires 6 transistors, SRAM has much larger memory cells and consequently lower density. The use of capacitors as storage cells in DRAM results in much smaller net memory cell size.

The advantages and disadvantages of DRAM memory can be summarized as follows. The major advantages are high density (capacity), cheaper cost per bit, and lower power consumption per bit. The disadvantage is that it must be refreshed periodically, due to the fact that the capacitor cell loses its charge; furthermore, while it is being refreshed, the data cannot be accessed. This is in contrast to SRAM's flip-flops, which retain data as long as the power is on, which do not need to be refreshed, and whose contents can be accessed at any time. Since 1970, the capacity of DRAM has exploded. After the 1K-bit (1024) chip came the 4K-bit in 1973, and then the 16K chip in 1976. The 1980s saw the introduction of 64K, 256K, and finally 1M and 4M memory chips. The 1990s saw 16M, 64M, 256M, and 1G-bit DRAM chips. By the time IBM came to the personal computer market, 16K-bit chips were widely used. Today, motherboards use 8M, 16M, 32M, and 64M chips. These will be discussed along with how they are used in the PC and compatibles. Keep in mind that when talking about IC memory chips, the capacity is always assumed to be in bits. Therefore, a 1M chip means 1 megabit and a 256K chip means a 256K-bit memory chip. However, when talking about the memory of a computer system, it is always assumed to be in bytes. For example, if one says that the IBM PC motherboard has 16M, it means 16M bytes of memory.

Packaging Issue in DRAM

In DRAM there is a problem of packing a large number of cells into a single chip with the normal number of pins assigned to addresses. For example, a 64K-bit chip (64Kx1) must have 16 address lines and 1 data line, requiring 16 pins to send in the address if the conventional method is used. This is in addition to V_{CC} power, ground, and read/write control pins. Using the conventional method of data access, the large number of pins defeats the purpose of high density and small packaging, so dearly cherished by IC designers. Therefore, to reduce the number of pins needed for addresses, multiplexing/demultiplexing is used. The method used is to split the address into half and send in each half of the address through the same pins, thereby requiring fewer address pins. Internally, the DRAM structure is divided into a square of rows and columns. The first half of the address is called the *row* and the second half is called the *column*. For example, in the case of DRAM of 64Kx1 organization, the first half of the address is sent in through the 8 pins A0–A7, and by activating RAS (row address strobe), the internal latches inside DRAM grab the first half of the address. After that, the second half of the address is sent in through the same pins and by activating CAS (column address strobe), the internal latches inside DRAM again latch this second half of the address. This results in using 8 pins for addresses plus RAS and CAS, for a total of 10 pins, instead of 16 pins that would be required without multiplexing. To access a bit of data from DRAM, both row and column addresses must be provided. For this concept to work, there must be a 2 by 1 multiplexer outside the DRAM circuitry and a demultiplexer inside every DRAM chip. Due to the complexities associated with DRAM interfacing (RAS, CAS, the need for multiplexer and refreshing circuitry), there are DRAM controllers designed to make DRAM interfacing much easier. However, many small microprocessor-based projects that do not require much RAM (usually less than 64K bytes) use SRAM instead of DRAM.

NV-RAM (Nonvolatile RAM)

While both DRAM and SRAM are volatile, there is a new type of RAM called NV-RAM, nonvolatile RAM. Like other RAMs, it allows the CPU to read and write to it, but when the power is turned off the contents are not lost, just like ROM. NV-RAM combines the best of RAM and ROM: the read and writability of RAM, plus the nonvolatility of ROM. To retain its contents, every NV-RAM chip internally is made of the following components:

1. It uses extremely power efficient (very, very low power consumption) SRAM cells built out of CMOS.

2. It uses an internal lithium battery as a backup energy source.

3. It uses an intelligent control circuitry. The main job of this control circuitry is to monitor the V_{CC} pin constantly to detect loss of the external power supply. If the power to the V_{CC} pin falls below out-of-tolerance conditions, the control circuitry switches automatically to its internal power source, the lithium battery. In this way, the internal lithium power source is used to retain the NV-RAM contents only when the external power source is off.

It must be emphasized that all three of these components are incorporated into a single IC chip, and for this reason nonvolatile RAM is a very expensive type of RAM as far as cost per bit is concerned. Offsetting the cost, however, is the fact that it can retain its contents up to ten years after the power has been turned off and allows one to read and write exactly the same as SRAM. See Table G.06-3 for NV-RAM parts made by Dallas Semiconductor.

Single In-Line Memory Modules (SIMMs)

When memory devices are combined to increase the word length, they are commonly packaged in a SIMM. A **SIMM** is essentially a small, rectangular-shaped PC board with a single row of input/output pins, as shown in Figure G.06-2. SIMMs are typically available in sizes from 1M x 8 to 8M x 32. The SIMM shown in Figure G.06-2 uses eight 4M x 1 bit memory chips to form a 4M x 8 bit memory module and is an example of word-length expansion.

FIGURE G.06-2
A 4M × 8 single-in-line memory module (SIMM).

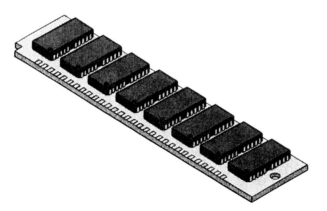

Dual In-Line Memory Module (DIMM)

The DIMM was created to fill the need of Pentium-class processors containing 64-bit data buses. A DIMM is like having two SIMMs side by side, and comes in 168-pin packages (more than twice that of a 72-pin SIMM). Ordinarily, SIMMs must be added in pairs on a Pentium motherboard to get the 64-bit bus width required by the Pentium.

Synchronous DRAM (SDRAM)

This type of RAM is very fast (up to 100 MHz operation) and is designed to synchronize with the system clock to provide high-speed data transfers.

Extended Data Out DRAM (EDO DRAM)

This type of DRAM is used with bus speeds at or below 66 MHz and is capable of starting a new access while the previous one is being completed. This ties in nicely with the bus architecture of the Pentium, which is capable of back-to-back pipelined bus cycles. Burst EDO (BEDO RAM) contains pipelining hardware to support pipelined burst transfers.

Video RAM (VRAM)

Video RAM is a special *dual-ported* RAM that allows two accesses at the same time. In a display adapter, the video electronics needs access to the VRAM (to display the Windows 95 desktop for example) and so does the processor (to open a new window on the desktop). This type of memory is typically local to the display adapter card.

Level-2 Cache

Cache is a special, high-speed memory capable of providing data within one clock cycle, and it is typically 10 times faster than regular DRAM. Although the processor itself contains a small amount of internal cache (8KB for instructions and 8KB for data in the original Pentium), you can add additional *level-2 cache* on the motherboard, between the CPU and main memory, as indicated in Figure G.06-3. Level-2 cache adds an additional 64KB to 2MB of external cache to complement the small internal cache of the processor. The basic operation of the cache is to speed up the average access time by storing copies of frequently accessed data.

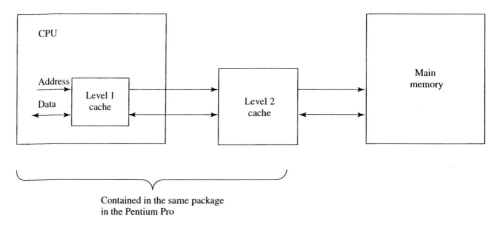

FIGURE G.06-3
Using cache in a memory system.

REVIEW QUESTIONS

True/False

1. RAM loses its information when power is turned off.

2. ROM loses its information when power is turned off.

Multiple Choice

3. A bit is stored in a dynamic RAM using a

    a. Logic element.

    b. Capacitor.

    c. Optical wavelet.

4. Motherboards typically use

    a. SIMMs or DIMMs.

    b. Level-2 cache.

    c. Both a and b.

5. Data is stored on a disk using

    a. Circular tracks of sectors.

    b. Records of 32-bits.

    c. Parallel groups of sectors called volumes.

G.07 Troubleshoot and repair types of microprocessor memory circuits

INTRODUCTION

Because memories can contain large numbers of storage cells, testing each cell can be a lengthy and frustrating process. Fortunately, memory testing is usually an automated process performed with a programmable test instrument or with the aid of software for in-system testing. Most microprocessor-based systems provide automatic memory testing as part of their system software.

ROM Testing

Since ROMs contain known data, they can be checked for the correctness of the stored data by reading each data word from the memory and comparing it with a data word that is known to be correct. One way of doing this is illustrated in Figure G.07-1. This process requires a reference ROM that contains the same data as the ROM to be tested. A special test instrument is programmed to read each address in both ROMs simultaneously and to compare the contents. A flowchart in Figure G.07-2 illustrates the basic sequence.

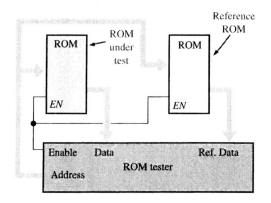

FIGURE G.07-1
Block diagram for a complete contents check of a ROM.

Checksum Method   Although the previous method checks each ROM address for correct data, it has the disadvantage of requiring a reference ROM for each different ROM to be tested. Also, a failure in the reference ROM can produce a fault indication.

In the checksum method, a number (the sum of the contents of all the ROM addresses) is stored in a designated ROM address when the ROM is programmed. To test the ROM, the contents of all the addresses except the checksum are added, and the result is compared with the checksum stored in the ROM. If there is a difference, there is definitely a fault. If the checksums compare, the ROM is most likely good. However, there is a remote possibility that a combination of bad memory cells could cause the checksums to compare.

FIGURE G.07-2
Flowchart for a complete contents check of a ROM.

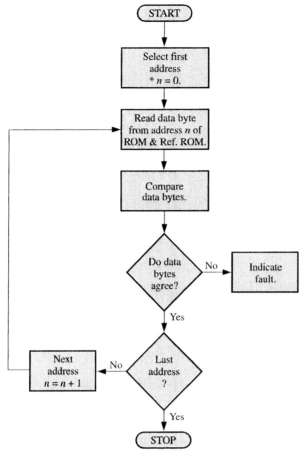

* *n* is the address number.

FIGURE G.07-3
Simplified illustration of a programmed ROM with the checksum stored at a designated address.

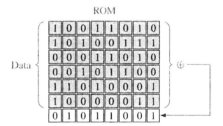

This process is illustrated in Figure G.07-3 with a simple example. The checksum in this case is produced by taking the sum of each column of data bits and discarding the carries. This is actually an XOR sum of each column. The flowchart in Figure G.07-4 illustrates the basic checksum test.

The checksum test can be implemented with a special test instrument, or it can be incorporated as a test routine in the built-in (system) software of microprocessor-based systems. In that case, the ROM test routine is automatically run on system start-up.

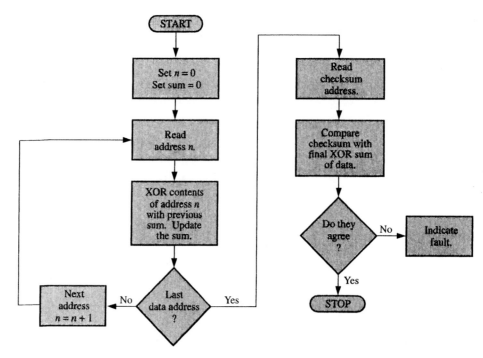

FIGURE G.07-4
Flowchart for a basic checksum test.

RAM Testing

To test a RAM for its ability to store both 0s and 1s in each cell, first 0s are written into all the cells in each address and then read out and checked. Next, 1s are written into all the cells in each address and then read out and checked. This basic test will detect a cell that is stuck in either a 1 state or a 0 state.

Some memory faults cannot be detected with the all-0s–all-1s test. For example, if two adjacent memory cells are shorted, they will always be in the same state, both 0s or both 1s. Also, the all-0s–all-1s test is ineffective if there are internal noise problems such that the contents of one or more addresses are altered by a change in the contents of another address.

The Checkerboard Pattern Test   One way to more fully test a RAM is by using a checkerboard pattern of 1s and 0s, as illustrated in Figure G.07-5. Notice that all adjacent cells have opposite bits. This pattern checks for a short between two adjacent cells, because if there is a short, both cells will be in the same state.

FIGURE G.07-5
The RAM checkerboard test pattern.

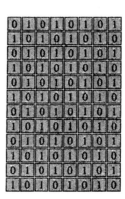

(a) (b)

After the RAM is checked with the pattern in Figure G.07-5(a), the pattern is reversed, as shown in part (b). This reversal checks the ability of all cells to store both 1s and 0s.

A further test is to alternate the pattern one address at a time and check all the other addresses for the proper pattern. This test will catch a problem in which the contents of an address are dynamically altered when the contents of another address change.

A basic procedure for the checkerboard test is illustrated by the flowchart in Figure G.07-6. The procedure can be implemented with the system software in microprocessor-based systems so that either the tests are automatic when the system is powered up or they can be initiated from the keyboard.

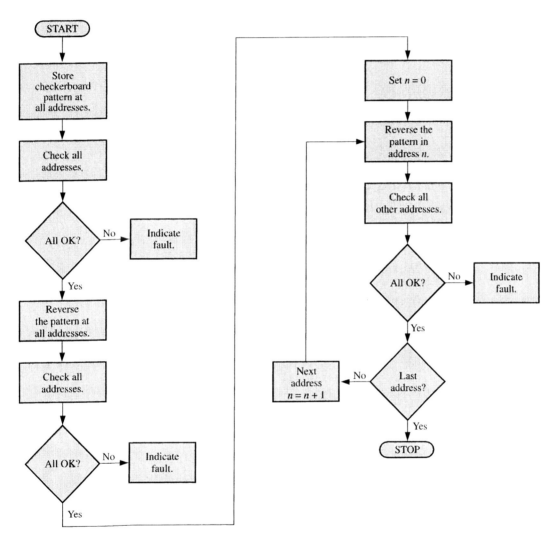

FIGURE G.07-6
Flowchart for basic RAM checkerboard test.

G.08 Understand principles and operations of microprocessor machine code and instruction sets

INTRODUCTION

In this section we will examine the machine language and instruction set of the Intel 80x86 family of microprocessors. Its use in the personal computer makes an understanding of its control language even more important.

DEVELOPING SOFTWARE FOR THE PERSONAL COMPUTER

To get the most use out of the power of the PC, we must understand how to control and use the hardware on the motherboard and the software capabilities of the processor. We will explore the software architecture of the 80x86 family by examining the instruction set and by looking at programming examples. What we will see is that the 80x86 processors speak a different language than we do.

Machine Language vs. Assembly Language

Our language is one of words and phrases. The 80x86 language is a string of 1s and 0s. For example, the instruction

```
ADD AX,BX
```

contains a word, **ADD,** that means something to us. Apparently, we are adding **AX** and **BX** together, whatever they are. So, even though we might be unfamiliar with the 80x86 instruction set, the instruction ADD AX,BX means something to us.

If we were instead given the binary string

```
0000  0001  1101  1000
```

or the hexadecimal equivalent

```
01  D8
```

and asked its meaning, we might be hard-pressed to come up with anything. We associate more meaning with ADD AX,BX than we do with 01 D8, which *is* the way the instruction is actually represented. All programs for the 80x86 will simply be long strings of binary numbers.

Because of the processor's internal decoders, different binary patterns represent different instructions. Here are a few examples to illustrate this point:

```
01 D8    ADD AX,BX    ;add BX to AX, result in AX
29 D8    SUB AX,BX    ;subtract BX from AX, result in AX
21 D8    AND AX,BX    ;AX equals AX AND BX
40       INC AX       ;add 1 to AX
4B       DEC BX       ;subtract 1 from BX
8B C3    MOV AX,BX    ;copy BX into AX
```

Can you guess the meaning of each instruction just by reading it? Do the hexadecimal codes for each instruction mean anything to you? What we see here is the difference between **machine language** and **assembly language.** The machine language for each instruction is represented by the hexadecimal codes. This is the binary language of the machine. The assembly language is represented by the wordlike terms that mean something to us. Putting groups of these wordlike instructions together is how a program is constructed.

INSTRUCTION TYPES

The 80x86 instruction set is composed of six main groups of instructions. Examining the instructions briefly here will give a good overall picture of the capabilities of the processor.

Data Transfer Instructions

Data transfer instructions are used to move data among registers, memory, and the outside world. Also, some instructions directly manipulate the stack, while others may be used to alter the flags.

The data transfer instructions are:

IN	Input byte or word from port
LAHF	Load AH from flags
LDS	Load pointer using data segment
LEA	Load effective address
LES	Load pointer using extra segment
MOV	Move to/from register/memory
OUT	Output byte or word to port
POP	Pop word off stack
POPF	Pop flags off stack
PUSH	Push word onto stack
PUSHF	Push flags onto stack
SAHF	Store AH into flags
XCHG	Exchange byte or word
XLAT	Translate byte

Arithmetic Instructions

These instructions make up the arithmetic group. Byte and word operations are available on almost all instructions. A nice addition are the instructions that multiply and divide. Previous 8-bit microprocessors did not include these instructions, forcing the programmer to write subroutines to perform multiplication and division when needed. Addition and subtraction of both binary and BCD operands are also allowed.

The arithmetic instructions are:

AAA	ASCII adjust for addition
AAD	ASCII adjust for division
AAM	ASCII adjust for multiply
AAS	ASCII adjust for subtraction
ADC	Add byte or word plus carry
ADD	Add byte or word

CBW	Convert byte or word
CMP	Compare byte or word
CWD	Convert word to double-word
DAA	Decimal adjust for addition
DAS	Decimal adjust for subtraction
DEC	Decrement byte or word by one
DIV	Divide byte or word (unsigned)
IDIV	Integer divide byte or word
IMUL	Integer multiply byte or word
INC	Increment byte or word by one
MUL	Multiply byte or word (unsigned)
NEG	Negate byte or word
SBB	Subtract byte or word and carry
SUB	Subtract byte or word

Bit Manipulation Instructions

Instructions capable of performing logical, shift, and rotate operations are contained in this group. Many common Boolean operations (AND, OR, NOT) are available in the logical instructions. These, as well as the shift and rotate instructions, operate on bytes or words.

The bit manipulation instructions are:

AND	Logical AND of byte or word
NOT	Logical NOT of byte or word
OR	Logical OR of byte or word
RCL	Rotate left through carry byte or word
RCR	Rotate right through carry byte or word
ROL	Rotate left byte or word
ROR	Rotate right byte or word
SAL	Arithmetic shift left byte or word
SAR	Arithmetic shift right byte or word
SHL	Logical shift left byte or word
SHR	Logical shift right byte or word
TEST	Test byte or word
XOR	Logical exclusive-OR of byte or word

String Instructions

String operations simplify programming whenever a program must interact with a user. User commands and responses are usually saved as ASCII strings of characters, which may be processed by the proper choice of string instruction.

The string instructions are:

CMPS	Compare byte or word string
LODS	Load byte or word string

MOVS	Move byte or word string
MOVSB (MOVSW)	Move byte string (word string)
REP	Repeat
REPE (REPZ)	Repeat while equal (zero)
REPNE (REPNZ)	Repeat while not equal (not zero)
SCAS	Scan byte or word string
STOS	Store byte or word string

Program Transfer Instructions

This group of instructions contains all jumps, loops, and subroutine (called **procedure**) and interrupt operations. The great majority of jumps are **conditional,** testing the processor flags before execution.

The program transfer instructions are:

CALL	Call procedure (subroutine)
INT	Interrupt
INTO	Interrupt if overflow
IRET	Return from interrupt
JA (JNBE)	Jump if above (not below or equal)
JAE (JNB)	Jump if above or equal (not below)
JB (JNAE)	Jump if below (not above or equal)
JBE (JNA)	Jump if below or equal (not above)
JC	Jump if carry set
JCXZ	Jump if CX equals zero
JE (JZ)	Jump if equal (zero)
JG (JNLE)	Jump if greater (not less or equal)
JGE (JNL)	Jump if greater or equal (not less)
JL (JNGE)	Jump if less (not greater or equal)
JLE (JNG)	Jump if less or equal (not greater)
JMP	Unconditional jump
JNC	Jump if no carry
JNE (JNZ)	Jump if not equal (not zero)
JNO	Jump if no overflow
JNP (JPO)	Jump if no parity (parity odd)
JNS	Jump if no sign
JO	Jump if overflow
JP (JPE)	Jump if parity (parity even)
JS	Jump if sign
LOOP	Loop unconditional
LOOPE (LOOPZ)	Loop if equal (zero)
LOOPNE (LOOPNZ)	Loop if not equal (not zero)
RET	Return from procedure (subroutine)

Processor Control Instructions

This last group of instructions performs small tasks that sometimes have profound effects on the operation of the processor. Many of the instructions manipulate the flags.

The processor control instructions are:

CLC	Clear carry flag
CLD	Clear direction flag
CLI	Clear interrupt enable flag
CMC	Complement carry flag
ESC	Escape to external processor
HLT	Halt processor
LOCK	Lock bus during next instruction
NOP	No operation
STC	Set carry flag
STD	Set direction flag
STI	Set interrupt enable flag
WAIT	Wait for $\overline{\text{TEST}}$ pin activity

ADDRESSING MODES

The 80x86 offers the programmer a wide number of choices when referring to a memory location. Many people believe that the number of **addressing modes** contained in a microprocessor is a measure of its power. If that is so, the 80x86 should be counted among the most powerful processors. Many of the addressing modes are used to generate a **physical address** in memory. A 20-bit address is formed by the sum of two 16-bit address values. One of the four segment registers will always supply the first 16-bit address. The second 16-bit address is formed by a specific addressing mode operation. The resulting 20-bit address points to one specific location in the processor's 1MB real-mode addressing space. We will see that there are a number of different ways the second part of the address may be generated.

Real-Mode Addressing Space

All addressing modes eventually create a physical address that resides somewhere in the 00000 to FFFFF addressing space of the processor. Figure G.08-1 shows a brief memory map of the real-mode addressing space, which is broken up into 16 blocks of 64KB each. Each 64KB block is called a **segment.** A segment contains all the memory locations that can be reached when a particular segment register is used. For example, if the data segment contains 0000, then addresses 00000 through 0FFFF can be generated when using the data segment. If, instead, register DS contains 1800, then the range of addresses becomes 18000 through 27FFF. It is important to see that a segment can begin on *any* 16-byte boundary. So, 00000, 00010, 00020, 035A0, 10800, and CCE90 are all acceptable starting addresses for a segment.

All together, 1,048,576 bytes can be accessed by the processor. This is commonly referred to as 1 **megabyte.** Small areas of the addressing space are reserved for special operations. At the very high end of memory, locations FFFF0 through FFFFF are assigned the role of storing the initial instruction used after a RESET operation. At the low end of memory, locations 00000 through 003FF are used to store the addresses for all 256 interrupts (although not all are commonly used in actual practice). This dedication of addressing

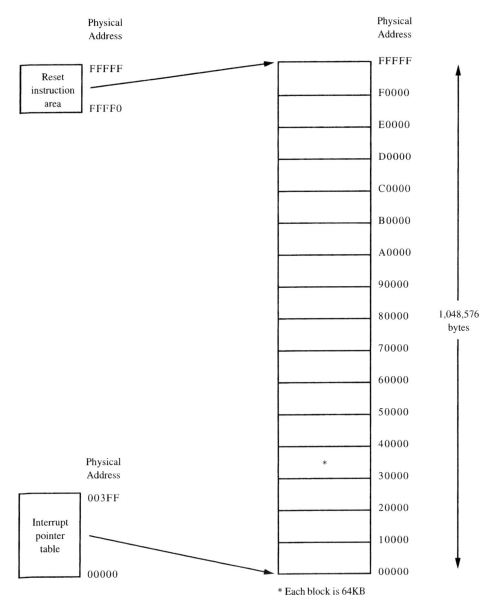

FIGURE G.08-1
Addressing space of the processor in real mode.

space is common among processor manufacturers, and may force designers to conform to specific methods or techniques when building systems around the 80x86. For instance, EPROM is usually mapped into high memory, so that the starting execution instructions will always be there at power-on.

Addressing Modes

The simplest addressing mode is known as **immediate.** Data needed by the processor is actually included in the instruction. For example:

```
MOV  CX,1024
```

contains the immediate data value 1024. This value is converted into binary and included in the code of the instruction.

When data must be moved between registers, **register** addressing is used. This form of addressing is very fast, because the processor does not have to access external memory (except for the instruction fetch). An example of register addressing is:

```
ADD  AL,BL
```

where the contents of registers AL and BL are added together, with the result stored in register AL. Notice that both operands are names of internal registers.

The programmer may refer to a memory location by its specific address by using **direct** addressing. Two examples of direct addressing are:

```
MOV  AX,[3000]
```

and

```
MOV  BL,COUNTER
```

In each case, the contents of memory are loaded into the specified registers. The first instruction uses square brackets to indicate that a memory address is being supplied. Thus, 3000 and [3000] are allowed to have two different meanings. 3000 means the number 3000, whereas [3000] means the number stored at memory location 3000. The second instruction uses the symbol name COUNTER to refer to memory. COUNTER must be defined somewhere else in the program for it to be used this way.

When a register is used within the square brackets, the processor uses **register indirect** addressing. For example:

```
MOV  BX,[SI]
```

instructs the processor to use the 16-bit quantity stored in the SI (source index) register as a memory address. A slight variation produces **indexed** addressing, which allows a small offset value to be included in the memory operand. Consider this example:

```
MOV  BX,[SI + 10]
```

The location accessed by the instruction is the sum of the SI register and the offset value 10.

When the register used is the base pointer (BP), **based** addressing is employed. This addressing mode is especially useful when manipulating data in large tables or arrays. An example of based addressing is:

```
MOV  CL,[BP + 4]
```

Including an index register (SI or DI) in the operand produces **based-indexed** addressing. The address is now the sum of the base pointer and the index register. An example might be:

```
MOV  [BP + DI],AX
```

When an offset value is also included in the operand, the processor uses **based-indexed with displacement** addressing. An example is:

```
MOV  DL,[BP + SI + 2]
```

Obviously, the 80x86 allows the base pointer to be used in many different ways.

Other addressing modes are used when string operations must be performed.

The processor is designed to access I/O ports, as well as memory locations. When **port** addressing is used, the address bus contains the address of an I/O port instead of a memory location. I/O ports may be accessed two different ways. The port may be specified in the operand field, as in:

```
IN  AL,80H
```

or indirectly, via the address contained in register DX:

```
OUT  DX,AL
```

Using DX allows a port range from 0000 to FFFF, or 65,536 individual I/O port locations. Only 256 (00 to FF) posts are allowed when the port address is included as an immediate operand.

Example G.08-1

Let us look at a practical application of the instructions and addressing modes just presented. The following is a subroutine designed to find the sum of 16 bytes stored in memory. It is not important at this time that you understand what each instruction does. We are simply trying to get a feel for what a source file might look like and what conventions to follow when we write our own programs.

```
        ORG  8000H
TOTAL:  MOV  AX,7000H   ;load address of data area
        MOV  DS,AX      ;init data segment register
        MOV  AL,0       ;clear result
        MOV  BL,16      ;init loop counter
        MOV  SI,0       ;init data pointer
ADDUP:  ADD  AL,[SI]    ;add data value to result
        INC  SI         ;increment data pointer
        DEC  BL         ;decrement loop counter
        JNZ  ADDUP      ;jump if counter not zero
        MOV  [SI],AL    ;save sum
        RET             ;and return
        END
```

The first line of source code contains a command that instructs the assembler to load its program counter with 8000H. The ORG (for origin) command is known as an assembler **pseudo-opcode,** a fancy name for a mnemonic that is understood by the assembler but not by the microprocessor. ORG does not generate any source code; it merely sets the value of the assembler's program counter. This is important when a section of code must be loaded at a particular place in memory. The ORG statement is a good way to generate instructions that will access the proper memory locations when the program is loaded into memory.

Hexadecimal numbers are followed by the letter H to distinguish them from decimal numbers. This is necessary since 8000 decimal and 8000 hexadecimal differ greatly in magnitude. For the assembler to tell them apart, we need a symbol that shows the difference. Some assemblers use $8000; others use &H8000. It is really a matter of whose software you purchase.

The second source line contains the major components normally used in a source statement. The label TOTAL is used to point to the address of the first instruction in the subroutine. ADDUP is also a label. Single-line assemblers do not allow the use of labels.

The opcode is represented by MOV and the operand field by AX,7000H. The order of the operands is <destination>, <source>. This indicates that 7000H is being MOVed into AX. So far we have three fields: label, opcode, and operand. The fourth field, if it is used, usually contains a comment explaining what the instruction is doing. Comments are preceded by a semicolon (;) to separate them from the operand field. In writing source code, you should follow the four-column approach. This will result in a more understandable source file.

The final pseudo-opcode in most source files is END. The END statement informs the assembler that it has reached the end of the source file. This is important, because many assemblers usually perform two passes over the source file. The first pass is used to determine the lengths of all instructions and data areas, and to assign values to all symbols (labels) encountered. The second pass completes the assembly process by generating the machine code for all instructions, usually with the help of the symbol table created in the first pass. The second pass also creates and writes information to the list and object files. The list file for our example subroutine looks like this:

```
1    8000                        ORG    8000H
2    8000  B8  0070  TOTAL:      MOV    AX,7000H
3    8003  8E  D8                MOV    DS,AX
4    8005  B0  00                MOV    AL,0
5    8007  B3  10                MOV    BL,16
6    8009  BE  0000              MOV    SI,0
7    800C  02  04    ADDUP:      ADD    AL,[SI]
8    800E  46                    INC    SI
9    800F  FE  CB                DEC    BL
10   8011  75  F9                JNZ    ADDUP
11   8013  88  04                MOV    [SI],AL
12   8015  CB                    RET
13                               END
```

Normally the comments would follow the instructions, but they have been removed for the purposes of this discussion.

The first column of numbers represents the original source line number.

The second column of numbers represents the memory addresses of each instruction. Notice that the first address matches the one specified by the ORG statement. Also notice that the ORG statement does not generate any code.

The third column of numbers is the machine code generated by the assembler. The machine codes are intermixed with data and address values. For example, B8 0070 in line 2 represents the instruction MOV AX,7000H, with the MOV instruction coded as B8 and the immediate word 7000H coded in byte-swapped form as 0070. In line 3, the machine codes 8E D8 represent MOV DS,AX. Neither of these 2 bytes are data or address values, as they were in line 2. Look for other data values in the instructions on lines 4 through 6. Line 5 makes an important point: the assembler will convert decimal numbers into hexadecimal (the 16 in the operand field has been converted into 10 in the machine code column).

Finally, another look at the list file shows that there are 1-, 2-, and 3-byte instructions present in the machine code. When an address or data value is used in an instruction, chances are good that you will end up with a 2- or 3-byte instruction (or possibly even more).

Following the code on each line is the text of the original source line. Having all of this information available is very helpful during the debugging process.

REVIEW QUESTIONS

True/False

1. Machine language and assembly language are the same thing.

2. An assembler converts from assembly language into machine language.

Multiple Choice

3. Which instruction is not a data transfer instruction?

    a. MOV.

    b. JMP.

    c. PUSH.

4. What is the order of the operands in an 80x86 instruction?

    a. Destination, source.

    b. Source, destination.

    c. Destination, destination.

5. What happens during execution of this instruction: MOV   AL,7?

    a. The value 7 is written into register AL.

    b. The value in register AL is written to location 7.

    c. The value in location 7 is written into register AL.

Microcomputers

H.01   Demonstrate an understanding of microcomputer operating systems

H.02   Demonstrate an understanding of essential microcomputer components

H.03   Demonstrate an understanding of microcomputer peripherals

H.04   Set up and configure a microcomputer using available operating systems and software packages

H.05   Troubleshoot and replace microcomputer peripherals

691

H.01 Demonstrate an understanding of microcomputer operating systems

INTRODUCTION

In this section we examine the operation of MS-DOS and the Windows 95 operating systems.

MS-DOS STRUCTURE

From Cold Boot to DOS Prompt

A cold boot occurs when the power switch of the PC is turned on. A warm boot is when the computer is reset using the CTRL-ALT-DEL keys. What happens from the moment the PC is turned on (cold boot) to the time the DOS prompt ">" appears on the screen? Figure H.01-1 on the facing page shows the major steps.

DOS Standard Device Names

DOS has reserved some names to be used for standard devices connected to the PC. Since they have special meaning for DOS, they should not be used for user-defined filenames or device names. They are listed in Table H.01-1.

TABLE H.01-1
DOS names for standard devices.

Device Name	Meaning
AUX	Auxiliary device (often another name referring to COM1)
CLOCK$	System clock (date and time)
COM1	Serial communication port #1
COM2	Serial communication port #2
COM3	Serial communication port #3
COM4	Serial communication port #4
CON	Console for system input (keyboard) and output (video)
LPT1	Line printer #1
LPT2	Line printer #2
LPT3	Line printer #3
LPT4	Line printer #4
NUL	NULL (an input and output that does not exist)
PRN	Printer attached to LPT1

FIGURE H.01-1
Boot process.

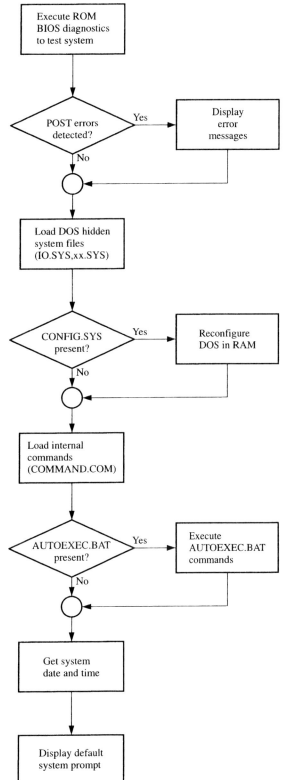

CONFIG.SYS and How It Is Used

Assume that you want to design and attach some kind of device such as a card reader to your PC. How is the PC notified of the presence of a card reader? CONFIG.SYS allows the user to incorporate new technology into the PC. CONFIG.SYS is an optional file which the PC reads and executes only during system boot. This is done right before the system loads COMMAND.COM. This optional file not only allows users to add a new nonstandard device such as a mouse and CD-ROM but also allows customizing DOS for a given system. Depending on the intended use of the PC, CONFIG.SYS can be modified to improve system performance dramatically. This is especially true for the way DOS allocates and uses the 640K of conventional memory. The setting of the parameters in the CONFIG.SYS file can have a dramatic impact on how much memory DOS leaves for application software. Note that the CONFIG.SYS file must be located in the root directory of the system disk.

AUTOEXEC.BAT and How It Is Used

AUTOEXEC.BAT is an optional file that contains a group (a batch) of DOS commands that are executed automatically when the PC is turned on. This saves the user from having to type in the commands each time the system is booted. The DOS commands contained in the batch file are executed in sequence without any external intervention from the user. After the PC is booted, CONFIG.SYS is loaded and executed, COMMAND.COM is loaded, and finally DOS searches for the AUTOEXEC.BAT file in the root directory of the system disk. If the file is present, it is loaded into memory and each command in the file is executed, one after the other. An example of a batch command is "prompt pg", which sets up the DOS prompt to display the current directory, as well as the ">". Another example is the command "win", which is placed in the AUTOEXEC.BAT file when the user wants the system to run Windows upon boot.

Types of DOS Commands

All DOS commands can be categorized according to when and where they can be invoked. The categories are internal, external, CONFIG.SYS, batch, and internal/external (internal or external). Each category is described next.

Internal DOS Commands After COMMAND.COM is loaded into the 640K conventional memory, it stays there, taking up precious memory. To keep the size of COMMAND.COM small, Microsoft includes only the most important DOS commands in COMMAND.COM. DOS commands included in COMMAND.COM are referred to as *internal* DOS commands. Commands such as COPY, DEL, TYPE, MD, and CD are internal commands. Remember that DOS is not case sensitive. Some examples are:

```
COPY PROG1.ASM PROG2.ASM    copies first file to second
DEL B:PROG1.ASM             deletes prog1.asm on drive b
TYPE PROG1.ASM              lists file to screen
MD PROGRAMS                 creates directory named programs
CD\TOM\PROGRAMS             changes current directory to
                            \tom\programs
```

External DOS Commands There are many DOS commands that are not part of COMMAND.COM and are provided as separate files on the DOS disk. These are called *external* DOS commands (external from the point of view of COMMAND.COM or one might say not residing in conventional memory). When an external DOS command is executed, it is loaded from disk into memory and executed. After that, it is released from memory. To use it again, it must be loaded from the disk again. When external DOS commands are used, the path must be specified; otherwise, it is assumed to be in the current directory.

Commands such as FORMAT, DISKCOPY, BACKUP, EDLIN, and UNDELETE are examples of external commands.

`FORMAT B:`	format diskette in drive b
`DISKCOPY A: B:`	copy diskette in drive a to disk in drive b
`BACKUP C:\SUE\PROGRAMS\*.* A:`	back up all files in that directory to disk a
`EDLIN PROG1.ASM`	edit file with DOS line editor program
`UNDELETE PROG1.ASM`	recover deleted file prog1.asm

WINDOWS 95

The Windows operating system has gone through many changes since it first appeared in the mid-1980s. It has evolved from a simple add-on to DOS to a multitasking, network-ready, object-oriented, user-friendly operating system. As the power of the underlying CPU running Windows has grown (from the initial 8086 and 8088 microprocessors through the Pentium II, as well as other microprocessors), so too have the features of the Windows operating system. For users familiar with Windows 3.x, the good news is that many of the operating system features are still there. For example, a left double-click is still used to launch an application. The purpose here is to familiarize you with many of the features that are new in Windows 95. Where possible, comparisons will be made to Windows 3.x to help you gain an appreciation for how things have changed.

The Desktop

Once Windows 95 has completely booted up, you may see a display screen similar to that shown in Figure H.01-2. This graphical display is called the *desktop,* because it resembles the desktop in an office environment. The desktop may contain various *folders* and *icons,*

FIGURE H.01-2
The Windows 95 desktop.

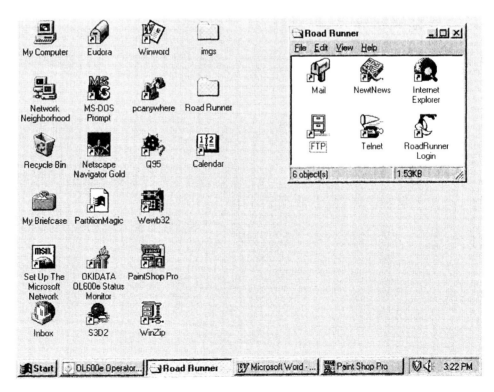

a *taskbar,* the *current time,* and possibly *open folders* containing other folders and icons. A folder is more than just a subdirectory. A folder can be shared across a network, and *cut* and *pasted* just like any other object. You can even e-mail a folder if you wish.

Typically, the bottom of the display will contain the *taskbar,* which contains the ever-present Start button, icons for all applications currently running or suspended, open desktop folders, and the current time. You can hide, resize, or move the taskbar to adjust the display area for applications. Simply left-clicking on an application's icon in the taskbar makes it the current application.

A new application may be launched by left double-clicking its desktop icon. The desktop may contain a picture, centered or tiled, called the *background image.* The desktop itself is an object that has its own set of properties. For example, you can control how many colors are available to display the desktop. Everything is controlled through the use of easily navigated pop-up menus.

Long File Names

File names in DOS were limited to eight characters with a three-letter extension (commonly called 8.3 notation). Since Windows 3.x ran on top of DOS, it, too, was limited to file names of the 8.3 variety, even though Program Manager allowed longer descriptive names on the program icons.

Windows 95 eliminates the short file name limitation by allowing up to 255 characters for a file name. As shown in Figure H.01-3, a *long file name* has two representations. One is compatible with older DOS applications (the old 8.3 notation using all uppercase characters). The other, longer representation is stored exactly as it was entered, with uppercase and lowercase letters preserved. To be compatible with older DOS applications, Windows 95 uses the first six characters of a long file name, followed in most cases by ~1. When two or more long file names appear in the same directory, Windows 95 will enumerate them (~1, ~2, etc.), as you can see in the directory listing of Figure H.01-3. To specify a long file name in a DOS command, use the abbreviated 8.3 notation, or enter the entire long file name surrounded by double quotes. For example, both of these DOS commands are identical in operation:

```
TYPE    WOWTHI~1.DOC
TYPE    "Wow This is a LONG filename.doc"
```

The great advantage of long file names is their ability to describe the contents of a file, without having to resort to cryptic abbreviations.

```
Volume in drive D is FIREBALLXL5
Volume Serial Number is 245F-15E6
Directory of D:\repair3e\e15

.                <DIR>        12-31-97 12:58a .
..               <DIR>        12-31-97 12:58a ..
LIST    TXT           0  01-06-98  2:01p list.txt
E15     DOC      31,232  01-06-98  2:01p e15.doc
WOWTHI~1 DOC     20,480  01-06-98  1:38p Wow This is a LONG filename.doc
WOWTHI~2 DOC     20,992  01-06-98  2:01p Wow This is LONG too.doc
        4 file(s)         72,704 bytes
        2 dir(s)      23,674,880 bytes free
```

FIGURE H.01-3
Two examples of long file names.

Context-Sensitive Menus

In many instances, right-clicking on an object (a program icon, a random location on the desktop, the taskbar) will produce a *context-sensitive menu* for the item. For example, right-clicking on a blank portion of the desktop produces the menu shown in Figure

FIGURE H.01-4
Context-sensitive desktop menu.

FIGURE H.01-5
Context-sensitive time/date menu.

H.01-4. Right-clicking on the time in the lower corner of the desktop generates a different menu, as indicated by Figure H.01-5. Note that the two example menus are different. This is what the "context-sensitive" term is all about. Windows 95 provides a menu tailored to the object you right-click on. This is a great improvement over Windows 3.x, which rarely did anything after right-clicking.

Windows Explorer

Windows 3.x provided two main applications that made life bearable: Program Manager and File Manager. In Windows 95, the services provided by these two applications, as well as many new features, are found in the new Windows Explorer program. Figure H.01-6 shows a typical Explorer window. The small box in the upper-left corner containing W95 (C:) indicates the current folder selected. Clicking the down arrow produces a list of folders to choose from. The two larger windows display, respectively, a directory tree of drive C: (folders only) and the contents of the currently selected folder (which also happens to be drive C:). Note the different icons associated with the files shown. Windows Explorer

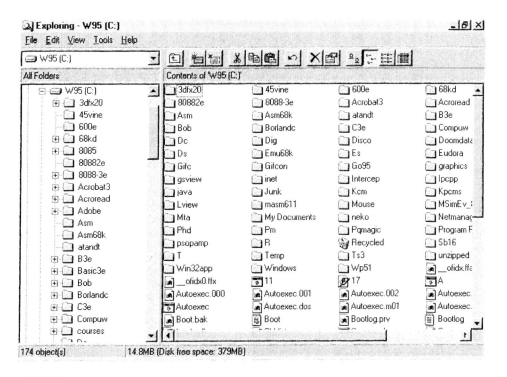

FIGURE H.01-6
Windows Explorer.

allows you to change the icon, or associate it with a different file. In general, as with Windows 3.x, double-clicking on a file or its icon opens the application associated with it.

Windows Explorer also lets you map network drives, search for a file or folder, and create new folders, among other things. It is truly one of the more important features of Windows 95.

The Registry

The Registry is the Windows 95 replacement for the SYSTEM.INI and WIN.INI configuration files used by Windows 3.x. The Registry is an internal operating system file maintained by Windows 95. As each application is installed, the installation program makes "calls" to the Registry to add configuration information, storing similar information to what was previously stored in the .INI files. In this way, the Registry is protected and therefore harder to corrupt. The Registry is accessed using the REGEDIT program as illustrated in Figure H.01-7. The Registry is nothing to fool around with simply because you feel like experimenting. A corrupt Registry can prevent Windows 95 from booting and could possibly require a complete reinstallation of Windows 95.

The Registry contains all the information Windows 95 knows about both the hardware and software installed on the computer.

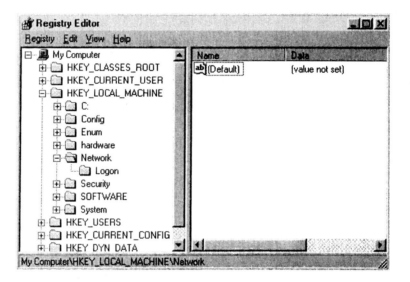

FIGURE H.01-7
REGEDIT window.

Networking

Windows 95 offers a major improvement in networking capabilities. Windows 3.1 had no built-in networking support and required network software to be loaded and maintained by DOS. This situation was improved slightly with the release of Windows for Workgroups 3.11, which provided limited networking via a network protocol called NetBEUI (NetBIOS Extended User Interface), which allows e-mail and file sharing in small peer-to-peer networks. NetBEUI is still available in Windows 95 (providing the network neighborhood feature, network drives, shared printers), along with other additions. Two important protocols have been added, PPP (point-to-point protocol) and TCP/IP (transmission control protocol/Internet protocol). PPP is used with a serial connection (such as a modem), and is the basis for the dial-up networking provided by Windows 95. TCP/IP is the protocol utilized by the Internet. You can use TCP/IP applications such as Netscape or Internet Explorer to browse the World Wide Web, connect to a remote computer and share files, send and receive e-mail, and much more. Typically, TCP/IP is used in conjunction with a network interface card for fast data transmission, although it can also be used over a modem connection (by encapsulating it inside the PPP protocol).

DOS

Yes, DOS is still a part of Windows. However, the role of DOS in Windows 95 is different in many ways from what was required under Windows 3.x. For example, Windows 3.x relied on the file system set up and maintained by DOS. This is commonly referred to as running "on-top" of DOS. Windows 95 does things differently, providing its own improved file system (long file names), and a *window* to run your DOS application in. You can open several DOS windows at the same time if necessary, and run different applications in each. Figure H.01-8 shows a typical DOS window.

Furthermore, in Windows 3.x, it is possible for a wayward application running in a DOS shell to completely hang the system. Applications are not allowed that much control in Windows 95. If a DOS application (or any other, for that matter) hangs up, all you need to do is press Ctrl-Alt-Del (oddly enough) to bring up the Close window. Figure H.01-9 illustrates the Close window.

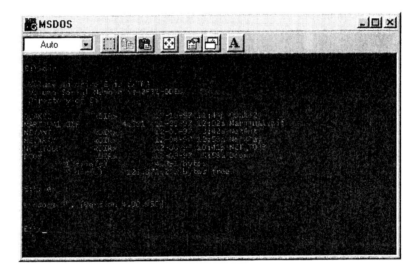

FIGURE H.01-8
DOS window.

FIGURE H.01-9
Close window.

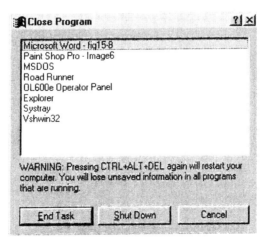

It is important to note that Windows 95 is always in control somewhere in the background, monitoring everything. A DOS application cannot hang the system like its Windows 3.x counterpart.

Recycle Bin

The Recycle Bin shown in Figure H.01-10(a) is a holding place for anything that is deleted in Windows 95. The nice thing about the Recycle Bin is that *you can get your files back* if you want to, using the Undelete option. Double-clicking on the Recycle bin icon brings up the window shown in Figure H.01-10(b). Any or all of the files shown in the window may be recovered. *Warning:* If you delete files while in DOS, they are not deposited in the Recycle Bin, and may be impossible to recover at a later time (with the old UNDELETE command).

Recycle Bin

(a) Program icon

FIGURE H.01-10
The Recycle Bin.

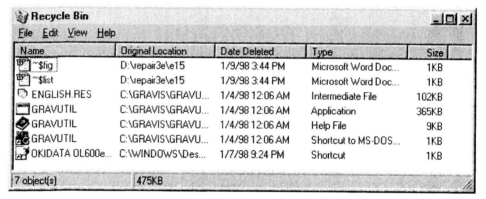

(b) Window

REVIEW QUESTIONS

True/False

1. A warm boot is performed the first time a computer is turned on.
2. Windows 95 runs on top of DOS.

Multiple Choice

3. The two files processed by DOS while booting are
   a. CONFIG.SYS and BOOT.BAT.
   b. CONFIG.SYS and AUTOEXEC.BAT.
   c. IO.SYS and AUTOEXEC.BAT.
4. The main user interface in the Windows 95 operating system is the
   a. Registry.
   b. Desktop.
   c. Taskbar.
5. Windows 95 comes with built-in
   a. Networking support.
   b. Improved help.
   c. Both a and b.

H.02 Demonstrate an understanding of essential microcomputer components

INTRODUCTION

The essential microcomputer components discussed in this section are the keyboard, video display, disk drive, memory, and CPU. These are the minimal requirements of a modern computing system.

The Microcomputer

When a microprocessor is connected to a memory unit, an input unit, and an output unit, you have a **microcomputer** as shown in Figure H.02-1. The CPU (microprocessor) communicates with the memory unit and the input and output units on the buses. Addresses are sent to the memory on the address bus, instructions and/or data are transferred between the CPU, memory, and input/output units on the data bus, and signals on the control bus coordinate all operations.

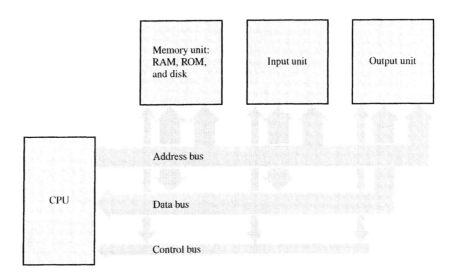

FIGURE H.02-1
Microcomputer block diagram.

Central Processing Unit (CPU) The microprocessor and associated support circuits make up the **CPU** which is sometimes called the MPU (microprocessing unit). Basically, the CPU addresses a memory location, obtains **(fetches)** a program instruction that is stored there, and carries out **(executes)** the instruction. After completing one instruction, the CPU moves on to the next one. This fetch and execute process is repeated until all of the instructions in a specific program have been executed. A simple example of an application program is a set of instructions stored in the memory as binary codes that directs the CPU to fetch a series of numbers also stored in the memory, add them, and store the sum back in the memory.

701

Memory Unit The memory unit typically consists of RAM, ROM, and disk. The RAM stores data and programs temporarily during processing. Data are numbers or other information in binary form, and programs are lists of instructions that tell the computer what to do. Because the RAM is generally a volatile memory, everything must be backed up in nonvolatile disk storage.

The ROM stores system programs such as BIOS (basic input/output system). These types of programs are used for handling video display graphics, printer communications, power-on self-test, and other routines for servicing peripherals and for general "housekeeping."

Input Unit The microcomputer receives information from the "outside world" via the input unit which, generally, can handle several external devices called **peripherals.** Examples of peripherals with which a computer communicates through the input unit are the keyboard and the mouse.

Output Unit The microcomputer sends information to the "outside world" via the output unit which, generally, can handle several peripherals. Examples of peripherals with which a computer communicates through the output unit are the video monitor and the printer. Some peripherals function as both input and output devices; examples are external disk drives, video monitors, and modems.

A Microcomputer System

For a microcomputer to accomplish a given task, it must communicate with the "outside world" by interfacing with people, sensing devices, or devices to be controlled. A typical microcomputer system is shown in Figure H.02-2. Usually, there is a keyboard for data entry, a video monitor, and a printer. A mouse is also used on most personal computers. Other types of peripherals such as an external disk drive, modem, scanner, graphics tablet, or voice input are often part of a system.

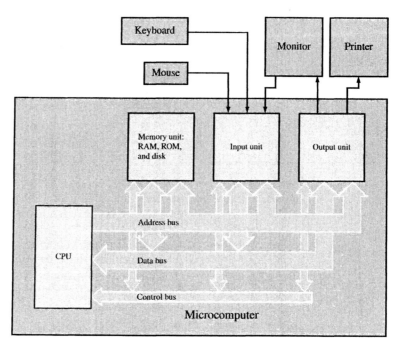

FIGURE H.02-2
A typical microcomputer system.

DISPLAY ADAPTERS AND MONITORS

The computer display system used by your computer consists of two separate but essential parts: the monitor and the video adapter card as shown in Figure H.02-3. Note from the figure that the monitor does not get its power from the computer; it has a separate power cord and its own internal power supply. Some monitors get their power from the computer instead.

The video adapter card (Figure H.02-3(b)) interfaces between the motherboard and the monitor. This card processes and converts data from the computer and allows you to see all the things you are used to seeing displayed on the monitor screen.

It is very important to realize that there are many different types of monitors and that each type of monitor essentially requires its own special video adapter card, as shown in Figure H.02-4 on the next page. Connecting a monitor to an adapter card not made for it can severely damage the monitor or adapter card, or both.

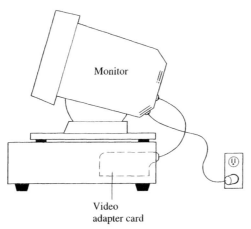

(a) Video adapter card with companion monitor

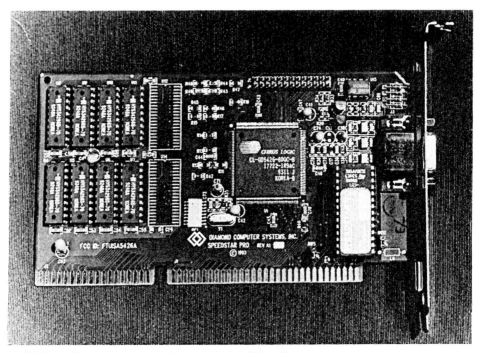

(b) SVGA graphics accelerator card (*photograph by John T. Butchko*)

FIGURE H.02-3
The two essential parts of a computer display system.

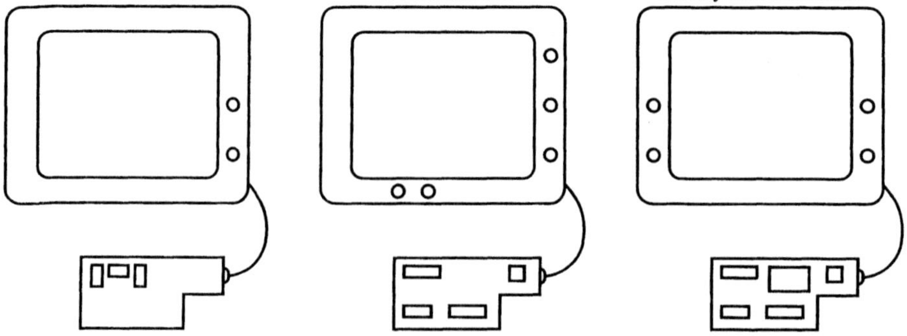

FIGURE H.02-4
Necessity of each computer monitor having its own matching adapter.

Types of Monitors

In order to understand the differences among the most common types of computer monitors, you must first understand the definitions of the terms used to describe them. Table H.02-1 lists the major terms used to distinguish one monitor from another.

Now that you know the definitions of some of the major terms used to distinguish one monitor from another, you can be introduced to the most common types of monitors in use today. Table H.02-2 lists the various types of monitors and their distinguishing characteristics.

TABLE H.02-1
Computer monitor terminology.

Term	Definition
Resolution	The number of pixels available on the monitor. A resolution of 640 by 480 means that there are 640 pixels horizontally and 480 pixels vertically.
Colors	The number of different colors that may be displayed at one time in the graphics mode. For some color monitors, more colors can be displayed in the text mode than in the graphics mode. This is possible because of the reduced memory requirements of the text mode.
Palette	A measure of the full number of colors available on the monitor. However, not all the available palette colors can be displayed at the same time (again, because of memory requirements). You can usually get a large number of colors with low resolution (fewer pixels) or a smaller number of colors (sometimes only one) with much higher resolution—again, because of memory limitations.
Display (digital or analog)	There are basically two different types of monitor displays, **digital** and **analog.** Some of the first computer monitors used poor-quality analog monitors. Then digital monitors, with their better overall display quality, became more popular. Now, however, the trend is back to analog monitors because of the increasing demand for high-quality graphics, where colors and shades can be varied continuously to give a more realistic appearance.

VGA (Video Graphics Array) Monitor The **VGA monitor** is one of the most popular color monitors; it provides high color resolution at a reasonable price. More and more software with graphics is making use of this type of monitor. The associated cards have a high scanning rate, resulting in less eye fatigue both in text and in graphics modes.

SVGA (Super VGA) Monitor Higher screen resolution and new graphics modes make the **SVGA monitor** even more popular than the VGA monitor.

TABLE H.02-2
Common types of computer monitors.

Type	Resolution*	Colors	Palette	Display
Monochrome composite	640 × 200	1	1	Analog
Color composite	640 × 200	4	4	Analog
Monochrome display	720 × 350	1	1	Digital
RGB (CGA)	640 × 200	4	16	Digital
EGA	640 × 350	16	64	Digital
PGA	640 × 480	Unlimited	Unlimited	Analog
VGA	640 × 480	256	262,144	Analog
SVGA	1280 × 1024	Varies	Varies	Digital/analog
Multiscan	varies	Unlimited	Unlimited	Digital/analog

*In general, the higher the resolution, the higher the scan frequency. For example, the typical scan frequencies of EGA and VGA monitors are 21.5 KHz and 31.5 KHz, respectively.

Multiscan Monitor   The **Multiscan monitor** was one of the first monitors that could be used with a wide variety of monitor adapter cards. Since this type of monitor can accommodate a variety of adapter cards, it is sometimes referred to as the *multidisplay* or *multi-sync* monitor.

Display Adapters

As previously stated, a computer monitor must be compatible with its adapter card. If it is not, damage to the monitor or adapter card, or both, could result.

VGA Adapter   The **VGA (video graphics array card)** was the fastest-growing graphics card in terms of popularity until the SVGA card became available. The VGA adapter card uses a 15-pin high-density pin-out, as shown in Figure H.02-5. The VGA 15-pin adapter can be wired to fit the standard nine-pin graphics adapter, as shown in Figure H.02-6.

SVGA Adapter   The Super VGA graphics interface uses the same connector that VGA monitors use. However, more display modes are possible with SVGA than with VGA.

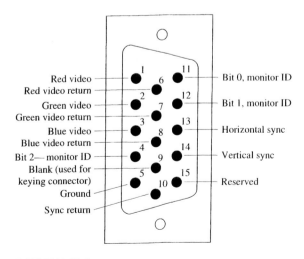

FIGURE H.02-5
Pin diagram for VGA adapter.

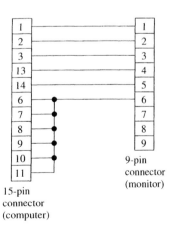

FIGURE H.02-6
Nine-pin adapter cable for VGA.

VESA The **VESA** (Video Electronics Standards Association) specification has been developed to guide the operation of new video cards and displays beyond VGA. New BIOS software that supports the VESA conventions is contained in an EPROM mounted on the display card. The software also supports the defined VESA video modes. Some of these new modes are 1024 by 768, 1280 × 1024, and 1600 × 1200, with up to 16 million possible colors.

Graphics Accelerator Adapters

A graphics accelerator is a video adapter containing a microprocessor designed specifically to handle the graphics processing work load. This eliminates the need for the system processor to handle the graphics information, allowing it to process other instructions (nongraphics related) instead.

Aside from the graphics processor, there are other features offered by graphics accelerators. These features include additional video memory, which is reserved for storing graphical representations, and a wide bus capable of moving 64 or 128 bits of data at a time. Video memory is also called VRAM and can be accessed much faster than conventional memory.

Many new multimedia applications require a *graphics accelerator* to provide the necessary graphics throughput in order to gain realism in multimedia applications. Table H.02-3 illustrates the settings available for supporting many different monitor types and refresh rates.

TABLE H.02-3
Computer monitor terminology.

Resolution	Colors	Memory	Refresh rates
640 × 480	256	2MB	60, 72, 75, 85
	65K	2MB	
	16M	2MB	
800 × 600	256	2MB	56, 60, 72, 75, 85
	65K	2MB	
	16M	2MB	
1024 × 768	256	2MB	43 (interlaced), 60, 72, 75, 85
	65K	2MB	
	16M	4MB	
1280 × 1024	256	2MB	43 (interlaced), 60, 75, 85
	65K	4MB	
	16M	4MB	

Most graphics accelerators are compatible with the new standards such as Microsoft DirectX which provide an application programming interface, or API, to the graphics subsystem. Usually, the graphics accelerators are also compatible with OpenGL for the Windows NT environment.

AGP Adapter

The Accelerated Graphics Port (AGP) is a new interface specification developed by Intel. The AGP adapter is based on PCI design but uses a special point-to-point channel so that the graphics controller can directly access the system main memory. The AGP channel is 32 bits wide and runs at 66 MHz. This provides a bandwidth of 266 MBps as opposed to the PCI bandwidth of 133 MBps.

TABLE H.02-4
AGP graphics mode.

Mode	Throughput (MBs)	Data Transfers per Cycle
1x	266	1
2x	533	2
3x	1066	4

AGP optionally supports two faster modes, with throughput of 533MB and 1.07GB. By sending either one (AGP 1X), two (AGP 2X) or four (AGP 4X) data transfers per clock cycle accomplishes these data rates. Table H.02-4 shows the different AGP modes. Other optional features include AGP texturing, sideband addressing, and pipelining. Each of these options provides additional performance enhancements.

AGP graphics support is provided by the new NLX motherboards, which also support the Pentium II microprocessor. It allows for the graphic subsystem to work much closer with the processor than previously available by providing new paths for data to flow among the processor, memory, and video memory. Figure H.02-7 shows this relationship.

AGP offers many advantages over traditional video adapters. You are encouraged to become familiar with the details of the AGP adapter.

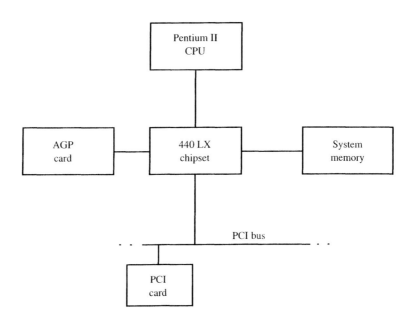

FIGURE H.02-7
AGP configuration.

The Floppy Disk

In order to store data on the disk, both sides are coated with magnetic materials. The principles behind the process of reading and writing (storing) digital data 0 and 1 on disks is the same as is used in any magnetic-based medium. Each side of the disk is organized into tracks and sectors as shown in Figure H.02-8. Tracks are organized as concentric circles and their number per disk varies from disk to disk, depending on the size and technology. Each track is divided into a number of sectors, and again the number of sectors per track varies, depending on the density of the disk and the version of the DOS operating system.

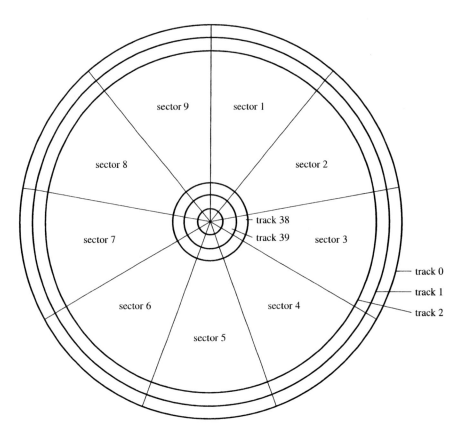

FIGURE H.02-8
Sectors and tracks of a 5 1/4" diskette.

Each sector stores 256 or 512 bytes of information, depending on the sector density. In what is commonly referred to as a *double-density* disk, the storage capacity of a single sector is 512 bytes. This format is supported by DOS 2.0 and higher. In addition to the number of tracks and sectors in the floppy disk, the physical size of the disk varies. Among the available sizes are 5 1/4 and 3 1/2 inches. The number of tracks and sectors, and the total capacity of double-sided floppy disks commonly used for the IBM PC and compatibles, are shown in Table H.02-5 along with the supporting DOS version. The total density of the various disk types can be verified by using 512 bytes for the sector density. For example, the 5 1/4" with 40 tracks and 9 sectors per track will be as follows:

```
40 tracks x 9 sectors per track = 360 sectors per disk side
360 sectors x 512 bytes per sector = 184,320 bytes per side
184,320 x 2 sides = 368,640 bytes per disk, or 360K bytes per
disk
```

TABLE H.02-5
Sectors and tracks for double-sided disks.

Diskette Type	Tracks per Side	Sectors per Track	Bytes	MS-DOS Version
5 1/4"	40	8	320K	1 and above
5 1/4"	40	9	360K	2 and above
5 1/4" high density	80	15	1.2M	3 and above
3 1/2"	80	9	720K	3.2 and above
3 1/2" high density	80	18	1.44M	3.3 and above

The sectors of a disk are grouped into clusters. Cluster size varies among formats, but a common size for floppy disks is 2 sectors per cluster. The file allocation table, or FAT, which will be discussed later, keeps track of what clusters are used to store which files.

Formatting Disks

Once a floppy disk is formatted, the computer can read from or write to that disk. *Formatting* organizes the sectors and tracks in a way that makes it possible for the disk controller to access the information on the disk. When a disk is formatted, a number of sectors are set aside for various functions and the remaining sectors are used to store the user's files. The formatting process sets aside a specific number of sectors for the boot record, directory, and FAT (file allocation table). It also copies some system files onto the disk if it was formatted with the "/s" option, which makes it a bootable disk.

Bootable and Nonbootable Disks

If a disk is formatted as a system disk (bootable), the first two files are IBMBIO.COM and IBMDOS.COM, which are followed by COMMAND.COM. The first two are hidden files; therefore, they will not be listed when DIR is used. However, if the disk is formatted as a nonbootable disk, it will not have those three files on it after it is formatted. The job of IBMBIO.COM is to provide low-level (hardware) communication (interface) between BIOS and DOS. The high-level (software) interface is provided by the IBMDOS.COM file. This is the section of DOS that contains INT 21H, among other things. Among the functions of COMMAND.COM is to provide the DOS prompt ">", read, interpret, and execute commands typed in by the user. The first two files have been given different names by MS-DOS from Microsoft: IO.SYS and MSDOS.SYS instead of IBMBIO.COM and IBMDOS.COM. Beginning with DOS 4.0, these three files no longer have to be the first directory entries and they can be located anywhere in the directory. The SYS command can be used to copy these files to a nonbootable disk to make it bootable.

HARD DISKS

Next we will look at the characteristics of hard disks and their organization with emphasis on performance factors such as access time and, finally, interfacing standards. The *hard disk,* referred to sometimes as *fixed disk,* or *winchester disk* in IBM literature, is judged according to three major criteria: capacity, access time (speed of accessing data), and interfacing standard. Before delving into each category, an explanation should be given for the use of different names such as fixed disk, winchester disk, and hard disk to refer to the same device. The term *hard disk* comes from the fact that it uses hard solid metal platters to store information instead of plastic as is the case in floppy disks. It is also called *fixed disk* because it is mounted (fixed) at a place on the computer and is not portable like the floppy disk (although some manufacturers make removable hard disks). Why is it also called the *winchester disk*? When IBM made the first hard disk for mainframes it was capable of storing 30 megabytes on each side and was therefore called a 30/30 disk. The 30/30 began to be called the winchester 30/30, after the rifle, and soon it came to be known simply as the winchester disk.

Hard Disk Capacity and Organization

One of the most important factors in judging a hard disk is its *capacity,* the number of bytes it can store. Capacity of hard disks ranges from 5 megabytes to many gigabytes (a gigabyte is 1024 megabytes). The 5 megabyte disk was used in the early days of the PC and is no longer made. At this time, when the capacity of hard disks is increasing to the gigabyte level, 2 to 8 gigabyte capacity disks are in common use for the Pentium computers.

Regardless of the capacity of the hard disk, they all use hard metal platters to store data. In general, the higher the number of platters, the higher the capacity of the disk. Just as in the floppy disk, both sides of each platter in the disk are coated with magnetic material. Likewise, it uses a storage scheme that divides the area into sectors and tracks just as the floppy disk does. There is one read/write head for each side of every platter, and these heads all move together. For example, a hard disk with 4 platters might have 8 read/write heads, one for each side, and they all move from the outer tracks into inner tracks by the same arm. Hard disks give rise to more complex organization and hence a new term: the *cylinder,* which consists of all the tracks of the same radius on each platter. Since all the read/write heads move together from track to track, it is logical to talk about cylinders in addition to tracks in the hard disk.

Why do all the heads move together? The answer is that it is too difficult and expensive to design a hard disk controller that controls the movement of so many different heads. In addition, it would prolong the access time since it must stop one head and then activate a different head continuously until it reaches the end of the file. Using the concept of the cylinder, all the tracks of the same radius are accessed at the same time, and if the end of the file is not reached, all the heads move together to the next track.

The number of read/write heads varies from one hard disk to another. The number is usually twice the number of platters. In some disks, one side of one platter is set aside for internal use and is not available for data storage by the user. Knowing the concepts of read/write heads and cylinders makes it possible to calculate the total number of tracks and the total capacity of the hard disk. The total capacity of a disk is calculated in the same way as it is for floppy disks:

number of tracks = number of cylinders × number of heads

HD capacity = number of tracks × number of sectors × sector density

Depending on the hard disk, often there are 17 to 36 sectors per track and 512 bytes for each sector.

Partitioning

Partitioning the disk is the process of dividing the hard disk into many smaller disks. This is done more frequently on disks larger than 2 gigabytes. For example, a given hard disk of 5 gigabytes capacity can be partitioned into three smaller logical disks with the DOS program FDISK. They are called *logical disks* since it is the same physical disk, but as far as DOS is concerned, they will be labeled disks C, D, and E. In the case of the 5 gigabyte disk, disk C will have 2GB, disk D 2GB, and the remaining 1GB are for disk E if the default partitioning mode is used. A hard disk can be divided into many logical disks of variable sizes with the names C, D, E, F, G, . . ., Z with no disk larger than 2 gigabytes.

After the hard disk has been partitioned, high-level formatting should be performed next. High-level formatting in the hard disk achieves exactly the same function as formatting a floppy disk. The C drive must be formatted with the system option (FORMAT C: /S) so that the system can boot from drive C. The remaining disks D through Z are in nonbootable format. The reason is that DOS always checks drive A first, then drive C for the system boot. In the absence of a bootable disk it goes automatically to the next drive until it finds one, and if it does not find any, it will display an appropriate message.

Clusters

In the 80x86 IBM PC, the sector size is always 512 bytes, but the size of the cluster varies among disks of various sizes. The cluster size is always a power of 2: 1, 2, 4, 8, and so on. The fact that a file of 1-byte size takes a minimum of 1 cluster is important and must be emphasized. This means that a number of small files on a disk with a large number of sectors per cluster will result in wasted space on the hard disk. Let's look at an example. In a hard disk with a cluster size of 16 sectors ($16 \times 512 = 8192$ bytes), storing a file of 26,000 bytes requires 4 clusters. The result is a waste of 6768 bytes since $4 \times 8192 = 32,768$ bytes, and $32,768 - 26,000$ bytes $= 6768$.

ESDI (Enhanced Small Device Interface) The ESDI standard was developed by a group of disk drive manufacturers in 1983. There are some differences between ESDI and ST412:

1. ESDI can achieve a data transfer rate of up to 20 Mbits per second in contrast to the 7.5 Mbits/second of the ST412.

2. With the same RPM as ST412, it can have more sectors per track. The number of sectors for ESDI can vary between the 20s and the 50s.

3. While in ST412 the defect information must be provided manually during low-level formatting, for ESDI the defect map is already stored on the drive.

4. In the ST412 standard, the number of cylinders, heads, and sectors is stored either in CMOS RAM of the system or in the ROM of the hard disk controller, in contrast to ESDI, where the configuration information is already provided and there is no need to store it externally.

IDE (Integrated Device Electronics) IDE is the standard for current PCs. In IDE, the controller is part of the hard disk. In other words, there is no need to buy a hard disk and a separate controller as is often the case for ST412. One of the reasons that the IDE drives have a better data transfer rate is the integration of many of the controller's functions into the drive itself with the use of VLSI chips. For example, in the ST412 standard the hard disk read/write heads would read the data and transfer it to the controller through the cable, and then the data is separated from the clock pulses by what is called data separator circuitry. By eliminating cable degradation, IDE and SCSI (discussed next) reach a much higher external data transfer rate.

SCSI (Small Computer System Interface) SCSI (pronounced "scuzzy") is one of the most widely used interface standards not only for high-performance IBMs and compatibles but also for non-80x86 computers by other manufacturers, such as Apple and Sun Micro. The main reason is that, unlike IDE, SCSI is the standard for all kinds of peripheral devices, not just hard disks. One can daisy chain up to 7 devices, such as CD-ROM, optical disk, tape drive, floppy disk drive, networks, and other I/O devices, using the SCSI standard. See Figure H.02-9.

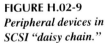

FIGURE H.02-9
Peripheral devices in SCSI "daisy chain."

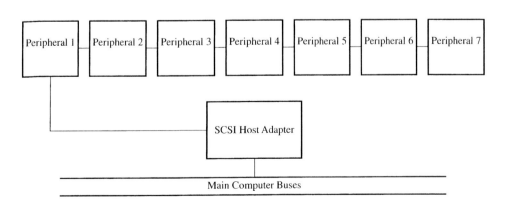

MEMORY

To understand interfacing memory to high-performance computers, the different types of available RAM must first be understood. Although SRAMs are fast, they are expensive and consume a lot of power due to the use of flip-flops in the design of the memory cell. At the opposite end of the spectrum is DRAM, which is cheaper but is slow (compared to CPU speed) and needs to be refreshed periodically. The refreshing overhead together with the long access time of DRAM is a major problem in the design of high-performance computers. The problem of the time taken for refreshing DRAM is minimal since it uses only a small percentage of bus time, but the solution to the slowness of DRAM is very involved. One common solution is using a combination of a small amount of SRAM, called *cache* (pronounced *cash*), along with a large amount of DRAM, thereby achieving the goal of near zero wait states.

Page Mode DRAM

The storage cells inside DRAM are organized in a matrix of N rows and N columns. In reading a given cell, the address for the row (A1 - An) is provided first and RAS is activated; then the address for the column (A1 - An) is provided and CAS is activated. In DRAM literature the term *page* refers to a number of column cells in a given row.

The idea behind page mode is that since in most situations memory locations are accessed consecutively, there is no need to provide both the row and column address for each location, as was the case in DRAM with standard timing. Instead, in page mode, first the row address is provided, RAS latches in the row address, and then the column addresses are provided and CAS toggles back and forth, latching in the column addresses until the last column of a given page is accessed. Then the address of the next row (page) is provided and the process is repeated. While the access time of the first cell is the standard access time using both row and column (t_{RAC}), in accessing the second cell on to the last cell of the same page (row), the access time is much shorter. This access time is often referred to as t_{CAC} (T of column access). In page mode DRAM, when we are in a given page, each successive cell can be accessed no faster than t_{PC} (page cycle time).

CACHE MEMORY

The most widely used memory design for high-performance CPUs implements DRAMs for main memory along with a small amount (compared to the size of main memory) of SRAM for cache memory. This takes advantage of the speed of SRAM and the high density and cheapness of DRAM. As mentioned earlier, to implement the entire memory of the computer with SRAM is too expensive and to use all DRAM degrades performance. Cache memory is placed between the CPU and main memory. See Figure H.02-10.

FIGURE H.02-10
CPU and its relation to various memories.

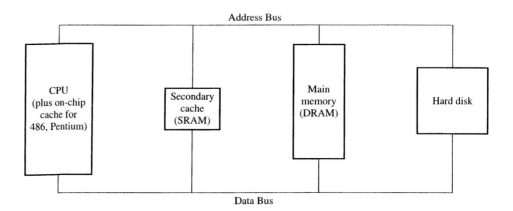

When the CPU initiates a memory access, it first asks cache for the information (data or code). If the requested data is there, it is provided to the CPU with zero wait states, but if the data is not in cache, the memory controller circuitry will transfer the data from main memory to the CPU while giving a copy of it to cache memory. In other words, at any given time the cache controller has knowledge of which information (code or data) is kept in cache; therefore, upon request for a given piece of code or data by the CPU, the address issued by the CPU is compared with the addresses of data kept by the cache controller. If they match (hit) they are presented to the CPU with zero weight states, but if the needed information is not in cache (miss), the cache controller along with the memory controller will fetch the data and present it to the CPU in addition to keeping a copy of it in cache for future reference. The reason a copy of data (or code) fetched from main memory is kept in the cache is to allow any subsequent request for the same information to result in a hit and provide it to the CPU with zero wait states. If the requested data is available in cache memory, it is called a *hit*; otherwise, if the data must be brought in from main memory, it is a *miss*.

In most computers with cache, the hit rate is 85% and higher. By combining SRAM and DRAM, cache memory's access time matches the memory cycle of the CPU. In the 80386/486 microprocessor with a frequency of 33 MHz and above, the use of cache is absolutely essential. For example, in the 33-MHz 80386-based computer with only a 60-ns read cycle time, only static RAM with an access time (cycle time) of 45 ns can provide the needed information to the CPU without inserting wait states. We have assumed that 15 ns $(60 - 45 = 15)$ is used for the delay associated with the address and data path. To implement the entire 16M of main memory of a 33-MHz 386/486 system with 45 ns SRAM is not only too expensive, but the power dissipation associated with such a large amount of SRAM would require a complex cooling system used only for expensive mini- and mainframe computers. The problem gets worse if we use faster processors.

It must be noted that when the CPU accesses memory, it is most likely to access the information in the vicinity of the same addresses, at least for a time. This is called the principle of *locality of reference*. In other words, even for a short program of 50 bytes, the CPU is accessing those 50 memory locations from cache with zero wait states. If it were not for this principle of locality and that the CPU accesses memory randomly, the idea of cache would not work. This implies that JMP and CALL instructions are bad for the performance of cache-based systems. The *hit rate,* the number of hits divided by the total number of tries, depends on the size of the cache, how it is organized (cache organization), and the nature of the program.

EDO AND SDRAM

In recent years the need for faster memory has led to the introduction of some very high-speed DRAMs. In this section we look at three of them: EDO (extended data-out), SDRAM (synchronous DRAM), and RDRAM (Rambus DRAM). In the mid-1990s, the speed of x86 processors went over 100 MHz and subsequently Intel and Digital Equipment began talking about 300–400 MHz CPUs. However, a major problem for these high-speed CPUs is the speed of DRAM. After all, cache has to be filled with information residing in main memory DRAM. Before we discuss some high-speed DRAMs, it needs to be noted that "300 MHz" CPU does not mean that its bus speed is also 300 MHz. For microprocessors over 100 MHz, the bus speed is often a fraction of the CPU speed. This is due to the expense and difficulty (e.g., crosstalk, electromagnetic interference) associated with the design of high-speed motherboards and the slowness of memory and logic gates. For example, in many150-MHz Pentium systems, the bus speed is only 66 MHz.

EDO DRAM: Origin and Operation

Earlier in this section we discussed page mode DRAM. It needs to be noted that page mode DRAM has been modified and now is referred to as *fast page mode* DRAM. Note that DRAM data books of the mid-1990s refer only to fast page DRAM (FPM DRAM) and not

TABLE H.02-6
70 ns 4M DRAM timing.

	FPM	EDO
Speed (ns)	70	70
t_{RAC} (ns)	70	70
t_{RC} (ns)	130	130
t_{PC} (ns)	40	30

Note: 256Kx16 DRAM from Micron Technology.

TABLE H.02-7
60, 50 ns 4M DRAM timing.

	FPM	EDO	EDO
Speed (ns)	60	60	50
t_{RAC} (ns)	60	60	50
t_{RC} (ns)	110	110	100
t_{PC} (ns)	35	25	20

Note: 256Kx16 DRAM from Micron Technology.

page mode. Tables H.02-6 and H.02-7 show typical DRAM timing for FPM and EDO DRAM.

SDRAM (Synchronous DRAM)

When the CPU bus speed goes beyond 75 MHz, even EDO is not fast enough. SDRAM is a memory for such systems. First, let us see why it is called synchronous DRAM. In all the traditional DRAMs (page mode, fast page, and EDO), CPU timing is not synchronized with DRAM timing, meaning that there is no common clock between the CPU and DRAM for reference. In those systems it is said that the DRAM is *asynchronous* with the micro-processor since the CPU presents the address to DRAM and memory provides the data in the master/slave fashion. If data cannot be provided on time the CPU is notified with the NOT READY signal. In response to NOT READY, the CPU inserts a wait state into its bus timing and waits until the DRAM is ready. In other words, the CPU bus timing is dependent upon the DRAM speed. This is not the case in synchronous DRAM. In systems with SDRAM, there is a common clock (called the *system clock*) that runs between the micro-processor and SDRAM. All bus activities (address, data, control) between the CPU and DRAM are synchronized with this common clock. That is, the common clock is the point of reference for both the CPU and SDRAM and there is no deviation from it and hence no waiting by the CPU.

REVIEW QUESTIONS

True/False

1. Only DRAM is required in a microcomputer system.
2. The resolution of a video display controls its memory requirements.

Multiple Choice

3. AGP stands for
   a. Active graphics processor.
   b. Accelerated graphics port.
   c. Advanced graphics port.

4. A hard drive may use a(n)
    a. IDE interface.
    b. SCSI interface.
    c. Both a or b.

5. Cache is a
    a. Special type of high speed memory.
    b. Cheaper form of memory than EDO RAM.
    c. Permanent secondary storage medium.

H.03 Demonstrate an understanding of microcomputer peripherals

INTRODUCTION

In this section we will review the operation of four of the most common peripherals used with a microcomputer system. These are printers, modems, CD-ROM drives, and sound cards.

PRINTERS

There are two fundamental types of printers used with personal computers: the **impact printer** and the **nonimpact printer.** The impact printer uses some kind of mechanical device to impart an impression to the paper through an inked ribbon. The nonimpact printer uses heat, a jet of ink, electrostatic discharge, or laser light. Nonimpact printers form printed images without making physical contact with the paper. These two types of printers are illustrated in Figure H.03-1.

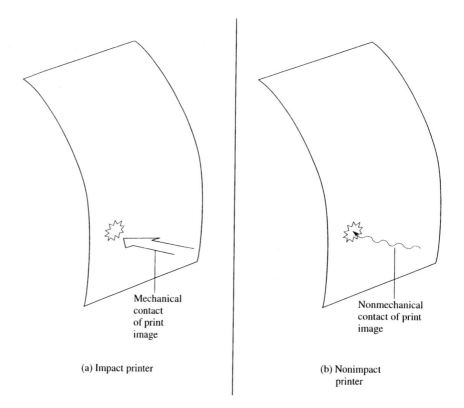

Mechanical contact of print image

(a) Impact printer

Nonmechanical contact of print image

(b) Nonimpact printer

FIGURE H.03-1
Two fundamental types of printers.

Impact Printers

The most common type of impact printer is the **dot-matrix printer.** The dot-matrix printer makes up its characters by means of a series of tiny mechanical pins that move in and out to form the various characters printed on the paper.

The Dot-Matrix Printer   The dot-matrix printer, one of the most popular types of printers, uses a mechanical printing head that physically moves across the paper to be printed. This mechanical head consists of tiny movable wires that strike an inked ribbon to form characters on the paper. There are two popular kinds of dot-matrix print heads. One consists of nine pins (the movable wires), and the other consists of 24 pins. A nine-pin print head is shown in Figure H.03-2.

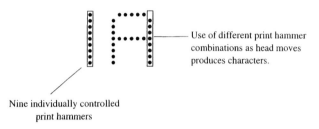

Use of different print hammer combinations as head moves produces characters.

Nine individually controlled print hammers

FIGURE H.03-2
Nine-pin dot-matrix print head.

The 24-pin dot-matrix printer is more expensive than the nine-pin model. However, because both types have modes of operation that allow for an *overstrike* of the image (with the head moving slightly and the image being struck again), the nine-pin model can produce close to what is known as letter-quality printing. The 24-pin model can produce an even sharper character when operated in the same overstrike mode. Because of the manner in which characters are formed in this type of printer, the printing of graphic images is possible.

Nonimpact Printers

The most popular nonimpact printers are the *ink-jet printer, bubble jet printer,* and the *laser printer.* The ink-jet printer uses tiny jets of ink that are electrically controlled. The laser printer uses a laser to form characters. The laser printer resembles an office photocopying machine.

Ink-Jet and Bubble Jet Printers   An **ink-jet printer** uses electrostatically charged plates to direct jets of ink onto paper. The ink is under pressure and is formed by a mechanical nozzle into tiny droplets that can be deflected to make up the required images on the paper. A bubble jet printer uses heat to form bubbles of ink. As the bubbles cool, they form the droplets applied to the paper. Ink-jet printers cost more than impact printers but are quieter and can produce high-quality graphic images.

Laser Printers   Through the operation of a laser and mirror (controlled by software), an electrical image is impressed on a photoreceptor drum. This drum picks up a powdered toner, which is then transferred to paper by electrostatic discharge. A second drum uses high temperature to bond this image to the paper. The result is a high-quality image capable of excellent characters and graphics.

Because of their high-quality output, laser printers find wide application in desktop publishing, computer-aided design, and other image-intensive computer applications. Laser printers are usually at the high end of the price range for computer printers.

TABLE H.03-1
Standard printable ASCII code.

Dec	Hex	Char	Dec	Hex	Char	Dec	Hex	Char
32	20		64	40	@	96	60	'
33	21	!	65	41	A	97	61	a
34	22	"	66	42	B	98	62	b
35	23	#	67	43	C	99	63	c
36	24	$	68	44	D	100	64	d
37	25	%	69	45	E	101	65	e
38	26	&	70	46	F	102	66	f
39	27	'	71	47	G	103	67	g
40	28	(	72	48	H	104	68	h
41	29	)	73	49	I	105	69	i
42	2A	*	74	4A	J	106	6A	j
43	2B	+	75	4B	K	107	6B	k
44	2C	,	76	4C	L	108	6C	l
45	2D	–	77	4D	M	109	6D	m
46	2E	.	78	4E	N	110	6E	n
47	2F	/	79	4F	O	111	6F	o
48	30	0	80	50	P	112	70	p
49	31	1	81	51	Q	113	71	q
50	32	2	82	52	R	114	72	r
51	33	3	83	53	S	115	73	s
52	34	4	84	54	T	116	74	t
53	35	5	85	55	U	117	75	u
54	36	6	86	56	V	118	76	v
55	37	7	87	57	W	119	77	w
56	38	8	88	58	X	120	78	x
57	39	9	89	59	Y	121	79	y
58	3A	:	90	5A	Z	122	7A	z
59	3B	;	91	5B	[	123	7B	{
60	3C	<	92	5C	\	124	7C	|
61	3D	=	93	5D	]	125	7D	}
62	3E	>	94	5E	^	126	7E	~
63	3F	?	95	5F	–			

The ASCII Code

Table H.03-1 lists all the printable characters for a standard printer. The code used to transmit this information is called the ASCII code. ASCII stands for American Standard Code for Information Interchange.

As you can see in Table H.03-1, each keyboard character is given a unique number value. For example, a space is number 32 (which is actually represented by the binary value 0010 0000 when transmitted from the computer to the printer). The number values that are less than 32 are used for controlling the operations of the printer. These are called **printer-control codes,** or simply **control codes.** These codes are shown in Table H.03-2.

The definitions of the control code abbreviations are as follows:

ACK	Acknowledge	GS	Group separator
BEL	Bell	HT	Horizontal tab
BS	Backspace	LF	Line feed
CAN	Cancel	NAK	Negative acknowledge
CR	Carriage return	NUL	Null
DC_1–DC_4	Device control	RS	Record separator
DEL	Delete idle	SI	Shift in
DLE	Data link escape	SO	Shift out

EM	End of medium	SOH	Start of heading
ENQ	Enquiry	SP	Space
EOT	End of transmission	STX	Start text
ESC	Escape	SUB	Substitute
ETB	End of transmission block	SYN	Synchronous idle
ETX	End text	US	Unit separator
FF	Form feed	VT	Vertical tab
FS	Form separator		

TABLE H.03-2
ASCII control codes.

Dec	Hex	Char	
0	0	^@	NUL
1	1	☺	SOH
2	2	●	STX
3	3	♥	ETX
4	4	♦	EOT
5	5	♣	ENQ
6	6	♠	ACK
7	7	•	BEL
8	8	◘	BS
9	9	○	HT
10	A	■	LF
11	B	♂	VT
12	C	♀	FF
13	D	♪	CR
14	E	♫	SO
15	F	○	SI
16	10	►	DLE
17	11	◄	DC$_1$
18	12	↕	DC$_2$
19	13	‼	DC$_3$
20	14	¶	DC$_4$
21	15	§	NAK
22	16	▬	SYN
23	17	↨	ETB
24	18	↑	CAN
25	19	↓	EM
26	1A	→	SUB
27	1B	←	ESC
28	1C	L	FS
29	1D	↔	GS
30	1E	▲	RS
31	1F	▼	US
32	20		SP

Extended ASCII Codes

If you set up your printer to act as a graphics printer (by setting the appropriate configuration; you must refer to the user manual that comes with the printer), you can extend the character set to include many other forms of printable characters. These characters are shown in Table H.03-3.

You can write these extended character codes to your printer by creating text files. To do this, hold down the ALT and SHIFT keys and type the number code into the numeric keypad on your keyboard. For example, to get the Greek letter Σ, simply press ALT-SHIFT-228; when you lift up on the CTRL-SHIFT keys, a Σ will appear on the monitor. If

TABLE H.03-3
Extended ASCII character set.

Dec	Hex	Char	Dec	Hex	Char	Dec	Hex	Char	Dec	Hex	Char
128	80	Ç	160	A0	á	192	C0	└	224	E0	α
129	81	ü	161	A1	í	193	C1	┴	225	E1	β
130	82	é	162	A2	ó	194	C2	┬	226	E2	Γ
131	83	â	163	A3	ú	195	C3	├	227	E3	π
132	84	ä	164	A4	ñ	196	C4	─	228	E4	Σ
133	85	à	165	A5	Ñ	197	C5	┼	229	E5	σ
134	86	å	166	A6	ª	198	C6	╞	230	E6	μ
135	87	ç	167	A7	º	199	C7	╟	231	E7	τ
136	88	ê	168	A8	¿	200	C8	╚	232	E8	φ
137	89	ë	169	A9	⌐	201	C9	╔	233	E9	θ
138	8A	è	170	AA	¬	202	CA	╩	234	EA	Ω
139	8B	ï	171	AB	½	203	CB	╦	235	EB	δ
140	8C	î	172	AC	¼	204	CC	╠	236	EC	∞
141	8D	ì	173	AD	¡	205	CD	=	237	ED	Ø
142	8E	Ä	174	AE	"	206	CE	╬	238	EE	∈
143	8F	Å	175	AF	"	207	CF	┴	239	EF	∩
144	90	É	176	B0	░	208	D0	╨	240	F0	≡
145	91	æ	177	B1	▒	209	D1	╤	241	F1	±
146	92	Æ	178	B2	▓	210	D2	╥	242	F2	≥
147	93	ô	179	B3	│	211	D3	╙	243	F3	≤
148	94	ö	180	B4	┤	212	D4	╘	244	F4	⌠
149	95	ò	181	B5	╡	213	D5	╒	245	F5	⌡
150	96	û	182	B6	╢	214	D6	╓	246	F6	÷
151	97	ù	183	B7	╖	215	D7	╫	247	F7	≈
152	98	ÿ	184	B8	╕	216	D8	╪	248	F8	°
153	99	Ö	185	B9	╣	217	D9	┘	249	F9	•
154	9A	Ü	186	BA	║	218	DA	┌	250	FA	·
155	9B	¢	187	BB	╗	219	DB	█	251	FB	√
156	9C	£	188	BC	╝	220	DC	▄	252	FC	η
157	9D	¥	189	BD	╜	221	DD	▌	253	FD	²
158	9E	Pt	190	BE	╛	222	DE	▐	254	FE	■
159	9F	ƒ	191	BF	┐	223	DF	▀	255	FF	

you include this in a text file (or do a PrintScreen), you can transfer it to the printer. The extended characters 176 through 223 are used for creating boxes, rectangles, and other shapes on the monitor or printer while it is still in the text mode. If you can't get these extended characters on the printer, it is either because you haven't set the printer to the IBM graphics mode or your printer simply can't perform the functions required by this mode.

MODEMS

The user of one computer may interact with another computer—which may be located thousands of miles away—as if it were sitting right in the same room, through the use of a modem. In order for computers to communicate in this manner, four items must be available, as shown in Figure H.03-3.

As shown in Figure H.03-3, there must be some kind of link between the computers. The most convenient link to use is the already established telephone system lines. Using

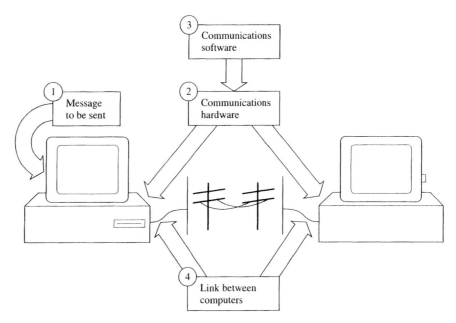

FIGURE H.03-3
Four items necessary for computer communications.

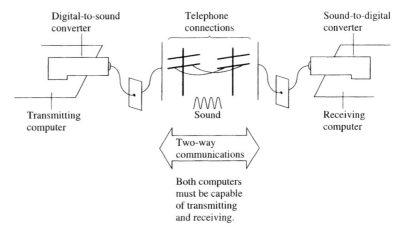

FIGURE H.03-4
Basic needs for the use of telephone lines in computer communications.

these lines and a properly equipped computer allows communications between any two computers that have access to a telephone. This becomes a very convenient and inexpensive method of communicating between computers.

There is, however, one problem. Telephone lines were designed for the transmission of the human voice, not for the transmission of digital data. Therefore, in order to make use of these telephone lines for transmitting computer data, the ONs and OFFs of the computer must first be converted to sound, sent over the telephone line, and then reconverted from sound to the ONs and OFFs that the computer understands. This concept is shown in Figure H.03-4.

The Modem

The word *modulate* means to change. Thus, an electronic circuit that changes digital data into sound data can be called a *modulator.* The word *demodulate* can be thought of as meaning "unchange," or restore to an original condition. Any electronic circuit that converts the sound used to represent the digital signals back to the ONs and OFFs understood by a computer can, therefore, be called a demodulator. Since each computer must be capable of both transmission and reception, each computer must contain an electrical circuit that can modulate as well as demodulate. Such a circuit is commonly called a modulator/demodulator, or **modem.**

For personal computers, a modem may be an internal or an external circuit—both perform identical functions.

The RS-232 Standard

The EIA (Electronics Industries Association) has published the EIA *Standard Interface Between Data Terminal Equipment Employing Serial Binary Data Interchange*—specifically, EIA-232-C. This is a standard defining 25 conductors that may be used in interfacing **data terminal equipment** (DTE, such as your computer) and **data communications equipment** (DCE, such as a modem) hardware. The standard specifies the function of each conductor, but it does not state the physical connector that is to be used. This standard exists so that different manufacturers of communications equipment can communicate with each other. In other words, the RS-232 standard is an example of an interface, essentially an agreement among equipment manufacturers on how to allow their equipment to communicate.

The RS-232 standard is designed to allow DTEs to communicate with DCEs. The RS-232 uses a DB-25 connector; the male DB-25 goes on the DTEs and the female goes on the DCEs. The RS-232 standard is shown in Figure H.03-5.

FIGURE H.03-5
The RS-232 standard.

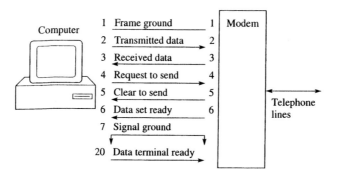

TABLE H.03-4
Standard baud rates.

Low Speed	High Speed
300	4800
600	9600
1200	14,400
2400	28,800

The RS-232 is a digital interface designed to operate at no more than 50 ft. with a 20,000-bit/s bit rate. The **baud,** named after J. M. E. Baudot, actually indicates the number of *discrete* signal changes per second. In the transmission of binary values, one such change represents a single bit. What this means is that the popular usage of the term baud has become the same as bits per second (bps). Table H.03-4 shows the standard set of baud rates.

Telephone Modem Setup

The most common problem with telephone modems is the correct setting of the software. There are essentially six distinct areas to which you must pay attention when using a telephone modem:

1. Port to be used
2. Baud rate

3. Parity

4. Number of data bits

5. Number of stop bits

6. Local echo ON or OFF

Most telephone modems have a default setting for each of these areas. However, as a user—and especially as a technician—you should understand what each of these areas means. You will have to consult the specific documentation that comes with the modem in order to see how to change any of these settings. For now it is important that you understand the idea behind each of these areas.

Port to Be Used The most common ports to be used are COM1 and COM2. Other possible ports are COM3 and COM4. The port you select from the communications software depends on the port to which you have the modem selected. On most communications software, once you set the correct port number, you do not need to set it again.

Baud Rate Typical values for the baud rate are 1200 baud and 2400 baud, which offer a comfortable reading rate when viewing received data. Again, this is a value that can be selected from the communications software menu. It is important that both computers be set at the same baud rate.

Parity Parity is a way of having the data checked. Normally, parity is not used. Depending on your software, there can be up to five options for the parity bit, as follows:

Space: Parity bit is always a 0.

Odd: Parity bit is 0 if there is an odd number of 1s in the transmission and is a 1 if there is an even number of 1s in the transmission.

Even: Parity bit is a 1 if there is an odd number of 1s in the transmission and is a 0 if there is an even number of 1s in the transmission.

Mark: Parity bit is always a 1.

None: No parity bit is transmitted.

Again, what is important is that both the sending and receiving units are set up to agree on the status of the parity bit.

Number of Data Bits The number of data bits to be used is usually set at 8. There are options that allow the number of data bits to be set at 7. It is important that both computers expect the same number of data bits.

Number of Stop Bits The number of stop bits used is normally 1. However, depending on the system, the number of stop bits may be 2. Stop bits are used to mark the beginning and the end of each character transmitted. Both computers must have their communications software set to agree on the number of stop bits used.

CD-ROM DRIVES AND SOUND CARDS

The term *multimedia* is now generally applied to personal computers equipped with CD-ROM drives and sound cards. Entire encyclopedias are now available on CD-ROM. Access the subject of spacecraft, and you get live-action video of an Apollo moon shot, complete with the accompanying audio. Hundreds of software packages are available on CD-ROM, with more appearing every day.

A computer is said to be **MPC (Multimedia PC) compliant** if it contains the following hardware:

☐ 386SX-16 processor (or better)

☐ 4MB of RAM (or more)

☐ 40MB hard drive (or more)

□ A color VGA display (or better)

□ A mouse

□ A single-spin CD-ROM (or faster)

Single-spin CD-ROM drives transfer data at a maximum rate of 150KB/sec. A double-spin CD-ROM drive transfers data at 300KB/sec, and so on.

A computer is **MPC-2 compliant** if it contains the following updated hardware:

□ 486SX-25 processor (or better)

□ 8MB of RAM (or more)

□ 40MB hard drive (or more)

□ A color VGA display (or better)

□ A mouse

□ A double-spin CD-ROM (or faster)

CD-ROM Operation

A CD-ROM stores binary information in the form of microscopic *pits* on the disk surface. The pits are so small that a CD-ROM typically stores over 650MB of data. This is equivalent to more than 430 1.44MB floppies. A laser beam is shone on the disk surface and either reflects (no pit) or does not reflect (pit), as you can see in Figure H.03-6.

FIGURE H.03-6
Reading data from a CD.

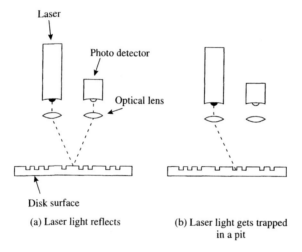

(a) Laser light reflects

(b) Laser light gets trapped in a pit

These two light states (reflection and no reflection) are easily translated into a binary 0 and a binary 1. Since the pits are mechanically pressed into a hard surface and only touched by light, they do not wear out or change as a result of being exposed to a magnetic surface.

Physical Layout of a Compact Disk

Figure H.03-7 shows the dimensions and structure of a compact disk. The pits previously described are put into the reflective aluminum layer when the disk is manufactured. Newer recordable CDs use a layer of gold instead of aluminum so that they can be written to using a low-power laser diode.

The High Sierra Format

The High Sierra format specifies the way the CD is logically formatted (tracks, sectors, directory structure, file name conventions). This specification is officially called ISO-9660 (international standards organization).

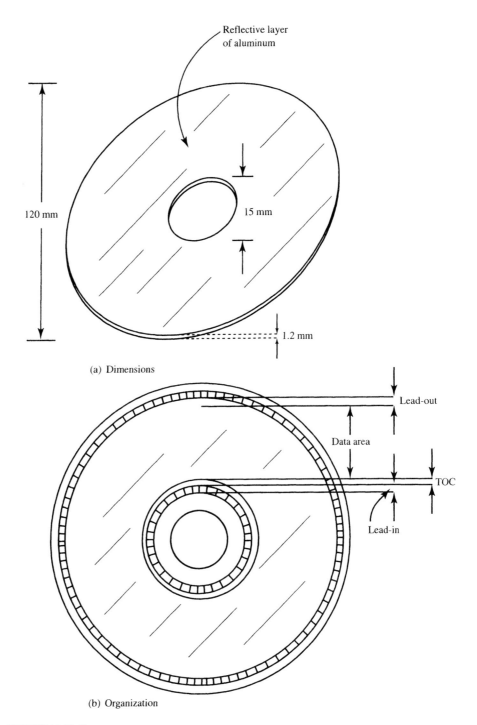

(a) Dimensions

(b) Organization

FIGURE H.03-7
Compact disk.

Photo CD

Developed by Kodak, the photo CD provides a way to store high-quality photographic images on a CD (using recordable technology) in the CD-I format. Each image is stored in several different resolutions, from 192 by128 to 3072 by 2048, using 24-bit color. This allows for around 100 images on one photo CD.

ATAPI

ATAPI stands for AT attachment packet interface. ATAPI is an improved version of the IDE hard drive interface, and utilizes *packets* of data during transfers. The ATAPI specification supports CD-ROM drives, hard drives, tape backup units, and plug-and-play adapters.

Sound Card Operation

Along with CD-ROM drives, sound cards for PCs have also increased in popularity. Currently, 16-bit sound cards are available that provide multiple audio channels and FM quality sound, and are compatible with the MIDI (Musical Instrument Data Interface) specification.

The basic operation of the sound card is shown in Figure H.03-8. Digital information representing samples of an analog waveform are inputted to a *digital-to-analog* converter, which translates the binary patterns into corresponding analog voltages. These analog voltages are then passed to a *low-pass filter* to smooth out the differences between the individual voltage samples, resulting in a continuous analog waveform. All of the digital/analog signal processing is done in a custom **digital signal processor** chip included on the sound card.

Sound cards also come with a microphone input that allows the user to record any desired audio signal.

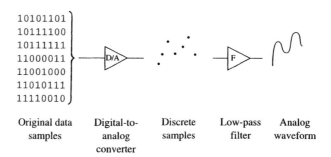

| 10101101 |
| 10111100 |
| 10111111 |
| 11000011 |
| 11001000 |
| 11010111 |
| 11110010 |

| Original data samples | Digital-to-analog converter | Discrete samples | Low-pass filter | Analog waveform |

FIGURE H.03-8
How binary data is converted into an analog waveform.

MIDI

MIDI stands for musical instrument digital interface. A MIDI-capable device (electronic keyboard, synthesizer) will use a MIDI-in and MIDI-out serial connection to send messages between a *controller* and a *sequencer.* The PC operates as the sequencer when connected to a MIDI device. MIDI messages specify the type of note to play and how to play it, among other things. Using MIDI, a total of 128 pitched instruments can generate 24 notes in 16 channels. This can be accomplished in a PC sound card by using frequency modulation or *wave table synthesis,* the latter method utilizing prerecorded samples of notes stored in a data table. The output of a note is controlled by several parameters. Figure H.03-9 illustrates the use of attack, decay, and release times to shape the output waveform envelope. Each of the four parameters can be set to a value from 0 to 15.

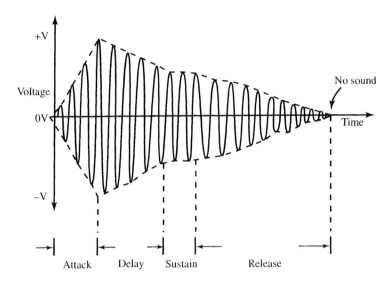

FIGURE H.03-9
Note envelope.

REVIEW QUESTIONS

True/False

1. Printers use the ASCII code for communication.
2. Modems are used over special, dedicated telephone lines.

Multiple Choice

3. Modems are
    a. Simplex devices.
    b. Duplex devices.
    c. Full duplex devices.
4. A CD-ROM drive typically stores
    a. 100 MB.
    b. 650 MB.
    c. 5100 MB.
5. A sound card may utilize
    a. Wave table synthesis.
    b. CD sampling synthesis.
    c. Attack-delay synthesis.

H.04 Set up and configure a microcomputer using available operating systems and software packages

INTRODUCTION

Setting up a machine for DOS operation is as simple as creating a partition on a hard drive (using FDISK), formatting it (using FORMAT), and then installing DOS from a set of floppy disks. Upgrading the same system to a Windows 95 system requires a working CD-ROM drive and the Windows 95 installation CD. The installation is typically a smooth process, with Windows 95 automatically determining the hardware capabilities of your system and loading the associated device drivers.

In this skill topic we will examine the specifics of installing new software in the Windows environment.

The process of installing software involves moving files from a floppy disk or CD-ROM to a suitable hard disk location (local or network). Every single file located on a computer disk has been installed, one way or another. The files actually consist of executable images, data files, initialization files, dynamic link libraries, and other custom files necessary for the computer or application to run. The files are placed in a directory structure determined by an application's installation program.

What types of software do we install? Actually, the operating system itself, plus every single application. Note that many applications register themselves with the operating system during the installation process. Specific application settings are stored in the Registry.

Existing Windows 3.x Software

Most data files and application programs are upward compatible. This means that Windows 3.x applications can be installed on Windows 95. Unfortunately, Windows 95 applications cannot be installed on a Windows 3.x computer. When a Windows 3.x computer is upgraded to Windows 95, all installed applications are also upgraded by placing information from the old .INI files to the new system registry.

Installing New Software from the Windows 95 CD-ROM

The Windows operating system can be configured in many different ways. There are likely to be many files that were not installed during the initial Windows installation. To see the current system configuration, double-click the Add/Remove Programs icon in the system Control Panel. Then select the Windows Setup tab as shown in Figure H.04-1. Just by looking at the window, it is apparent what components are installed.

Check boxes along the left margin of the component window identify three situations. First, if a box is checked and the inside of the box is white, all components of the category have been installed. If a box is checked and the inside of the box is gray, only some of the components from that particular category are installed. The "Details" button is used to show the specific status of each component. If the Windows operating system components are changed, the computer will request the Windows 95 CD-ROM or floppy installation media to copy the additional files.

FIGURE H.04-1
Windows Setup menu.

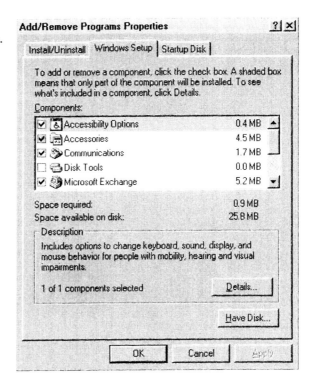

The Windows 95 files are stored in .CAB files on the CD-ROM drive. These files contain the Windows 95 operating system components in compressed form for distribution. Each file is extracted from the .CAB file using a special procedure built into the system.

Getting the Latest Updates from Microsoft

As improvements are made in the Windows operating system, they are posted on the World Wide Web. Figure H.04-2 shows the downloads page for support drives, patches, and service packs from Microsoft. As you can see, each category contains specific applications such as Word, Exchange, and the different Microsoft operating systems.

Installing Third Party Software

Application software is usually installed with a custom software installation wizard. The first step in performing a software installation is identifying which installation program to run. Figure H.04-3 shows the Run window with the path specified to a setup file. When the "OK" button is selected, the Setup Wizard begins the installation process. Figure H.04-4 shows the "Norton Utilities for Windows 95 Setup Wizard" installation screen. The installation program asks for the user name and company. This information is usually displayed by the application each time it is run. Each window has option buttons allowing the program user to move back and forward through the screens presented during the installation.

The software producers do not actually sell their code, or executables—they license them. The license allows the user to run the product within the scope of the license agreement. It is impossible to install the software without agreeing with the license terms, as shown in Figure H.04-5 on page 731.

During an installation, the user is given choices about how the Setup Wizard is to install or reconfigure the application software. A complete installation usually installs all components of an application, and a custom installation may allow for one or more

FIGURE H.04-2
Operating system update from the Web.

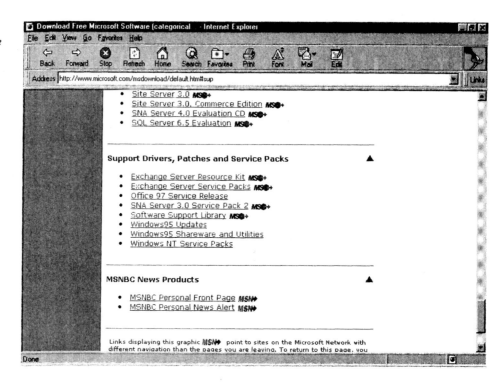

FIGURE H.04-3
Location and name of the installation file.

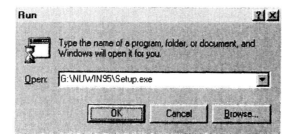

FIGURE H.04-4
Installation setup wizard.

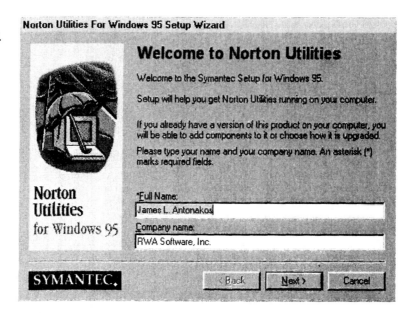

FIGURE H.04-5
License agreement.

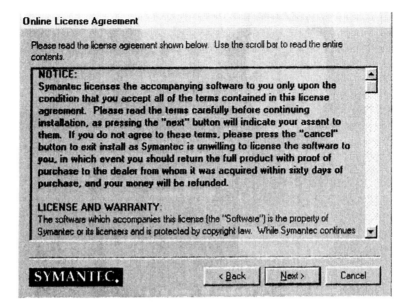

FIGURE H.04-6
Installation options.

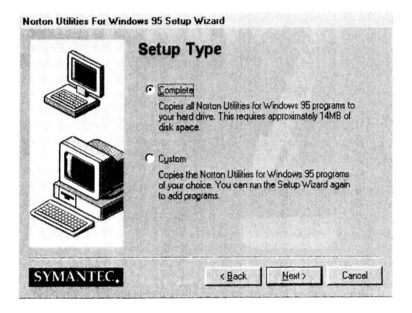

components of a product to be installed individually. Figure H.04-6 shows a typical Setup option window indicating that a complete installation is to be performed. The installation proceeds by clicking the "Next" button.

Some applications will search for a previous installation of the application to identify which particular components are installed. When the user is prompted to confirm the program location, it will display the current installation locations. Figure H.04-7 shows how a custom location can be selected as the installation directory. The directory will be created when the "Next" button is clicked.

The installation process continues with the Setup Wizard asking about how the new application software is to be configured, such as adding special features to the Recycle Bin, as shown in Figure H.04-8 and whether or not to run the system doctor when the system boots as shown in Figure H.04-9. These options are usually set to a default answer. When in doubt, use the default responses.

Eventually, the Setup Wizard begins the process of copying files from the installation media and the selected installation location. The progress meter shown in Figure

FIGURE H.04-7
Selecting a custom storage location.

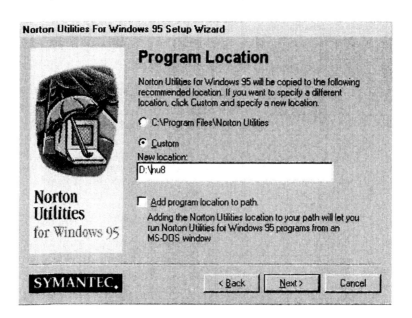

FIGURE H.04-8
More installation options.

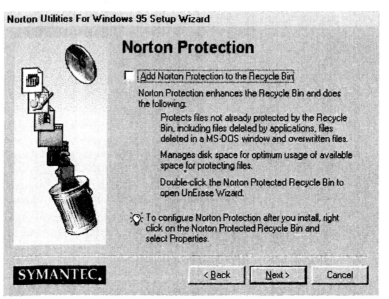

FIGURE H.04-9
More installation options.

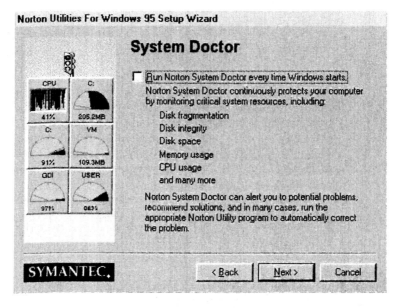

FIGURE H.04-10
File being copied to the hard disk.

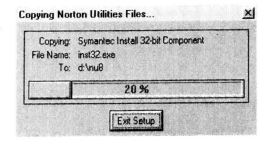

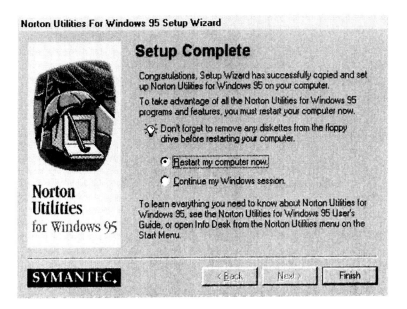

FIGURE H.04-11
Installation processes require a system reboot.

H.04-10 shows a percentage of the how close the installation is from being complete and the current file being processed.

When the installation process completes, it may require the system to be rebooted. This allows for new drivers to be loaded during system initialization. Figure H.04-11 shows a congratulatory message for successfully completing the installation.

REVIEW QUESTIONS

True/False

1. The Windows 95 CD-ROM is required to install new software.

2. The World Wide Web can be used to update software.

Multiple Choice

3. Most Windows 95 applications use a

   a. Setup script.

   b. Setup wizard.

   c. Setup shell.

4. During an installation, you may need to
    a. Reboot to make the required changes.
    b. Change the default install directory.
    c. Both a and b.
5. Before installing DOS, the hard drive must be
    a. Interleaved.
    b. Activated.
    c. Partitioned.

H.05 Troubleshoot and replace microcomputer peripherals

INTRODUCTION

In this section we examine several methods for troubleshooting printer, modem, tape and CD-ROM drive, and sound card problems.

Printer Problems

Most printer problems are caused by software. What this means is that the software does not match the hardware of the printer. This is especially true when printing graphics. When there is a hardware problem with printers, it is usually the interface cable that goes from the printer adapter card to the printer itself. This is illustrated in Figure H.05-1.

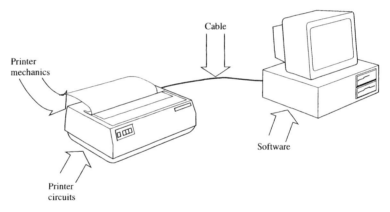

FIGURE H.05-1
Problem areas in computer printers.

Printer Cables   The interface cable is used to connect the printer to the computer. Previously, there was limited communication between the printer and computer. The computer received a few signals from the printer such as the on-line or off-line indicator, the out-of-paper sensor, and the print buffer status. As long as the printer was sending the correct signals to the computer, the computer would continue to send data.

Advances in printer technology now require a two-way communication between the computer and printer. As a result of these changes, a new bidirectional printer cable is required to connect most new printers to the computer. The bidirectional cables may or may not adhere to the new IEEE standard for Bidirectional Parallel Peripheral Interface. The IEEE 1284 Bitronic printer cable standard requires 28 AWG construction, a Hi-flex jacket, and dual shields for low EMI emissions. The conductors are twisted into pairs with varying left-hand lays to reduce possible cross talk.

Check the requirements for each printer to determine the proper cable type. Figure H.05-2 shows two popular parallel printer cable styles. Many different lengths of printer cables are available. It is usually best to use the shortest cable possible as this can reduce the possibility of communication errors.

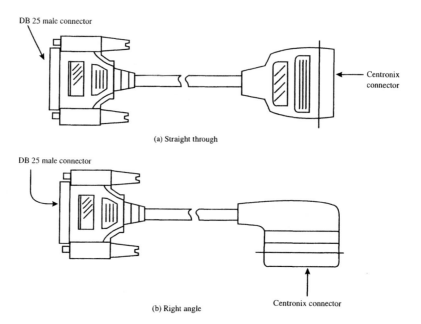

FIGURE H.05-2
Typical printer cables.

Printer Hardware A printer requires periodic maintenance. This includes vacuuming out the paper chaff left inside the printer. A soft dry cloth should be used to keep the paper and ribbon paths clean. It is a good idea to use plastic gloves when cleaning a printer, because the ink is usually difficult to remove from the skin. With dot-matrix printers, be careful of the print heads, which can get quite hot after extended use. Make sure you do not turn the print platen rollers when the power is on. The reason is that a stepper motor is engaged when power is applied. This little motor is trying to hold the print platen roller in place. If you force it to move, you could damage the stepper motor.

Laser Hardware Essentially, laser printers require very little maintenance. If you follow the instructions that come with the printer, the process of changing the cartridge (after about 3500 copies) also performs the required periodic maintenance on the printer.

When using a laser printer, remember that such a machine uses a large amount of electrical energy and this produces heat. So make sure that the printer has adequate ventilation and a good source of reliable electrical power. This means that you should not use an electrical expansion plug from your wall outlet to run your computer, monitor, and laser printer. Doing so may overload your system.

When shipping a laser printer, be sure to remove the toner cartridge. If you don't remove it, there is the possibility that it could open up and spill toner (a black powder) over the inside of the printer, causing a mess that is difficult to clean up.

Testing Printers

When faced with a printer problem, first determine from the customer if the printer ever worked at all or if this is a new installation that never worked. If it is a new installation and has never worked, a careful reading of the manual that comes with the printer is usually required to make sure that the device is compatible with the printer adapter card. Table H.05-1 lists some of the most direct methods for troubleshooting a computer printer.

TABLE H.05-1
Printer troubleshooting methods.

Checks	Comments
Check if printer is plugged in and turned on.	The printer must have external AC power to operate.
Check if printer is on-line and has paper.	Printers must be *on-line,* meaning that their control switches have been set so that they will print (check the instruction manual). Some printers will not operate if they do not have paper inserted.
Print a test page.	Select the "Print Test Page" option as shown in Figure H.05-3. Confirm that the page printed properly. See Figure H.05-4. Figure H.05-5 shows the printer test page output.
Do a printer self-test.	Most printers have a self-test mode. In this mode, the printer will repeat its character set over and over again. You must refer to the documentation that comes with the printer to see how this is done.
Do a PrintScreen.	If the printer self-test works, then with some characters on the computer monitor, hold down the SHIFT key and press the PRINTSCRN key at the same time. What is on the monitor should now be printed. Do not use a program (such as a word-processing program) because the software in the program may not be compatible with the printer.
Exchange printer cable.	If none of the tests above work, the problem may be in the printer cable. At this point, you should swap the cable with a known good one.
Replace the printer adapter card.	Try replacing the printer adapter card with a known good one. Be sure to refer to the printer manual to make sure you are using the correct adapter card.
Check parameters for a serial interface.	If you are using a serial interface printer from a serial port, make sure you have the transmission rate set correctly, along with the parity, number of data bits, and number of stop bits. Refer to the instruction manual that comes with the printer and use the correct form of the DOS MODE command.
Check the configuration settings.	Check all of the configuration settings available on the printer.
Check the software installation.	When software is installed (such as word-processing and spreadsheet programs), the user may have had the wrong printer driver installed (the software that actually operates the printer from the program).

FIGURE H.05-3
"Print Test Page" option.

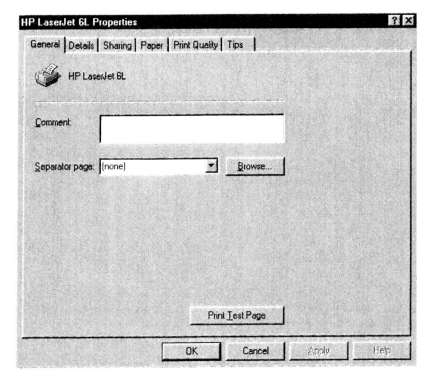

FIGURE H.05-4
Printer test page confirmation.

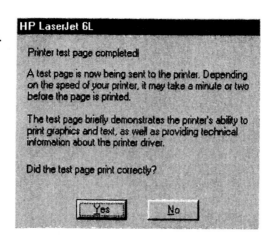

Windows 95
Printer Test Page

Congratulations!

If you can read this information, you have correctly installed your HP LaserJet 6L.

The information below describes your printer driver and port settings.

```
Printer name:    HP LaserJet 6L
Printer model:   HP LaserJet 6L
Driver name:     HPW.Drv
Driver version:  4.00
Color support:   No
Port name:       LPT1:
Data format:     EMF
```

This is the end of the printer test page.

FIGURE H.05-5
Printer test page output.

MODEM PROBLEMS

Table H.05-2 lists some of the most common problems encountered in telephone modems. As you will see, most of the problems are software related.

Other common problems encountered involve very simple hardware considerations. For example, telephone modems usually come with two separate telephone line connectors.

The purpose of the phone input is to connect a telephone, not the output line from the modem, to the telephone wall jack. The phone input is simply a convenience. It allows the telephone to be used without having to disconnect a telephone line from the computer to the wall telephone jack. If you mistakenly connect the line from the wall telephone jack to the phone input, you will be able to dial out from your communications software, but your system will hang up on you. Make sure that the telephone line that goes to the telephone wall jack comes from the *line output* and not the *phone output* jack of your modem.

Another common hardware problem is a problem in your telephone line. This can be quickly checked by simply using your phone to get through to the other party. If you can't do this, then neither can your computer.

A problem that is frequently encountered in an office or school building involves the phone system used within the building. You may have to issue extra commands on your

TABLE H.05-2
Common telephone modem problems.

Symptom	Possible Cause(s)
Can't connect.	Usually this means that your baud rates or numbers of data bits are not matched. This is especially true if you see garbage on the screen, especially the { character.
Can't see input.	You are typing in information but it doesn't appear on the monitor screen. However, if the person on the other side can see what you are typing, it means that you need to turn your local echo on. In this way, what you type will be echoed back to you, and you will see it on your monitor screen.
Get double characters.	Here you are typing information and getting double characters. This means that if you type HELLO, you get HHEELLLLOO; at the same time, what the other computer is getting appears normal. This means that you need to turn your local echo off. In this way, you will not be echoing back the extra character. With some systems *half-duplex* refers to local echo on, whereas *full duplex* refers to local echo off.

software in order to get your call out of the building. In this case you need to check with your telecommunications manager or the local phone company.

Sometimes your problem is simply a noisy line. This may have to do with your communications provider or it may have to do with how your telephone line is installed. You may have to switch to a long-distance telephone company that can provide service over more reliable communication lines. Or, you may have to physically trace where your telephone line goes from the wall telephone jack. If this is an old installation, your telephone line could be running in the wall right next to the AC power lines. If this is the case, you need to reroute the phone line.

CD-ROM DRIVE AND SOUND CARD PROBLEMS

One of the most common reasons a new CD-ROM drive or sound card does not work has to do with the way its interrupts and/or DMA channels are assigned.

Figure H.05-6 shows the location of the sound card in the hardware list provided by Device Manager. The AWE-32 indicates that the sound card is capable of advanced wave effects using 32 voices.

Figure H.05-7 shows the interrupt and DMA assignments for the sound card. Typically, interrupt 5 is used (some network interface cards also use interrupt 5), as well as DMA channels 1 and 5. If the standard settings do not work, you need to experiment until you find the right combination.

TAPE BACKUP PROBLEMS

Many problems associated with hard drive backups are usually a result of the connection between the computer and the tape drive. The parallel port is used to communicate with tape devices (since the data can be transferred faster using parallel data lines). The parallel port in newer computers can be configured in BIOS in different ways, such as standard or EPP. If the setting is incorrect, the tape device will not work properly. It may be necessary to modify the BIOS setting to change the parallel port from EPP to standard or vice versa. Many times, the documentation provided by the manufacturer can provide the answer to this problem as well as many other types of common problems.

FIGURE H.05-6
Selecting the sound card.

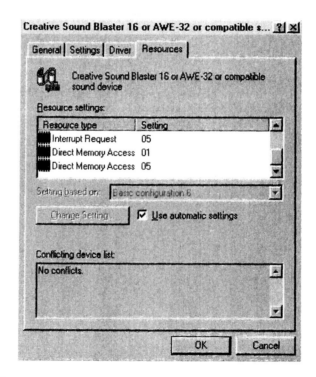

FIGURE H.05-7
Sound card settings.

REVIEW QUESTIONS

True/False

1. The length of a printer cable may cause problems with the printer.

2. Windows 95 provides a test-page feature for testing a printer.

Multiple Choice

3. If modem-to-modem communication is not possible, the reason may be

   a. Incorrect modem settings.

   b. An error-free phone line.

   c. Both a and b.

4. CD-ROM and sound cards use

   a. Interrupts.

   b. DMA channels.

   c. Both a and b.

5. EPP is a

   a. Modem standard.

   b. Serial port standard.

   c. Parallel port standard.

Answers to Review Questions

A General

A.01: 1. F, 2. T, 3. B, 4. C, 5. A
A.02: 1. F, 2. F, 3. C, 4. A, 5. B
A.03: 1. F, 2. F, 3. C, 4. C, 5. C
A.04: 1. T, 2. F, 3. B, 4. B, 5. C
A.05: 1. F, 2. T, 3. C, 4. B, 5. A
A.06: 1. F, 2. F, 3. C, 4. C, 5. C
A.07: 1. F, 2. F, 3. C, 4. A, 5. B
A.08: None
A.09: None
A.10: 1. T, 2. F, 3. C, 4. A, 5. A
A.11: None
A.12: 1. T, 2. T, 3. B, 4. C, 5. C

B DC Circuits

B.01: 1. T, 2. T, 3. C. 4. A, 5. C
B.02: 1. T, 2. F, 3. B, 4. B, 5. B
B.03: 1. F, 2. T, 3. B, 4. C, 5. B
B.04: 1. F, 2. F, 3. A, 4. A, 5. A
B.05: 1. T, 2. T, 3. C, 4. C, 5. A
B.06: 1. T, 2. T, 3. B, 4. A, 5. C
B.07: 1. F, 2. F, 3. B, 4. B, 5. A
B.08: 1. T, 2. T, 3. B, 4. B, 5. C
B.09: None
B.10: None
B.11: 1. F, 2. T, 3. A, 4. C, 5. B
B.12: None
B.13: None
B.14: 1. F, 2. T, 3. B, 4. B. 5. C
B.15: None
B.16: None
B.17: 1. F, 2. T, 3. B, 4. A, 5. B
B.18: 1. T, 2. F, 3. C, 4. B, 5. A
B.19: None
B.20: None
B.21: 1. F, 2. T, 3. C, 4. B, 5. C
B.22: None

B.23: None
B.24: 1. T, 2. T, 3. A, 4. C, 5. A

C AC Circuits

C.01: 1. T, 2. F, 3. B, 4. A, 5. C
C.02: 1. T, 2. F, 3. C, 4. B, 5. C
C.03: 1. F, 2. F, 3. A, 4. C, 5. A
C.04: None
C.05: 1. F, 2. T, 3. B, 4. B, 5. A
C.06: 1. T, 2. T, 3. C, 4. C, 5. C
C.07: 1. F, 2. T, 3. B, 4. C, 5. A
C.08: 1. T, 2. T, 3. B, 4. B, 5. B
C.09: None
C.10: None
C.11: 1. T, 2. F, 3. A, 4. A, 5. B
C.12: None
C.13: None
C.14: 1. F, 2. T, 3. B, 4. A, 5. A
C.15: 1. T, 2. T, 3. C, 4. B, 5. B
C.16: None
C.17: None
C.18: 1. T, 2. F, 3. B, 4. A, 5. C
C.19: None
C.20: None
C.21: 1. T, 2, T, 3. A, 4. C, 5. B
C.22: None
C.23: None
C.24: 1. F, 2. T, 3. B, 4. A, 5. B
C.25: None
C.26: None
C.27: 1. T, 2. F, 3. A, 4. B, 5. A
C.28: None
C.29: None
C.30: 1. T, 2. F, 3. C, 4. B, 5. C
C.31: 1. F, 2. T, 3. A, 4. C, 5. C
C.32: None

D Discrete Solid State Devices

D.01:   1. T,   2. F,   3. B,   4. A,   5. A
D.02:   1. T,   2. T,   3. A,   4. A,   5. A
D.03:   1. T,   2. T,   3. A,   4. C,   5. B
D.04:   1. F,   2. F,   3. C,   4. C,   5. C
D.05:   1. T,   2. F,   3. B,   4. C,   5. B
D.06:   1. F,   2. F,   3. A,   4. B,   5. B
D.07:   None
D.08:   None
D.09:   1. T,   2. T,   3. B,   4. A,   5. C
D.10:   None
D.11:   None
D.12:   1. T,   2. F,   3. B,   4. B,   5. B
D.13:   None
D.14:   None
D.15:   1. F,   2. T,   3. B,   4. B,   5. C
D.16:   None
D.17:   None

E Analog Circuits

E.01:   1. F,   2. T,   3. A,   4. A,   5. B
E.02:   None
E.03:   None
E.04:   1. F,   2. T,   3. C,   4. B,   5. C
E.05:   None
E.06:   None
E.07:   1. T,   2. T,   3. B,   4. C,   5. A
E.08:   None
E.09:   None
E.10:   1. T,   2. T,   3. B,   4. B,   5. C
E.11:   None
E.12:   None
E.13:   1. F,   2. F,   3. C,   4. A,   5. C
E.14:   None
E.15:   None
E.16:   1. T,   2. T,   3. C,   4. B,   5. C
E.17:   None
E.18:   1. T,   2. T,   3. B,   4. C,   5. B
E.19:   None
E.20:   1. T,   2. T,   3. A,   4. B,   5. B
E.21:   None
E.22:   1. F,   2. T,   3. A,   4. B,   5. A
E.23:   None
E.24:   1. T,   2. T,   3. C,   4. B,   5. A
E.25:   None

E.26:   None
E.27:   1. T,   2. F,   3. B,   4. B,   5. A
E.28:   None
E.29:   1. F,   2. T,   3. B,   4. B,   5. A
E.30:   1. F,   2. F,   3. A,   4. B,   5. A

F Digital Circuits

F.01:   1. F,   2. F,   3. A,   4. B,   5. A
F.02:   1. T,   2. F,   3. B,   4. C,   5. C
F.03:   1. T,   2. T,   3. A,   4. B,   5. B
F.04:   None
F.05:   1. F,   2. T,   3. B,   4. C,   5. C
F.06:   None
F.07:   None
F.08:   1. T,   2. F,   3. C,   4. B,   5. B
F.09:   None
F.10:   None
F.11:   1. F,   2. T,   3. B,   4. C,   5. A
F.12:   None
F.13:   None
F.14:   1. F,   2. F,   3. A,   4. A,   5. C
F.15:   None
F.16:   None
F.17:   1. T,   2. T,   3. C,   4. B,   5. C
F.18:   None
F.19:   None
F.20:   1. F,   2. T,   3. C,   4. B,   5. B
F.21:   None
F.22:   1. T,   2. T,   3. B,   4. C,   5. C
F.23:   None
F.24:   1. T,   2. F,   3. B,   4. A,   5. C
F.25:   None
F.26:   None
F.27:   None
F.28:   1. T,   2. T,   3. B,   4. C,   5. A
F.29:   None
F.30:   1. T,   2. T,   3. C,   4. C,   5. C
F.31:   None
F.32:   1. T,   2. T,   3. A,   4. B,   5. A
F.33:   None

G Microprocessors

G.01:   1. F,   2. F,   3. A,   4. C,   5. C
G.02:   None

G.03: 1. F, 2. F, 3. C, 4. B, 5. A
G.04: 1. T, 2. T, 3. C, 4. B, 5. C
G.05: 1. T, 2. T, 3. C, 4. C, 5. A
G.06: 1. T, 2. F, 3. B, 4. C, 5. A
G.07: None
G.08: 1. F, 2. T, 3. B, 4. A, 5. A

H Microcomputers

H.01: 1. F, 2. F, 3. B, 4. B, 5. C
H.02: 1. F, 2. T, 3. B, 4. C, 5. A
H.03: 1. T, 2. F, 3. C, 4. B, 5. A
H.04: 1. F, 2. T, 3. B, 4. A, 5. C
H.05: 1. T, 2. T, 3. A, 4. C, 5. C